Lecture Notes in Computer Science   **14448**

The series Lecture Notes in Computer Science (LNCS), including its subseries Lecture Notes in Artificial Intelligence (LNAI) and Lecture Notes in Bioinformatics (LNBI), has established itself as a medium for the publication of new developments in computer science and information technology research, teaching, and education.

LNCS enjoys close cooperation with the computer science R & D community, the series counts many renowned academics among its volume editors and paper authors, and collaborates with prestigious societies. Its mission is to serve this international community by providing an invaluable service, mainly focused on the publication of conference and workshop proceedings and postproceedings. LNCS commenced publication in 1973.

Biao Luo · Long Cheng · Zheng-Guang Wu ·
Hongyi Li · Chaojie Li
Editors

Neural
Information Processing

30th International Conference, ICONIP 2023
Changsha, China, November 20–23, 2023
Proceedings, Part II

 Springer

Editors
Biao Luo 🆔
Central South University
Changsha, China

Zheng-Guang Wu 🆔
Zhejiang University
Hangzhou, China

Chaojie Li 🆔
UNSW Sydney
Sydney, NSW, Australia

Long Cheng 🆔
Chinese Academy of Sciences
Beijing, China

Hongyi Li 🆔
Guangdong University of Technology
Guangzhou, China

ISSN 0302-9743   ISSN 1611-3349 (electronic)
Lecture Notes in Computer Science
ISBN 978-981-99-8081-9   ISBN 978-981-99-8082-6 (eBook)
https://doi.org/10.1007/978-981-99-8082-6

This Springer imprint is published by the registered company Springer Nature Singapore Pte Ltd.
The registered company address is: 152 Beach Road, #21-01/04 Gateway East, Singapore 189721, Singapore

Paper in this product is recyclable.

Preface

Welcome to the 30th International Conference on Neural Information Processing (ICONIP2023) of the Asia-Pacific Neural Network Society (APNNS), held in Changsha, China, November 20–23, 2023.

The mission of the Asia-Pacific Neural Network Society is to promote active interactions among researchers, scientists, and industry professionals who are working in neural networks and related fields in the Asia-Pacific region. APNNS has Governing Board Members from 13 countries/regions – Australia, China, Hong Kong, India, Japan, Malaysia, New Zealand, Singapore, South Korea, Qatar, Taiwan, Thailand, and Turkey. The society's flagship annual conference is the International Conference of Neural Information Processing (ICONIP). The ICONIP conference aims to provide a leading international forum for researchers, scientists, and industry professionals who are working in neuroscience, neural networks, deep learning, and related fields to share their new ideas, progress, and achievements.

ICONIP2023 received 1274 papers, of which 256 papers were accepted for publication in Lecture Notes in Computer Science (LNCS), representing an acceptance rate of 20.09% and reflecting the increasingly high quality of research in neural networks and related areas. The conference focused on four main areas, i.e., "Theory and Algorithms", "Cognitive Neurosciences", "Human-Centered Computing", and "Applications". All the submissions were rigorously reviewed by the conference Program Committee (PC), comprising 258 PC members, and they ensured that every paper had at least two high-quality single-blind reviews. In fact, 5270 reviews were provided by 2145 reviewers. On average, each paper received 4.14 reviews.

We would like to take this opportunity to thank all the authors for submitting their papers to our conference, and our great appreciation goes to the Program Committee members and the reviewers who devoted their time and effort to our rigorous peer-review process; their insightful reviews and timely feedback ensured the high quality of the papers accepted for publication. We hope you enjoyed the research program at the conference.

October 2023

Biao Luo
Long Cheng
Zheng-Guang Wu
Hongyi Li
Chaojie Li

Organization

Honorary Chair

Weihua Gui   Central South University, China

Advisory Chairs

Jonathan Chan	King Mongkut's University of Technology Thonburi, Thailand
Zeng-Guang Hou	Chinese Academy of Sciences, China
Nikola Kasabov	Auckland University of Technology, New Zealand
Derong Liu	Southern University of Science and Technology, China
Seiichi Ozawa	Kobe University, Japan
Kevin Wong	Murdoch University, Australia

General Chairs

Tingwen Huang	Texas A&M University at Qatar, Qatar
Chunhua Yang	Central South University, China

Program Chairs

Biao Luo	Central South University, China
Long Cheng	Chinese Academy of Sciences, China
Zheng-Guang Wu	Zhejiang University, China
Hongyi Li	Guangdong University of Technology, China
Chaojie Li	University of New South Wales, Australia

Technical Chairs

Xing He	Southwest University, China
Keke Huang	Central South University, China
Huaqing Li	Southwest University, China
Qi Zhou	Guangdong University of Technology, China

Local Arrangement Chairs

Wenfeng Hu	Central South University, China
Bei Sun	Central South University, China

Finance Chairs

Fanbiao Li	Central South University, China
Hayaru Shouno	University of Electro-Communications, Japan
Xiaojun Zhou	Central South University, China

Special Session Chairs

Hongjing Liang	University of Electronic Science and Technology, China
Paul S. Pang	Federation University, Australia
Qiankun Song	Chongqing Jiaotong University, China
Lin Xiao	Hunan Normal University, China

Tutorial Chairs

Min Liu	Hunan University, China
M. Tanveer	Indian Institute of Technology Indore, India
Guanghui Wen	Southeast University, China

Publicity Chairs

Sabri Arik	Istanbul University-Cerrahpaşa, Turkey
Sung-Bae Cho	Yonsei University, South Korea
Maryam Doborjeh	Auckland University of Technology, New Zealand
El-Sayed M. El-Alfy	King Fahd University of Petroleum and Minerals, Saudi Arabia
Ashish Ghosh	Indian Statistical Institute, India
Chuandong Li	Southwest University, China
Weng Kin Lai	Tunku Abdul Rahman University of Management & Technology, Malaysia
Chu Kiong Loo	University of Malaya, Malaysia

| Qinmin Yang | Zhejiang University, China |
| Zhigang Zeng | Huazhong University of Science and Technology, China |

Publication Chairs

Zhiwen Chen	Central South University, China
Andrew Chi-Sing Leung	City University of Hong Kong, China
Xin Wang	Southwest University, China
Xiaofeng Yuan	Central South University, China

Secretaries

| Yun Feng | Hunan University, China |
| Bingchuan Wang | Central South University, China |

Webmasters

| Tianmeng Hu | Central South University, China |
| Xianzhe Liu | Xiangtan University, China |

Program Committee

Rohit Agarwal	UiT The Arctic University of Norway, Norway
Hasin Ahmed	Gauhati University, India
Harith Al-Sahaf	Victoria University of Wellington, New Zealand
Brad Alexander	University of Adelaide, Australia
Mashaan Alshammari	Independent Researcher, Saudi Arabia
Sabri Arik	Istanbul University, Turkey
Ravneet Singh Arora	Block Inc., USA
Zeyar Aung	Khalifa University of Science and Technology, UAE
Monowar Bhuyan	Umeå University, Sweden
Jingguo Bi	Beijing University of Posts and Telecommunications, China
Xu Bin	Northwestern Polytechnical University, China
Marcin Blachnik	Silesian University of Technology, Poland
Paul Black	Federation University, Australia

Anoop C. S.	Govt. Engineering College, India
Ning Cai	Beijing University of Posts and Telecommunications, China
Siripinyo Chantamunee	Walailak University, Thailand
Hangjun Che	City University of Hong Kong, China
Wei-Wei Che	Qingdao University, China
Huabin Chen	Nanchang University, China
Jinpeng Chen	Beijing University of Posts & Telecommunications, China
Ke-Jia Chen	Nanjing University of Posts and Telecommunications, China
Lv Chen	Shandong Normal University, China
Qiuyuan Chen	Tencent Technology, China
Wei-Neng Chen	South China University of Technology, China
Yufei Chen	Tongji University, China
Long Cheng	Institute of Automation, China
Yongli Cheng	Fuzhou University, China
Sung-Bae Cho	Yonsei University, South Korea
Ruikai Cui	Australian National University, Australia
Jianhua Dai	Hunan Normal University, China
Tao Dai	Tsinghua University, China
Yuxin Ding	Harbin Institute of Technology, China
Bo Dong	Xi'an Jiaotong University, China
Shanling Dong	Zhejiang University, China
Sidong Feng	Monash University, Australia
Yuming Feng	Chongqing Three Gorges University, China
Yun Feng	Hunan University, China
Junjie Fu	Southeast University, China
Yanggeng Fu	Fuzhou University, China
Ninnart Fuengfusin	Kyushu Institute of Technology, Japan
Thippa Reddy Gadekallu	VIT University, India
Ruobin Gao	Nanyang Technological University, Singapore
Tom Gedeon	Curtin University, Australia
Kam Meng Goh	Tunku Abdul Rahman University of Management and Technology, Malaysia
Zbigniew Gomolka	University of Rzeszow, Poland
Shengrong Gong	Changshu Institute of Technology, China
Xiaodong Gu	Fudan University, China
Zhihao Gu	Shanghai Jiao Tong University, China
Changlu Guo	Budapest University of Technology and Economics, Hungary
Weixin Han	Northwestern Polytechnical University, China

Xing He	Southwest University, China
Akira Hirose	University of Tokyo, Japan
Yin Hongwei	Huzhou Normal University, China
Md Zakir Hossain	Curtin University, Australia
Zengguang Hou	Chinese Academy of Sciences, China
Lu Hu	Jiangsu University, China
Zeke Zexi Hu	University of Sydney, Australia
He Huang	Soochow University, China
Junjian Huang	Chongqing University of Education, China
Kaizhu Huang	Duke Kunshan University, China
David Iclanzan	Sapientia University, Romania
Radu Tudor Ionescu	University of Bucharest, Romania
Asim Iqbal	Cornell University, USA
Syed Islam	Edith Cowan University, Australia
Kazunori Iwata	Hiroshima City University, Japan
Junkai Ji	Shenzhen University, China
Yi Ji	Soochow University, China
Canghong Jin	Zhejiang University, China
Xiaoyang Kang	Fudan University, China
Mutsumi Kimura	Ryukoku University, Japan
Masahiro Kohjima	NTT, Japan
Damian Kordos	Rzeszow University of Technology, Poland
Marek Kraft	Poznań University of Technology, Poland
Lov Kumar	NIT Kurukshetra, India
Weng Kin Lai	Tunku Abdul Rahman University of Management & Technology, Malaysia
Xinyi Le	Shanghai Jiao Tong University, China
Bin Li	University of Science and Technology of China, China
Hongfei Li	Xinjiang University, China
Houcheng Li	Chinese Academy of Sciences, China
Huaqing Li	Southwest University, China
Jianfeng Li	Southwest University, China
Jun Li	Nanjing Normal University, China
Kan Li	Beijing Institute of Technology, China
Peifeng Li	Soochow University, China
Wenye Li	Chinese University of Hong Kong, China
Xiangyu Li	Beijing Jiaotong University, China
Yantao Li	Chongqing University, China
Yaoman Li	Chinese University of Hong Kong, China
Yinlin Li	Chinese Academy of Sciences, China
Yuan Li	Academy of Military Science, China

Yun Li	Nanjing University of Posts and Telecommunications, China
Zhidong Li	University of Technology Sydney, Australia
Zhixin Li	Guangxi Normal University, China
Zhongyi Li	Beihang University, China
Ziqiang Li	University of Tokyo, Japan
Xianghong Lin	Northwest Normal University, China
Yang Lin	University of Sydney, Australia
Huawen Liu	Zhejiang Normal University, China
Jian-Wei Liu	China University of Petroleum, China
Jun Liu	Chengdu University of Information Technology, China
Junxiu Liu	Guangxi Normal University, China
Tommy Liu	Australian National University, Australia
Wen Liu	Chinese University of Hong Kong, China
Yan Liu	Taikang Insurance Group, China
Yang Liu	Guangdong University of Technology, China
Yaozhong Liu	Australian National University, Australia
Yong Liu	Heilongjiang University, China
Yubao Liu	Sun Yat-sen University, China
Yunlong Liu	Xiamen University, China
Zhe Liu	Jiangsu University, China
Zhen Liu	Chinese Academy of Sciences, China
Zhi-Yong Liu	Chinese Academy of Sciences, China
Ma Lizhuang	Shanghai Jiao Tong University, China
Chu-Kiong Loo	University of Malaya, Malaysia
Vasco Lopes	Universidade da Beira Interior, Portugal
Hongtao Lu	Shanghai Jiao Tong University, China
Wenpeng Lu	Qilu University of Technology, China
Biao Luo	Central South University, China
Ye Luo	Tongji University, China
Jiancheng Lv	Sichuan University, China
Yuezu Lv	Beijing Institute of Technology, China
Huifang Ma	Northwest Normal University, China
Jinwen Ma	Peking University, China
Jyoti Maggu	Thapar Institute of Engineering and Technology Patiala, India
Adnan Mahmood	Macquarie University, Australia
Mufti Mahmud	University of Padova, Italy
Krishanu Maity	Indian Institute of Technology Patna, India
Srimanta Mandal	DA-IICT, India
Wang Manning	Fudan University, China

Piotr Milczarski	Lodz University of Technology, Poland
Malek Mouhoub	University of Regina, Canada
Nankun Mu	Chongqing University, China
Wenlong Ni	Jiangxi Normal University, China
Anupiya Nugaliyadde	Murdoch University, Australia
Toshiaki Omori	Kobe University, Japan
Babatunde Onasanya	University of Ibadan, Nigeria
Manisha Padala	Indian Institute of Science, India
Sarbani Palit	Indian Statistical Institute, India
Paul Pang	Federation University, Australia
Rasmita Panigrahi	Giet University, India
Kitsuchart Pasupa	King Mongkut's Institute of Technology Ladkrabang, Thailand
Dipanjyoti Paul	Ohio State University, USA
Hu Peng	Jiujiang University, China
Kebin Peng	University of Texas at San Antonio, USA
Dawid Połap	Silesian University of Technology, Poland
Zhong Qian	Soochow University, China
Sitian Qin	Harbin Institute of Technology at Weihai, China
Toshimichi Saito	Hosei University, Japan
Fumiaki Saitoh	Chiba Institute of Technology, Japan
Naoyuki Sato	Future University Hakodate, Japan
Chandni Saxena	Chinese University of Hong Kong, China
Jiaxing Shang	Chongqing University, China
Lin Shang	Nanjing University, China
Jie Shao	University of Science and Technology of China, China
Yin Sheng	Huazhong University of Science and Technology, China
Liu Sheng-Lan	Dalian University of Technology, China
Hayaru Shouno	University of Electro-Communications, Japan
Gautam Srivastava	Brandon University, Canada
Jianbo Su	Shanghai Jiao Tong University, China
Jianhua Su	Institute of Automation, China
Xiangdong Su	Inner Mongolia University, China
Daiki Suehiro	Kyushu University, Japan
Basem Suleiman	University of New South Wales, Australia
Ning Sun	Shandong Normal University, China
Shiliang Sun	East China Normal University, China
Chunyu Tan	Anhui University, China
Gouhei Tanaka	University of Tokyo, Japan
Maolin Tang	Queensland University of Technology, Australia

Shu Tian	University of Science and Technology Beijing, China
Shikui Tu	Shanghai Jiao Tong University, China
Nancy Victor	Vellore Institute of Technology, India
Petra Vidnerová	Institute of Computer Science, Czech Republic
Shanchuan Wan	University of Tokyo, Japan
Tao Wan	Beihang University, China
Ying Wan	Southeast University, China
Bangjun Wang	Soochow University, China
Hao Wang	Shanghai University, China
Huamin Wang	Southwest University, China
Hui Wang	Nanchang Institute of Technology, China
Huiwei Wang	Southwest University, China
Jianzong Wang	Ping An Technology, China
Lei Wang	National University of Defense Technology, China
Lin Wang	University of Jinan, China
Shi Lin Wang	Shanghai Jiao Tong University, China
Wei Wang	Shenzhen MSU-BIT University, China
Weiqun Wang	Chinese Academy of Sciences, China
Xiaoyu Wang	Tokyo Institute of Technology, Japan
Xin Wang	Southwest University, China
Xin Wang	Southwest University, China
Yan Wang	Chinese Academy of Sciences, China
Yan Wang	Sichuan University, China
Yonghua Wang	Guangdong University of Technology, China
Yongyu Wang	JD Logistics, China
Zhenhua Wang	Northwest A&F University, China
Zi-Peng Wang	Beijing University of Technology, China
Hongxi Wei	Inner Mongolia University, China
Guanghui Wen	Southeast University, China
Guoguang Wen	Beijing Jiaotong University, China
Ka-Chun Wong	City University of Hong Kong, China
Anna Wróblewska	Warsaw University of Technology, Poland
Fengge Wu	Institute of Software, Chinese Academy of Sciences, China
Ji Wu	Tsinghua University, China
Wei Wu	Inner Mongolia University, China
Yue Wu	Shanghai Jiao Tong University, China
Likun Xia	Capital Normal University, China
Lin Xiao	Hunan Normal University, China

Qiang Xiao	Huazhong University of Science and Technology, China
Hao Xiong	Macquarie University, Australia
Dongpo Xu	Northeast Normal University, China
Hua Xu	Tsinghua University, China
Jianhua Xu	Nanjing Normal University, China
Xinyue Xu	Hong Kong University of Science and Technology, China
Yong Xu	Beijing Institute of Technology, China
Ngo Xuan Bach	Posts and Telecommunications Institute of Technology, Vietnam
Hao Xue	University of New South Wales, Australia
Yang Xujun	Chongqing Jiaotong University, China
Haitian Yang	Chinese Academy of Sciences, China
Jie Yang	Shanghai Jiao Tong University, China
Minghao Yang	Chinese Academy of Sciences, China
Peipei Yang	Chinese Academy of Science, China
Zhiyuan Yang	City University of Hong Kong, China
Wangshu Yao	Soochow University, China
Ming Yin	Guangdong University of Technology, China
Qiang Yu	Tianjin University, China
Wenxin Yu	Southwest University of Science and Technology, China
Yun-Hao Yuan	Yangzhou University, China
Xiaodong Yue	Shanghai University, China
Paweł Zawistowski	Warsaw University of Technology, Poland
Hui Zeng	Southwest University of Science and Technology, China
Wang Zengyunwang	Hunan First Normal University, China
Daren Zha	Institute of Information Engineering, China
Zhi-Hui Zhan	South China University of Technology, China
Baojie Zhang	Chongqing Three Gorges University, China
Canlong Zhang	Guangxi Normal University, China
Guixuan Zhang	Chinese Academy of Science, China
Jianming Zhang	Changsha University of Science and Technology, China
Li Zhang	Soochow University, China
Wei Zhang	Southwest University, China
Wenbing Zhang	Yangzhou University, China
Xiang Zhang	National University of Defense Technology, China
Xiaofang Zhang	Soochow University, China
Xiaowang Zhang	Tianjin University, China

Contents – Part II

Theory and Algorithms

Theory and Algorithms

Distributed Nash Equilibrium Seeking of Noncooperative Games with Communication Constraints and Matrix Weights

Shuoshuo Zhang[1], Jianxiang Ren[1], Xiao Fang[1(✉)], and Tingwen Huang[2]

[1] School of Mathematics, Southeast University, Nanjing 211189, China
{zhangshuoshuo,renjianxiang,fangxiao}@seu.edu.cn
[2] Department of Science, Texas A&M University at Qatar, Doha, Qatar
tingwen.huang@qatar.tamu.edu

Abstract. Distributed Nash equilibrium seeking is investigated in this paper for a class of multi-agent systems under intermittent communication and matrix-weighted communication graphs. Different from most of the existing works on distributed Nash equilibrium seeking of noncooperative games where the players (agents) can communicate continuously over time, the players considered in this paper are assumed to exchange information only with their neighbors during some disconnected time intervals while the underlying communication graph is matrix-weighted. A distributed Nash equilibrium seeking algorithm integrating gradient strategy and leader-following consensus protocol is proposed for the noncooperative games with intermittent communication and matrix-weighted communication graphs. The effect of the average intermittent communication rate on the convergence of the distributed Nash equilibrium seeking algorithm is analyzed, and a lower bound of the average intermittent communication rate that ensures the convergence of the algorithm is given. The convergence of the algorithm is established by means of Lyapunov stability theory. Simulations are presented to verify the proposed distributed Nash equilibrium seeking algorithm.

Keywords: Nash equilibrium · Noncooperative game · Intermittent communication · Multi-agent system

1 Introduction

In recent years, there has been increasing interest in the distributed coordination control and decision-making of multi-agent systems [1–3]. The advancement of distributed coordination and autonomous decision-making technology for multi-agent systems have emerged as a core focus in the field of artificial intelligence.

This work was supported by in part the National Key Research and Development Program of China under Grant No. 2022YFA1004702 and in part by the General Joint Fund of the Equipment Advance Research Program of Ministry of Education under Grant No. 8091B022114.

In practice, agents in systems may encounter conflicts of interest, and game theory provides feasible solutions to address these issues. By utilizing sophisticated mathematical models, game theory provides a theoretical foundation for decision-making for agents situated within environments characterized by conflict and confrontation. As networked games continue to develop, distributed Nash equilibrium seeking of noncooperative games has a wide range of applications, including economic dispatching in electric power systems and cooperative task assignment in multi-unmanned systems.

Recently, significant efforts have been devoted to the study of Nash equilibrium seeking methods in noncooperative games for multi-agent systems, resulting in many profound results established [4–9]. Gradient method was used for finding differential Nash equilibrium in continuous-time games in [4]. Distributed Nash equilibrium seeking of multi-cluster games with consistency-constraint was studied in [5], where the networked noncooperative games and distributed optimization were put in a unified framework. In [6], a study was conducted on the online solution of team games by combining cooperative control, game theory, and reinforcement learning, and a cooperative policy iteration algorithm was proposed for graphical games, which guarantees convergence to the cooperative Nash equilibrium when all agents simultaneously update their policies. For a two-network zero-sum game under the undirected communication topology, a distributed saddle-point strategy was synthesized and its convergence to the Nash equilibrium was established for a group of functions with strict concavity-convexity and local Lipschitz continuity in [7]. For the distributed aggregative game problem, two types of algorithms, namely, a synchronous and an asynchronous distributed algorithms were proposed and the convergence of the algorithms to Nash equilibrium was established in [8]. In [9], a continuous-time distributed Nash equilibrium seeking algorithm was developed in virtue of a leader-following consensus protocol and gradient strategy. Furthermore, the work of [9] was extended in [10], where the communication graph of players is switching.

It is important to note that many of the findings regarding multi-agent systems' pursuit of Nash equilibrium are based on the common assumption that agents continuously exchange information. The assumption that each agent can share information with its neighbors without any communication restrictions is often insufficient in real-world scenarios. In some cases, agents can only communicate with their neighbors at disconnected time intervals due to potential communication restrictions and cyber-attacks [11]. [12] provided a solution to the intermittent communication in wireless sensor networks. [13] studied resource allocation problem under limited bandwidth constraints in communication. In [14], a new consensus protocol was developed that utilizes globally synchronized intermittent local information feedback. This protocol ensures that the states of agents can achieve consensus in an exponential manner. However, for the noncooperative game problems of multi-agent systems under intermittent communication, the distributed Nash equilibrium seeking has not been studied yet.

Matrix weight has been extensively used in network analysis and optimization problems. An efficient matrix weighted low rank approximation algorithm was proposed in [15], which can be used to process large-scale data. [16] introduced the concept of a matrix-weighted graph as a means to study the synchronization problem of linear systems. The coupling between two systems is characterized by different output matrices in this approach. [17] studied the distributed robust optimization of networked agent systems where the communication graph of agents randomly switches among some matrix-weighted and connected graphs. In this paper, we consider the distributed Nash equilibrium seeking problem of multi-agent system under intermittent communication and matrix-weighted communication graph. By using Lyapunov stability theory, it is theoretically proved that the states of multi-agent system can converge to Nash equilibrium under the proposed algorithm provided that the average intermittent communication rate is greater than a threshold. The theoretical result is demonstrated through simulations.

Notations. Let $\mathbb{R}^{N \times N}$ be $N \times N$-dimensional real matrix space and $\mathbb{R}^N$ be N-dimensional real vector space. Symbol $\mathbb{N}$ is natural number set. Let I_N $(0_{N \times N})$ be $N \times N$-dimensional identity (zero) matrix and 1_N (0_N) be N-dimensional column vector with all elements being 1 (0). For symmetric matrix $M \in \mathbb{R}^{N \times N}$, $\lambda_{\max}(M)$ and $\lambda_{\min}(M)$ represents the maximum and minimum eigenvalue of M. $A > 0$ indicates that matrix $A \in \mathbb{R}^{n \times n}$ is a positive definite matrix. $\otimes$ is the Kronecker product.

2 Problem Formulation

Consider a noncooperative game involving N players (or agents). The set of players is $\mathcal{N} = \{1, ..., N\}$. Let $x = \left[x_1^T, ..., x_N^T\right]^T \in \mathbb{R}^{Nn}$ denote all players' actions with $x_i \in \mathbb{R}^n$ being the action of player i. The payoff function of player i is denoted by $f_i(x)$. The objective of player i is as follows,

$$\max_{x_i \in \Omega_i} f_i(x) = f_i(x_i, x_{-i}), \quad \forall i \in \mathcal{N}. \tag{1}$$

where $x_{-i} = \left[x_1^T, x_2^T, ..., x_{i-1}^T, x_{i+1}^T, ..., x_N^T\right]^T$ and Ω_i is the constraint set of the action of player i. This paper considers the case that the players' actions are unconstrained, that is, $\Omega_i = \mathbb{R}^n$, $\forall i \in \mathcal{N}$.

An action profile $x^* = \left(x_i^*, x_{-i}^*\right)$ is called Nash equilibrium if [18]

$$f_i\left(x_i^*, x_{-i}^*\right) \geq f_i\left(x_i, x_{-i}^*\right), \quad \forall i \in \mathcal{N}. \tag{2}$$

The communication graph of the players is denoted by undirected graph $\mathcal{G} = \{\mathcal{N}, \mathcal{E}, \mathcal{A}\}$, where $\mathcal{E}$ denotes the set of edges, $\mathcal{A} = \left\{\Gamma_{ij} \in \mathbb{R}^{n \times n} | \Gamma_{ij} = \Gamma_{ij}^T > 0, \sum_j \Gamma_{ij} = I_n\right\}$ denotes the set of matrix weights. Communication is achievable between players i and j if $(i, j) \in \mathcal{E}$. The matrix-weighted Laplacian matrix $L = [l_{ij}]_{Nn \times Nn}$ of $\mathcal{G}$ is defined as

$$l_{ij} = \begin{cases} -\Gamma_{ij}, & i \neq j, \\ \sum\limits_{k=1,k\neq i}^{N} \Gamma_{ik}, & i = j. \end{cases} \qquad (3)$$

Lemma 1 [10]. *There is at least one eigenvalue 0 for the Laplacian matrix L of undirected graph $\mathcal{G}$, and all other eigenvalues of L are positive. 0 is a simple eigenvalue of Laplacian matrix L if and only if the undirected graph $\mathcal{G} = \{\mathcal{N}, \mathcal{E}, \mathcal{A}\}$ is connected.*

Moreover, we consider the noncooperative game with intermittent communication, that is, players only communicate with their neighbors during some disconnected time intervals. Let T represent the set of time intervals during which the players can communicate with each other, and $\bar{T}$ represent the set of time intervals during which the players cannot communicate with each other. Note that $T \cup \bar{T} = [0, \infty)$. We assume that there are non-overlapping time intervals $[t_k, t_{k+1})$, $k \in \mathbb{N}$, $t_0 = 0$, and $\bigcup\limits_k [t_k, t_{k+1}) = [0, \infty) = T \cup \bar{T}$. It can be reasonably speculated that the convergence to Nash equilibrium of distributed seeking algorithms under intermittent communication will be affected by the ratio of communication time to total time, so we define ζ_k as the lebesgue measure of $\{t | t \in [t_k, t_{k+1}) \cap T\}$, and $\rho_k = t_{k+1} - t_k$. Then, the average intermittent communication rate of time interval $[t_k, t_{k+1})$ is given by

$$a_k = \frac{\zeta_k}{\rho_k}. \qquad (4)$$

The following assumptions are made regarding communication graph and payoff functions.

Assumption 1. *$\mathcal{G}$ is connected.*

Assumption 2. *The players' payoff functions $f_i(x)$, $\forall i \in \mathcal{N}$, are $\mathcal{C}^2$ with respect to x. And for each player $i \in \mathcal{N}$, the payoff function $f_i(x_i, x_{-i})$ is concave in x_i, for given x_{-i}.*

Assumption 3. *Define $G(x) = \left[\frac{\partial f_1(x)}{\partial x_1}^T, \frac{\partial f_2(x)}{\partial x_2}^T, ..., \frac{\partial f_N(x)}{\partial x_N}^T \right]^T$. Then there is a constant $\eta > 0$ such that $\forall x, z \in \mathbb{R}^{Nn}$,*

$$(x - z)^T (G(x) - G(z)) \leq -\eta \|x - z\|^2. \qquad (5)$$

Remark 1. By Assumption 2 and Assumption 3, it can be obtained that the Nash equilibrium of the noncooperative game exists and is unique [8,19].

Remark 2. By Assumption 2, the following properties about $G(x)$ can be obtained:

(1)

$$G(x^*) = 0_{Nn}, \qquad (6)$$

where x^* is the Nash equilibrium.

(2) $\forall x, y \in \mathbb{R}^{Nn}$, there is some Lipschitz constant $l^g > 0$ such that

$$\|G(x) - G(y)\| \leq l^g \|x - y\|, \tag{7}$$

and l^g depends on the Lipschitz coefficients of $\frac{\partial f_i(x)}{\partial x_i}$, $i \in \mathcal{N}$.

3 Main Results

3.1 Distributed Nash Equilibrium Seeking Algorithm Under Intermittent Communication

Motivated by [9,14] and [17], a distributed Nash equilibrium seeking algorithm adapted to intermittent communication and matrix weights will be presented in this subsection.

In a non-cooperative game, each player's payoff is influenced by the choices made by other players and they do not have access to information about players who are not their immediate neighbors. Therefore, to seek Nash equilibrium in a distributed manner, let each player make a local estimation on the global action vector x. Specifically, let $y_{ij} \in \mathbb{R}^n$ denote player i's estimation on player j's action, and $y_i = \left[y_{i1}^T, y_{i2}^T, ..., y_{iN}^T\right]^T \in \mathbb{R}^{Nn}$ denote player i's estimation on the global action vector x. Then player i updates its action as follows:

$$\dot{x}_i = k_i \frac{\partial f_i}{\partial x_i}(y_i), \tag{8}$$

where $k_i = \delta \bar{k}_i$ where $\bar{k}_i > 0$ is fixed parameter and $\delta > 0$ is a small parameter to be designed.

Inspired by the consensus-based method proposed in [9] and considering that there are time intervals when players cannot communicate with each other, design the following update law for $y_{ij}, i, j \in \mathcal{N}$,

$$\dot{y}_{ij} = -\left(\sum_{k=1}^{N} \Gamma_{ik}(y_{ij} - y_{kj}) + \Gamma_{ij}(y_{ij} - x_j)\right), \ t \in T,$$
$$\dot{y}_{ij} = 0_n, \ t \in \bar{T}, \tag{9}$$

where $\Gamma_{ij}, i, j \in \mathcal{N}$ is the matrix weight between player i and player j.

Remark 3. If $T = [0, \infty)$, then players can communicate with each other all the time.

3.2 Convergence Analysis

The subsequent analysis demonstrates that the Nash equilibrium seeking strategy, defined in (8) and (9), leads to the convergence of players' actions to the Nash equilibrium.

For convenience, define

$$y = [y_1^T, y_2^T, ..., y_N^T]^T, \ J(y) = \left[\frac{\partial f_1}{\partial x_1}(y_1)^T, \frac{\partial f_2}{\partial x_2}(y_2)^T, ..., \frac{\partial f_N}{\partial x_N}(y_N)^T\right]^T,$$

$$h_1(x) = -\left[(\Gamma_{11}x_1)^T, (\Gamma_{12}x_2)^T, ..., (\Gamma_{1N}x_N)^T, (\Gamma_{21}x_1)^T, (\Gamma_{22}x_2)^T, ..., (\Gamma_{NN}x_N)^T\right]^T,$$

$$B_0 = diag\{\Gamma_{ij}\}, \ i,j \in \mathcal{N}, \ \bar{k} = diag\{\bar{k}_i I_n\}, \ i \in \mathcal{N}, \ H = L \otimes I_N + B_0.$$

where $diag\{\Gamma_{ij}\}$ $(diag\{\bar{k}_i I_n\})$ is a block diagonal matrix with the diagonal elements are respectively $\Gamma_{11}, ..., \Gamma_{1N}, \Gamma_{21}, ..., \Gamma_{NN}$ $(\bar{k}_1 I_n, ..., \bar{k}_N I_n)$. Then, the merged form of (8) is

$$\dot{x} = \delta \bar{k} J(y), \tag{10}$$

and the merged form of (9) is

$$\dot{y} = -Hy - h_1(x), \ t \in T,$$
$$\dot{y} = 0_{N^2 n}, \ t \in \bar{T}, \tag{11}$$

where L is the Laplacian matrix. According to Assumption 1, Lemma 1 and the definition of B_0, it can be obtained that H is a positive definite matrix, implying that $-H$ is Hurwitz. Then there are symmetric positive definite matrices Q_1, P_1 such that

$$Q_1 = P_1 H + H^T P_1. \tag{12}$$

Remark 4. By Assumption 2, it can be obtained that $\forall x, y \in \mathbb{R}^{N^2 n}$, there is some constant $l > 0$ such that

$$\|J(x) - J(y)\| \le l\|x - y\|, \tag{13}$$

where l can be regarded as a local Lipschitz constant that depends on the Lipschitz coefficients of $\frac{\partial f_i(x)}{\partial x_i}$, $i \in \mathcal{N}$.

Theorem 1. *Suppose that Assumptions 1-3 hold, and players update their actions according to (8) and (9). Assuming the existence of a sequence of time intervals $[t_k, t_{k+1})$, $k \in \mathbb{N}$, $t_0 = 0$ which are uniformly bounded and non-overlapping. Then the Nash equilibrium x^* is exponentially stable under the proposed Nash equilibrium seeking algorithm (8) and (9) if the following conditions are satisfied:*

(1)

$$0 < \delta < \frac{4(c - c^2)\eta\lambda_{\min}(Q_1)}{4c\eta r_2 + r_1^2},$$

(2)

$$a_k \ge \frac{\gamma_1 \lambda_{\max}(B_2)}{\gamma_2 \lambda_{\min}(B_1) + \gamma_1 \lambda_{\max}(B_2)},$$

where a_k is defined as (4),

$$\gamma_1 = \max\{\frac{c}{2}\lambda_{\max}(\bar{k}^{-1}), (1-c)\lambda_{\max}(P_1)\}, \gamma_2 = \min\{\frac{c}{2}\lambda_{\min}(\bar{k}^{-1}), (1-c)\lambda_{\min}(P_1)\}.$$

$$B_1 = \begin{pmatrix} c\eta & -\frac{r_1}{2} \\ -\frac{r_1}{2} & -r_2 + \frac{(1-c)\lambda_{\min}(Q_1)}{\delta} \end{pmatrix}, B_2 = \begin{pmatrix} -c\eta & \frac{r_1}{2} \\ \frac{r_1}{2} & r_2 \end{pmatrix}.$$

$$r_1 = cl_1 + 2\sqrt{N}(1-c)l_1^g \left\|\bar{k}\right\| \left\|P_1\right\|, \quad r_2 = 2\sqrt{N}(1-c)l_1 \left\|\bar{k}\right\| \left\|P_1\right\|.$$

Q_1, P_1 *are defined as (12), $c \in (0,1)$ is a constant, η is defined in Assumption 3, and l_1, l_1^g are Lipschitz constants.*

Proof. Define $y^q = \left[y_{11}^q{}^T, y_{12}^q{}^T, \ldots, y_{1N}^q{}^T, y_{21}^q{}^T, \ldots, y_{2N}^q{}^T, \ldots, y_{NN}^q{}^T\right]^T$, $y_{ij}^q = x_j, \forall i, j \in \mathcal{N}$, and $\bar{y} = y - y^q$, $\bar{x} = x - x^*$, where x^* is the Nash equilibrium. It is not difficult to get that $G(x) = J(y^q)$. For $\bar{y}$, according to (11), one can obtain that when $t \in T$,

$$\begin{aligned}\dot{\bar{y}} &= \dot{y} - \dot{y}^q \\ &= -(H(\bar{y} + y^q) + h_1(x)) - \dot{y}^q \\ &= -H\bar{y} - \dot{y}^q, \end{aligned} \tag{14}$$

and when $t \in \bar{T}$, $\dot{\bar{y}} = \dot{y} - \dot{y}^q = -\dot{y}^q$.

According to (10), one can obtain that

$$\begin{aligned}\left\|\frac{dx}{dt}\right\| &= \left\|\delta\bar{k}J(y)\right\| = \left\|\delta\bar{k}J(y) - \delta\bar{k}G(x^*)\right\| \\ &= \left\|\delta\bar{k}J(y) - \delta\bar{k}J(y^q) + \delta\bar{k}G(x) - \delta\bar{k}G(x^*)\right\| \\ &\leq \delta\left\|\bar{k}\right\|(l_1\left\|\bar{y}\right\| + l_1^g\left\|\bar{x}\right\|). \end{aligned} \tag{15}$$

where $G(x^*) = 0_{Nn}$ and $G(x) = J(y^q)$ are used and $l_1 > 0$ ($l_1^g > 0$) is Lipschitz constant satisfying that $\|J(y) - J(y^q)\| \leq l_1\|\bar{y}\|$ ($\|G(x) - G(x^*)\| \leq l_1^g\|\bar{x}\|$).

Define a Lyapunov function candidate as

$$V = \frac{c}{2}\bar{x}^T\bar{k}^{-1}\bar{x} + (1-c)\bar{y}^T P_1\bar{y}, \tag{16}$$

where $c \in (0,1)$ is a constant and P_1 is defined as (12).

Define $z = [\bar{x}^T, \bar{y}^T]^T$. Then $\gamma_2\|z\|^2 \leq V \leq \gamma_1\|z\|^2$, where $\gamma_1 = \max\{\frac{c}{2}\lambda_{\max}(\bar{k}^{-1}), (1-c)\lambda_{\max}(P_1)\}$, $\gamma_2 = \min\{\frac{c}{2}\lambda_{\min}(\bar{k}^{-1}), (1-c)\lambda_{\min}(P_1)\}$.

For $t \in [t_k, t_{k+1}) \cap T$, $k \in \mathbb{N}$, differentiating V along the trajectories of (8) and (9) gives

$$\begin{aligned}\dot{V} &= \delta c\bar{x}^T J(y) + (1-c)\dot{\bar{y}}^T P_1\bar{y} + (1-c)\bar{y}^T P_1\dot{\bar{y}} \\ &= \delta c\bar{x}^T G(x) + \delta c\bar{x}^T(J(y) - G(x)) \\ &\quad + (1-c)(-H\bar{y} - \dot{y}^q)^T P_1\bar{y} + (1-c)\bar{y}^T P_1(-H\bar{y} - \dot{y}^q) \\ &= \delta c\bar{x}^T(G(x) - G(x^*)) + \delta c\bar{x}^T(J(y) - J(y^q)) \\ &\quad + (1-c)(-H\bar{y} - \dot{y}^q)^T P_1\bar{y} + (1-c)\bar{y}^T P_1(-H\bar{y} - \dot{y}^q) \\ &\leq \delta cl_1\left\|\bar{x}\right\|\left\|\bar{y}\right\| - \delta c\eta\left\|\bar{x}\right\|^2 - (1-c)\bar{y}^T Q_1\bar{y} - 2(1-c)\bar{y}^T P_1\dot{y}^q, \end{aligned} \tag{17}$$

where $G(x^*) = 0_{Nn}$ and $G(x) = J(y^q)$ are used in the derivation of the third equality of (17), Assumption 3 and (12) are used in the derivation of the last inequality of (17) and $l_1 > 0$ is Lipschitz constant satisfying that $\|J(y) - J(y^q)\| \le l_1\|\bar{y}\|$. By the positive-definiteness of the matrix Q_1 and (15), one can further get that

$$
\begin{aligned}
\dot{V} \le & - \delta c\eta\|\bar{x}\|^2 + \delta cl_1 \|\bar{x}\| \|\bar{y}\| \\
& - (1-c)\lambda_{\min}(Q_1)\|\bar{y}\|^2 - 2(1-c)\bar{y}^T P_1\left(\frac{\partial(y^q)^T}{\partial x}\right)^T \frac{dx}{dt} \\
\le & - \delta c\eta\|\bar{x}\|^2 + \delta(cl_1 + 2\sqrt{N}(1-c)l_1^g \|\bar{k}\| \|P_1\|) \|\bar{x}\| \|\bar{y}\| \\
& + 2\delta\sqrt{N}(1-c)l_1 \|\bar{k}\| \|P_1\| \|\bar{y}\|^2 - (1-c)\lambda_{\min}(Q_1)\|\bar{y}\|^2.
\end{aligned}
\tag{18}
$$

Define $r_1 = cl_1 + 2\sqrt{N}(1-c)l_1^g \|\bar{k}\| \|P_1\|$, $r_2 = 2\sqrt{N}(1-c)l_1 \|\bar{k}\| \|P_1\|$, $B_1 = \begin{pmatrix} c\eta & -\frac{r_1}{2} \\ -\frac{r_1}{2} & -r_2 + \frac{(1-c)\lambda_{\min}(Q_1)}{\delta} \end{pmatrix}$. By condition (1), we get that B_1 is positive definite. Then $\dot{V} \le -\delta\lambda_{\min}(B_1)\|z\|^2 \le -\frac{\delta\lambda_{\min}(B_1)}{\gamma_1}V$.

For $t \in [t_k, t_{k+1}) \cap \bar{T}$, $k \in \mathbb{N}$, differentiating V along the trajectories of (8) and (9) gives

$$
\begin{aligned}
\dot{V} = & \delta c\bar{x}^T J(y) + (1-c)\dot{\bar{y}}^T P_1\bar{y} + (1-c)\bar{y}^T P_1\dot{\bar{y}} \\
= & \delta c\bar{x}^T G(x) + \delta c\bar{x}^T (J(y) - G(x)) \\
& + (1-c)(-\dot{y}^q)^T P_1\bar{y} + (1-c)\bar{y}^T P_1 (-\dot{y}^q) \\
= & \delta c\bar{x}^T (G(x) - G(x^*)) + \delta c\bar{x}^T (J(y) - J(y^q))) \\
& + (1-c)(-\dot{y}^q)^T P_1\bar{y} + (1-c)\bar{y}^T P_1 (-\dot{y}^q) \\
\le & - \delta c\eta \|\bar{x}\|^2 + \delta cl_1 \|\bar{x}\| \|\bar{y}\| - 2(1-c)\bar{y}^T P_1\dot{y}^q.
\end{aligned}
\tag{19}
$$

Similar to (18) gives

$$
\begin{aligned}
\dot{V} \le & - \delta c\eta\|\bar{x}\|^2 + \delta(cl_1 + 2\sqrt{N}(1-c)l_1^g \|\bar{k}\| \|P_1\|) \|\bar{x}\| \|\bar{y}\| \\
& + 2\delta\sqrt{N}(1-c)l_1 \|\bar{k}\| \|P_1\| \|\bar{y}\|^2.
\end{aligned}
\tag{20}
$$

Define $B_2 = \begin{pmatrix} -c\eta & \frac{r_1}{2} \\ \frac{r_1}{2} & r_2 \end{pmatrix}$. Obviously, $\lambda_{\max}(B_2) > 0$, then $\dot{V} \le \delta\lambda_{\max}(B_2)\|z\|^2 \le \frac{\delta\lambda_{\max}(B_2)}{\gamma_2}V$.

Given the analysis above, we obtain that $V(t_1) \le V(0)e^{-\delta\Delta_0}$, $\Delta_0 = \frac{\lambda_{\min}(B_1)}{\gamma_1}\zeta_0 - \frac{\lambda_{\max}(B_2)}{\gamma_2}(\rho_0 - \zeta_0)$. According to condition (2), we have $\Delta_0 > 0$. By recursion, for any integer $k \in \mathbb{N}$,

$$
V(t_{k+1}) \le V(0)e^{-\sum_{j=0}^{k}\delta\Delta_j},
\tag{21}
$$

where $\Delta_j = \frac{\lambda_{\min}(B_1)}{\gamma_1}\zeta_j - \frac{\lambda_{\max}(B_2)}{\gamma_2}(\rho_j - \zeta_j)$, $j = 0, 1, 2, \ldots, k$. According to condition (2), we have $\Delta_j > 0$. For any given $t > 0$, there exists a positive integer s such that $t_{s+1} < t \le t_{s+2}$. Furthermore, since $[t_k, t_{k+1})$, $k \in \mathbb{N}$ is uniformly bounded and non-overlapping, we may let $\kappa = \min_{i \in \mathbb{N}} \Delta_i > 0$, $\rho_{\max} = \max_{i \in \mathbb{N}} \rho_i$. Thus, it follows that

$$
\begin{aligned}
V(t) &\leqslant V\left(t_{s+1}\right) e^{\rho_{\max} \delta \frac{\lambda_{\max}(B_2)}{\gamma_2}} \\
&\leqslant e^{\rho_{\max} \delta \frac{\lambda_{\max}(B_2)}{\gamma_2}} V(0) e^{-\sum_{j=0}^{s} \delta \Delta_j} \\
&\leqslant e^{\rho_{\max} \delta \frac{\lambda_{\max}(B_2)}{\gamma_2}} V(0) e^{-\delta(s+1)\kappa} \\
&\leqslant e^{\kappa \delta + \rho_{\max} \delta \frac{\lambda_{\max}(B_2)}{\gamma_2}} V(0) e^{-\frac{\kappa \delta}{\rho_{\max}} t},
\end{aligned}
\tag{22}
$$

that is $V(t) \leqslant K_0 e^{-K_1 t} V(0)$, $\forall t > 0$, where $K_0 = e^{\kappa \delta + \rho_{\max} \delta \frac{\lambda_{\max}(B_2)}{\gamma_2}}$ and $K_1 = \frac{\kappa \delta}{\rho_{\max}}$. Recall that $\gamma_2 \|z\|^2 \leq V \leq \gamma_1 \|z\|^2$, then we have $\|z(t)\|^2 \leq \frac{V}{\gamma_2} \leq \frac{K_0}{\gamma_2} e^{-K_1 t} V(0) \leq \frac{\gamma_1}{\gamma_2} K_0 e^{-K_1 t} \|z(0)\|^2$, then $\|z(t)\| \leq \sqrt{\frac{\gamma_1}{\gamma_2} K_0} e^{-\frac{K_1}{2} t} \|z(0)\|$.

Let $M_r(t) = [(x - x^*)^T, (y - y^q(x^*))^T]^T$, where $y^q(x^*) = 1_N \otimes x^*$, we have $\|M_r(t)\| \leq \|z(t)\| + \|y^q(x) - y^q(x^*)\| \leq R_1 \|z(t)\| \leq R_1 \sqrt{\frac{\gamma_1}{\gamma_2} K_0} e^{-\frac{K_1}{2} t} \|z(0)\|$

$\leq \quad R_1 \sqrt{\frac{\gamma_1}{\gamma_2} K_0} e^{-\frac{K_1}{2} t} (\|M_r(0)\| \quad + \quad \|y^q(x^*) - y^q(x(0))\|) \quad \leq$

$R_2 \sqrt{\frac{\gamma_1}{\gamma_2} K_0} e^{-\frac{K_1}{2} t} \|M_r(0)\|$, for some positive constants R_1 and R_2. Hence, the conclusion is derived.

4   Numerical Example

In this section, a noncooperative game of 10 players is considered. The communication topology for the players is depicted in Fig. 1 and matrix weights are given by

$$
\Gamma_{ij} = \begin{cases} \frac{1}{2} I_2, & (i,j) \in \mathcal{E}, \\ 0_{2\times 2}, & otherwise. \end{cases}
\tag{23}
$$

The players' payoff functions are

$$
f_i(x) = -\frac{1}{2} \sum_{j=1}^{N} (x_i - x_j)^T \Gamma_{ij}(x_i - x_j) - \frac{1}{2}\|x_i + 4\|^2, \quad \forall i \in \mathcal{N},
\tag{24}
$$

where $x_i \in \mathbb{R}^2$.

Through theoretical analysis, the Nash equilibrium is $x_i^* = [-4, -4]^T$, $\forall i \in \mathcal{N}$. The assumptions of Theorem 1 are all satisfied. According to Theorem 1, when parameter δ and average intermittent communication rate a_k satisfy condition (1) and condition (2) respectively, Nash equilibrium is exponentially stable. The simulation is performed in the time interval $[0, 10)$, which is divided into equal parts of length δ and the time intervals that players can communicate with each other are randomly selected from all of the sub–intervals based on the given average intermittent communication rate. Denote the state error $e_i(t) = \|x_i(t) - x_i^*\|$, $i \in \mathcal{N}$. The trajectories of the state errors are plotted in Fig. 2 ($a_k = 0$, $\forall k \in \mathbb{N}$), Fig. 3 ($a_k = 0.01$, $\forall k \in \mathbb{N}$), Fig. 4 ($a_k = 0.6$, $\forall k \in \mathbb{N}$).

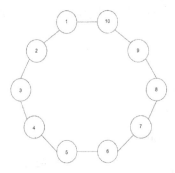

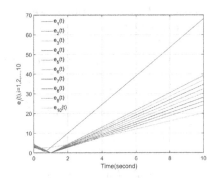

Fig. 1. Communication topology for the players in the numerical examples.

Fig. 2. Plot of $e_i(t), i \in \mathcal{N}$ when $a_k = 0, \forall k \in \mathbb{N}$

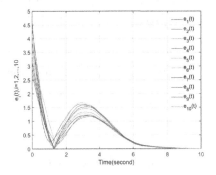

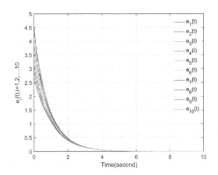

Fig. 3. Plot of $e_i(t), i \in \mathcal{N}$, when $a_k = 0.01, \forall k \in \mathbb{N}$

Fig. 4. Plot of $e_i(t), i \in \mathcal{N}$, when $a_k = 0.6, \forall k \in \mathbb{N}$

When the average intermittent communication rate is 0, all the errors diverge to infinity eventually, which indicates the actions of the players diverge. When the average intermittent communication rate is 0.01, the errors zigzag down to zero, which indicates the actions of the players converge to the Nash equilibrium in a zigzag way, while when the average intermittent communication rate is 0.6, the errors decay to zero monotonically, which indicates the actions of the players converge to the Nash equilibrium. These simulations verify theoretical results.

5   Conclusion

Distributed Nash equilibrium seeking of noncooperative games under intermittent communication and matrix-weighted communication graph is considered in this paper. A communication threshold is proposed to make sure the states of players can converge to Nash equilibrium. Through Lyapunov method, it is theoretically proved that the states of players exponentially converge to Nash equilibrium under the proposed algorithm when parameter δ and average intermittent communication rate a_k satisfy certain conditions. Finally, theoretical results are validated through numerical simulations.

References

1. Wen, G., Yu, W., Lv, Y., Wang, P.: Cooperative Control of Complex Network Systems with Dynamic Topologies, 1st edn. CRC Press, Boca Raton (2021)
2. Zhou, J., Wen, G., Lv, Y., Yang, T., Chen, G.: Distributed resource allocation over multiple interacting coalitions: a game-theoretic approach. arXiv:2210.02919 (2022)
3. Zhou, J., Lv, Y., Wen, C., Wen, G.: Solving specified-time distributed optimization problem via sampled-data-based algorithm. IEEE Trans. Netw. Sci. Eng. **9**(4), 2747–2758 (2022)
4. Ratliff, L., Burden, S., Sastry, S.: On the characterization of local Nash equilibria in continuous games. IEEE Trans. Autom. Control **61**(8), 2301–2307 (2016)
5. Zhou, J., Lv, Y., Wen, G., Lü, J., Zheng, D.: Distributed Nash equilibrium seeking in consistency-constrained multi-coalition games. IEEE Trans. Cybern. **53**(6), 3675–3687 (2023)
6. Vamvoudakis, K., Lewis, F., Hudas, G.: Multi-agent differential graphical games: online adaptive learning solution for synchronization with optimality. Automatica **48**(8), 1598–1611 (2012)
7. Gharesifard, B., Cortes, J.: Distributed convergence to Nash equilibria in two-network zero-sum games. Automatica **49**(6), 1683–1692 (2013)
8. Koshal, J., Nedić, A., Shanbhag, U.V.: Distributed algorithms for aggregative games on graphs. Oper. Res. **64**(3), 680–704 (2016)
9. Ye, M., Hu, G.: Distributed Nash equilibrium seeking by a consensus based approach. IEEE Trans. Autom. Control **62**(9), 4811–4818 (2017)
10. Ye, M., Hu, G.: Distributed Nash equilibrium seeking in multiagent games under switching communication topologies. IEEE Trans. Cybern. **48**(11), 3208–3217 (2018)
11. Wen, G., Wang, P., Lv, Y., Chen, G., Zhou, J.: Secure consensus of multi-agent systems under denial-of-service attacks. Asian J. Control **25**(2), 695–709 (2023)
12. Kobo, H.I., Abu-Mahfouz, A.M., Hancke, G.P.: A survey on software-defined wireless sensor networks: challenges and design requirements. IEEE Access **5**, 1872–1899 (2017)
13. Guo, H., Wang, H.: A resource allocation algorithm for multimedia communication systems with limited bandwidth. Int. J. Commun. Syst. **22**(4), 399–414 (2009)
14. Wen, G., Duan, Z., Yu, W., Chen, G.: Consensus in multiagent systems with communication constraints. Int. J. Robust Nonlinear Control **22**(2), 170–182 (2012)
15. Krahmer, F., Ward, R., Wang, Y.: Efficient matrix weighted low-rank approximation. IEEE Trans. Signal Process. **64**(3), 749–762 (2016)
16. Tuna, S.E.: Synchronization under matrix-weighted Laplacian. Automatica **73**, 76–81 (2016)
17. Wen, G., Zheng, W., Wan, Y.: Distributed robust optimization for networked agent systems with unknown nonlinearities. IEEE Trans. Autom. Control **68**, 5230–5244 (2022). https://doi.org/10.1109/TAC.2022.3216965
18. Nash, J.: Non-cooperative games. Ann. Math. **54**(4), 286–295 (1951)
19. Rosen, J.B.: Existence and uniqueness of equilibrium points for concave N-person games. Econometrica **33**(3), 520–534 (1965)

Accelerated Genetic Algorithm with Population Control for Energy-Aware Virtual Machine Placement in Data Centers

Zhe Ding[1], Yu-Chu Tian[1(✉)], Maolin Tang[1], You-Gan Wang[2], Zu-Guo Yu[3], Jiong Jin[4], and Weizhe Zhang[5,6]

[1] School of Computer Science, Queensland University of Technology, Brisbane, QLD 4001, Australia
zhe.ding@hdr.qut.au, {y.tian,m.tang}@qut.edu.au

[2] Institute for Learning Sciences and Teacher Education, Australian Catholic University, Brisbane, QLD 4000, Australia
you-gan.wang@acu.edu.au

[3] Key Laboratory of Intelligent Computing and Information Processing of the Ministry of Education of China, and Hunan Key Laboratory for Computation and Simulation in Science and Engineering, Xiangtan University, Xiangtan 411105, China
yuzuguo@aliyun.com

[4] School of Science, Computing and Engineering Technologies, Swinburne University of Technology, Melbourne, VIC 3122, Australia
jiongjin@swin.edu.au

[5] School of Cyberspace Science, Harbin Institute of Technology, Harbin 150001, China
wzzhang@hit.edu.cn

[6] The Cyberspace Security Research Center, Peng Cheng Laboratory, Shenzhen, China

Abstract. Energy efficiency is crucial for the operation and management of cloud data centers, which are the foundation of cloud computing. Virtual machine (VM) placement plays a vital role in improving energy efficiency in data centers. The genetic algorithm (GA) has been extensively studied for solving the VM placement problem due to its ability to provide high-quality solutions. However, GA's high computational demands limit further improvement in energy efficiency, where a fast and lightweight solution is required. This paper presents an adaptive population control scheme that enhances gene diversity through population control, adaptive mutation rate, and accelerated termination. Experimental results show that our scheme achieves a 17% faster acceleration and 49% fewer generations compared to the standard GA for energy-efficient VM placement in large-scale data centers.

Keywords: Data center · energy efficiency · virtual machine · genetic algorithm · population

B. Luo et al. (Eds.): ICONIP 2023, LNCS 14448, pp. 14–26, 2024.
https://doi.org/10.1007/978-981-99-8082-6_2

1   Introduction

Data centers are essential for supporting the growing demand for cloud services worldwide. However, this increased reliance on data centers has led to a surge in electricity consumption. In the United States alone, data centers consumed 78 billion kWh of energy in 2020, necessitating the construction of 50 additional large power plants each year to meet the demand [5]. Therefore, improving the energy efficiency of data centers has become crucial.

Among various factors, the placement of virtual machines (VMs) onto physical machines (PMs) plays a significant role in enhancing energy efficiency for large-scale data centers [1]. Improved VM placement strategies have shown energy savings of over 20% in recent reports [1], which motivates our research of power-aware VM placement techniques to further reduce energy consumption.

The First Fit Decreasing (FFD) and Best Fit Decreasing (BFD) algorithms are commonly used heuristics for VM placement. FFD effectively handles general bin-packing problems and has been adopted for energy optimization in VM placement [6]. BFD, another heuristic algorithm, also addresses bin-packing problems based on resource utilization ratios.

The Genetic Algorithm (GA) has been explored as meta-heuristics for optimized resource management in data centers [10]. It offers higher-quality solutions than FFD or BFD, albeit with increased execution time, due to searching a larger solution space.

However, GA suffers from slow execution [4], making its computation, such as crossover and mutation, time-consuming for energy-efficient data centers. Thus, there is a need to accelerate GA for real-world data center applications.

Efforts have been made to tune GA parameters for improved performance [9]. An adaptive GA (AGA) has been developed, optimizing parameters such as population size, crossover rate, mutation rate, generation gap, scaling window, and selection strategy [9]. Another recent work focuses on improving GA's computation for virtual resource management in data centers [8], using the concept of decrease-and-conquer and an initial feasible solution derived from FFD. However, all these studies primarily use on static parameters in GA.

This paper presents our progress in accelerating GA for energy-efficient VM placement in data centers. Our main contribution is an adaptive population control scheme that enhances gene diversity. It is achieved through dynamic population control, adaptive mutation rates, and accelerated termination.

2   Insights into GA Computation

In our previous study of the computation for GA's fitness function (FF), we have the following equation:

$$T_{fit} = \frac{1}{2 * f * u_{cpu}} * N_{pop}^2 * N_{gen} * N_{ic\text{-}fit1} \tag{1}$$

where T_{fit} stands for computation time for a single FF, f for CPU frequency, u_{CPU} for CPU utilization, N_{pop} for population size, N_{gen} for the amount of executed generations (iterations), and $N_{ic\text{-}fit1}$ for the amount of CPU instructions for a single FF computation.

According to Eq. (1), the value of the term $\frac{1}{2*f*u_{cpu}}$ is constant, which is determined by the system configuration. Thus, N_{pop}, N_{gen} and N_{fit1} are dominant factors for the fitness computation.

On the other hand, if we are able to calculate and save the fitness value of a VM placement plan as soon as it is generated, we do not need to calculate it again during tournament. Thus, in the ith generation, we have $N_{fit-new}$ and $N_{ic\text{-}fit1-new}$ as the new calculation of N_{fit1} and $N_{ic\text{-}fit1}$, respectively:

$$N_{fit-new} = \sum_{i=1}^{N_{gen}} N_{pop-i} \qquad (2)$$

where N_{pop-i} stands for the population size of the ith generation.

Then, considering both Eqs. (2) and (1), we have:

$$N_{ic-fit1-new} = \frac{f * T_{fit-new} * u_{cpu}}{\sum_{i=1}^{N_{gen}} N_{pop-i}} \qquad (3)$$

From Eq. (3), $T_{fit-new}$ can be expressed as:

$$T_{fit-new} = \frac{1}{f * u_{cpu}} * N_{ic-fit1-new} * \sum_{i=1}^{N_{gen}} N_{pop-i} \qquad (4)$$

According to Eqs. (1) and (4), $T_{fit-new} < T_{fit}$ when any N_{pop-i} is never far greater than N_{pop}.

3   Adaptive Population Control

3.1   Basic Architecture

Enriching GA's gene diversity has several advantages, such as a higher chance of achieving greater optimality and avoiding local-optimal traps. These advantages significantly enhance the final results of the GA, as confirmed by our experimental findings presented later in this paper. However, as the population size grows, practical drawbacks arise, particularly increased computational demand.

Therefore, we have two objectives for the population size in our GA: 1) Enlarge the population size to enhance gene diversity and achieve optimal solutions, and 2) Limit the population size before the computational demand becomes excessively high. To fulfill these objectives, we introduce the concept of Environmental Carrying Capacity (ECC) into our GA to control its population. ECC represents the relationship between convergence and population size in our GA.

To illustrate the relationship of ECC, consider the following example phases. When the GA starts, it is still far from reaching the optimal solution or convergence. In other words, the population has not reached the limit of ECC in terms of evolution or group size. At this stage, the population is about to grow and evolve. As the GA gets closer to the optimal solution and convergence, the population approaches the limit of ECC. Evolution becomes more challenging, and the group size stops growing. Finally, when the GA terminates with the optimal solution, the population, or civilization, has surpassed ECC. As a result, the population decreases with minimal evolution. These example phases are illustrated in Fig. 1.

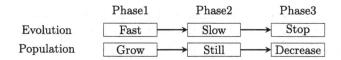

Fig. 1. The architecture of our population control scheme inspired by ECC. The three 'phases' are automatically merged by our scheme in GA, not manually separated.

3.2   Managing the Population Size in GA

To quantify ECC in our GA, we use a parameter g to represent the quotient of current generation (f_c) divided by the fitness value of previous generation (f_p):

$$g = f_c / f_p \tag{5}$$

Then, we use it to control the population of our GA. In the ith generation, current population size (N_{pop-i}) is designed to gain a chance to grow or decrease, which is determined by ECC or g. Firstly, we determine the amount of individuals (as VM-placement plans), which is about to populate ($\lceil \rceil$ stands for the ceiling integer):

$$N_{pop-i\ add} = \lceil N_{pop-(i-1)}/g_i \rceil - N_{pop-(i-1)} \tag{6}$$

where g_i refers to the g value in ith generation. Since g_i is positive and its maximum is 1, $N_{pop-i\ add}$ must be an natural number.

Secondly, for each individual that is about to join population, the chance is to be determined if this individual is really about to be born (like the chance against fetal mortality):

$$P(N_{pop-i\ grow}) = f_c / f_0 \tag{7}$$

where f_0 stands for the fitness value of the very first generation. In addition, when an individual confirms to be born by Eq. (7), it will be granted an additional mutation process to further enrich gene diversity.

Thirdly, if the population has not evolved at all, there is a chance that GA may eliminate a worst plan from the population:

$$P(N_{pop-i\ decrease}) = R_{m-i} * lg(N_{nig}) \tag{8}$$

where R_{m-i} stands for the mutation rate in the ith generation, and N_{nig} stands for the number of successive generations where population has not evolved since last successful evolution.

Combining Eqs. (6)–(8), we finally have:

$$N_{pop-i} = N_{pop-(i-1)} + N_{pop-i\ grow} - N_{pop-i\ decrease} \tag{9}$$

N_{pop-i} stands for the population size for the next generation. A new group of young individuals are now ready to their life in our GA.

3.3 Updating Systematical Parameters Dynamically

Updating Mutation Rate Dynamically. Among GA's genetic operators, mutation is the operator which directly affect genes in every individual. It forces one individual to update its gene, 'randomly'. Thus, mutation may enrich the gene diversity by mutating gene into something which never appears in the population. For example, in our previous experiments, GA works up to 1% greater with 1.5% more mutation rate.

In addition, notice that $P(N_{pop-i\ decrease})$ is determined by the mutation rate in current generation (as R_{m-i}). Therefore, we have designed an mutation rate which dynamically updates while the population evolving. We have R_{m-i} updated dynamically as follows:

$$R_{m-i} = R_{m-(i-1)} * [1 + (1 - g_i) * lg(i)] \tag{10}$$

In Eq. (10), i is monotonously increasing, with $lg(i)$ following $O(\log n)$ trend, and $(1 - g_i)$ is always positive. Thus, R_{m-i} is monotonously increasing as well.

Updating the Condition for Termination Dynamically. GA's termination is often judged by N_{nig}, which is introduced in Eq. (8). When N_{nig} grows beyond a given threshold N_{t-base}, GA automatically terminates and outputs its final solution. However, N_{nig} refers to the generations with no improvement. Thus, if N_t is adjusted and reduced, less generations will be wasted by judging GA's termination. Therefore, we design N_{t-i}, which dynamically updates along with GA's generations, to reduce the 'junk time' from GA. N_{t-i} is relevantly inferior to N_{t-base}. N_{t-i} is designed as follow:

$$N_{t-i} = N_{t-base}^2 / N_{pop-i} \tag{11}$$

3.4 Population Control Scheme

Our population control scheme is summarized in Algorithm 1. Initially, Algorithm 1 collects f_c, f_p, and $N_{pop-(i-1)}$. Then, in Line 1, it calculates g as the fundamental parameter for subsequent processing. Lines 2 to 7 manage the populating process, determining $N_{pop-i-add}$ as the populating size. The process copies

and mutates the best individual to finalize the populating step. Following populating, Lines 8 and 9 handle the depopulating process, removing the worst individual when $N_{nig} > 0$. Line 10 updates the mutation rate R_{m-i}. Line 11 updates
N_{t-i}, the accelerating termination parameter in our GA. Finally, Algorithm 1
outputs the updated population for the next generation.

To provide an intuitive representation of the population-control scheme in our
GA, we execute a sample GA run. Figure 2 illustrates the evolution of N_{pop} and
R_m in this GA run. This sample GA run follows our experimental configuration
simulating a large-scale data center, with the pattern being set as P5.

Algorithm 1: Population control for our new GA

Input: Current population
Output: Population for the next generation
Initialize: Get f_c, f_p, and $N_{pop-(i-1)}$
1 Calculate g by Eq. (5);
2 Calculate $N_{pop-i-add}$ by Eq. (6);
3 **foreach** $N_{pop-i-add}$ **do**
4   **if** *Random chance by Eq. (7)* **then**
5   Find and copy the best plan in current generation;
6   Mutate the copied plan;
7   Add the copied plan into population;

8 **if** $N_{nig} > 0$ *and Random chance by Eq. (8)* **then**
9   Remove the worst plan from population;
10 Update R_{m-i} by Eq. (10);
11 Update N_{t-i} by Eq. (11);
12 **return** *Population for the next generation*;

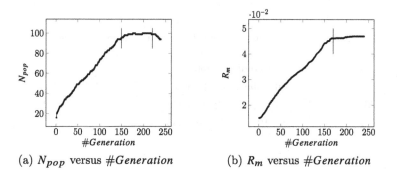

(a) N_{pop} versus #Generation   (b) R_m versus #Generation

Fig. 2. A large-scale GA run, which is integrated with Algorithm 1 and P5.

4 Simulation Experiments

4.1 Experimental Design

Our experiments aim to evaluate the effectiveness of the presented alternative features of GA, especially for energy-efficient VM placing problems in data centers. All 5 features presented will be tested: improved data structure in Eq. (4), enlarging population size in Eqs. (6) and (7), limiting population size in Eq. (8), adjusting mutation rate in Eq. (10), and accelerating termination in Equation (11). Six patterns are designed for experiments, namely P0–P5, as listed in Table 1. P0 is the original GA.

Google's Cluster-Usage Traces [7] are used as the input data set. As the data set is huge, only a small part of the logs is extracted from the traces for demonstration. Our Data configurations are tabulated in Table 2. The medium-scale data input represents a private, university-level cloud data center, running about two thousand VMs in 500 to 600 PMs. The large-scale scenario simulates a large, public cloud data center, running thousands of VMs in over 1,000 PMs.

Table 1. Pattern configuration. "✓" indicates that the this pattern has been integrated with corresponding feature, "·" means no this pattern.

Feature	Pattern					
	P0	P1	P2	P3	P4	P5
Improved data structure	·	✓	✓	✓	✓	✓
Adding population size	·	·	✓	✓	✓	✓
Reducing population size	·	·	·	✓	✓	✓
Dynamic mutation rate	·	·	·	·	✓	✓
Faster termination	·	·	·	·	·	✓

Table 2. The input configuration from Google's data set.

Scale	Input configuration			
	Start time	Initial job ID	Initial task ID	#Items
Medium scale	$3.717 * 10^{11}$	6289303978	424	2242
Large scale	$1.8 * 10^9$	4028922835	429	5365

Algorithm 2: Task assignment [2,3]

Input: All tasks (applications)
Output: A plan of application assignments to VMs
Initialize: An empty VM set
1 **foreach** *task in all given tasks* **do**
2   Assign this task to a VM of proper size;
3   Append this VM into the VM set;

4 Convert the VM set to the assignment plan;
5 **return** *an application assignment plan*;

It is worth mentioning that Google's cluster traces only record tasks, not VMs. To use Google's data set for VM-placement research, it is necessary to assign these tasks to VMs prior to placing VMs to PMs. The task-assignment Algorithm 2 is a replicate of the work in [2,3]. It assigns each incoming task to a VM of an appropriate proper size. The VMs in our experiments are designed with fixed sizes, which are specified in Table 3. The use of fixed VM sizes is a common practice in data center management, as in Amazon's cloud.

In addition, the FF computation from our previous study is given in Algorithm 3. Experiments in this paper are integrated with this FF to initially accelerate corresponding GAs.

After all tasks are assigned into VMs, we execute our GA for VM placement to PMs. Our GA implies tournament selection strategy. In addition, we use an First-Fit solution as one individual in the GA's first generation for better initialization [8].

Algorithm 3: FF computation of fitness N_{apm}

Input: A VM-placement plan, including the corresponding PM list
Output: Fitness value N_{apm} (as an indicator of $\mathbb{E}$)
Initialize: An empty PM counter N_{apm}
1 **foreach** *Planned VM* **do**
2   Update corresponding active PM status;

3 **foreach** *PM on the PM list* **do**
4   **if** *PM utilization* $\neq 0$ **then**
5         $N_{apm} \leftarrow N_{apm} + 1$;

6 **return** N_{apm} *as this plan's fitness value*;

Table 3. Six types of VMs with normalized CPU capacity.

CPU type	Huge	Large	Medium	Normal	Small	Tiny
CPU capacity	0.45	0.30	0.15	0.10	0.045	0.015

Our simulation experiments are conducted on a desktop computer. The computer is equipped with Intel Core 2 Q6700, 3.4 GHz CPU, and 16 GB DDR4 2666 MHz RAM. It runs Windows 10 Professional operating system. The input of our experiments is a comma separated-value (CSV) file extracted from Google's Cluster-usage Traces. The outputs are a placement plan, the energy consumption of the data center, the number of GA generations, GA execution time, and the average execution time for each GA generation.

4.2 Results for Large-Scaled Data Center

In terms of #Generation, as shown in Fig. 3 (c), P0 and P1 require similar generations because they have the same genetic operators. P2–P5 require much fewer generations than P0 and P1 due to the population control features. Nevertheless, P5 costs least generations, empowered by an accelerated termination. P3 costs slightly more generations due to the fact that limiting population size alone does not help convergence.

On #PMs, as shown in Fig. 3 (a), all 6 patterns show minor differences. This is due to the fact that GA has been very close to the optimal, and even a 0.1% improvement on average will be significant.

Regarding the computation time, significant differences are observed among the patterns, as depicted in Fig. 3 (b). P1 demonstrates faster performance compared to P0, due to our improved data structure. P2 exhibits the slowest execution time due to its large population size. However, by limiting the population

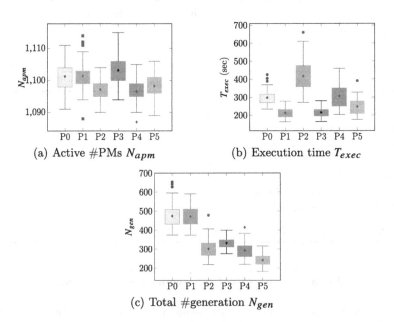

(a) Active #PMs N_{apm} (b) Execution time T_{exec}

(c) Total #generation N_{gen}

Fig. 3. Box plots of experimental results for a large-scaled data center. The configurations of P0–P5 are presented in Table 1.

size, P3 achieves considerably faster results than P2. It is worth noting that as the mutation rate increases, P4 and P5 experience slower execution times due to the additional computational overhead associated with mutation in each generation. Notably, P5 outperforms P4 in terms of speed, owing to its accelerated termination mechanism. Overall, P1, P3, and P5 remain faster than P0.

The numerical comparisons presented in Table 4 support these observations. They indicate that with the incorporation of all our enhancements, P5 achieves a speedup of 17% compared to P0, with 49% fewer generations and a 0.28% improvement in energy savings. Furthermore, P4 achieves the highest energy savings while maintaining a similar computation time as P0, with 38% fewer generations. Lastly, by solely improving the data structure, P1 achieves a speedup of 29% compared to P0.

Table 4. Experimental results for the large-scaled data center. All values are calculated as percentage compared with P0 on average.

Feature	Pattern				
	P1	P2	P3	P4	P5
Computation time	−29%	40%	−28%	3%	−17%
#Generations	−1%	−37%	−30%	−38%	−49%
#PMs	0.01%	−0.37%	0.17%	−0.43%	−0.28%

4.3   Results for Medium-Scaled Data Center

The following figures and tables indicates our experimental results in our medium-scaled data center.

From the perspective of the number of generations (#Generation), as depicted in Fig. 4 (c), P0 and P1 require a similar number of generations since P1 uses the same genetic operators as P0. In contrast, P2 to P5 demonstrate significantly fewer generations compared to P0 and P1, benefiting from our population control features. However, P4 and P5 require slightly more generations. This is attributed to the fact that within the medium-scale input, population size and gene diversity have a greater impact on convergence compared to just accelerating termination.

Regarding #PMs, as shown in Fig. 4 (a), all six patterns exhibit minor differences. This is because the GA has been very close to the optimal solution, and even a 0.1% improvement in the average value becomes significant.

In terms of computation time, as illustrated in Fig. 4 (b), significant differences can be observed among the patterns. P1 demonstrates faster performance than P0, due to our improved data structure. P2 and P3 exhibit the slowest execution times due to their large population sizes. However, as the mutation rate increases, P4 and P5 become faster due to the accelerated convergence mentioned earlier. Overall, P1, P4, and P5 remain faster than P0.

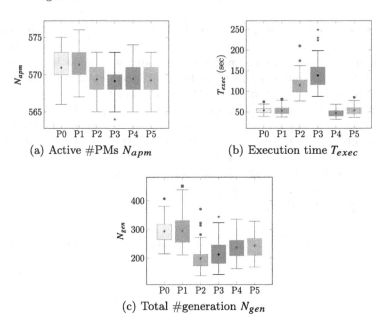

(a) Active #PMs N_{apm}

(b) Execution time T_{exec}

(c) Total #generation N_{gen}

Fig. 4. Box plots of experimental results for a large-scaled data center. Configuration of P0–P5 is presented in Table 1.

The numerical comparisons shown in Table 5 provide further insights. It can be observed that with the incorporation of all our enhancing features, P5 achieves the same computational speed as P0, while requiring 21% fewer generations and achieving a 0.3% improvement in energy savings. Moreover, P3 achieves the best energy savings but requires a significantly longer execution time.

Table 5. Experimental results for the medium-scaled data center. All values are calculated as percentage compared with P0 on average.

Feature	Pattern				
	P1	P2	P3	P4	P5
Computation time	−1.5%	114%	158%	−13%	0.1%
#Generations	−0.3%	−47%	−38%	−24%	−21%
#PMs	0.2%	−0.27%	−0.32%	−0.26%	−0.3%

Table 6. Average execution time in each generation for the large-scaled data center. All values are calculated as percentage compared with P0 on average.

Scale	Pattern				
	P1	P2	P3	P4	P5
Large scale	−28%	121%	3%	66%	63%
Medium scale	−1%	−37%	−30%	−38%	−49%

4.4 Additional Computation Demand in Each Generation

In addition to the experimental perspectives discussed earlier (such as N_{apm}, N_g, and T), we also examine the computational demand in each generation of our tested GAs. Several factors can influence the computation demand in each generation, including data structure, population size, and mutation rate. Here we provide a deeper understanding of GA's computation and discusses potential solutions to further enhance GA's convergence.

Table 6 presents the average duration of each generation for both scales, categorized by our six patterns. It compares the average duration of generations for P1 to P5 with that of P0. P1 exhibits a faster computation time than P0 in each generation due to its improved data structure. On the other hand, P2 to P5 are slower, in different degrees, primarily due to our population control scheme. The introduction of additional individuals or a higher mutation rate can result in increased computational demand for GAs.

Lastly, it is worth noting that increasing the computational demand in GA does not necessarily guarantee better results. For instance, let us consider P2 as an example. It takes 40% longer than P4 or P5 to execute. However, as shown in Table 4, P2 does not achieve significant improvements in terms of energy perspective (N_{apm}), which is the primary optimization objective in this paper. This highlights the significance of our approaches to depopulate, increase mutation rate, and accelerate termination in order to achieve the desired results.

5 Conclusion

A population-controlling scheme has been presented in this paper for GA computation of energy-efficient VM placement in large-scale cloud data centers. While improving the quality of solution in terms of energy savings in data centers, our approach is shown to run 17% faster with 49% fewer generations compared to the standard GA for VM placement. The GA acceleration is achieved through the integration of adaptive population control, mutation rate adjustment, and accelerated termination in a unified framework. Heuristics developed from extensive experiments have been provided for managing population scale, mutation rate, and termination configuration, resulting in significant enhancements in computational efficiency.

Acknowledgements. This work was supported in part by the Australian Research Council (ARC) through the Discovery Project Scheme under Grant DP220100580 and Grant DP160104292, and the Industrial Transformation Training Centres Scheme under Grant IC190100020.

References

1. Alharbi, F., Tian, Y.C., Tang, M., Zhang, W.Z., Peng, C., Fei, M.: An ant colony system for energy-efficient dynamic virtual machine placement in data centers. Expert Syst. Appl. **120**, 228–238 (2019)
2. Ding, Z., Tian, Y.C., Tang, M.: Efficient fitness function computation of genetic algorithm in virtual machine placement for greener data centers. In: 2018 IEEE 16th International Conference on Industrial Informatics (INDIN), pp. 181–186, Porto, Portugal, 18–20 July 2018
3. Ding, Z., Tian, Y.C., Tang, M., Li, Y., Wang, Y.G., Zhou, C.: Profile-guided three-phase virtual resource management for energy efficiency of data centers. IEEE Trans. Industr. Electron. **67**(3), 2460–2468 (2020)
4. Elsayed, S., Sarker, R., Coello Coello, C.A.: Fuzzy rule-based design of evolutionary algorithm for optimization. IEEE Trans. Cybern. **49**(1), 301–314 (2019)
5. Kumar, S., Pandey, M.: Energy aware resource management for cloud data centers. Int. J. Comput. Sci. Inf. Secur. **14**(7), 844 (2016)
6. Liu, Z., Xiang, Y., Qu, X.: Towards optimal CPU frequency and different workload for multi-objective VM allocation. In: 2015 12th Annual IEEE Consumer Communications and Networking Conference (CCNC), pp. 367–372 (2015)
7. Reiss, C., Wilkes, J., Hellerstein, J.L.: Google cluster-usage traces: format+schema. Google Inc., White Paper, pp. 1–14 (2011)
8. Sonkiln, C., Tang, M., Tian, Y.C.: A decrease-and-conquer genetic algorithm for energy efficient virtual machine placement in data centers. In: IEEE 15th International Conference on Industrial Informatics (INDIN 2017). Eden, Germany, 24–26 July 2017
9. Srinivas, M., Patnaik, L.M.: Adaptive probabilities of crossover and mutation in genetic algorithms. IEEE Trans. Syst. Man Cybern. **24**(4), 656–667 (1994)
10. Wu, G., Tang, M., Tian, Y.-C., Li, W.: Energy-efficient virtual machine placement in data centers by genetic algorithm. In: Huang, T., Zeng, Z., Li, C., Leung, C.S. (eds.) ICONIP 2012. LNCS, vol. 7665, pp. 315–323. Springer, Heidelberg (2012). https://doi.org/10.1007/978-3-642-34487-9_39

A Framework of Large-Scale Peer-to-Peer Learning System

Yongkang Luo[1] , Peiyi Han[1,2](✉) , Wenjian Luo[1,2] , Shaocong Xue[1] ,
Kesheng Chen[1] , and Linqi Song[3]

[1] Guangdong Provincial Key Laboratory of Novel Security Intelligence Technologies,
School of Computer Science and Technology, Harbin Institute of Technology,
Shenzhen 518055, Guangdong, China
hanpeiyi@hit.edu.cn
[2] Peng Cheng Laboratory, Shenzhen 518055, Guangdong, China
[3] Department of Computer Science, City University of Hong Kong, Hong Kong,
China

Abstract. Federated learning (FL) is a distributed machine learning
paradigm in which numerous clients train a model dispatched by a central server while retaining the training data locally. Nonetheless, the failure of the central server can disrupt the training framework. Peer-to-peer
approaches enhance the robustness of system as all clients directly interact with other clients without a server. However, a downside of these
peer-to-peer approaches is their low efficiency. Communication among
a large number of clients is significantly costly, and the synchronous
learning framework becomes unworkable in the presence of stragglers. In
this paper, we propose a semi-asynchronous peer-to-peer learning system (P2PLSys) suitable for large-scale clients. This system features a
server that manages all clients but does not participate in model aggregation. The server distributes a partial client list to selected clients that
have completed local training for local model aggregation. Subsequently,
clients adjust their own models based on staleness and communicate
through a secure multi-party computation protocol for secure aggregation. Through our experiments, we demonstrate the effectiveness of
P2PLSys for image classification problems, achieving a similar performance level to classical FL algorithms and centralized training.

Keywords: Federated learning · Semi-asynchronous learning ·
Peer-to-peer learning system

This study is supported by the National Key R&D Program of China (Grant
No. 2022YFB3102100), Shenzhen Fundamental Research Program (Grant No.
JCYJ20220818102414030), the Major Key Project of PCL (Grant No. PCL2022A03,
PCL2023AS7-1), Guangdong Provincial Key Laboratory of Novel Security Intelli-
gence Technologies (No. 2022B1212010005), Shenzhen Science and Technology Pro-
gram (Grant No. ZDSYS20210623091809029, RCBS20221008093131089).

B. Luo et al. (Eds.): ICONIP 2023, LNCS 14448, pp. 27–41, 2024.
https://doi.org/10.1007/978-981-99-8082-6_3

1   Introduction

Deep neural networks (DNNs) with exceptional performance typically require a large amount of available data. Consequently, obtaining sufficient data for high-performance deep learning networks is a critical challenge. Traditional training approaches are centralized, involving the collection of data from numerous applications. However, with the introduction of privacy constraints and regulations, data sharing has become increasingly restricted.

Federated Learning (FL) is a promising paradigm in distributed machine learning (ML) that facilitates learning without sharing the actual datasets [1]. FL establishes a distributed framework that consists of a centralized server and numerous clients, thus overcoming data sharing limitations while still supporting efficient learning [2,3]. The server needs to perform multiple operations such as client selection, update collection, model aggregation and global model broadcast [4].

Each client trains the model locally using their own private dataset for a certain number of epochs. Subsequently, the server selects specific clients to aggregate their updated models and broadcasts the aggregated model to all clients. The core idea of FL is to "bring the code to the data, instead of the data to the code" [5], ensuring privacy while maintaining effective learning within the distributed framework.

FL can be classified into asynchronous and synchronous frameworks [6]. The synchronous FL approach is often inefficient, as it requires the server to wait for all clients to complete local training before model aggregation. Due to device heterogeneity and network instability, the presence of stragglers is often unavoidable, which significantly impairs algorithm performance [7]. To mitigate the influence of stragglers, asynchronous FL (AFL) has been proposed [8]. Specifically, the server aggregates local models from clients immediately without waiting for all clients to complete their processes. However, stale local models based on outdated global models must be adjusted before aggregation to balance staleness tolerance and convergence rate [7,9,10]. In addition, semi-asynchronous FL is introduced as a hybrid of synchronous FL and AFL [11], where the server stores the local models and aggregates them periodically.

FL workflows can generally be divided into two categories: centralized aggregation and peer-to-peer approaches [12]. In centralized topology, a server coordinates all training nodes, aggregating and distributing models to and from the peer nodes. However, the central server can often be unreliable, and a server failure could lead to the entire federated learning system collapsing. Decentralized federated learning, which does not require a central server, is an area worth investigating. In decentralized FL, there is no centralized server, and instead, each peer node aggregates the model by connecting to one or more nodes. A serverless, peer-to-peer federated learning framework has been proposed to address these concerns [13]. Swarm Learning (SL), which operates without a centralized server and involves selecting a leader node to coordinate all other nodes, has also gained significant research interest [14].

In both FL and SL, model parameters or gradients are directly transmitted, implying that the client's model is visible to the server or leader client. This visibility raises privacy concerns due to the potential presence of malicious or honest-but-curious clients [15]. To address this issue, participants can jointly determine a common model by averaging all associated model weights without sharing the actual weight values [16]. The secure multi-party computation protocol is utilized to calculate the aggregated model in a peer-to-peer manner, resulting in secure aggregation that upholds the privacy of each client's model [17].

Compared to traditional centralized FL, decentralized FL offers improved robustness. However, the communication between peer nodes can be costly, particularly when the number of nodes is large in decentralized FL. For instance, the piecewise secure multi-party summation protocol can significantly increase communication costs and demand more network resources in a P2PL paradigm [18]. Additionally, synchronous frameworks are acceptable when the number of peer nodes is small. However, when dealing with a large number of peer nodes, the emergence of lagging nodes can greatly reduce training efficiency. Therefore, it is necessary to design an efficient large-scale P2PL system framework that features reduced memory requirements, more efficient communication protocols, and asynchronous training. Especially, the server cannot be used to store and aggregate all parameters when we are training a large model. Instead, in P2PLSys, the server only plays a management and scheduler role. GPUs can communication and exchange model information with each other. The P2PLSys algorithm can become a more practical solution for various machine learning scenarios.

In this work, we propose P2PLSys, a semi-asynchronous learning framework designed to mitigate the effects of stragglers in large-scale peer nodes. Specifically, a central server is utilized to record essential information, the IDs of nodes, in order to reduce memory pressure on clients. However, the central server does not store any model information and is only responsible for scheduling clients. When a subset of clients has completed training, the corresponding nodes' IDs are downloaded from the clients and local aggregation. Models and data remain stored within the nodes, and model aggregation is restricted to clients. We employ an efficient secure multi-party protocol for secure aggregation. The main contributions of this paper are listed as follows.

- We propose a P2PLSys framework for large-scale clients. P2PLSys is a semi-asynchronous framework designed to improve the training efficiency of peer-to-peer approaches. Furthermore, we propose a model adjustment method based on staleness and a secure aggregation method based on secure multi-party computation protocols.
- Our experiments demonstrate high accuracy and efficiency in the models of each node under practical settings, showcasing the promise of our proposed framework in addressing the challenges of large-scale, decentralized learning.

The remainder of this paper is organized as follows. In Sect. 2, we introduce some related work. Then, P2PLSys is described in detail including the architecture and overall process in Sect. 3. Meanwhile, we introduce the modules of the server and peer nodes. In Sect. 4, we verify the performance of P2PLSys in two datasets. Finally, Sect. 5 comes to a conclusion about this paper.

2 Preliminaries

In this section, we begin by briefly introducing the basic concept of P2PL paradigm. Next, we present a simple secure multi-party computation protocol for secure aggregation. Finally, we describe weight adjustment methods used in asynchronous FL, as we apply a similar idea in our P2PLSys framework.

2.1 P2PL Paradigm

The P2PL paradigm is employed to achieve truly decentralized learning without a server or leader node different from FL or SL. Suppose there are K peer nodes in the P2PL paradigm. Once the local training of all peer nodes is completed, the K peer nodes aggregate their models by communicating each other directly. Every node can communicate with all other nodes to get their model information. However, it dangerous and inefficient. For secure aggregation, the secure multi-party computation protocol is always utilized in P2PL paradigm [16,19,20]. The secure multi-party computation protocols can protect the model information, ensuring that models remain invisible among all peer nodes. However, the most pressing issue to be addressed in the P2PL paradigm is its inefficiency.

2.2 Secure Multi-party Computation

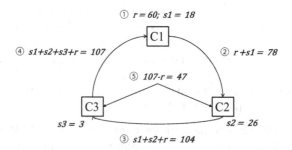

Fig. 1. The basic secure multi-party summation

Secure multi-party computation aims to compute an arbitrary function on private data across multiple participants while keeping their data hidden [21]. We introduce a simple protocol illustrated in Fig. 1. Each client maintains their own values $s1$, $s2$ and $s3$. They would like to calculate the sum of these values without revealing their own values to other clients. First, client $C1$ generates a random value r and sends the sum of r and $s1$ to client $C2$. $C2$ adds its value $s2$ to the sum and sends it to the next client. $C1$ calculates the sum of all clients' values by subtracting r from the summation value received from client $C3$. Lastly, the total sum is sent to all other clients.

The secure multi-party computing protocols for rings are efficient. Although this type of protocol may be vulnerable to client-side collusion, the large number of clients in the P2PLSys environment and the uncertainty of clients at each exchange make such risks tolerable. We introduce another secure multi-party computation protocol. Each peer node splits their model into K segments using a random segments generation method. Then, each of the K peer nodes sends $K - 1$ segments to the other peer nodes separately. Next, every node sums their own segment along with the $K - 1$ segments received from others. Finally, each node obtains the aggregated model after receiving the sums from all other peer nodes [18]. These protocols are deemed acceptable in the P2PLSys setting.

2.3 Asynchronous Federated Learning

In AFL, the most significant challenge is the impact of stragglers, whose models exhibit staleness. To reduce the impact of staleness, weighted aggregation methods are commonly employed. We briefly introduce some weight adjustment methods utilized in AFL [8].

$$w_t = (1 - \alpha_t)w_{t-1} + \alpha_t w_{new}, \tag{1}$$

where w_{t-1} is the global model parameters of the $t - 1$ epoch, w_{new} is the new local model parameters from a client and α_t is a staleness parameter.

$$\alpha_t = \alpha \times s(t - \tau) \tag{2}$$

where α is a mixing hyperparameter, τ is the time stamp of the local model from the client, t is the current time and the crucial problem is to design the weighting function $s(t - \tau)$. [8] proposed a polynomial and hinge weighting function to measure the staleness. The polynomial strategy is shown as follow.

$$s_a(t - \tau) = (t - \tau + 1)^{-a} \tag{3}$$

The hinge strategy is computed as Eq. 4.

$$s_a(t - \tau) = \begin{cases} 1, & \text{if } t - \tau <= b \\ \frac{1}{a(t-\tau+b)+1}, & \text{otherwise} \end{cases} \tag{4}$$

In [22], the staleness function is defined as follows.

$$s_a(t - \tau) = (e/2)^{-(t-\tau)} \tag{5}$$

We draw inspiration from the weight adjustment concept used in asynchronous federated learning. However, the most significant difference in P2PLSys is the absence of a global model. Instead, models are merged among a subset of clients. As a result, it is crucial to devise new methods for measuring model staleness and adjusting the aggregation approaches among clients to effectively address the unique challenges presented by the P2PLSys framework.

3 P2PL System

In this section, we first introduce the architecture of P2PLSys, which includes a server and multiple peer nodes. Next, we outline the modules of the server and peer nodes. Lastly, we describe the general workflow of P2PLSys.

3.1 The Architecture of P2PLSys

P2PLSys includes a server and a large number of peer nodes, as illustrated in Fig. 2. The server is designed to record essential information, such as IDs of peer nodes. When local training is complete, peer nodes can download other nodes' IDs from the server to aggregate models. Importantly, the server only maintains information about the peer nodes but does not participate in model aggregation, ensuring model privacy. Furthermore, multi-party computation protocols are employed to protect privacy among peer nodes.

Given the large number of peer nodes in the system, a synchronous model aggregation approach is impractical. To improve training efficiency, we utilize an asynchronous algorithm. However, the main challenge in asynchronous training is model staleness. We address this issue by employing weighted aggregation, which improves overall model performance. Specifically, the server measures model staleness and calculates weight parameters. Peer nodes then adjust the model parameters according to these weight parameters and aggregate the weighted model, creating a robust and efficient asynchronous training process.

3.2 The Overall Process of P2PLSys

P2PLSys is a semi-asynchronous learning framework designed for large-scale peer nodes. The specific process is illustrated in Fig. 2. Each client continuously performs local training and sends a signal to the server upon completing the training. Once the server receives signals from a sufficient number of clients, it schedules PN peer nodes for model aggregation. The P2PLSys process is divided into four steps:

Step 1: The server maintains an $NList$ and $SArchive$, waiting for signals from clients indicating the completion of training. $NList$ is the clients list, including the IDs of peers nodes and other necessary information. $SArchive$ is the archive of clients model updated times which is used to adjust model weights to mitigate the impact of model staleness.

Step 2: When the server receives a signal from a peer node that local training is complete, it updates the corresponding $SArchive$. The server selects PN peer nodes and calculates the model weight when the number of received signals is greater than PN. Model weight design focuses on model staleness and sample quantity. The IDs and corresponding weights of PN nodes are then distributed among them.

Step 3: The peer node adjusts the model and communicates with other nodes using the ID list, reducing storage pressure on the peer nodes. During the communication process, the server can synchronously schedule other nodes. The peer node employs a secure multi-party computation protocol to perform secure model aggregation.

Step 4: Peer nodes update the parameter model using the aggregated model. Once PN nodes complete a local secure aggregation, the process returns to **Step 1**.

The most significant difference in P2PLSys is that even though the server exists, it does not participate in any model aggregation. All clients' model information remains invisible to the server. The server only records necessary client information to reduce storage pressure. Additionally, the semi-asynchronous learning framework is an innovative feature of P2PLSys. Unlike general asynchronous federated learning, there is no global model for the server; instead, each client maintains a model. The model adjustment based on staleness relies solely on certain local clients, termed **local model adjustment**. To protect model information among clients, secure multi-party computation protocols are used for communication.

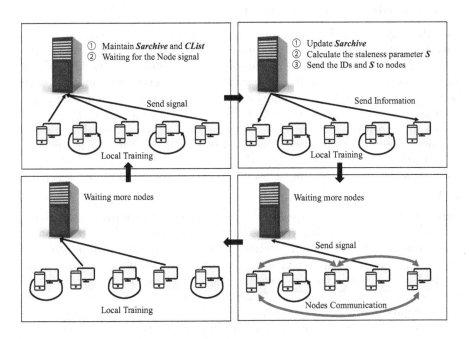

Fig. 2. The overall process of P2PLSys

Algorithm 1. Monitoring

Input: The peer nodes list $NList$, the SMC parameters PN;
Output: The PN nodes which have finished local training;
1: $NSet = \{\}$;
2: **while** Ture **do**
3:    Received the signal of client k finished local training;
4:    $SArc(k) = SArc(k) + 1$ and $NSet = NSet \cup \{k\}$;
5:    **if** $NSet.size() \geq PN$ **then**
6:       $NSet_{PN}$: randomly select PN nodes from $NSet$;
7:       $NSet = NSet - NSet_{PN}$;
8:       Perform **Scheduler**($NSet_{PN}$);
9:    **end if**
10: **end while**

3.3 The Modules of the Server

We introduce the modules of server, including the management, monitor and scheduler.

Management: In the P2PLSys, the server maintains two lists. One is $NList$, the clients list. The other one is $SArchive$. The peer nodes must report personal information to server when they join the P2PLSys. Then, the server will update $NList$ and $SArchive$ to add the new peer nodes to the P2PLSys.

Monitoring: The goal of monitoring module is to detect whether peer nodes have finished training their models. Due to varying computing power across peer nodes, the time required to complete local training differs for each node. The system contains a large number of peer nodes, so the server must monitor which nodes have completed local training for model aggregation. The monitoring operation is performed continuously within the system. The monitoring operation is shown in Algorithm 1, where $NSet$ represents the peer nodes which finished the local training. The server will update the $SArc(k)$ by adding one and $NSet$ when receives the signal that peer node k finished the local training. The server will randomly select PN peer nodes which have trained the model selected to perform scheduler operation.

Scheduler: The primary purpose of the scheduler algorithm is to assign the ID list and weighted parameters based on model staleness. The specific process is shown in Algorithm 2. S represents the weight of the nodes' models based on staleness. We calculate the staleness weight of each node in Line 4. Unlike asynchronous federated learning, the server only needs to dispatch PN nodes to aggregate their models. We measure the staleness of each client among the PN nodes and consider the aggregation as a local aggregation without taking other nodes into account. We then calculate the staleness parameter using two methods, termed **Exp-S** and **Unif-S**, both of which are described in Sect. 4.1. Meanwhile, in order to aggregate the models, the server sends the total number of dataset samples for all PN peer nodes. This way, the server ensures accurate weight parameters and efficient model aggregation while considering staleness-related factors (Fig. 3).

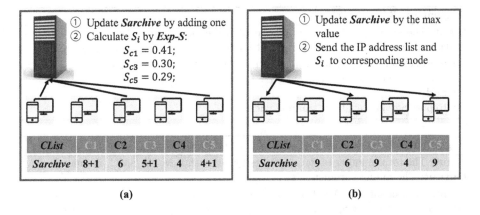

Fig. 3. The Scheduler operation of the server.

Algorithm 2. Scheduler($NSet_{PN}$)

Input: The clients list $NSet_{PN}$, the archive of clients model updated times $SArchive$, the ID list of clients $IDList$;

Output: The weight of nodes model S;

1: $NList_{PN} = \{\}$;
2: **for** Each client k in $NSet_{PN}$ **do**
3:     Calculate the staleness of client k S_k;
4:     $SArchive(k) = Max(SArchive(NSet_{PN}))$;
5:     $NList_{PN} = NList_{PN} \cup NList_k$;
6: **end for**
7: Send $(S_k, NList_{PN})$ to each client k in $NSet_{PN}$;
8: Perform **NodesComm**($NSet_{PN}, NList_{PN}$);

3.4   The Modules of Peer Nodes

We will briefly introduce the modules of peer nodes including Local-Training and NodesComm.

Local-Training: The local training of peer nodes is concisely demonstrated in Algorithm 3. Each peer node performs E epochs to update their model [1]. Once the local training is completed, the peer node sends a signal to the server indicating the conclusion of its training process. This approach ensures the server stays informed about the progress of each node and can coordinate the scheduling and aggregation processes accordingly.

Communication: When the server performs the scheduler operation, the PN peer nodes aggregate their models using the secure multi-party computation protocol. Specifically, the nodes obtain the staleness parameter S and the ID list of other nodes $IDList$. First, each node $k \in NSet_{PN}$ adjusts its model using the staleness parameter S_k and the total number of data samples. Second, the PN nodes aggregate their models using the secure multi-party computation protocol,

---

Algorithm 3. LocalTrain

Input: The number of local training epoch E, the local dataset D;
Output: The weight parameter of client model;
1: **for** Each local iteration i from 1 to E **do**
2:     $w_i = w_{i-1} - \gamma * \partial_{w_{i-1}}(w_{i-1}; D)$
3: **end for**
4: Report Server that local training ends;

Algorithm 4. NodesComm($NSet_{PN}, NList_{PN}$)

Input: The total number of data samples for all clients N, the SMC parameters PN;
Output: The model parameters θ_T;
1: **for** Each client k in $NSet_{PN}$ **do**
2:     $w_k^{new} = N_k/N * S_i * w_k$;
3: **end for**
4: Perform Secure Multi-Party Summation Protocol to aggregate w^{new};

---

as described in Sect. 2.2. Finally, each node updates its model by aggregating other models, ensuring that the overall system effectively shares information while preserving privacy and handling staleness.

4 Experiments

In this section, we empirically evaluate the performance of the P2PLSys algorithm. First, we introduce the experimental design. Next, we describe the datasets and network architectures used in the evaluation. Finally, we demonstrate the performance of P2PLSys on two popular datasets under various settings, showcasing the effectiveness of our proposed system in addressing decentralized learning challenges. Our code has been put in https://github.com/MiLab-HITSZ/2023LuoP2PLSys.

4.1 Experimental Design

We apply P2PLSys to train convolutional neural networks [23] for image classification with n peer nodes. To simulate the varying computing power of each node, we randomly initialize a value within the range of [1-T] for each node, representing the time it takes to complete a local training. We set different T values to verify the stability of P2PLSys. Pytorch, a popular deep learning library, is used for our experiments. We set a global timer to simulate the asynchronous settings, and the parameter settings are displayed in Table 1. The baseline algorithm is FedAvg [22], which implements a synchronous method. For FedAvg, we randomly select 10 clients to update and aggregate. We also consider SGD as another baseline for comparison purposes.

Table 1. The settings of parameters

Parameters	Settings	Description
n	100	Number of nodes
T	10/20	Maximum value of staleness
PN	10	Number of aggregated nodes
$NLocal$	2K/40K	Number of local training

We propose two kinds of method to calculate the staleness S_k. We term the first kind method as **Unif-S** showing as follow.

$$s_k^U = \frac{SArc(k) - Min(SArc(PN)) + 1}{\sum_j^{PN} SArc(j) - Min(SArc(PN)) + 1} \quad (6)$$

where the $SArc(k)$ indicates the value of node k in $SArc$. We calculate s_k^U according to the proportion of $SArc(k)$ in sum of $SArc(PN)$.

We term the second kind method as **Exp-S** showed as follow.

$$s_k^E = \frac{e^{SArc(k)-Min(SArc(PN))+1}}{\sum_j^{PN} e^{SArc(j)-Min(SArc(PN))+1}} \quad (7)$$

where we do an exponential operation on the staleness value. This method aims to increase the weight of the newer models. Other methods are also possible, but we will only briefly present these two in this discussion

4.2 Dataset and Neural Network Architectures

We evaluate our algorithm using the MNIST [24] and CIFAR-10 [25] datasets which are popular datasets. MNIST contains 60K training images and 10K testing images. Each picture in MNIST consists of 28×28 pixels, depicting handwritten digits ranging from 0 to 9. CIFAR-10 contains 60K RGB images of objects from 10 classes, divided into 50K training samples and 10K testing samples.

In the experiments, we employ convolutional neural networks (CNN) [23] for MNIST, which consist of three convolution layers, three max-pooling layers, and two fully connected layers. Due to its complexity, we use a more advanced network, ResNet18 [26], for CIFAR-10. For all experiments, the training dataset is divided into n parts for n nodes, with training data independently and identically distributed, maintaining the same number of samples across different nodes. The testing data is used to evaluate global classification accuracy. The models of different nodes are initialized with the same architecture and parameters.

4.3 MNIST Classification

In this experiment, we test P2PLSys with various stale parameters and aggregation methods. Since the model performance of each nodes differs, we employ three metrics, the maximum, minimum, and average prediction accuracy values, to measure the performance of P2PLSys. Additionally, we utilize two kinds of local model adjustment. As shown in Fig. 4, we compare P2PLSys with FedAvg and SGD on the MNIST dataset.

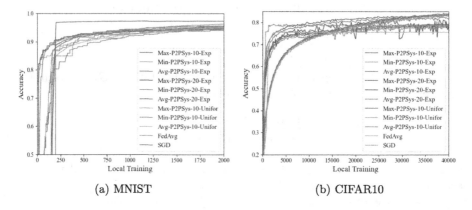

(a) MNIST   (b) CIFAR10

Fig. 4. The accuracy of MNIST and CIFAR10.

Analysis of Settings: We assign two values, 10 and 20, to indicate the range of clients' computing power. A higher value suggests that P2PLSys must be more robust. In Fig. 4 (a), we observe that the larger value converges as quickly as the smaller value, and their performance is similar. This result implies that P2PLSys demonstrates strong robustness to clients with varying computing power. Then, we compare the **Exp-S** and **Unif-S** methods. **Exp-S** converges faster than **Unif-S**, aligning with our analysis since the exponential operation increases the weight of the latest model, which in turn improves convergence speed.

Comparison with SGD and FedAvg: We notice that P2PLSys converges as fast as FedAvg, but slightly slower than SGD. Particularly, during the initial stage of training, the maximum value of P2PLSys and FedAvg accuracy is similar. In the later training period, the performance of both methods is not significantly different. Compared to SGD, the performance of P2PLSys and FedAvg is moderately worse, as they converge slower than SGD. However, their final performance is comparable to that of SGD.

4.4   CIFAR10 Classification

Analysis of Settings: In Fig. 4 (b), we observe that the larger value converges as quickly as the smaller value, with similar performance. This indicates that P2PLSys demonstrates strong robustness to clients with various computing power. Then, we compare the **Exp-S** and **Unif-S** methods. We notice that **Exp-S** converges faster than **Unif-S**, which aligns with our analysis since the operation of the index increases the weight of the latest model, improving the convergence speed.

Comparison with SGD and FedAvg: We can see that P2PLSys converges as quickly as FedAvg but slightly slower than SGD. Specifically, the maximum value of P2PLSys and FedAvg accuracy is similar in the initial stage of training. In the late training period, the performance does not differ significantly during the initial stage of training. Compared to SGD, the performance of P2PLSys and FedAvg is slightly worse, as they converge slower than SGD. However, their final performance is similar to that of SGD, indicating the effectiveness of P2PLSys as an alternative approach.

Overall, the P2PLSys algorithm has achieved performance not inferior to SGD and FedAvg. On the one hand, nodes with fast training speed do not need to wait for stragglers, and can be fully trained. On the other hand, stragglers also have the opportunity to exchange model information with other nodes. It makes the overall performance of the system well.

5   Conclusion

In this paper, we present a semi-asynchronous peer-to-peer learning system, P2PLSys, designed for large-scale clients in decentralized training. In contrast to traditional FL systems with servers, P2PLSys adopts a peer-to-peer approach and semi-asynchronous framework. Additionally, it employs a secure multi-party computation protocol for secure aggregation. P2PLSys is developed to address the problem of non-robustness arising from the presence of servers in traditional federated learning and the issue of low learning efficiency in peer-to-peer learning. Our experimental results demonstrate that P2PLSys achieves performance similar to FedAvg and SGD under various experimental settings, showcasing the promise of our proposed system in effectively handling decentralized learning challenges. In future work, we can not only consider the stragglers, but also consider the issue of data heterogeneity. Our framework has the ability to be well compatible with data heterogeneity issues.

References

1. McMahan, B., Moore, E., Ramage, D., Hampson, S., Aguera y Arcas, B.: Communication-efficient learning of deep networks from decentralized data. In: Artificial Intelligence and Statistics, pp. 1273–1282. PMLR (2017)
2. Sarkar, S., Agrawal, S., Gadekallu, T.R., Mahmud, M., Brown, D.J.: Privacy-preserving federated learning for pneumonia diagnosis. In: International Conference on Neural Information Processing, pp. 345–356. Springer, Singapore (2022). https://doi.org/10.1007/978-981-99-1648-1_29
3. Teng, L., et al.: Flpk-bisenet: federated learning based on priori knowledge and bilateral segmentation network for image edge extraction. IEEE Trans. Netw. Serv. Manag. (2023)

4. Alazab, M., et al.: Federated learning for cybersecurity: concepts, challenges, and future directions. IEEE Trans. Indust. Inf. **18**(5), 3501–3509 (2021)

5. Bonawitz, K., et al.: Towards federated learning at scale: system design. Proc. Mach. Learn. Syst. **1**, 374–388 (2019)

6. Xu, C., Qu, Y., Xiang, Y., Gao, L.: Asynchronous federated learning on heterogeneous devices: a survey. arXiv preprint arXiv:2109.04269 (2021)

7. Chen, Y., Ning, Y., Slawski, M., Rangwala, H.: Asynchronous online federated learning for edge devices with non-IID data. In: Proceedings of the 2020 IEEE International Conference on Big Data (Big Data), pp. 15–24. IEEE (2020)

8. Xie, C., Koyejo, S., Gupta, I.: Asynchronous federated optimization. arXiv preprint arXiv:1903.03934 (2019)

9. Shi, G., Li, L., Wang, J., Chen, W., Ye, K., Xu, C.Z.: Hysync: hybrid federated learning with effective synchronization. In: 2020 IEEE 22nd International Conference on High Performance Computing and Communications; IEEE 18th International Conference on Smart City; IEEE 6th International Conference on Data Science and Systems (HPCC/SmartCity/DSS), pp. 628–633. IEEE (2020)

10. Zhou, C., Tian, H., Zhang, H., Zhang, J., Dong, M., Jia, J.: Tea-fed: time-efficient asynchronous federated learning for edge computing. In: Proceedings of the 18th ACM International Conference on Computing Frontiers, pp. 30–37 (2021)

11. Xu, C., Qu, Y., Xiang, Y., Gao, L.: Asynchronous federated learning on heterogeneous devices: a survey. arXiv preprint arXiv:2109.04269 (2021)

12. Rieke, N., et al.: The future of digital health with federated learning. NPJ Digit. Med. **3**(1), 1–7 (2020)

13. Roy, A.G., Siddiqui, S., Pölsterl, S., Navab, N., Wachinger, C.: Braintorrent: a peer-to-peer environment for decentralized federated learning. arXiv preprint arXiv:1905.06731 (2019)

14. Warnat-Herresthal, S., et al.: Swarm learning for decentralized and confidential clinical machine learning. Nature **594**(7862), 265–270 (2021)

15. Kairouz, P., et al.: Advances and open problems in federated learning. Found. Trends® Mach. Learn. **14**(1–2), 1–210 (2021)

16. Wink, T., Nochta, Z.: An approach for peer-to-peer federated learning. In: Proceedings of the 51st Annual IEEE/IFIP International Conference on Dependable Systems and Networks Workshops (DSN-W), pp. 150–157. IEEE (2021)

17. Zapechnikov, S.: Secure multi-party computations for privacy-preserving machine learning. Procedia Comput. Sci. **213**, 523–527 (2022)

18. Luo, Y., Zhiyun, X., Huang, L.: Secure multi-party statistical analysis problems and their applications. Comput. Eng. Appl. **41**(24), 141–143 (2005)

19. Kanagavelu, R., et al.: Two-phase multi-party computation enabled privacy-preserving federated learning. In: Proceedings of the 20th IEEE/ACM International Symposium on Cluster, Cloud and Internet Computing (CCGRID), pp. 410–419. IEEE (2020)

20. Mugunthan, V., Polychroniadou, A., Byrd, D., Balch, T.H.: SMPAI: secure multi-party computation for federated learning. In: Proceedings of the NeurIPS 2019 Workshop on Robust AI in Financial Services (2019)

21. Ranbaduge, T., Vatsalan, D., Christen, P.: Secure multi-party summation protocols: are they secure enough under collusion? Trans. Data Priv. **13**(1), 25–60 (2020)

22. Chen, Y., Sun, X., Jin, Y.: Communication-efficient federated deep learning with layerwise asynchronous model update and temporally weighted aggregation. IEEE Trans. Neural Netw. Learn. Syst. **31**(10), 4229–4238 (2019)

23. Jmour, N., Zayen, S., Abdelkrim, A.: Convolutional neural networks for image classification. In: Proceedings of the 2018 International Conference on Advanced Systems and Electric Technologies, pp. 397–402 (2018)
24. LeCun, Y., Bottou, L., Bengio, Y., Haffner, P.: Gradient-based learning applied to document recognition. Proc. IEEE **86**(11), 2278–2324 (1998)
25. Krizhevsky, A., et al.: Learning multiple layers of features from tiny images (2009)
26. He, K., Zhang, X., Ren, S., Sun, J.: Deep residual learning for image recognition. In: Proceedings of the IEEE Conference on Computer Vision and Pattern Recognition, pp. 770–778 (2016)

Optimizing 3D UAV Path Planning: A Multi-strategy Enhanced Beluga Whale Optimizer

Chen Ye[1], Wentao Wang[2], Shaoping Zhang[1], and Peng Shao[1(✉)]

[1] School of Computer and Information Engineering, Jiangxi Agricultural University, Nanchang 330045, China
pshao@whu.edu.cn
[2] College of Software, Nankai University, Tianjin 300350, China

Abstract. The goal of 3D UAV path planning problem is to assist the UAV in planning a flight path with the lowest total overhead cost. In this paper, we present a novel approach to address the problem by incorporating flight distance, threat cost, flight altitude and path smoothness constraints into a comprehensive cost function. The current popular metaheuristic algorithm is utilized to solve for the closest globally optimal UAV flight path. To overcome the challenges of local optima and slow convergence associated with the conventional Beluga Whale Optimizer (BWO), this paper proposes a modified beluga whale optimizer (OGGBWO) based on random opposition-based learning strategy, adaptive Gauss variational operator and elitist group genetic strategy. Extensive experiments conducted on the CEC2022 test set and four distinct terrain scenarios of varying complexity demonstrate that the OGGBWO algorithm outperforms classical and state-of-the-art metaheuristics. It achieves superior optimization performance across all 12 CEC2022 test functions and exhibits exceptional convergence in generating flight paths with the lowest total cost function in diverse terrain scenarios.

Keywords: Path Planning · UAV · Beluga Whale Optimizer

1 Introduction

With the advancement of unmanned aerial vehicle (UAV) technology, UAVs have become a vital tool for various path planning tasks, including logistics [1], smart agriculture [2], rescue operations [3], and target tracking [4]. The UAV path planning problem is to design an optimal flight path for an UAV to reach a specific target or complete a mission, provided that the origin and destination are determined. In the process of generating candidate paths, considerations include not only the path length, but also the flight environment (including terrain, obstacles, weather, etc.), fuel costs and mission requirements, among many other factors. Therefore, the flight path planning problem for UAVs is complex and multi-constrained optimization problem. Several approaches, including graph search algorithms, artificial potential field methods and metaheuristics, have been employed to assist in UAV path planning [5–7].

© The Author(s), under exclusive license to Springer Nature Singapore Pte Ltd. 2024
B. Luo et al. (Eds.): ICONIP 2023, LNCS 14448, pp. 42–54, 2024.
https://doi.org/10.1007/978-981-99-8082-6_4

The problem of UAV 3D path planning is a complex optimization problem that involves various factors such as flight altitude, UAV climbing cost, and safety considerations. The main challenge in this field is to minimize unnecessary flight time and distance while ensuring compliance with altitude and other constraints. Metaheuristic algorithms have emerged as a promising approach to solving optimization problems, as they can quickly converge to near-optimal solutions. In a recent study by Song et al. [6], a novel approach combining cuckoo search algorithm with compact parallelism techniques was presented, aiming at avoiding collisions with obstacles in the optimized path of the UAV to the target location. Experiments demonstrate that the compactness of the method and the parallel communication strategy are more effective in reducing the memory footprint of the unmanned robot and improving accuracy. Another study by Phung et al. [7] introduced a spherical vector-based particle swarm optimization algorithm (SPSO) for tackling multiple threats to UAV path planning in complex environments. Through experiments, which involve real UAV operations, it is demonstrated that the use of ball vectors to represent the position and velocity of particles can seek the optimal path.

A new population-based metaheuristic algorithm, called beluga whale optimization (BWO), was proposed by Zhong et al. [8] in 2022. BWO algorithm has been applied in property prediction [9], power dispatch [10], intrusion detection [11] and other real-world optimization problems with good results. To address the shortcomings of slow convergence of BWO and the tendency to be attracted to local extremes on complex problems, we present a novel algorithm named beluga whale optimization based on random opposition-based learning strategy, adaptive Gauss variational operator and elitist group genetic strategy (OGGBWO).

2 Related Work

2.1 3D Path Planning Model for UAVs

The objective of 3D UAV path planning is to identify the optimal flight path for a UAV, considering a multitude of intricate flight constraints. The flight path of the UAV is divided into n path nodes, so the flight path through n path nodes is denoted as X_i. Each path node is denoted by coordinates as $N_j^i = \left(x_j^i, y_j^i, z_j^i\right)$. For feasible path X_i, the total cost function can be modelled as follows:

$$F(X_i) = \sum_{k=1}^{4} b_k F_k(X_i) \tag{1}$$

where $F_k(X_i)$ means four individual cost function; b_k represents weighting factor for the kth cost function. The cost function more aptly describes the 3D UAV path planning problem mathematically, which provides the fitness function for subsequent use of metaheuristics to optimize the 3D UAV path planning problem. The four constraints involved are detailed below:

Distance Cost. For the UAV to operate efficiently, the planned path needs to be optimal in some criterion. We therefore choose to minimize the path length. The path cost is primarily determined by the flight distance covered by the UAV, spanning from its initial starting point to the destination. Thus, the Euclidean distance between two adjacent path nodes is defined as $\left\| \overrightarrow{N_j^i N_{j+1}^i} \right\|$, and the distance cost F_1 is calculated using the following equation:

$$
\begin{cases}
F_1(X_i) = \sum_{j=1}^{n-1} \left\| \overrightarrow{N_j^i N_{j+1}^i} \right\|, \\
\left\| \overrightarrow{N_j^i N_{j+1}^i} \right\| = \sqrt{\left(x_{j+1}^i - x_j^i\right)^2 + \left(y_{j+1}^i - y_j^i\right)^2 + \left(z_{j+1}^i - z_j^i\right)^2}.
\end{cases} \tag{2}
$$

Threat Cost. In performing missions, the UAV should avoid obstacles encountered in its path as much as possible. In this study, all obstacles are considered as cylinders and represented by the set K. Let C_k be the centroid of the obstacle and R_k the radius. Figure 1 illustrates the obstacle projection of the UAV through a given path $\left\| \overrightarrow{N_j^i N_{j+1}^i} \right\|$.

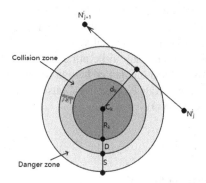

Fig. 1. Projection of a drone passing through an obstacle.

d_k is the straight-line distance from the UAV flight path to the centre of the obstacle C_k. D and S represent the diameter of the UAV and its hazard distance to the collision zone, respectively, which in concert with R_k determine the division of the collision, hazard and safety zones. The threat cost F_2 from passing through the set of obstacles K is calculated using the following equation:

$$
\begin{cases}
F_2(X_i) = \sum_{j=1}^{n-1} \sum_{k=1}^{K} T_k\left(\overrightarrow{N_j^i N_{j+1}^i}\right), \\
T_k\left(\overrightarrow{N_j^i N_{j+1}^i}\right) = \begin{cases} 0, & \text{if } d_k > S + D + R_k \\ (S + D + R_k) - d_k, & \text{if } D + R_k < d_k \leq S + D + R_k \\ \infty, & \text{if } d_k \leq D + R_k \end{cases}
\end{cases} \tag{3}
$$

Altitude Constraint. Flight altitude constraints are essential to increase the efficiency of UAVs in performing their tasks. By limiting the flight altitude of the UAV to a given range, the altitude H_j^i is kept at an average altitude. The altitude cost associated with waypoint P_j^i is calculated as Eq. (4). The smaller the difference between the flight altitude and the average altitude, the lower the altitude cost. In contrast, when the UAV flies into a domain where the height constraint is violated, the altitude cost will then be considered infinite.

$$
\begin{cases}
F_3(X_i) = \sum\limits_{j=1}^{n} H_j^i \\
H_j^i = \begin{cases} \left| h_j^i - \frac{(h_{max}+h_{min})}{2} \right|, & \text{if } h_{min} \leq h_j^i \leq h_{max} \\ \infty, & \text{otherwise.} \end{cases}
\end{cases}
\tag{4}
$$

where h_j^i indicates the flight altitude of the UAV from the ground; h_{min} and h_{max} represent the minimum and maximum altitude.

Smoothness Constraint. Reducing the UAV's turning and climbing rates is essential for generating feasible paths. Figure 2 shows a three-dimensional diagram of the UAV from node N_j^i via N_{j+1}^i to N_{j+2}^i. $N_j'^i$, $N_{j+1}'^i$ and $N_{j+2}'^i$ are the projection points of these three path nodes in the XOY plane respectively. The projection vectors $\overrightarrow{N_j'^i N_{j+1}'^i}$ and $\overrightarrow{N_{j+1}'^i N_{j+2}'^i}$ form the turning angle $\varnothing_j^i$, which can be calculated using the following equation:

$$
\phi_j^i = \arctan\left(\frac{\left\| \overrightarrow{N_j^i N_{j+1}^i} \times \overrightarrow{N_{j+1}'^i N_{j+2}'^i} \right\|}{\overrightarrow{N_j^i N_{j+1}^i} \cdot \overrightarrow{N_{j+1}'^i N_{j+2}'^i}} \right)
\tag{5}
$$

where the projection $\overrightarrow{N_j'^i N_{j+1}'^i}$ is solved using the relationship between the unit vector $\overrightarrow{k}$ in the z-axis direction and the path segment $\overrightarrow{N_j^i N_{j+1}^i}$ by the following equation:

$$
\overrightarrow{N_j'^i N_{j+1}'^i} = \overrightarrow{k} \times \left(\overrightarrow{N_j^i N_{j+1}^i} \times \overrightarrow{k} \right)
\tag{6}
$$

The climbing angle ψ_j^i, between the vectors $\overrightarrow{N_j'^i N_{j+1}'^i}$ and $\overrightarrow{N_{j+1}'^i N_{j+2}'^i}$, reflects the climb rate of the UAV. It is computed by:

$$
\psi_j^i = \arctan\left(\frac{z_{j+1}^i - z_j^i}{\left\| \overrightarrow{N_j'^i N_{j+1}'^i} \right\|} \right)
\tag{7}
$$

The smooth cost is then given by the following equation:

$$
F_4(X_i) = a_1 \sum_{j=1}^{n-2} \phi_j^i + a_2 \sum_{j=1}^{n-1} \left| \psi_j^i - \psi_{j-1}^i \right|
\tag{8}
$$

where a_1 and a_2 are respectively the penalty coefficients of the turning and climbing angles.

The 3D UAV path planning method is based on a spherical coordinate system and is based on the basic principle that each feasible path X_i is encoded into a set of vectors from the origin to the end point. Each vector specifically describes the movement of the UAV as it flies from one path node to the next. In a spherical coordinate system, these vectors are specifically composed of these three components $\rho \in (0, path\_length)$, $\theta \in (-\pi/2, \pi/2)$ and $\emptyset \in (-\pi, \pi)$, which represent the flight amplitude, flight elevation and azimuth of the UAV respectively. A feasible flight path Ω_i with N waypoints can be represented by the following 3N-dimensional spherical vector:

$$\Omega_i = (\rho_{i1}, \psi_{i1}, \phi_{i1}, \rho_{i2}, \psi_{i2}, \phi_{i2}, \cdots, \rho_{iN}, \psi_{iN}, \phi_{iN}), N = n - 2 \tag{9}$$

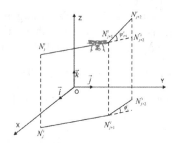

Fig. 2. UAV flight path map in 3D coordinate system.

2.2 Beluga Whale Optimization

The Beluga Whale algorithm models the three phases of exploration, exploitation, and whale fall within its mathematical framework, as described in the literature [8]. The balance factor B_f and whale fall probability W_f are introduced to select the current appropriate population renewal method, which are calculated as follows, respectively:

$$B_f = B_0 \cdot \left(1 - \frac{It}{MaxIt}\right) \tag{10}$$

$$W_f = 0.1 - 0.05 \cdot \frac{It}{MaxIt} \tag{11}$$

where B_0 is a random value in the range (0,1) that changes in each iteration; It is the current number of iterations; $MaxIt$ is the maximum number of iterations.

When $B_f > 0.5$, the algorithm is in the exploration phase; when $B_f \leq 0.5$, the algorithm is in the exploitation phase; when $B_f < W_f$, the algorithm is in the whale fall phase. As the number of iterations t increases, the fluctuation range of B_f narrows from (0,1) to (0,0.5). After the transition between the exploration and exploitation phases, the decision regarding the occurrence of whale fall behavior is determined by comparing the probability of whale fall W_f with the equilibrium factor B_f. When W_f exceeds B_f, the algorithm enters the whale fall phase. The detailed mathematical model for each stage is described as follows:

Exploration Phase. This phase garners inspiration from swimming in pairs of belugas. The new position $X_{i,j}^{t+1}$ for the ith beluga whale on the jth dimension is generated by (the following equation:

$$\begin{cases} X_{i,j}^{t+1} = X_{i,P_j}^t + \left(X_{r,P_1}^t - X_{i,P_j}^t\right)(1+r_1)\sin(2\pi r_2), j = even \\ X_{i,j}^{t+1} = X_{i,P_j}^t + \left(X_{r,P_1}^t - X_{i,P_j}^t\right)(1+r_1)\cos(2\pi r_2), j = odd \end{cases} \tag{12}$$

where t is the current iteration; $p_j (j = 1, 2, \cdots, d)$ is a random number selected from d dimension; X_{i,p_j}^t and X_{r,p_1}^t are the current position for the ith and r^{th} beluga whale on the $p_j{}^{th}$ dimension, and r is a randomly selected beluga whale; r_1 and r_2 are random number between (0,1), which are applied to enhance the randomness in this phase.

Exploitation Phase. The exploitation phase of the BWO algorithm incorporates behavioral characteristics inspired by the foraging behaviors of belugas. To augment global convergence during this phase, the algorithm introduces the Lévy flight strategy into the prey predation model. The following equation denotes the expression for generating a random number that obeys the Lévy distribution:

$$L_f = 0.05 \times \frac{\mu \cdot \sigma}{|v|^{1/\beta}} \tag{13}$$

$$\sigma = \left\{ \frac{\Gamma(1+\beta) \times \sin(\pi\beta/2)}{\Gamma[(1+\beta)/2] \times \beta \times 2^{(\beta-1)/2}} \right\}^{1/\beta} \tag{14}$$

where μ and v are both random numbers obeying the normal distribution; the value of β is usually 1.5. The mathematical model of beluga whale using the Lévy flight strategy to hunt its prey is as follows:

$$X_i^{t+1} = r_3 \cdot X_{best}^t - r_4 \cdot X_i^t + S \cdot L_f \cdot \left(X_r^t - X_i^t\right) \tag{15}$$

where r_3 and r_4 are both random numbers between (0,1); L_f is a random number obeying the Lévy distribution; and S denotes the random jump degree, applied for measuring the strength of the Lévy flight, which is described by the following equation:

$$S = 2 \cdot r_4 \cdot \left(1 - \frac{It}{MaxIt}\right) \tag{16}$$

Whalefall Phase. To ensure that the population size remains constant, the position update formula is devised based on the beluga's current position x_i^t and the falling length step x_{step}^t:

$$X_i^{t+1} = r_5 \cdot X_i^t - r_6 \cdot X_r^t + r_7 \cdot X_{step}^t \tag{17}$$

$$X_{step}^t = e^{\frac{-C \cdot It}{MaxIt}} \cdot (Ub - Lb) \tag{18}$$

where r_5, r_6 and r_7 are all random numbers between (0,1); Lb and Ub are the lower and upper bounds respectively; x_r^t is the position of individual r chosen at random from the current population; C is the step factor, related to the population size N and the previously mentioned whale fall probability W_f, calculated as follows:

$$C = 2N \times W_f \tag{19}$$

3   Multi-strategy Enhanced Beluga Whale Optimizer

3.1   Elitist Group Genetic Strategy

The genetic algorithm (GA) [12] has been known for the ability to quickly identify the range where the global optimum may reside. In response to BWO's premature converge, genetic operators and strategy specific to BWO are employed alternately with a certain probability to update the population. A random number between (0,1) is generated and the BWO's strategy is executed when rand < 0.5; otherwise, the genetic operator. This allows for optimal utilization of GA's strong global search capability.

The three whales with the best fitness values in the population are selected as the elite group g_{i1}, g_{i2}, g_{i3} and compared with the remaining other whales $g_{j1}, \cdots, g_{jm}$. If whale i has a better fitness value than whale j, whale i is genetically crossed with whale j. Each genetic crossover process will produce two new individuals $NewX_j^i$ and $NewX_j^{i+1}$, and the equation for gorilla genetic hybridization is as follows:

$$NewX_j^i = (r_8 + 1) \cdot X_j^{gi} + (1 - r_9) \cdot X_j^{gj}, i \in [1, n], j \in [1, d] \qquad (20)$$

$$NewX_j^{i+1} = (r_8 + 1) \cdot X_j^{gj} + (1 - r_9) \cdot X_j^{gi}, i \in [1, n], j \in [1, d] \qquad (21)$$

where X_j^{gi} and X_j^{gj} denote gene j of the X^{gi} and X^{gj}, respectively. r_8 and r_9 is a random number in the range [0,2].

In contrast, whale i and whale j are genetically mutated. Each mutation process also produces two new whale individuals $NewX_j^i$ and $NewX_j^{i+1}$. The equation for whale gene mutation is as follows:

$$NewX_j^i = X_j^{gi} + \delta \cdot \gamma, i \in [1, n], j \in [1, d] \qquad (22)$$

$$NewX_j^{i+1} = X_j^{gj} + \delta \cdot \gamma, i \in [1, n], j \in [1, d] \qquad (23)$$

$$\delta = 0.1 \cdot (Ub - Lb) \qquad (24)$$

where γ is a random value that conforms to a normal distribution.

3.2   Adaptive Gauss Variational Operator

The adaptive Gauss variance operator [13] is introduced in the exploration phase. The Gauss variation employs a Gauss distribution to perturb the position, and the probability density function of the Gauss distribution is shown in Eq. (25).

$$Gauss : f(x) = \frac{1}{\sqrt{2\pi}\sigma} e^{-\frac{(x-\mu)^2}{2\sigma^2}} \qquad (25)$$

where μ denotes the expected value and σ^2 denotes the variance. When $\mu = 0$ and $\sigma = 1$, the equation denotes the Gauss normal distribution.

The beluga whale swimming equation is improved by introducing a Gaussian variant to the following equation:

$$X(t+1) = X(t)(1 + kGsuss(0, 1)) \tag{26}$$

where $X(t)$ refers to the position before mutation; $X(t+1)$ denotes the position after mutation; k in the mutation operator is a factor controlling the step size for the purpose of better coordinating exploration and exploitation capabilities, as calculated by:

$$\begin{cases} k = 1 - It \times a^{\frac{1}{MaxIt}} \\ a = 2 - It \times \frac{2}{MaxIt} \end{cases} \tag{27}$$

It can be seen from Eq. (27): k keeps getting smaller dynamically as the evolutionary iteration increases. The fact indicates that the step size factor allows the algorithm to generate large perturbations in the early part of the iteration, while the perturbations become smaller in the later part. This can provide a favorable premise for effectively balancing the global and local search capabilities of the BWO algorithm.

3.3 Random Opposition-Based Learning Strategy

Opposition-based Learning (OBL) is a strategy for improvement in the field of group intelligence optimization proposed by Tizhoosh et al. [14] in 2005. Central to this approach is the computation of the inverse solution, predicated upon the current feasible solution in the population. By comparing objective function values of the current and inverse solutions, the superior one proceeds to the next iteration. The mathematical expression for OBL is as follows:

Theorem 1. Let $P(x_1, x_2, \cdots, x_D)$ be a point in D-dimensional space, where $x_i \in [a_i, b_i]$ and $i = 1, 2, \cdots, D$. The opposite point of $\widehat{P}(\hat{x}_1 \hat{x}_2, \cdots, \hat{x}_D)$ is defined by:

$$\hat{x}_i = a_i + b_i - \hat{x}_i \tag{28}$$

While OBL strategy generates a reverse solution that can extend the search, the distance between the current solution and its reverse solution is a certain value and lacks randomness. If both solutions deviate from the global optimum, local optima are still encountered. To tackle this, the random opposition-based learning (ROBL) strategy is proposed. This novel approach adds randomness to inverse solution calculation, as depicted in the formula below:

$$x_i^* = a_i + b_i - r_{10} \times \hat{x}_i \tag{29}$$

where r_{10} denotes a random number taken from 0 and 1.

At the end of the algorithm, incorporating the ROBL strategy remarkably enhances the population diversity. The fitness of the oppositional solutions using the ROBL strategy is compared with the original solutions. Opposing solutions with higher fitness replace original ones if superior.

4 Validation Experiments

In this subsection, CEC2022 test functions [15] are utilized to comprehensively inves-tigate the overall performance of GGOBWO. F_1 are single-peaked functions based on rotation matrices, which are single-peaked functions based on rotation matrices, which are utilized to measure the development capability and convergence of the algorithm. F_2-F_5 are multi-peaked functions based on translation and rotation matrices, which have a large quantity of local optima and test the local search capabilities. F_6-F_8 and F_9-F_{12} are hybrid and composite functions, respectively. Their complex function constructions present a formidable challenge for algorithms searching for the global optimal solution. We selected five representative algorithms, whale optimization algorithm (WOA) [16],

Table 1. CEC2022 test function values for each algorithm.

Function	Performance	WOA	SCSO	POA	DBO	BWO	OGGBWO
F_1	Mean	2.75e+04	3.53e+04	8.02e+03	2.08e+04	6.08e+04	**1.07e+03**
	Std	1.23e+04	1.24e+04	2.76e+03	7.05e+03	1.92e+04	**4.87e+02**
F_2	Mean	5.74e+02	1.47e+03	6.43e+02	4.92e+02	2.07e+03	**4.68e+02**
	Std	5.47e+01	3.15e+02	1.14e+02	5.95e+01	2.83e+02	**2.43e+01**
F_3	Mean	6.65e+02	6.83e+02	6.49e+02	6.37e+02	6.79e+02	**6.03e+02**
	Std	1.04e+01	1.07e+01	1.05e+01	1.55e+01	7.32e+00	**1.47e+00**
F_4	Mean	9.28e+02	9.66e+02	8.83e+02	9.03e+02	9.72e+02	**8.63e+02**
	Std	3.60e+01	1.03e+01	1.31e+01	2.80e+01	**8. 06e+00**	1.77e+01
F_5	Mean	3.83e+03	3.28e+03	2.20e+03	1.98e+03	3.64e+03	**1.66e+03**
	Std	8.66e+02	4.32e+02	**2.23e+02**	5.93e+02	2.38e+02	5.64e+02
F_6	Mean	9.43e+05	3.21e+08	1.02e+06	3.92e+05	1.11e+09	**4.93e+03**
	Std	1.52e+06	2.75e+08	4.77e+06	1.18e+06	4.05e+08	**3.61e+03**
F_7	Mean	2.19e+03	2.21e+03	2.11e+03	2.12e+03	2.22e+03	**2.04e+03**
	Std	4.55e+01	2.63e+01	4.08e+01	5.71e+01	3.66e+01	**1.62e+01**
F_8	Mean	2.29e+03	2.44e+03	2.28e+03	2.30e+03	2.28e+03	**2.23e+03**
	Std	5.61e+01	1.22e+02	6.56e+01	8.08e+01	**1.76e+01**	2.22e+01
F_9	Mean	2.56e+03	2.75e+03	2.56e+03	2.51e+03	3.02e+03	**2.48e+03**
	Std	3.60e+01	1.29e+02	3.87e+01	3.17e+01	1.38e+02	**1.40e-01**
F_{10}	Mean	4.45e+03	4.45e+03	3.89e+03	3.09e+03	3.48e+03	**2.65e+03**
	Std	1.18e+03	1.91e+03	8.88e+02	1.02e+03	8.43e+02	**1.98e+02**
F_{11}	Mean	3.37e+03	7.61e+03	4.68e+03	3.16e+03	7.99e+03	**2.95e+03**
	Std	1.92e+02	5.90e+02	8.15e+02	5.04e+02	4.09e+02	**9.44e+01**
F_{12}	Mean	3.28e+03	3.55e+03	3.25e+03	3.25e+03	3.54e+03	**3.20e+03**
	Std	5.87e+01	1.40e+02	5.25e+01	5.45e+01	7.23e+01	**3.02e+01**

sand cat swarm optimization algorithm (SCSO) [17], peafowl optimization algorithm (POA) [18], dung beetle optimizer (DBO) [19] and BWO, as the comparison algorithms. They are all new metaheuristics that have emerged in recent years.

To enable a comprehensive comparison of algorithm performance, the parameters of all algorithms are set according to the original literature [8, 16–19]. To avoid chance errors, for each test function, all algorithms are simulated in 30 independent times and the mean and standard deviation of the test results are recorded. Each experiment is terminated when the maximum number of iterations (set to 1000) is reached. Table 1 summarizes the performance of each algorithm on CEC2022, and the best results for the corresponding functions are shown in bold.

5   OGGBWO Algorithm for 3D UAV Path Planning

In an endeavor to gauge the efficacy of OGGBWO algorithm proposed in the present study for a real-world problem, we employ it to obtain viable paths for UAVs in 3D terrain environments. A total of four different terrain scenarios are designed in this paper, with progressively increasing complexity. The obstacle information data for each terrain is listed in Table 2. OGGBWO and other comparative algorithms are all spherical vector-based planning methods for search in the optimization process. Each searched path is encoded by the algorithm into a set of vector groups consisting of magnitude, elevation and azimuth angles and searched in the configuration space and a flight path with the lowest total cost function is planned. The model parameters involved in the 3D UAV path planning problem and the experimental parameters are shown in Table 3.

The population size of each metaheuristic algorithm is set to 100 and the maximum number of iterations is set to 200. To avoid chance influencing the experimental results,

Table 2. Obstacle data in the four terrain scenarios.

Scenario	Obstacle coordinates	Obstacle radius	–	–	Obstacle radius
1	(500, 500,100)	80	2	(300,450,150)	80
	(700,400,150)	100		(700,450,150)	80
	(710,220,100)	60		(500,450,150)	80
	–	–		(700,265,80)	70
3	(650,520,150)	70	4	(200,200,100)	60
	(400,500,150)	80		(400,200,80)	70
	(500,350,150)	70		(530,350,150)	80
	(710,680,80)	80		(400,500,180)	70
	(600,200,150)	80		(580,700,130)	70
	(800,300,100)	50		(620,550,150)	50
	–	–		(770,400,80)	80
	–	–		(420,700,80)	50

the comparison experiment is repeated 30 times for each terrain scenario. The mean and standard deviation of the experimental results are also recorded in Table 4.

Table 3. Experimental parameters for UAV path planning.

Parameter significance	Parameter
Number of path nodes n	12
The medium weight parameters of the smoothness cost function	$a_1 = 1, a_2 = 1$
The weighting parameters of each cost function in Eq. (1)	$b_1 = 5, b_2 = 1, b_3 = 10, b_4 = 1$
UAV flight height limitation	100 m–200 m
Drone diameter	1 m
Safe distance between UAV and obstacle	1 m

Table 4. Cost function values for each algorithm in terrain scenarios.

Scenario	Performance	WOA	SCSO	POA	DBO	BWO	OGGBWO
1	Mean	4992.1576	6820.8589	4695.2355	4758.7134	4690.5099	**4685.3127**
	Std	254.5019	269.0055	11.6933	83.4159	4.5391	**3.6273**
2	Mean	5287.1284	7011.1156	4797.0948	4926.8160	4778.5108	**4736.0970**
	Std	235.7259	421.0403	41.0735	80.8828	27.1687	**6.9502**
3	Mean	5961.3775	7727.0729	5223.2256	5304.9985	5302.2075	**5211.7422**
	Std	450.5779	427.0001	**81.9074**	169.2191	94.6612	172.1102
4	Mean	5606.4208	7771.7513	4892.5040	5310.0170	4901.6182	**4742.7510**
	Std	497.7995	483.7660	266.7570	267.2909	207.6780	**169.1825**

From the data in Table 4, the flight paths generated using the OGGBWO algorithm in four different scenarios all have the smallest fitness function values. This indicates that the OGGBWO algorithm proposed in this paper can find a path for the UAV with the lowest flight cost, which not only has a relatively short distance, but also satisfies constraints such as altitude restrictions during flight. The universality of the OGGBWO algorithm allows it to find a better flight path in both simple and complex scenarios. In scenario 1, the difference between the OGGBWO algorithm and the second ranked BWO algorithm is approximately 42 units. In complex scenario 4, the advantage of the OGGBWO algorithm is better shown, with a difference of approximately 150 units between the OGGBWO algorithm and the second ranked POA algorithm. From a stability point of view, the standard deviation of the LCGARO algorithm is the smallest in scenarios 1, 2 and 4. In scenario 3, the OGGBWO algorithm ranks third in terms of stability and is not too far from the POA algorithm with the best standard deviation at this point. To compare the performance of OGGBWO more clearly with other algorithms in finding the optimal path, the flight trajectory of the UAV avoiding obstacles was plotted. Figure 3(a)-(d) and

3(e)-(h) show the top view and the corresponding 3D diagrams respectively, where the solid circle and the pentagram represent the flight start and end points respectively.

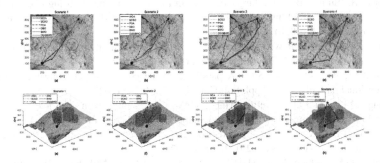

Fig. 3. Top view of the planned path for each algorithm and the corresponding 3D view.

6 Conclusion

To solve the problem that the beluga whale optimizer tends to fall into local optimality and converge slowly when solving the UAV path planning problem, this paper combines three strategies and proposes a new algorithm OGGBWO. First, the CEC2022 test function is used as a benchmark to measure the advantages and disadvantages of the OGGBOW algorithm relative to existing algorithms and to evaluate its generalization capability and effectiveness on different problems. The experimental results show that OGGBOW has a more robust and stable optimization performance than WOA, SCSO, POA, DBO and BWO. In addition, OGGBWO is thoroughly compared with other algorithms in benchmark scenarios of four different complexities. The results of the comparison experiments show that OGGBWO always minimizes the total cost function of the UAV flight and achieves the best quality path in all cases. Our future work will focus on designing precise flight constraints based on specific application scenarios, such as UAV irrigation. We will further develop the optimized performance and applicability of OGGBOW, evaluating its performance under different benchmark functions.

Patents. This research is supported by the National Natural Science Foundation of China (No. 71863018 and No. 71403112), Jiangxi Provincial Social Science Planning Project (No. 21GL12) and Technology Plan Projects of Jiangxi Provincial Education Department (No. GJJ200424).

References

1. Moshref-Javadi, M., Winkenbach, M.: Applications and research avenues for drone-based models in logistics: a classification and review. Expert Syst. Appl. **177**, 114854 (2021)
2. Abbas, A., et al.: Drones in plant disease assessment, efficient monitoring, and detection: a way forward to smart agriculture. Agronomy **13**(6), 1524 (2023)

3. McRae, J.N., Gay, C.J., Nielsen, B.M., Hunt, A.P.: Using an unmanned aircraft system (Drone) to conduct a complex high altitude search and rescue operation: a case study. Wilderness Environ. Med. **30**(3), 287–290 (2019)

4. Bhagat, S., Sujit, P.B.: UAV target tracking in urban environments using deep reinforcement learning. In: 2020 International Conference on Unmanned Aircraft Systems (ICUAS), pp. 694–701. IEEE (2020)

5. McLain, T.W., Beard, R.W.: Coordination variables, coordination functions, and cooperative timing missions. J. Guid. Control. Dyn. **28**(1), 150–161 (2005)

6. Song, P.C., Pan, J.S., Chu, S.C.: A parallel compact cuckoo search algorithm for three-dimensional path planning. Appl. Soft Comput. **94**, 106443 (2020)

7. Phung, M.D., Ha, Q.P.: Safety-enhanced UAV path planning with spherical vector-based particle swarm optimization. Appl. Soft Comput. **107**, 107376 (2021)

8. Zhong, C., Li, G., Meng, Z.: Beluga whale optimization: a novel nature-inspired metaheuristic algorithm. Knowl.-Based Syst. **251**, 109215 (2022)

9. Alsoruji, G.S., Sadoun, A.M., Abd Elaziz, M., Al-Betar, M.A., Abdallah, A.W., Fathy, A.: On the prediction of the mechanical properties of ultrafine grain Al-TiO2 nanocom-posites using a modified long-short term memory model with beluga whale optimizer. J. Market. Res. **23**, 4075–4088 (2023)

10. Hassan, M.H., Kamel, S., Jurado, F., Ebeed, M., Elnaggar, M.F.: Economic load dispatch solution of large-scale power systems using an enhanced beluga whale optimizer. Alex. Eng. J. **72**, 573–591 (2023)

11. Raj, M.G., Pani, S.K.: Hybrid feature selection and BWTDO enabled DeepCNN-TL for intrusion detection in fuzzy cloud computing. Soft Comput. 1–20 (2023) https://doi.org/10.1007/s00500-023-08573-3

12. Goldberg, D.E., Holland, J.H.: Genetic algorithms and machine learning. Mach. Learn. **3**(2), 95–99 (1988)

13. Salgotra, R., Singh, U.: Application of mutation operators to flower pollination algorithm. Expert Syst. Appl. **79**, 112–129 (2017)

14. Tizhoosh, H.R.: Opposition-based learning: a new scheme for machine intelligence. In: International Conference on Computational Intelligence for Modelling, Control and Automation and International Conference on Intelligent Agents, Web Technologies and Internet Commerce (CIMCA-IAWTIC'06), vol. 1, pp. 695–701. IEEE (2005)

15. Biedrzycki, R., Arabas, J., Warchulski, E.: A version of NL-SHADE-RSP algorithm with midpoint for CEC 2022 single objective bound constrained problems. In: 2022 IEEE Congress on Evolutionary Computation (CEC), pp. 1–8. IEEE (2022)

16. Mirjalili, S., Lewis, A.: The whale optimization algorithm. Adv. Eng. Softw. **95**, 51–67 (2016)

17. Seyyedabbasi, A., Kiani, F.: Sand Cat swarm optimization: a nature-inspired algorithm to solve global optimization problems. Eng. Comput. **39**, 1–25 (2022)

18. Wang, J., et al.: Novel phasianidae inspired peafowl (Pavo muticus/cristatus) optimization algorithm: design, evaluation, and SOFC models parameter estimation. Sustain. Energy Technol. Assess. **50**, 101825 (2022)

19. Xue, J., Shen, B.: Dung beetle optimizer: a new meta-heuristic algorithm for global optimization. J. Supercomput.Supercomput. **79**(7), 7305–7336 (2023)

Interactive Attention-Based Graph Transformer for Multi-intersection Traffic Signal Control

Yining Lv[1,2], Nianwen Ning[1,2(✉)], Hengji Li[1,2], Li Wang[1,2], Yanyu Zhang[1,2], and Yi Zhou[1,2]

[1] School of Artificial Intelligence, Henan University, Zhengzhou 450046, China
{lyn,lihengji,wangli123,zyy,zhouyi}@henu.edu.cn
[2] International Joint Research Laboratory for Cooperative Vehicular Networks of Henan, Zhengzhou 450046, China
nnw@henu.edu.cn

Abstract. With the exponential growth in motor vehicle numbers, urban traffic congestion has become a pressing issue. Traffic signal control plays a pivotal role in alleviating the problem. In modeling multi-intersection, most studies focus on communication with regional intersections. They rarely consider the cross-regional. To address the above limitation, we construct an interactive attention-based graph transformer network for traffic signal control (GTLight). Specifically, the model considers correlations between cross-regional intersections using an interactive attention mechanism. In addition, the model designs a phase-timing optimization algorithm to solve the problem of overestimation of Q-value in signal timing strategies. We validate the effectiveness of GTLight on different traffic datasets. Compared to the recent graph-based reinforcement learning method, the average travel time is improved by 28.16%, 26.56%, 25.79%, 26.46%, and 19.59%, respectively.

Keywords: Traffic signal control · Cross-regional intersections · Graph transformer network · Interactive attention mechanism · Phase-timing

1 Introduction

The explosion of traffic data brings us into the era of big data. According to the statistics, nearly 99 h and \$88 billion per year are lost per capita in the United States due to traffic congestion [1]. Real-time signal control is an effective way to alleviate traffic congestion and reduce traffic flow pressure.

Early researchers applied statistical models to study traffic signal timing strategies. These algorithms can impose severe labor burdens. Adaptive traffic control systems (ATCS), such as SCOOT [2] and SCATS [3], can adjust the signal timing based on the current traffic conditions. However, these algorithms have high computational complexity. Data-based traffic signal control schemes are implemented by developing data analysis techniques. In [4], the maximum throughput control (MTC) method maximizes capacity by dynamically greedily

B. Luo et al. (Eds.): ICONIP 2023, LNCS 14448, pp. 55–67, 2024.
https://doi.org/10.1007/978-981-99-8082-6_5

adjusting the signal timing. Reinforcement learning (RL) based traffic signal control methods, such as [5–7], can predict real-time traffic status dynamically. To enhance the learning capability of RL, many researchers combine deep neural networks (DNNs) with RL. As in the [8] algorithm, it utilizes deep reinforcement learning (DRL) for traffic signal control at a single intersection. The algorithm is ensured to be stable by a target network and an empirical replay mechanism. When modeling multi-intersection scenarios, researchers assign an agent to each intersection, such as [9–11]. These algorithms overcome the scalability problems of single-agent and improve the accuracy of traffic signal control.

However, the existing algorithms have several limitations. **First, higher-order spatial dependence.** Collaboration of the target intersection only with neighboring intersections can reduce the effective control strategy. The agent needs to sense the traffic in higher-order neighborhoods. Most existing methods do not consider cross-regional intersections cooperation. **Second, global optimal strategy.** Equal consideration of neighboring intersections with different characteristics can result in a lack of coordination between upstream and downstream controllers. The globally optimal strategy cannot be achieved.

Inspired by Ashish et al. [12], we combine the transformer with graph neural networks (GNNs) to create an interactive attention-based graph transformer network for traffic signal control (GTLight). It considers the correlation between cross-regional intersections and solves the higher-order spatial dependency problem. Moreover, the model designs a phase-timing optimization algorithm to address the issue of Q-value overestimation. Unlike the original transformer, our attention mechanism is interactive. It is used to distinguish the impact of different intersections. The main contributions of this paper are summarized:

- We construct a GTLight that aggregates the global state information of the road network. It learns multiple dependency patterns through an interactive attention mechanism. Different relevant information is captured from various potential subspaces. The problem of higher-order spatial dependency between intersections is solved.
- A phase-timing optimization algorithm is devised. It incorporates the idea of a dual network structure. A new DNN is added to calculate the optimal action. The original network evaluates the Q-value of the subsequent actions. Decouples the Q-values used for action selection and evaluation to solve the Q-value overestimation problem.
- Different experiments are conducted on synthetic and real-world datasets. The results show that GTLight outperforms the baseline algorithms in average travel time and can achieve faster convergence.

The rest of the paper is organized as follows. Section 2 reviews the existing traffic signal control methods. Section 3 describes the graph reinforcement learning (GRL) setup. Section 4 introduces the specific implementation details of the GTLight. Section 5 performs the experimental analysis. Section 6 summarizes the work and future perspectives of this study.

2   Related Work

2.1   Traffic Signal Control Based on DRL

DRL has been applied in various intelligence fields. Most researchers used DRL to optimize the traffic signal timing strategy. [13] solely focused on the state information of individual agents without interacting with other agents. [14] proposed the multi-agent A2C (MA2C) algorithm for traffic signal control. It took into account the first-order neighboring state information for each agent. In the study of [15], the author introduced that neighboring agents share local state and traffic information to control multi-intersection using Q-learning networks. [16] integrated RL and game theory, where an agent played with its neighbors and learned the best response strategy. Maximum pressure was employed as a reward for dynamic network traffic signal control in the study conducted by [17]. These methods directly connected neighboring agents and treated the impacts of adjacent intersections equally. However, traffic conditions are constantly changing. Each adjacent intersection has a different impact on the target intersection. Therefore, all adjacent intersections cannot be considered at the same level.

2.2   Traffic Signal Control Based on GRL

In recent years, the researchers applied GNNs [18] in traffic signal control research to solve the limitation of DRL. [19] used a variational learning method to model complex dependencies between different variables using probabilistic GNNs. [20] created a traffic light adjacency graph by considering the spatial relationships of the road network. [21] used graph convolutional neural networks (GCNNs) to model and predict traffic states. [22,23] utilized graph attention networks (GAT) to aggregate features of adjacent intersections. [24] proposed a meta-learning spatial-temporal GAT that enabled the model to learn traffic signal control tasks through weight updates. These models have demonstrated the significance of incorporating varying degrees of neighboring intersections. However, they cannot model information about features beyond the "field of perception". The spatial attributes of the road network are divided into regional and cross-regional spaces. Traffic flows in different regions are interrelated. Information on traffic conditions across the road network needs to be considered.

3   Problem Description

The multi-intersection traffic signal control problem is modeled from the perspective of GRL. Specifically, the state features of the target intersection are first obtained. Then, the spatial dependency is captured using GNNs. Finally, the signal timing strategy is obtained by the DRL. The process is modelled as $MDP = < \mathcal{S}, \mathcal{O}, \mathcal{A}, \mathcal{P}, \mathcal{R}, \pi, \gamma >$.

State Space $\mathcal{S}$ and Observation Space $\mathcal{O}$. Each agent i observes a portion of the global state $s \in \mathcal{S}$ as its observation $o \in \mathcal{O}$. The observed state at time t

is denoted as $o_i^t = [\boldsymbol{n}, \boldsymbol{p}]$. $\boldsymbol{n}$ represents the number of vehicles in each lane, and $\boldsymbol{p}$ is the current phase value.

Action $\mathcal{A}$. The agent i makes its next-time decision by choosing an action a_i from the set of available actions A_i. We use eight variable signal phases with the same set of actions for each agent, i.e., $A = \{a_1, a_2, a_3, a_4, a_5, a_6, a_7, a_8\}$.

Reward $\mathcal{R}$. It is the immediate reward that the agent receives from the environment at each time t. $r_i^t = -\sum lq_{i,l}^t$ is defined as the reward of the model. $q_{i,l}^t$ denotes the queue length on the incoming lane l at time t.

Transfer Probability $\mathcal{P}$. It portrays the dynamic characterization of the environment. $p(s_{t+1}|s_t, a_t)$ denotes the probability that state s_t to the next state s_{t+1} when a_t is taken.

Policy π. It represents the mapping of states to actions.

Discount Factor γ. It is used to moderate the near and far-term impact and takes a value in the $(0,1]$ range.

4 Methodology

4.1 Framework of GTLight

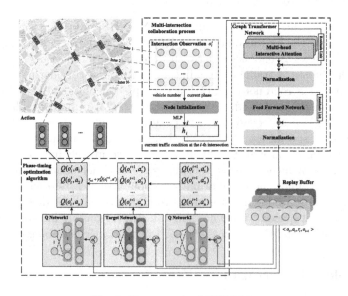

Fig. 1. Framework of GTLight.

The framework of GTLight is shown in Fig. 1. It is divided into two main parts: a multi-intersection cooperation process and a phase-timing optimization algorithm. Specifically, the agent acquires information on the observed state of each intersection and performs node initialization. Then, the spatial structure information of the road network is learned using the graph transformer network.

Finally, the signal timing strategy is determined using a phase timing optimization algorithm. We will describe these modules in detail.

4.2 Node Initialization Module

At the moment t, we obtain the original observation state information o_i^t from intersection i. It is encoded by a layer of multi-layer perceptron (MLP) with a rectified linear unit (RELU) [25]. Then, the initial representation of the node is obtained, as shown in Eq. (1):

$$h_i = Dense(o_i^t) = \sigma(o_i W_e + b_e), \tag{1}$$

where $Dense$ is a fully connected layer. $o_i^t \in \mathbb{R}^d$ is the state observation of intersection i at time t. Weight matrix $W_e \in \mathbb{R}^{d \times m}$, bias vector $b_e \in \mathbb{R}^m$. The hidden state $h_i \in \mathbb{R}^m$ denotes the current traffic state at the i-th intersection.

4.3 Graph Transformer Network Module

The module integrates the state features of adjacent intersections and extracts higher-order hidden states. The multi-head interactive attention mechanism calculation process is shown in Fig. 2.

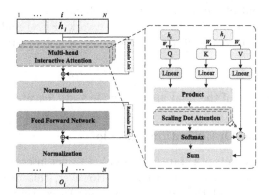

Fig. 2. Computational process of multi-head interactive attention mechanism.

First, three potential subspaces are trained for each node based on the vectors generated from the current observation state. Q is the query subspace, K is the key subspace and V is the value subspace. As shown in Eq (2):

$$Q = h_i W_q, K = h_j W_k, V = h_j W_v, \tag{2}$$

where $W_q \in \mathbb{R}^{m \times n}$, $W_k \in \mathbb{R}^{m \times n}$, $W_v \in \mathbb{R}^{m \times n}$ are the weight matrices of Q, K, V respectively. $j \in \mathcal{N}_i$ is a neighboring intersection of target intersection i. The function of the three trainable weight matrices is to extract features.

The attention weights of each pair of target intersection i and neighboring intersections j are calculated. The status information of the target intersection is updated. The equation is as follows:

$$h_i = Attention(Q, K, V) = softmax(QK^T)V, \tag{3}$$

where $Attention(.)$ denotes the attention weights between each pair of nodes. The aim is to differentiate the importance of the intersections.

Multi-head interactive attention is used to capture information from different subspaces. The equation is as follows:

$$head_h = Attention(QW_h^Q, KW_h^K, VW_h^V), \tag{4}$$

$$MultiHeadAttention = Concat(head_1, ..., head_H)W^O, \tag{5}$$

where W_h^Q, W_h^K, W_h^V are the weight of the h-th attention head learnable. W^O is the dimensionality reduction projection. Multiple dependency patterns can be learned using the multi-head interactive attention mechanism and capture richer information about the intersection.

The model uses two residual connections to solve the problem of difficulty in training multi-layer neural networks. As shown in Eq. (6):

$$z_i = h_i + MultiHeadAttention(h_i), \tag{6}$$

where z_i is the output of the residual layer, which enhances the fit of the model.

The proposed model uses a feedforward network (FFN) and batch normalization to improve the representation of the nodes. As shown in Eq. (7):

$$h_i' = BN(FFN(BN(z_i))), \tag{7}$$

where h_i' is the output of the final updated node state feature.

4.4 Phase-Timing Optimization Module

The prediction of road network state features h_i' at target intersections using deep Q-network (DQN) can achieve good results. However, this method sometimes overestimates the value of the behavior and fails to converge to the optimal value function.

Therefore, a new DNN is added to the DQN in the phase-timing optimization module. We input the next moment state into this neural network to get the action with the maximum Q-value. The purpose is to reduce the correlation of the Q-network during iteration. The original network evaluates the Q-value of the subsequent actions. Improve the stability of the algorithm by decoupling the Q-values used for action selection and evaluation to address Q-value overestimation. The Q-value is calculated as follows:

$$Q = \phi(h_i'), \tag{8}$$

where ϕ is a two-layer fully connected layer with RELU activation.

To prevent overestimation of the Q-value, a new objective is defined as:

$$y_t = r_{t+1} + \gamma Q(s_{t+1}, argmax_{a'} Q'(s_{t+1}, a'; \theta_t); \theta'_t), \quad (9)$$

where θ and θ' for action selection and evaluation respectively.

The loss function for optimizing the current policy is as follows:

$$\mathcal{L}(\theta) = \frac{1}{T} \sum_{t=1}^{T} \sum_{i=1}^{N} (y_t - Q(s_t, a_t, \theta_t))^2, \quad (10)$$

where $s_t = \{o_1^t, o_2^t, ..., o_i^t\}$. T represents the total simulation time per episode. N denotes the number of intersections.

The training procedure of GTLight is presented in Algorithm 1.

Algorithm 1: GTLight Training Process

Input: Set of target intersections $\mathcal{I}_N$ and adjacency matrix A;
Output: Optimized parameter θ;
1: Initialize network parameters θ, green time T_g and simulation time T;
2: **for** epoch $= 1$ **to** max-epochs **do**
3:    Reset the environment;
4:    **for** $t = 1, 2, ..., \frac{T}{T_g}$ **do**
5:       Encode the observed state as hidden representation h_i;
6:       Learning spatial dependency h_i' by Eq. (7);
7:       Compute the Q-value of the agent i at time t by Eq. (8);
8:       Select random action a_t or $a_t = argmax_a Q_i^t(s_t, a)$;
9:       Take action a_t, receive reward r_t and next state o_i^{t+1};
10:      Store transition (o_t, a_t, r_t, o_{t+1}) in reply buffer $\mathcal{D}$;
11:   **end for**
12:   **for** each target intersection $\mathcal{I}_i$ in $\mathcal{I}_N$ **do**
13:      Sample a random mini-batch of S samples from $\mathcal{D}$;
14:      Update evaluating network by Eq. (10);
15:      Update parameters $\theta = (1 - \beta)\theta' + \beta\theta$;
16:   **end for**
17: **end for**

---

5 Experiments

We validate the effectiveness of GTLight using different datasets and use CityFlow[1] traffic simulation software to build traffic environment. It can simulate the movement of each vehicle and supports large-scale traffic signal control.

---

[1] http://cityflow-project.github.io.

5.1 Experimental Setup

Parameter Setting. Firstly, the green, yellow, and red signal times are set to 10 s, 3 s, and 2 s, respectively. Secondly, the simulation duration is 3600 s per round for 100 episodes. Each set is trained 100 times. The training batch is 20. The discount factor is 0.8. The sample size is 1000. The empirical replay buffer size is 10000. Thirdly, the optimization algorithm is the RMSprop [26].

Synthetic Dataset. The dataset is synthesized by analyzing two real-world traffic in Hangzhou and Jinan. It consists of 16 four-way intersections and 300m road segments. We test with four configs (config1 is a peak arrival rate of 0.388, config2 is a slow arrival rate of 0.388, config3 is a peak arrival rate of 0.416, and config4 is a slow arrival rate of 0.416). Set the left-turn rate at 10%, the straight-through rate at 60%, and the right-turn rate at 30%.

Real-World Dataset. The dataset is 16 intersections in Gudang Street, Hangzhou, China. It is captured by intersection cameras. We take the number of vehicles passing through the intersections as traffic flow. The duration is one hour.

5.2 Comparison Algorithms

To validate the effectiveness of GTLight, we compare different baseline methods. Each model is learned without any pre-training parameters to ensure fairness.

Traffic Signal Control Methods on Traditional

FixedTime [27]: The model cycles traffic signals in a fixed phase sequence.

SOTL [28]: The model manually adjusts the threshold for the number of waiting vehicles to change the signal phase.

The comparison algorithms are chosen to verify that data analysis techniques can improve the validity of our model.

Traffic Signal Control Based on DRL

PressLight [17]: The model uses the DQN algorithm with stress as a reward.

LitLight [13]: The model relates RL to classical transportation theory.

The comparison algorithms are chosen to verify that graph structure enhances the ability of the model to capture information.

Traffic Signal Control Based on GRL

CoLight [29]: The model uses GAT to communicate between intersections.

SwarmCoLight [30]: The model is a hierarchical and decentralized multi-agent GRL method.

The comparison algorithms are chosen to verify the effect of considering higher-order road network information on the model performance.

5.3 Evaluation Indicator

We use the **average travel time** as an evaluation indicator for the model. It is the average time in seconds spent by all vehicles entering and exiting each intersection in each round. Lower time cost represents better performance.

5.4   Experimental Results

Overall Analysis. We compare the proposed and baseline methods on synthetic and real-world datasets. The performance of each model in average travel time is shown in Table 1. We train the models on an Ubuntu 20.04 LTS server. Our code is available at Github.[2] We obtain the following intuitive observations:

Table 1. The performance of the synthetic and real-world datasets is compared in average travel time.

Methods		Synthetic				Real-word
		config1	config2	config3	config4	hangzhou
Traditional	FixedTime	923.67	875.35	826.03	902.53	806.28
	SOTL	1466.06	1511.38	1508.10	1460.93	1287.22
DRL	PressLight	745.37	718.18	704.28	775.04	584.79
	LitLight	404.86	433.62	489.01	539.24	480.66
GRL	CoLight	514.60	483.80	525.11	504.79	521.34
	SwarmCoLight	466.80	503.17	491.03	533.99	536.45
Ours	**GTLight**	**335.34**	**369.54**	**364.37**	**392.70**	**431.35**

Our GTLight outperforms all comparative algorithms, achieving the shortest travel times. Specifically, the traditional traffic signal control methods (Fixed-Time, SOTL) rely heavily on oversimplified assumptions. They are prone to fail under dynamic traffic conditions and therefore perform poorly. PressLight does not consider the different degrees of impact of adjacent intersections on the target intersection. LitLight does not consider the impact of downstream lanes. Compared with PressLight and LitLight, our GTLight fully considers the spatial state information between adjacent intersections and achieves some improvements in travel time. Moreover, compared with CoLight, the performance of GTLight improved by 34.83%, 23.62%, 30.61%, 22.21%, and 17.26% on different datasets. Compared with SwarmColight, GTLight improved by 28.16%, 26.56%, 25.79%, 26.46%, and 19.59%, respectively. The reason is that CoLight and SwarmCoLight are concerned with the impact of neighboring intersections around a small area. However, considering only a small range of neighboring intersections can limit the scope of cooperation at the target intersection. In relative terms, our GTLight captures global road network state information using a relational enhanced interactive attention mechanism.

Trend Analysis. We compare the corresponding learning curves of the proposed GTLight with the baseline methods, as shown in Fig. 3. The traditional traffic signal control methods are not shown in the figure because they do not

---

[2] https://github.com/ddlyn/GTLight.

have a training process. As can be seen from the figure, GTLight achieves faster convergence and better performance than the other methods. Specifically, PressLight uses only the decentralized DQN algorithm, which suffers from instability. Therefore, it is difficult to achieve optimal performance even after a long training period. Although LitLight shows good learning performance, it does not consider the relevant information of downstream intersections. Compared with CoLight and SwarmCoLight, GTLight combines the idea of a two-layer network to improve stability. It also considers information on the state of cross-regional intersections to obtain optimal strategies more quickly.

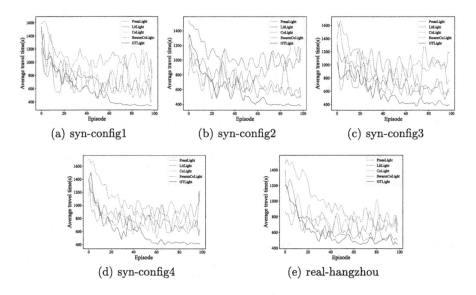

(a) syn-config1   (b) syn-config2   (c) syn-config3

(d) syn-config4   (e) real-hangzhou

Fig. 3. Average travel time for different models with different datasets.

Ablation Experiment. The ablation study of the GTLight on different datasets is shown in Fig. 4. Our GTLight consistently outperforms all variants of the model. Specifically, GTLight-GD uses the traditional attention mechanism and DQN. GTLight-G uses the traditional attention mechanism and phase-timing optimization algorithm. GTLight-D uses the graph transformer network and DQN. The traditional attention mechanism focuses only on a small area of neighboring intersections. It does not consider the correlation of traffic flow between intersections in different areas. In addition, the optimal action of the DQN algorithm is chosen based on the parameters of the target Q-network. It leads to an overestimation situation. In our GTLight, the graph transformer network focuses on the traffic flow at intersections in different regions. The optimal action selection is based on the parameters of the currently updated Q-network. The overestimation situation is reduced to a certain extent.

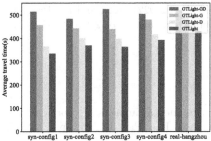
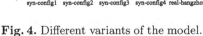
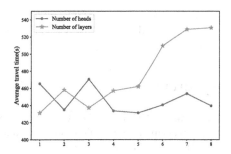

Fig. 4. Different variants of the model.   **Fig. 5.** Effect of different numbers.

5.5 Study of GTLight

Effect of Attentional Head Number. We test the effect of attention head numbers on GTLight using the real-world dataset, as shown in Fig. 5. The model achieves the shortest travel time using five attention heads. However, when the number of attention heads is sufficient, the model pays sufficient attention to the node information. More than five heads may introduce some noise effects, making the performance degraded.

Effect of Graph Transformer Network Layer Number. We use the real-world dataset to evaluate the effect of graph transformer layer number on GTLight, as shown in Fig. 5. The traditional transformer model contains n-layer of the network. However, training becomes more challenging as the depth of network layers increases. While normalization techniques can alleviate this problem, difficulties still exist. The experimental results show that our model performs better with one and three layers, but one layer is even better.

6 Conclusion

In this paper, a GTLight model is constructed for multi-intersection traffic signal control. It uses a graph transformer network to solve the higher-order spatial dependence problem. In addition, a phase-timing optimization algorithm is designed to solve the overestimation of the Q-value. We conduct performance, ablation, and parametric experiments on different datasets. Experiments have shown that GTLight performs well on the evaluation metrics. The proposed model has practical implications for improving urban road traffic conditions and enhancing the travel experience for residents. In our further research, we aim to improve the safety and disturbance resistance of the model.

Acknowledgements. This work was supported by National Natural Science Foundation of China (No. 62176088), the Key Science and Technology Research Project of

Henan Province of China (Grant No. 22102210067, 222102210022), and the Program for Science and Technology Development of Henan Province (No. 212102210412 and 202102310198).

References

1. Fan, Z., Harper, C.D.: Congestion and environmental impacts of short car trip replacement with micromobility modes. Transp. Res. Part D: Transp. Environ. **103**, 103173 (2022)
2. Robertson, D.I., Bretherton, R.D.: Optimizing networks of traffic signals in real time-the SCOOT method. IEEE Trans. Veh. Technol. **40**(1), 11–15 (1991)
3. Luk, J.: Two traffic-responsive area traffic control methods: SCAT and SCOOT. Traff. Eng. Control **25**(1) (1984)
4. Liao, X.C., Qiu, W.J., Wei, F.F., Chen, W.N.: Combining traffic assignment and traffic signal control for online traffic flow optimization. In: Tanveer, M., Agarwal, S., Ozawa, S., Ekbal, A., Jatowt, A. (eds.) ICONIP 2022. CCIS, vol. 1793, pp. 150–163. Springer, Singapore (2023). https://doi.org/10.1007/978-981-99-1645-0_13
5. Noaeen, M., et al.: Reinforcement learning in urban network traffic signal control: a systematic literature review. Exp. Syst. Appl. **199**, 116830 (2022)
6. Wei, H., Zheng, G., Yao, H., Li, Z.: Intellilight: a reinforcement learning approach for intelligent traffic light control. In: Proceedings of the 24th ACM SIGKDD International Conference on Knowledge Discovery and Data Mining, pp. 2496–2505 (2018)
7. Kuang, L., Zheng, J., Li, K., Gao, H.: Intelligent traffic signal control based on reinforcement learning with state reduction for smart cities. ACM Trans. Internet Technol. **21**(4), 1–24 (2021)
8. Li, L., Lv, Y., Wang, F.Y.: Traffic signal timing via deep reinforcement learning. IEEE/CAA J. Automat. Sinica **3**(3), 247–254 (2016)
9. Wu, T., et al.: Multi-agent deep reinforcement learning for urban traffic light control in vehicular networks. IEEE Trans. Veh. Technol. **69**(8), 8243–8256 (2020)
10. Ying, Z., Cao, S., Liu, X., Ma, Z., Ma, J., Deng, R.H.: PrivacySignal: privacy-preserving traffic signal control for intelligent transportation system. IEEE Trans. Intell. Transp. Syst. **23**(9), 16290–16303 (2022)
11. Tan, T., Bao, F., Deng, Y., Jin, A., Dai, Q., Wang, J.: Cooperative deep reinforcement learning for large-scale traffic grid signal control. IEEE Trans. Cybernet. **50**(6), 2687–2700 (2020)
12. Vaswani, A., et al.: Attention is all you need. In: Proceedings of the 31st International Conference on Neural Information Processing Systems, pp. 6000–6010 (2017)
13. Zheng, G., et al.: Diagnosing reinforcement learning for traffic signal control. arXiv preprint arXiv:1905.04716 (2019)
14. Chu, T., Wang, J., Codecà, L., Li, Z.: Multi-agent deep reinforcement learning for large-scale traffic signal control. IEEE Trans. Intell. Transp. Syst. **21**(3), 1086–1095 (2020)
15. Arel, I., Liu, C., Urbanik, T., Kohls, A.G.: Reinforcement learning-based multi-agent system for network traffic signal control. IET Intel. Transp. Syst. **4**(2), 128–135 (2010)

16. El-Tantawy, S., Abdulhai, B., Abdelgawad, H.: Multiagent reinforcement learning for integrated network of adaptive traffic signal controllers (MARLIN-ATSC): methodology and large-scale application on downtown toronto. IEEE Trans. Intell. Transp. Syst. **14**(3), 1140–1150 (2013)
17. Wei, H., et al.: PressLight: learning max pressure control to coordinate traffic signals in arterial network. In: Proceedings of the 25th ACM SIGKDD International Conference on Knowledge Discovery and Data Mining, pp. 1290–1298 (2019)
18. Zhang, M., Wu, S., Yu, X., Liu, Q., Wang, L.: Dynamic graph neural networks for sequential recommendation. IEEE Trans. Knowl. Data Eng. **35**(5), 4741–4753 (2023)
19. Zhong, T., Xu, Z., Zhou, F.: Probabilistic graph neural networks for traffic signal control. In: Proceedings of the IEEE International Conference on Acoustics, Speech and Signal Processing, pp. 4085–4089 (2021)
20. Wang, Y., Xu, T., Niu, X., Tan, C., Chen, E., Xiong, H.: STMARL: a spatio-temporal multi-agent reinforcement learning approach for cooperative traffic light control. IEEE Trans. Mob. Comput. **21**(6), 2228–2242 (2022)
21. Zeng, Z.: GraphLight: graph-based reinforcement learning for traffic signal control. In: Proceedings of the IEEE International Conference on Computer and Communication Systems, pp. 645–650 (2021)
22. He, L., Li, Q., Wu, L., Wang, M., Li, J., Wu, D.: A spatial-temporal graph attention network for multi-intersection traffic light control. In: Proceedings of the IEEE International Joint Conference on Neural Networks, pp. 1–8 (2021)
23. Wu, L., Wang, M., Wu, D., Wu, J.: DynSTGAT: dynamic spatial-temporal graph attention network for traffic signal control. In: Proceedings of the 30th ACM International Conference on Information and Knowledge Management, pp. 2150–2159 (2021)
24. Wang, M., Wu, L., Li, M., Wu, D., Shi, X., Ma, C.: Meta-learning based spatial-temporal graph attention network for traffic signal control. Knowl. Based Syst. **250**, 109166 (2022)
25. Xu, B., Wang, N., Chen, T., Li, M.: Empirical evaluation of rectified activations in convolutional network. arXiv preprint arXiv:1505.00853 (2015)
26. Zou, F., Shen, L., Jie, Z., Zhang, W., Liu, W.: A sufficient condition for convergences of ADAM and RMSPROP. In: Proceedings of the IEEE/CVF Conference on Computer Vision and Pattern Recognition, pp. 11119–11127 (2019)
27. Koonce, P., Rodegerdts, L.: Traffic signal timing manual. Tech. rep., United States. Federal Highway Administration (2008)
28. Cools, S.B., Gershenson, C., D'Hooghe, B.: Self-organizing traffic lights: a realistic simulation. In: Advances in Applied Self-Organizing Systems, pp. 45–55 (2013)
29. Wei, H., et al.: CoLight: learning network-level cooperation for traffic signal control. In: Proceedings of the 28th ACM International Conference on Information and Knowledge Management, pp. 1913–1922 (2019)
30. Yi, Y., Li, G., Wang, Y., Lu, Z.: Learning to share in multi-agent reinforcement learning. In: Proceedings of the 36th Annual Conference on Neural Information Processing Systems, pp. 1–13 (2022)

PatchFinger: A Model Fingerprinting Scheme Based on Adversarial Patch

Bo Zeng, Kunhao Lai, Jianpeng Ke, Fangchao Yu, and Lina Wang[✉]

Key Laboratory of Aerospace Information Security and Trusted Computing, Ministry of Education, School of Cyber Science and Engineering, Wuhan University, Wuhan, China
{bobozen,kh.lai,kejianpeng,fangchao,lnwang}@whu.edu.cn

Abstract. As deep neural networks (DNNs) gain great popularity and importance, protecting their intellectual property is always the topic. Previous model watermarking schemes based on backdoors require explicit embedding of the backdoor, which changes the structure and parameters. Model fingerprinting based on adversarial examples does not require any modification of the model, but is limited by the characteristics of the original task and not versatile enough. We find that adversarial patch can be regarded as an inherent backdoor and can achieve the output of specific categories injected. Inspired by this, we propose PatchFinger, a model fingerprinting scheme based on adversarial patch which is applied to the original samples as a model fingerprinting through a specific fusion method. As a model fingerprinting scheme, PatchFinger does not sacrifice the accuracy of the source model, and the characteristics of the adversarial patch make it more flexible and highly robust. Experimental results show that PatchFinger achieves an ARUC value of 0.936 in a series of tests on the Tiny-ImageNet dataset, which exceeds the baseline by 19%. When considering average query accuracy, PatchFinger gets 97.04% outperforming the method tested.

Keywords: Deep Neural Network · Intellectual Property Protection · Model Fingerprinting · Adversarial Patch

1 Introduction

In the past decades, DNNs have shown superior performance in computer vision [8], intelligent healthcare [6], autonomous driving [4] and other fields. However, to obtain high-quality DNNs, it is necessary to allocate expensive resources and considerable time to the training process. Unfortunately, there has been an increasing number of cases where these trained models have been illegally copied and misappropriated [22]. These illegal activities pose severe threats to the intellectual property of the creators. Therefore, protecting the intellectual property rights of DNNs should be taken into urgent consideration.

In order to protect the intellectual property of DNNs, there are currently two mainstream methods: model watermarking and model fingerprinting. Model

© The Author(s), under exclusive license to Springer Nature Singapore Pte Ltd. 2024
B. Luo et al. (Eds.): ICONIP 2023, LNCS 14448, pp. 68–80, 2024.
https://doi.org/10.1007/978-981-99-8082-6_6

watermarking requires to embed the watermark into the model during training. The current methods that achieve good results are based on backdoors but they inevitably sacrifice the accuracy of the model, and cannot protect a model that has already been trained.

Unlike model watermarking, model fingerprinting is based on the internal attributes of the model to bind it to the owner. After attackers steal the model, they may try to destroy the model fingerprinting to hinder ownership verification. However, some model fingerprinting methods [3,14] use adversarial samples as fingerprints which perform poor against such conditions.

The primary challenge with watermarking technology is its modification of the model during deployment, whereas task-specific fingerprint technologies are not robust enough to handle different scenarios adequately. To solve the above issues, we design a model fingerprinting scheme named PatchFinger based on adversarial patch [2]. We discover that adversarial patch has the backdoor property that it can output the designed label after insertion. Moreover due to its innate characteristics, adversarial patches are robust enough to different attacks. Inspired by this, we utilize adversarial patch in the model fingerprint scheme, achieving the same effect as a backdoor-based watermarking scheme without compromising the accuracy of the model. The patch exhibits versatility and adaptability since it can added to any original image and maintain its adversarial properties under different backgrounds, lighting, and rotations, working exceptionally well under various transformations.

Specifically, PatchFinger consists of two phases: patch injection and fingerprint generation. In the patch injection phase, a specific fusion method is used to inject the patch into the original image, generating a patched image input for the subsequent phase. In the fingerprint generation phase, all models are divided into positive and negative models. The patched image is recognized as a specific category by all positive models but as the original category by all negative models. By optimizing the loss of positive and negative models, as well as the restriction of adversarial patch, query images that can be used for fingerprinting are selected.

To summarize, we mainly make the following contributions:

- We propose PatchFinger, to our best knowledge, which first introduces adversarial patch to model fingerprinting. It combines the advantages of adversarial patch and fingerprints, resulting in high robustness without altering the model.
- In the fingerprint generation algorithm, we involve a mask to control the embedding position when injecting adversarial patch, rather than directly rotating or scaling, making it more flexible and better adapted to images.
- The experimental results show that PatchFinger has strong robustness against attacks such as model modifications, model extractions, and input modifications. Compared with previous methods, PatchFinger achieved an ARUC value of 0.936 and an average query accuracy of 97.04%, outperforming the baseline.

2   Related Work

2.1   Model Watermarking

DNN model watermarking can be classified into two types: parameter-based methods [13,21] and query-based methods [1,12,24]. Parameter-based DNN watermarking alters the weight distribution compromising the accuracy of the model. Compared to parameter-based schemes, query-based ones are known as dynamic watermarking techniques. Le Merrer et al. first utilize dynamic watermarking [12], using the adversarial samples of the DNN model to embed the watermark. However, Adi et al. [1] claim that such methods heavily depend on the transferability of adversarial samples across different architectures. According to Zhang et al. [24], they add information to the backdoor sample pattern in the watermarking dataset to embed the watermark.

2.2   Model Fingerprinting

Unlike model watermarking techniques, model fingerprinting maintains the source model unchanged. Jia et al. propose PoL [10], which extracts model fingerprints from the static properties of the model. However, Zhang et al. [16] point out that PoL cannot resist to the fingerprint forgery attack and they use adversarial samples to overcome it.

Besides the static properties of the model, some methods [3,14] construct fingerprints dynamically by building query sets as model fingerprintings. Cao et al. [3] claim that each DNN model has a unique decision boundary which can be exploited using model fingerprinting to obtain particular labels. Lukas et al. [14] indicate IPGuard is not robust to model extraction attacks and introduce Conferrable adversarial examples to strengthen their scheme. In order to improve fingerprint recognition, MetaV [18] uses model preparation techniques to generate test cases and compare fingerprints. MetaFinger [23] proposes DNN enhancement, such that trained models have increased diversity by randomly injecting Gaussian noise into their model parameters. In our approach, in addition to using advanced fingerprint methods, we first use adversarial patch as fingerprints, which have high robustness against various attacks.

3   Problem Definition

3.1   Threat Model

The threat model considered in this paper includes a model owner who trains and deploys the source model, and an attacker who steals the source model through illegal copying.

The objective of the model owner is to verify whether a suspicious model is a copy of the source model and to correctly identify stolen models, while recognizing other unrelated models from trusted third parties as negative models.

The owner has white-box access to the source model (such as model parameters and training data) and black-box access to the suspicious model.

The attacker's goal is to evade ownership verification and ensure that the suspicious model's resemblance to the source model's performance avoids detection as a stolen model. The attacker is assumed to own the access to some data with the same distribution of training and can modify the model. The attacker can also use some techniques to modify and reject abnormal query examples.

3.2   Fingerprinting DNN Models

The DNN model fingerprintings can be divided into two stages: fingerprint extraction and fingerprint verification. The two stages are shown as follows:

Fingerprint Extraction. Given a DNN model m, the owner can obtain a fingerprint F_m through a function $F_m = Extract(m)$.

Fingerprint Verification. For a suspicious DNN model m_s, verification function $Verify(F_m, m_s)$ outputs 0 and 1, where 1 represents that the model is illegally obtained.

3.3   Adversarial Techniques

Based on the above threat model, DNN models may be subjected to different types of attacks. Common attacks include model modifications, model extractions, as well as input modifications. Below, we provide a brief explanation of attack types and specific settings are detailed in Sect. 5.1.

Model Modification. Model Modifications convert the source model to a theft model, with the same functionality as much as possible, including fine-tuning, weight pruning, weight noising, adversarial training and so on.

Model Extraction. Model extractions recover the internal information or functionality of the source model without directly accessing the source model, including distillation, knockoff and so on.

Input Modification. Input modifications aim to modify the inputs to resist adversarial samples, and it can also be used to block the verification of model ownership.

4   Method

From Fig. 1, the overall process consists of two parts: patch injection and fingerprint generation. As shown on the left side of the figure, in our patch injection part, the size of the adversarial patch is the same as that of the original image. Unlike the traditional method of flipping, scaling, and locating, we select a mask to determine the embedding rule. As shown on the right side, during the fingerprint generation, the samples that have been embedded with the adversarial patch are designated as a specific category. After optimization using positive and negative models, the final model fingerprint image is generated. The specific details are as follows:

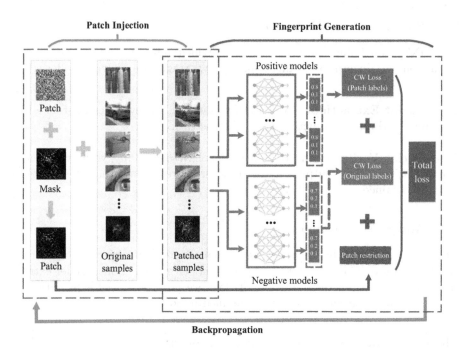

Fig. 1. PatchFinger framework based on adversarial patch.

4.1   Patch Injection

Considering the previous patch injection method in [2]

$$x' = A(x, \Delta, l, t) \tag{1}$$

where $A(\cdot)$ represents the function that injects the patch into the original image, x represents the original image, Δ represents the patch, l represents the patch embedding position, and t represents the operation of flipping or scaling the patch. We can see just simple changes are performed on the patch, so the patch embedded in this way is limited by its own shape and needs to be generated in advance. To overcome these limitations, we have made some modifications to the above definition, and the modified equation is shown as Eq. 2.

$$x' = A(x, \Delta, m) \qquad (2)$$
$$x'_{i,j,e} = (1 - m_{i,j}) \cdot x_{i,j,e} + m_{i,j} \cdot \Delta_{i,j,e} \qquad (3)$$

where x and Δ keep the same, but m represents a two-dimensional matrix of the mask, which determines the proportion of the patch. And i, j, e in Eq. 3 represent the indexes of the columns, rows and channels of the image, respectively.

The mask value is in the range of $[0, 1]$. When $m_{i,j} = 1$, the pixel at that position in the original image is completely covered by the adversarial patch and remain unchanged otherwise. By using a continuous mask, not only is the mask specific to the dataset, but also makes it more appropriate to integrate the adversarial patch into the original image. By injecting the patch, the image x' with the embedded adversarial patch can be obtained.

4.2   Fingerprint Generation

After obtaining the patched samples, the final loss is reduced by optimizing the samples to obtain fingerprint samples. Algorithm 1 explains the process of generating the fingerprint samples. First, the generated samples are transformed

---

Algorithm 1. Fingerprint generation

Input: Adversarial patch Δ, mask m, input image X, original label Y_{true}, patch label Y_{target}, learning rate η, iteration number N_e, loss control factor λ, positive example model M_p, negative example model M_n

Output: Query set Q

1: $\Delta = \text{random.uniform}()$
2: $m = \text{random.uniform}()$
3: **for** $e = 1$ to N_e **do**
4:    $X' = A(X, \Delta, m)$
5:    $X' = \text{transform}(X')$
6:    $L_{pos} = \frac{1}{|M_p|} \sum_{i=1}^{|M_p|} CW\left(M_p^i(X'), Y_{target}\right)$
7:    $L_{neg} = \frac{1}{|M_n|} \sum_{i=1}^{|M_n|} CW(M_n^i(X'), Y_{true})$
8:    $Loss = L_{pos} + L_{neg} + \lambda|m|$
9:    $\Delta = \Delta - \eta \nabla Loss$
10:   $m = m - \eta \nabla Loss$
11: **end for**
12: $Q = \text{screen}(X, Y_{true}, Y_{target}, M_p, M_n)$
13: **return** Q

---

before the input model to reduce overfitting (line 5). In this paper, three losses in Eq. 4 are calculated (lines 6–8).

$$Loss = L_{pos} + L_{neg} + \lambda|m| \qquad (4)$$

The first part is the loss of all positive models. For the patched sample with the specified label Y_{target}, the target for all positive models is to predict the label of the image X' as Y_{target}, as shown in Eq. 5.

$$L_{pos} = \frac{1}{|M_p|} \sum_{i=1}^{|M_p|} CW \left(M_p^i(X'), Y_{target} \right) \tag{5}$$

The second part is the loss of all negative models, which should recognize the patched sample as a normal image. Therefore, their goal is to predict the label of the image X' as Y_{true}, which is the original correct label of the sample, as shown in Eq. 6.

$$L_{neg} = \frac{1}{|M_n|} \sum_{i=1}^{|M_n|} CW(M_n^i(X'), Y_{true}) \tag{6}$$

where $CW(\cdot)$ represents the Carlini-Wagner loss [5].

The third part is the constraint on the adversarial patch. The adversarial patch should be as small as possible, that is, it should modify only a limited part of the image. In this paper, the L_1 norm $\lambda|m|$ of the mask m is used to constrain the size of the patch. The smaller the λ, the smaller the constraint on the mask m and the larger the patch size, resulting in a higher success rate for the specified classification. We use the Adam optimizer to optimize the above problem.

Finally, the samples are filtered (line 12), and not all samples are correctly classified by the positive and negative models. Here, we only selects the target images where all positive models classify the image as Y_{target} and all negative models classify the image as Y_{true} and add them to the query set.

5 Experiments

5.1 Experimental Setup

Dataset. In this paper, 50 categories are randomly selected from Tiny-ImageNet [11]. The datasets are divided into two parts for positive and negative models. The details of the attacks to the models are shown in the Table 1. Positive models use the ResNet18 [8] structure as the source model and employ 6 types of attacks to generate 48 different models (including the source model). The negative models also use the ResNet18 structure as the source model, but are trained on different datasets and undergo the same attacks.

Experimental Environment. All experiments are conducted on a Windows server with an Intel(R) Core(TM) i7-6700K CPU and 12 GB NVIDIA TITAN X GPU. Unless otherwise specified, we always set the number of fingerprint samples for both PatchFinger and the baseline to 100 to achieve fair comparison. The learning rate η in Algorithm 1 is set to 0.001 and the number of iterations is 1000. The λ that constrains the patch size is set to 150.

Table 1. Construction of positive and negative models.

Types	Attacks	Num	Production process
Positive Models	Training	1	Train a ResNet18 model [8] as the source model
	Fine-tuning	4	Fine-tuning with each mode in [1]
	Weight Pruning	9	Prune the model weights with $p = 0.1, 0.2, ..., 0.9$ [7], and then fine-tune
	Weight Noising	9	Perturb the model weights with Gaussian noise [23], and then fine-tune
	Adversarial Training	9	Generate adversarial examples for $r = 0.1, 0.2, ..., 0.9$ of the fine-tuning data [14], and fine-tune with augmented datasets.
	Distillation	8	Distill [9] a model for each architecture.
	Knockoff	8	Knockoff [17] each architecture to steal models.
Negative Models	The same as the above	48	Execute the same attack as the positive model to generate a mirrored negative model

5.2 Performance Metrics

This section summarizes the evaluation metrics commonly used to evaluate DNN intellectual property protection methods.

- **Fidelity.** The fingerprint needs to make no influences to the accuracy of the source model.
- **Effectiveness.** The verification method can correctly identify the illegal model.
- **Robustness.** The fingerprint should be able to withstand various attacks.
- **Uniqueness.** The verification method should not produce false positives for models that are not illegal.

PatchFinger is compared with three other advanced schemes, namely IPGuard [3], MetaV [18], and MetaFinger [23] on the Tiny-ImageNet dataset. All compared methods maintain the accuracy without loss and achieve the desired level of effectiveness. Therefore, we focus on robustness and uniqueness properties.

5.3 Overall Results of Baseline Testing for PatchFinger

We compare PatchFinger with three other model fingerprinting approaches on multiple models by ARUC curve proposed in [3]. Figure 2 shows the ARUC values for all methods, where ρ represents the threshold of the matching rate. A higher ARUC means the model can possess both robustness and uniqueness simultaneously. We can see PatchFinger achieves the highest ARUC value (0.936), surpassing the second-ranked MetaFinger method by 19% (0.153).

5.4 Specific Results of Adversarial Techniques

The specific query accuracy of each method under different attacks in accord with Table 1 is shown in this section.

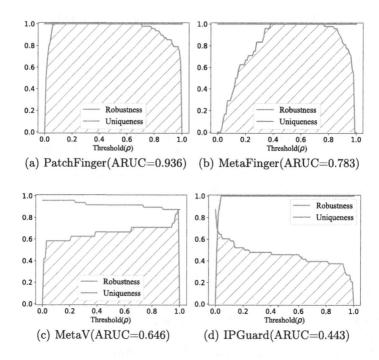

(a) PatchFinger(ARUC=0.936)   (b) MetaFinger(ARUC=0.783)

(c) MetaV(ARUC=0.646)   (d) IPGuard(ARUC=0.443)

Fig. 2. ARUC values of all methods on the settings in Table 1

In Table 2, there shows the query accuracy of four methods on four different fine-tuning modes in [3] on the Tiny-ImageNet. As we can see, all methods can achieve almost 100% query accuracy when just fine-tuning the model(FTLL, FTAL). But the performances of pre-trained models in RTLL and RTAL will be worse. Unlike other schemes, PatchFinger can maintain such query accuracy under all fine-tuning modes.

Table 2. Query accuracy of all methods under different fine-tuning modes.

Methods	FTLL	FTAL	RTLL	RTAL
IPGuard	100.00%	93.00%	66.00%	61.00%
MetaV	100.00%	100.00%	0	0
MetaFinger	99.00%	99.00%	89.00%	84.00%
PatchFinger	100.00%	100.00%	**100.00%**	**100.00%**

Figure 3 shows the results of the other 5 kinds of attacks. From Fig. 3(a) and (b), we can see that our scheme is extremely robust against weight pruning and weight noising, with almost no loss of query accuracy. Figure 3(c) and (d) show the query accuracy of positive and negative models after adding adversarial samples to the training data at different ratios. On the whole, PathFinger and MetaFinger are more stable at a good performance level with no sharp changes. When the rate is less than 0.5, PathFinger outperforms MetaFinger in both positive and negative models, while their performances are similar otherwise.

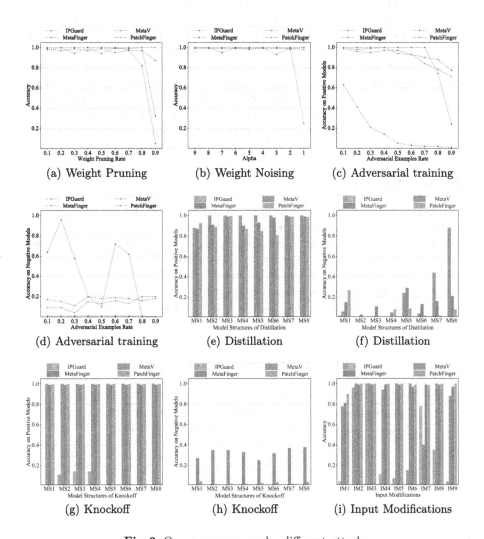

Fig. 3. Query accuracy under different attacks

Figure 3(e), (f), (g) and (h) show the experimental results of the model extraction attacks, where we use MobileNetV2 [19], ResNet18 [8], ResNet34, ResNet50, ShuffleNetV2 [15], VGG13_BN [20], VGG16_BN and VGG19_BN as model stuructures represented by (MS1, ..., MS8) in turn. We only select ResNet18, ResNet34, VGG13_BN, and VGG16_BN as training models in PatchFinger, but its performances on other structures are still good. Except for IPGuard, the other three methods are task-independent. The model extraction attacks make large changes to the model, so the query accuracy of IPGuard is very low. For the other three methods, all perform well in positive models, but both MetaFinger and MetaV perform worse than PatchFinger on negative models.

Input modifications are simple and effective ways to defend against adversarial samples and can also be used to hinder the verification of model ownership by pre-processing the inputs. We test the performances of all approaches on Tiny-ImageNet under a range of modification attacks, including Gaussian blur, Gaussian noise, universal noise, horizontal flipping, random translation, random cropping, bit depth reduction, JPEG compression and R&P represented by (IM1, ..., IM9) in turn. Figure 3(i) shows the performance results that almost all methods experience a certain degree of performance drop. But still PatchFinger performs the best in all cases, and the lowest query accuracy is 90%.

5.5   Average Accuracy

In summary, the average query accuracy of our proposed method PatchFinger and three comparison methods on all the aforementioned attacks is shown in Table 3. PatchFinger shows the best performance on the positive models as evidenced by its 97.04% query accuracy, thereby demonstrating its strong robustness. IPGuard has the lowest query accuracy on the negative models, but its accuracy on the positive models is not comparable to ours. As for two other schemes perform similarly on the positive models, our scheme outperforms them on the negative models a lot, demonstrating its good uniqueness.

Table 3. The average query accuracy

Average Accuracy	IPGuard	MetaV	MetaFinger	PatchFinger
Positive models	43.75%	91.39%	96.02%	**97.04%**
Negative models	**0.56%**	21.92%	21.08%	7.96%

6   Conclusion

We propose PatchFinger, a model fingerprinting scheme based on adversarial patch. PatchFinger achieves the same effect as the model watermarking scheme based on backdoors without compromising model accuracy. Moreover, the inherent characteristics of the adversarial patch make the scheme highly robust. PatchFinger comprises two stages: patch injection and fingerprint generation. The patch injection stage inserts an adversarial patch into the original image, and the fingerprint generation stage optimizes the loss to select a query set that meets the final conditions. Experimental results demonstrate that PatchFinger exhibits strong robustness when exposed to various attacks. Additionally, its query accuracy outperforms similar fingerprint schemes significantly. Moreover, the scheme has very low accuracy in identifying unrelated models exhibiting good uniqueness as well.

Acknowledgements. This work was supported in part by the National Natural Science Foundation of China under No. 62372334; in part by the National Key Research and Development Program of China under No. 2020YFB1805400.

References

1. Adi, Y., Baum, C., Cisse, M., Pinkas, B., Keshet, J.: Turning your weakness into a strength: watermarking deep neural networks by backdooring. In: USENIX Security Symposium, pp. 1615–1631 (2018)
2. Brown, T.B., Mané, D., Roy, A., Abadi, M., Gilmer, J.: Adversarial patch. arXiv preprint arXiv:1712.09665 (2017)
3. Cao, X., Jia, J., Gong, N.Z.: IPGuard: protecting intellectual property of deep neural networks via fingerprinting the classification boundary. In: Proceedings of the 2021 ACM Asia Conference on Computer and Communications Security, pp. 14–25 (2021)
4. Cao, Y., et al.: Adversarial sensor attack on LiDAR-based perception in autonomous driving. In: Proceedings of the 2019 ACM SIGSAC Conference on Computer and Communications Security, pp. 2267–2281 (2019)
5. Carlini, N., Wagner, D.: Towards evaluating the robustness of neural networks. In: 2017 IEEE Symposium on Security and Privacy (SP), pp. 39–57. IEEE (2017)
6. Esteva, A., et al.: Dermatologist-level classification of skin cancer with deep neural networks. Nature **542**(7639), 115–118 (2017)
7. Han, S., Pool, J., Tran, J., Dally, W.: Learning both weights and connections for efficient neural network. In: Advances in Neural Information Processing Systems, vol. 28 (2015)
8. He, K., Zhang, X., Ren, S., Sun, J.: Deep residual learning for image recognition. In: Proceedings of the IEEE Conference on Computer Vision and Pattern Recognition, pp. 770–778 (2016)
9. Hinton, G., Vinyals, O., Dean, J.: Distilling the knowledge in a neural network. arXiv preprint arXiv:1503.02531 (2015)
10. Jia, H., et al.: Proof-of-learning: Definitions and practice. In: 2021 IEEE Symposium on Security and Privacy (SP), pp. 1039–1056. IEEE (2021)
11. Le, Y., Yang, X.: Tiny ImageNet visual recognition challenge. CS231n **7**(7), 3 (2015)
12. Le Merrer, E., Perez, P., Trédan, G.: Adversarial frontier stitching for remote neural network watermarking. Neural Comput. Appl. **32**, 9233–9244 (2020)
13. Liu, H., Weng, Z., Zhu, Y.: Watermarking deep neural networks with greedy residuals. In: ICML, pp. 6978–6988 (2021)
14. Lukas, N., Zhang, Y., Kerschbaum, F.: Deep neural network fingerprinting by conferrable adversarial examples. arXiv preprint arXiv:1912.00888 (2019)
15. Ma, N., Zhang, X., Zheng, H.-T., Sun, J.: ShuffleNet V2: practical guidelines for efficient CNN architecture design. In: Ferrari, V., Hebert, M., Sminchisescu, C., Weiss, Y. (eds.) Computer Vision – ECCV 2018. LNCS, vol. 11218, pp. 122–138. Springer, Cham (2018). https://doi.org/10.1007/978-3-030-01264-9_8

16. Maini, P., Yaghini, M., Papernot, N.: Dataset inference: ownership resolution in machine learning. arXiv preprint arXiv:2104.10706 (2021)
17. Orekondy, T., Schiele, B., Fritz, M.: Knockoff Nets: stealing functionality of black-box models. In: Proceedings of the IEEE/CVF Conference on Computer Vision and Pattern Recognition, pp. 4954–4963 (2019)
18. Pan, X., Yan, Y., Zhang, M., Yang, M.: MetaV: a meta-verifier approach to task-agnostic model fingerprinting. In: Proceedings of the 28th ACM SIGKDD Conference on Knowledge Discovery and Data Mining, pp. 1327–1336 (2022)
19. Sandler, M., Howard, A., Zhu, M., Zhmoginov, A., Chen, L.C.: MobileNetV2: inverted residuals and linear bottlenecks. In: Proceedings of the IEEE Conference on Computer Vision and Pattern Recognition, pp. 4510–4520 (2018)
20. Simonyan, K., Zisserman, A.: Very deep convolutional networks for large-scale image recognition. arXiv preprint arXiv:1409.1556 (2014)
21. Wang, T., Kerschbaum, F.: RIGA: covert and robust white-box watermarking of deep neural networks. In: Proceedings of the Web Conference 2021, pp. 993–1004 (2021)
22. Yan, M., Fletcher, C., Torrellas, J.: Cache telepathy: leveraging shared resource attacks to learn DNN architectures. In: USENIX Security Symposium (2020)
23. Yang, K., Wang, R., Wang, L.: MetaFinger: fingerprinting the deep neural networks with meta-training. In: 31st International Joint Conference on Artificial Intelligence, IJCAI 2022 (2022)
24. Zhang, J., et al.: Protecting intellectual property of deep neural networks with watermarking. In: Proceedings of the 2018 on Asia Conference on Computer and Communications Security, pp. 159–172 (2018)

Attribution of Adversarial Attacks
via Multi-task Learning

Zhongyi Guo[1], Keji Han[1], Yao Ge[1], Yun Li[1,2]([✉]), and Wei Ji[3]

[1] School of Computer Science, Nanjing University of Posts and Telecommunications,
Nanjing 210023, China
{1221045710,1016041119,2020070131}@njupt.edu.cn
[2] Jiangsu Key Laboratory of Big Data Security and Intelligent Processing,
Nanjing 210023, China
liyun@njupt.edu.cn
[3] School of Communications and Information Engineering, Nanjing University of
Posts and Telecommunications, Nanjing 210003, China
jiwei@njupt.edu.cn

Abstract. Deep neural networks (DNNs) can be easily fooled by adversarial examples during inference phase when attackers add imperceptible perturbations to original examples. Many works focus on adversarial detection and adversarial training to defend against adversarial attacks. However, few works explore the tool-chains behind adversarial examples, which is called Adversarial Attribution Problem (AAP). In this paper, AAP is defined as the recognition of three signatures, i.e., *attack algorithm*, *victim model* and *hyperparameter*. Existing works transfer AAP into a single-label classification task and ignore the relationship among above three signatures. Actually, there exists owner-member relationship between attack algorithm and hyperparameter, which means hyperparameter recognition relies on the result of attack algorithm classification. Besides, the value of hyperparameter is continuous, hence hyperparameter recognition should be regarded as a regression task. As a result, AAP should be considered as a multi-task learning problem rather than a single-label classification problem or a single-task learning problem. To deal with above problems, we propose a multi-task learning framework named Multi-Task Adversarial Attribution (MTAA) to recognize above three signatures simultaneously. It takes the relationship between attack algorithm and the corresponding hyperparameter into account and uses the uncertainty weighted loss to adjust the weights of three recognition tasks. The experimental results on MNIST and ImageNet show the feasibility and scalability of the proposed framework.

Keywords: Deep neural network · Adversarial attack attribution · Multi-task learning

1 Introduction

In the past two decades, deep neural networks (DNNs) have shown outstanding performance across various tasks in computer vision, such as image classifica-

B. Luo et al. (Eds.): ICONIP 2023, LNCS 14448, pp. 81–94, 2024.
https://doi.org/10.1007/978-981-99-8082-6_7

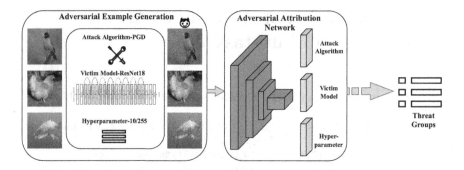

Fig. 1. Overview of Adversarial Attribution.

tion [1] and object detection [2], etc. Nevertheless, DNNs are demonstrated to be easily fooled by adversarial examples [3], which is accomplished during inference phase by adding imperceptible perturbations to examples. Some classic adversarial attack algorithms include Fast Gradient Sign Method (FGSM) [4], Projected Gradient Descent (PGD) [5] and Carlini & Wagner (C&W) [6].

Extensive efforts focus on adversarial detection [7] and adversarial training [8], while few works study the Adversarial Attribution Problem (AAP), which is an important part of Reverse Engineering of Deceptions (RED) [9–11]. According to the assertion of Defense Advanced Research Projects Agency (DARPA), RED is aimed at "developing techniques that automatically reverse engineer the tool-chains behind attacks such as multimedia falsification, adversarial machine learning attacks, or other information deception attacks." With the same purpose of RED, as shown in Fig. 1, AAP concentrates on better understanding the hidden signatures used to generate adversarial examples, i.e., *attack algorithm*, *victim model* and *hyperparameter*. Then aid in attribution to a particular threat group, which is a critical step in formulating a response and holding the attackers accountable.

Adversarial attribution can help defenders to seize the clues about the originator of the attack, their goals, and provide insight into the most effective defense algorithm against corresponding attacks [9]. The following works make a preliminary exploration of AAP. Ref. [13] investigates the attribution of attack types using fewer training samples by self-supervised learning. Ref. [14] primarily explores the attributability of attack algorithm, victim model, hyperparameter and norm using a self-built 11-layer neural network. However, they consider these four signatures separately and ignore the relationship among them. In addition, they conduct experiments on small-scale datasets like MNIST and CIFAR-10, thus there is a lack of diversity in the dataset. Ref. [15] explores the attribution of attack algorithm and victim model using the structure of ResNet50's feature extractor plus a Multilayer Perceptron(MLP) classifier. Ref. [16] explores the attribution of attack algorithm and hyperparameter using ResNet50 as backbone. Unfortunately, Ref. [13,15,16] only study one or two signatures recognition and Ref. [14] consider hyperparameter values at large intervals, e.g., 0.03, 0.1, 0.2

for maximum perturbation ε in FGSM L_∞ and 0.01, 0.1, 1.0 for confidence κ in C&W L_2. Thus above works lack the integrity of signature recognition. What is more, all of these works transfer AAP into single-label classification problem, i.e., combine Attack Algorithm+Victim Model+Hyperparameter together to form one label, such as PGD+ResNet18+10/255. When the number of attack algorithms and victim models increase, the combination explosion problem appears. That is, the rapid growth of Attack Algorithm+Victim Model+Hyperparameter classes make single-label classification difficult. Last but not least, the relationship among these signatures is neglected in these works. Overall, it is urgent to propose a unified and extensible framework for adversarial attribution to deal with more signatures and alleviate the combination explosion issue.

In this paper, to figure out AAP, we explore the relationship among three signatures and propose a multi-task learning framework called MTAA. MTAA contains a perturbation extraction module, an adversarial-only extraction module and a classification and regression module. It takes the relationship between attack algorithm and the corresponding hyperparameter into account and uses the uncertainty weighted loss to adjust the weights of three recognition tasks.

We summarize the main contributions as follows:

- To solve Adversarial Attribution Problem (AAP), we propose a Multi-Task Adversarial Attribution (MTAA) method. MTAA recognizes three signatures simultaneously in just one unified model, explores the relationship among them and alleviates the combination explosion problem.
- A feature-level adversarial perturbation extractor is proposed to improve the performance of MTAA.
- The experimental comparison among MTAA, single-label baseline and single-task baseline illustrates the scalability and computational efficiency of MTAA.

2   Related Work

Adversarial Attack was first discovered by Szegedy [3], who reveals the vulnerability of deep learning model that attackers can manipulate its predictions by adding visually imperceptible perturbations to images. Recently, large amounts of adversarial attack algorithms spring out. The most representative attack algorithms among them are gradient-based attacks like one-shot FGSM [4] and iterative PGD [5], as well as optimization-based attack like C&W [6].

FGSM [4] crafts adversarial example with the sign of gradient in regard to ground truth label and can be formulated as:

$$x' = x + \varepsilon \cdot \text{sign}\left(\nabla_x \ell(h(x;\theta), u)\right) \tag{1}$$

where x is a clean example, u is its label and x' is an adversarial example. $h(\cdot)$ is the victim model whose parameter is θ. $\ell(\cdot)$ is the loss function. $\nabla_x(\cdot)$ is gradient of $\ell(\cdot)$ with respect to x. sign$(\cdot)$ is the gradient sign function. ε is the hyperparameter that controls the attack intensity.

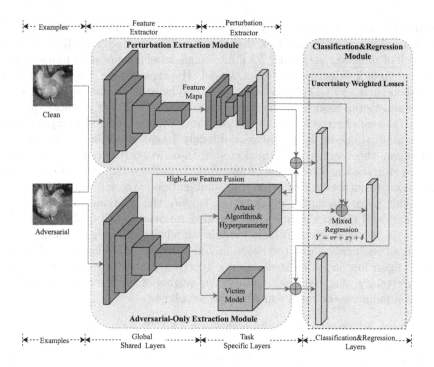

Fig. 2. Overall architecture of Multi-Task Adversarial Attribution.

PGD [5] can be seen as the iterative version of FGSM and is formulated as:

$$x'_{t+1} = \text{Clip}_{x,\varepsilon}\left(x'_t + \alpha \cdot \text{sign}\left(\nabla_x \ell(h(x;\theta), u)\right)\right) \tag{2}$$

where x'_t is an adversarial example in *step t*, u is its label. $h\left(\cdot\right)$ is the victim model whose parameter is θ. $\ell\left(\cdot\right)$ is the loss function. $\text{Clip}_{x,\varepsilon}\left(\cdot\right)$ performs clipping at attack intensity ε. α is the step-size in each attack iteration.

C&W [6] computes the adversarial perturbation by solving the following optimisation problem:

$$\min \|\rho\|_p + e \cdot a(x + \rho), \quad \text{s.t. } x + \rho \in [0, 1]^m. \tag{3}$$

where ρ is the perturbation to be optimized. e is a suitably chosen constant. $a\left(\cdot\right)$ is an objective function satisfying $h(x + \rho)=l$ if and only if $a(x + \rho) \leq 0$, in which $h\left(\cdot\right)$ is the victim model and l is the target label.

3   Methodology

In Sect. 3.1, we propose the architecture of Multi-Task Adversarial Attribution (MTAA) framework. In Sect. 3.2, we present the overall loss function weighted by homoscedastic uncertainty.

3.1 Multi-task Adversarial Attribution (MTAA)

Overall Architecture. As shown in Fig. 2, the architecture of multi-task learning framework for AAP can be divided into three parts: (1) perturbation extraction module (2) adversarial-only extraction module (3) classification&regression module.

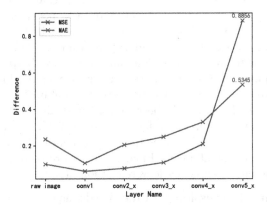

Fig. 3. The MSE and MAE between feature maps of clean and adversarial examples on ImageNet. The 'conv' means different conv blocks of ResNet101.

Perturbation Extraction Module. The perturbation extraction module leverages information from both clean and adversarial examples. We use Auto-Encoder (AE) as perturbation extractor. AE learns effective representations of a set of data in an unsupervised manner. With an encoder R and a decoder G, AE is forced to minimize the reconstruction error $\|x - G(R(x))\|_2^2$ for each input sample x. However, learning the background information is unhelpful for adversarial attribution, it has also been proved by [18] that building a flow density estimator on latent representation (feature maps) works better than on the raw image. On the other hand, we find that the difference between the feature maps of the original images and adversarial examples becomes larger with deeper layers, as shown in Fig. 3. Thus we add a ResNet101 feature extractor (or other DNNs) before AE to obtain latent representation (feature maps, *fm* for short) of adversarial and corresponding clean examples, which will help AE better learn the pixel difference between them. So we optimize AE by minimizing the loss function:

$$\mathcal{L}_{MSE_1} = \frac{1}{n}\sum_{j=1}^{n}\left(x_j^{fm'} - G(R(x_j^{fm'})) - x_j^{fm}\right)^2. \tag{4}$$

the objective of function (4) is to let feature-level perturbation $x_j^{fm''}$ equal to $x_j^{fm'} - x_j^{fm} = G(R(x_j^{fm'}))$, where the feature maps of jth adversarial example is $x_j^{fm'}$ and corresponding clean example is x_j^{fm} with total number n. We train AE to learn manifolds of adversarial perturbation as the augmented feature.

Adversarial-Only Extraction Module. The adversarial-only extraction module only takes adversarial examples as input. Adversarial examples first pass global shared layers that leverage ResNet101's feature extractor (or other DNNs) to learn the shared representation of three signatures. Then task specific layers learn task-specific representations for attack and hyperparameter as well as victim model separately. Note that the structure of task specific layers are the final feature extraction layers of ResNet101. We also use a high-low feature fusion in learning representation of attack and hyperparameter signatures. The high-low feature fusion is to concatenate high and low features maps from layer near the output and input of ResNet101, respectively. Generally, low level features have small receptive field, thus they capture partial/detailed information, which helps the recognition of L_∞ attacks because L_∞ constraint perturbation on each pixel. While high level feature have large receptive field, so that they capture integral/rich information, which helps the recognition of L_2 attacks because L_2 constraint perturbation on all pixels. Finally, the feature vectors learnt from task specific layers are sent to classification&regression module.

Classification and Regression Module. We define attack algorithm and victim model attribution as two classification tasks. With regard to two classification layers, we use a fully connection layer. We optimize these two classification tasks by minimizing two cross-entropy losses:

$$\mathcal{L}_{CE_1} = \frac{1}{n} \sum_{j} \sum_{b_1=1}^{m_1} Q_j^{b_1} \log \left(P_j^{b_1} \right), \tag{5}$$

$$\mathcal{L}_{CE_2} = \frac{1}{n} \sum_{j} \sum_{b_2=1}^{m_2} Q_j^{b_2} \log \left(P_j^{b_2} \right). \tag{6}$$

where $\mathcal{L}_{CE_1}$ and $\mathcal{L}_{CE_2}$ are the loss function of attack algorithm and victim model classification, respectively. The number of adversarial examples x_j' is n and Q_j^b is an indicator function that judges whether x_j''s label is the same as b. m_1 and m_2 are the number of attack algorithm and victim model labels, respectively. P_j^b estimates the probability that x_j' belongs to label b with a softmax function.

We define hyperparameter attribution as a mixed regression task. For the dependency of attack algorithm and hyperparameter, we concatenate the result of attack algorithm classifier with extracted feature as the input of hyperparameter regression. Formally, the regression problem is expressed as:

$$Y = v\tau + z\gamma + \delta. \tag{7}$$

where Y is predicted value of hyperparameter, v is the features extracted by task specific layers for attack and hyperparameter as well as perturbation extraction module, τ is the weight of v, z is the logits of attack classification layer, γ is the weight of z and δ is stochastic noise with mean 0 and variance σ to explain the measurement error of the data itself.

We optimize this regression task by minimizing the mean square error (MSE) loss:

$$\mathcal{L}_{MSE_2} = \frac{1}{m_3} \sum_j^{m_3} \left(\hat{Y}_j - \overline{Y}_j\right)^2. \tag{8}$$

where $\hat{Y}_j$ is the estimated value of x_j''s hyperparameter and $\overline{Y}_j$ is ground truth. m_3 is the number of hyperparameter values.

3.2 Uncertainty Weighted Losses

Inspired by [19], the performance of multi-task learning model is strongly rely on weight between different tasks. As a result, we choose the standard uncertainty weighted losses, which leverage homoscedastic uncertainty that is not dependent on input data but dependent on task uncertainty. Following the steps of conducting maximum likelihood inference, we first describe the probabilistic model of regression tasks and classification tasks as (9) and (10), respectively:

$$p\left(y|f^W(x), \sigma_1\right) = N\left(f^W(x), \sigma_1^2\right), \tag{9}$$

$$p\left(y|f^W(x), \sigma_2\right) = softmax\left(\frac{1}{\sigma_2^2} f^W(x)\right). \tag{10}$$

where $f^W(x)$ is the output of a multi-task learning model with parameters W and input x. $N(\cdot)$ is the Gaussian likelihood accompanied by an observation noise parameter σ_1 that captures how much noise we have in the outputs. The classification likelihood is scaled by σ_2^2 to meet a Boltzmann distribution with the logits of model through a *Softmax* function.

The second step is factorizing over the outputs, which is illustrated in Eq. (11):

$$p\left(y_1, \ldots, y_T|f^W(x)\right) = p\left(y_1|f^W(x)\right) \ldots p\left(y_T|f^W(x)\right). \tag{11}$$

where $y_1, \ldots, y_T$ are model's outputs with T tasks, such as attack algorithm and victim model classification as well as hyperparameter regression in AAP.

The third step is taking log of likelihood function to conduct *maximum likelihood* inference. The inference of regression tasks and classification tasks are given in Eq. (12) and (13), respectively:

$$\log p\left(y|f^W(x), \sigma_1\right) \propto -\frac{1}{2\sigma_1^2} \left\|y - f^W(x)\right\|^2 - \log \sigma_1, \tag{12}$$

$$\log p\left(y|f^W(x), \sigma_2\right) = \frac{1}{\sigma_2^2} f_c^W(x) - \log \sum_{c'} exp\left(\frac{1}{\sigma_2^2} f_{c'}^W(x)\right). \tag{13}$$

where $f_c^W(x)$ is the c'th component of the vector $f^W(x)$.

In our case, by adding uncertainty weighted losses to overall loss function and following *maximum likelihood* inference, we can formally describe the combined loss function as follows:

$$\mathcal{L}\left(W, \sigma_1, \sigma_2, \sigma_3\right) \tag{14}$$

$$= -\log p\left(y_1, y_2, y_3 = c | f^W(x)\right) \tag{15}$$

$$= -\log N\left(y_1; f^W(x), \sigma_1^2\right) \cdot softmax\left(y_2 = c; f^W(x), \sigma_2\right) \tag{16}$$

$$\cdot softmax\left(y_3 = c; f^W(x), \sigma_3\right) \tag{17}$$

$$= \frac{1}{2\sigma_1^2}\left\|y_1 - f^W(x)\right\|^2 + \log\sigma_1 - \log p\left(y_2 = c | f^W(x), \sigma_2\right) - \log p\left(y_3 = c | f^W(x), \sigma_3\right) \tag{18}$$

$$= \frac{1}{2\sigma_1^2}\mathcal{L}_1(W) + \frac{1}{\sigma_2^2}\mathcal{L}_2(W) + \frac{1}{\sigma_3^2}\mathcal{L}_3(W) + \log\sigma_1 \tag{19}$$

$$+ \log\frac{\sum_{c'} exp\left(\frac{1}{\sigma_2^2}f_{c'}^W(x)\right)}{\left(\sum_{c'} exp\left(f_{c'}^W(x)\right)\right)^{\frac{1}{\sigma_2^2}}} + \log\frac{\sum_{c'} exp\left(\frac{1}{\sigma_3^2}f_{c'}^W(x)\right)}{\left(\sum_{c'} exp\left(f_{c'}^W(x)\right)\right)^{\frac{1}{\sigma_3^2}}} \tag{20}$$

$$\approx \frac{1}{2\sigma_1^2}\mathcal{L}_1(W) + \frac{1}{\sigma_2^2}\mathcal{L}_2(W) + \frac{1}{\sigma_3^2}\mathcal{L}_3(W) + \log\sigma_1 + \log\sigma_2 + \log\sigma_3 \tag{21}$$

where $\mathcal{L}_1(W) = \left\|y_1 - f^W(x)\right\|^2$ represents the MSE loss of hyperparameter regression task, $\mathcal{L}_2(W) = -log\left(Softmax\left(y_2, f^W(x)\right)\right)$ represents the cross entropy loss of attack algorithm classification task and victim model classification's cross entropy loss is $\mathcal{L}_3(W) = -log\left(Softmax\left(y_3, f^W(x)\right)\right)$. The third step in the equation transformation uses the approximation in (12) and (13). We can minimize the combined loss function by optimizing the parameters W, σ_1, σ_2 and σ_3. In order to simplify the optimization objective, the explicit simplifying assumption $\frac{1}{\sigma^2}\sum_{c'} exp\left(\frac{1}{\sigma^2}f_{c'}^W(x)\right) \approx \left(\sum_{c'} exp\left(f_{c'}^W(x)\right)\right)^{\frac{1}{\sigma^2}}$ is used in the last approximate transition. σ_1, σ_2, σ_3 measures the uncertainty of each task, which means higher scale value causes lower contribution of loss function. The three scales are regulated by the last three *log* terms in the formula, which penalizes the objective when values are too high.

4   Experiments

4.1   Experimental Setup

Dataset: The attribution scenario we consider is shown in Table 1. ε in FGSM and PGD is maximum perturbation. For PGD, α is step size and *step* is attack iteration number. For C&W, κ is confidence of attack, C is parameter for box-constraint and *step* is attack iteration number. For hyperparameter regression task, we focus on recognizing maximum perturbation ε for FGSM and PGD ranging from $10/255$ to $200/255$ with step size $10/255$ while confidence κ for

C&W ranging from 5% to 100% with step size 5%. We leverage shuffled MNIST training dataset that contains 55000 images and ImageNet validating dataset which contains 50000 images, along with attack algorithms tool box Cleverhans [12] to generate adversarial examples. For each Attack Algorithm+Victim Model+Hyperparameter class, we generate 160 examples for training and 20 examples for testing.

Table 1. The attribution scenario of adversarial attribution.

Attack Algorithm	Hyperparameter	Victim Model
FGSM(L_∞)	ε: **10/255-200/255(10/255)**	InceptionV3, ResNet18, ResNet50, VGG16, VGG19
PGD(L_2)	ε: **10/255-200/255(10/255)**α: 10/255*step*: 100	
C&W(L_2)	κ: **5%–100%(5%)***C*: 50*step*: 500	

Implementation Details: This paper uses PyTorch for training and inferencing on a computer with 2 Intel Xeon Platinum 8255C 2.50 GHz * 32 CPUs and 43 GB memory, 2 NVIDIA RTX3090 GPUs. As to the multi-task learning architecture, for perturbation extraction module we utilize pretrained ResNet50 and ResNet101 to extract feature maps of MNIST and ImageNet, respectively, and Auto-Encoder to extract perturbation; for adversarial-only extraction module we utilize pretrained ResNet50/ ResNet101 as global shared layers and two ResNet50's/ResNet101's last bottlenecks as task specific layers for MNIST and ImageNet, respectively. Adam [17] is employed with cosine annealing LR schedule whose initial learning rate $\beta = 0.001$, weight decay $\pi = 0.001$ and mini-batch size= 512. Besides, in order to unify the measurement scale of FGSM and PGDs' ε and C&W's κ in hyperparameter regression task, we magnify the attack intensity labels for FGSM and PGD 255 times.

Evaluation Metrics: The accuracy is used to evaluate attack algorithm classification task and victim model classification task. The Root Mean Square Error (RMSE) is used to evaluate hyperparameter regression task. We measure the *multi-task learning performance* Δ_{MTL} as in [20], i.e., the multi-task performance of model f is the average per-task drop in performance w.r.t. the single-task baseline B:

$$\Delta_{MTL} = \frac{1}{T} \sum_{k=1}^{T} (-1)^{o_k} (M_{f,k} - M_{B,k})/M_{B,k}. \qquad (22)$$

where T is the total number of tasks, $M_{f,k}$ is the value of metric M of task k on multi-task model f, $M_{B,k}$ is the value of metric M of task k on single-task model B, i.e., accuracy for two classification tasks and RMSE for one regression task. $o_k = 1$ if a lower value means better performance for metric M_k of task k, and 0 otherwise. The single-task performance is measured for a fully-converged model that uses the same backbone network only to perform that task.

In addition to a performance evaluation, we also consider the model resources, i.e., number of parameters and FLOPS, when comparing the MTAA.

4.2   Attribution Experimental Results

In this section, we compare the performance of our MTAA with corresponding single-task baseline, i.e., train individual DNN for each three task, that rely on the same backbone. The results are shown in Table 2 and 3. Note that [14] and [15] treat AAP as single-label classification task, i.e., combine attack algorithm, victim model and hyperparameter together to form one label. However, in order to compare our MTAA with these two works, we reproduce the backbone of these two works and treat them as single-task baseline. For FGSM and PGD attack, the RMSE is actually $6.04/255$ on MINST and $6.79/255$ on ImageNet because we magnify ε's labels 255 times to unify measurement scale.

The *multi-task learning performance* Δ_{MTL} achieves 1.17% and 3.93% on MNIST and ImageNet, respectively, which means MTAA improves over the single-task baseline. For two classification tasks, MTAA achieve 100% and 99.88% accuracy on MNIST, as well as 99.78% and 97.84% accuracy on ImageNet, respectively. For regression task, the RMSE results of MTAA are better than single-task baseline on both datasets, which means our framework captures the relationship between attack algorithm and the corresponding hyperparameter. Besides, MTAA reduces the required amount of resources, i.e., fewer number of parameters and FLOPS. So we should consider AAP as a multi-task learning problem rather than a single-task learning problem.

Table 2. Results ($\%/RMSE$) of [14] and [15] as single-task baseline and MTAA on MNIST.

	Backbone	Model	FLOPS (G)	Params (M)	Attack Algorithm (%)↑	Victim Model (%)↑	Hyperparameter (RMSE)↓	Δ_{MTL} (%)↑
Single-Task	Self-built	[14]	-	-	95.36	93.47	9.64	+0.00
	ResNet-50	[15]	0.97	71	100	99.81	6.46	+0.00
MTL	ResNet-50	MTAA	0.77	48	100	99.88	6.04	+1.17

Table 3. Results ($\%/RMSE$) of [14,15] and ResNet101 as single-task baseline and MTAA on ImageNet.

	Backbone	Model	FLOPS (G)	Params (M)	Attack Algorithm (%)↑	Victim Model (%)↑	Hyperparameter (RMSE)↓	Δ_{MTL} (%)↑
Single-Task	Self-built	[14]	-	-	88.54	83.22	12.97	+0.00
	ResNet-50	[15]	12	71	97.43	93.25	7.93	+0.00
	ResNet-101		24	128	98.68	94.72	7.33	+0.00
MTL	ResNet-50	MTAA	9	48	99.68	96.95	7.32	+4.66
	ResNet-101	MTAA	21	108	99.78	97.84	6.79	+3.93

4.3   Scalability of MTAA

We highlight the scalability of MTAA from the aspect of *Attribution Scenario*. [14] and [15] treat AAP as a single-label classification task, so as the number of attack algorithms and victim models increased, the combination explosion problem appears. That is, the rapid growth of Attack Algorithm+Victim

Model+Hyperparameter classes make single-label classification difficult. To discuss this problem, we conduct experiments on 2 attack algorithms, 3 victim models, 20 hyperparameter values and 3 attacks, 5 victims, 20 hyperparameter values with hyperparameter setting in Table 1, respectively. Therefore for the single-label classification baseline, there are $2 \times 3 \times 20 = 120$ classes and $3 \times 5 \times 20 = 300$ classes, respectively. The results are shown in Table 4 and 5. Note that MTAA uses ResNet50 and ResNet101 as backbone on MINST and ImageNet, respectively. To ensure the fairness of comparison, we also add column 'ResNet101' in Table 5 to make the backbone of single-label classification the same as MTAA.

The accuracy of single-label classifier ([14] and [15]) dramatically drop when we increase the number of attack algorithms and victim models. Besides, it is interesting that single-label classifier's accuracy reduce to minimum when facing PGD and C&W. We think it is because PGD and C&Ws' hyperparameter is harder to be recognized than that of FGSM. However, our MTAA performs stably on two classification tasks when we increase the number of attack algorithms and victim models, and the performance on regression task fluctuates in a small range. Therefore we should consider AAP as a multi-task learning problem rather than a single-label classification problem.

Table 4. Results ($\%/RMSE$) for 2 attack algorithms, 3 victim models and 3 attack algorithms, 5 victim models on MNIST.

Attack Algorithms	Victim Models	[14]	[15]	MTAA		
		Single-Label Classification		Attack Algorithm	Victim Model	Hyperparameter
FGSM PGD C&W	InceptionV3 ResNet18 ResNet50 VGG16 VGG19	57.53	64.15	100	99.88	6.04
FGSM PGD	InceptionV3 ResNet18 VGG16	83.74	92.21	100	100	3.72
FGSM C&W	InceptionV3 ResNet18 VGG16	55.76	68.33	100	99.96	5.71
FGSM PGD	InceptionV3 ResNet50 VGG19	85.54	91.71	100	100	2.06
PGD C&W	InceptionV3 ResNet50 VGG19	50.06	56.62	100	99.96	7.02

4.4 Ablation Study

To validate the effectness of different components in adversarial-only extractor module in MTAA, such as global shared layers (GSL), weight of loss and task specific layers (TSL), some ablation experiments are conducted and the results

Table 5. Results (%/$RMSE$) for 2 attack algorithms, 3 victim models and 3 attack algorithms, 5 victim models on ImageNet.

Attack Algorithms	Victim Models	[14]	[15]	ResNet101	MTAA		
		Single-Label Classification			Attack Algorithm	Victim Model	Hyperparameter
FGSM PGD C&W	InceptionV3 ResNet18 ResNet50 VGG16 VGG19	50.71	59.73	61.22	99.78	97.84	6.79
FGSM PGD	InceptionV3 ResNet18 VGG16	71.26	78.53	82.96	99.97	98.93	5.96
FGSM C&W	InceptionV3 ResNet18 VGG16	49.98	59.42	62.33	99.97	98.24	7.76
FGSM PGD	InceptionV3 ResNet50 VGG19	63.76	71.32	84.07	99.93	98.69	6.02
PGD C&W	InceptionV3 ResNet50 VGG19	39.98	47.21	48.04	99.97	98.21	8.01

are shown in Table 6 and 7. From the second row in Table 6 and 7, we can obtain that a suitable weight balancing method greatly influences the performance of AAP, especially in victim model classification. The experimental results in third row of both tables show that it is better to deploy TSL for different tasks that can further extracts task-specific features. According to experimental results in the forth row of both tables, we can conclude that a deeper network (ResNet18 to ResNet50 for MNIST and ResNet50 to ResNet101 for ImageNet) for GSL performs better because it has stronger feature extraction capability. We also provide the ablation experimental results for perturbation extractor (PE) module in MTAA in last row of both tables, which indicates the addition of PE improve the attribution performance for AAP.

Table 6. Results (%/$RMSE$) of ablation study on different model architecture of MTAA on MNIST.

Architecture of MTAA	Attack Algorithm	Victim Model	Hyperparameter
ResNet18+simple add loss	98.92	84.72	7.88
ResNet18+Uncertainty loss weight	99.54	97.84	7.21
ResNet18+Uncertainty loss weight+TSL	99.79	98.37	7.02
ResNet50+Uncertainty loss weight+TSL	99.94	99.26	6.42
ResNet50+Uncertainty loss weight+TSL+PE	**100**	**99.88**	**6.04**

Table 7. Results ($\%/RMSE$) of ablation study on different model architecture of MTAA on ImageNet.

Architecture of MTAA	Attack Algorithm	Victim Model	Hyperparameter
ResNet50+simple add loss	98.12	70.82	8.7
ResNet50+Uncertainty loss weight	99.3	95.98	7.9
ResNet50+Uncertainty loss weight+TSL	99.48	96.1	7.81
ResNet101+Uncertainty loss weight+TSL	99.61	97.13	7.25
ResNet101+Uncertainty loss weight+TSL+PE	**99.78**	**97.84**	**6.79**

5 Conclusion

Adversarial Attribution Problem (AAP) is a vital part in Reverse Engineering of Deceptions to recognize the signatures behind the adversarial examples, such as *attack algorithm, victim model* and *hyperparameter*. In view of existing works neglect the relationship among the above signatures while lacking a scalable and unified framework, we propose a multi-task learning framework accompanied by uncertainty weighted loss to solve this problem efficiently and scalably. The experimental results on MNIST and ImageNet show MTAA has the-state-of-art performance.

Acknowledgements. This work was partially supported by National Natural Science Foundation of China (No. 61772284).

References

1. He, K., Zhang, X., Ren, S., Sun, J.: Deep residual learning for image recognition. In: CVPR, pp. 770–778 (2016)
2. Girshick, R., Donahue, J., Darrell, T., Malik, J.: Rich feature hierarchies for accurate object detection and semantic segmentation. In: CVPR, pp. 580–587 (2014)
3. Szegedy, C., et al.: Intriguing properties of neural networks. In: ICLR (2014)
4. Goodfellow, I.J., Shlens, J., Szegedy, C.: Explaining and harnessing adversarial examples. In: ICLR (2015)
5. Madry, A., Makelov, A., Schmidt, L., Tsipras, D., Vladu, A.: Towards deep learning models resistant to adversarial attacks. In: ICLR (2018)
6. Carlini, N., Wagner, D, Towards evaluating the robustness of neural networks. In: S&P, pp. 39–57 (2017). https://doi.org/10.1109/SP.2017.49
7. Qin, Y., Frosst, N., Sabour, S., Raffel, C., Cottrell, G., Hinton, G.: Detecting and diagnosing adversarial images with class-conditional capsule reconstructions. In: ICLR (2020)
8. Tramèr, F., Kurakin, A., Papernot, N., Goodfellow, I., Boneh, D., McDaniel, P.: Ensemble adversarial training: attacks and defenses. In: ICLR (2018)
9. DARPA Artificial Intelligence Exploration (AIE) Opportunity and DARPA-PA-19-02-09 and Reverse Engineering of Deceptions (RED)
10. Gong, Y., et al.: Reverse engineering of imperceptible adversarial image perturbations. In: ICLR (2022)

11. Thaker, D., Giampouras, P., Vidal, R.: Reverse Engineering ℓ_p attacks: a block-sparse optimization approach with recovery guarantees. In: ICML, pp. 21253–21271 (2022)
12. Papernot, N., et al.: Technical report on the CleverHans v2.1.0 adversarial examples library. arXiv preprint arXiv:1610.00768 (2016)
13. Moayeri, M., Feizi, S, Sample efficient detection and classification of adversarial attacks via self-supervised embeddings. In: ICCV, pp. 7677–7686 (2021)
14. Dotter, M., Xie, S., Manville, K., Harguess, J., Busho, C.: Rodriguez, M.: Adversarial attack attribution: discovering attributable signals in adversarial ML attacks. In: RSEML Workshop at AAAI (2021)
15. Li, Y.: Supervised Classification on Deep Neural Network Attack Toolchains. Northeastern University (2021)
16. Nicholson, D.A., Emanuele, V.: Reverse engineering adversarial attacks with fingerprints from adversarial examples. arXiv preprint arXiv:2301.13869 (2023)
17. Kingma, D. P., Ba, J, Adam: a method for stochastic optimization. In: ICLR (2015)
18. Zhang, H., Li, A., Guo, J., Guo, Y.: Hybrid models for open set recognition. In: Vedaldi, A., Bischof, H., Brox, T., Frahm, J.-M. (eds.) ECCV 2020. LNCS, vol. 12348, pp. 102–117. Springer, Cham (2020). https://doi.org/10.1007/978-3-030-58580-8_7
19. Kendall, A., Gal, Y., Cipolla, R.: Multi-task learning using uncertainty to weigh losses for scene geometry and semantics. In: CVPR, pp. 7482–7491 (2018)
20. Maninis, K.K., Radosavovic, I., Kokkinos, I.: Attentive single-tasking of multiple tasks. In: CVPR, pp. 1851–1860 (2019)

A Reinforcement Learning Method for Generating Class Integration Test Orders Considering Dynamic Couplings

Yanru Ding[1,2], Yanmei Zhang[1], Guan Yuan[1(✉)], Yingjie Li[1], Shujuan Jiang[1], and Wei Dai[2]

[1] School of Computer Science and Technology, China University of Mining and Technology, Xuzhou 221116, China
{yrding,ymzhang,yuanguan,yingjieli,shjjiang}@cumt.edu.cn
[2] Artificial Intelligence Research Institute, China University of Mining and Technology, Xuzhou 221116, China
weidai@cumt.edu.cn

Abstract. In recent years, with the rapid development of artificial intelligence, reinforcement learning has made significant progress in various fields. However, there are still some challenges when applying reinforcement learning to solve problems in software engineering. The generation of class integration test orders is a key challenge in object-oriented program integration testing. Previous research mainly focused on static couplings and neglected dynamic couplings, leading to inaccurate cost measurement of class integration test orders. In this paper, we propose a reinforcement learning method to generate class integration test orders considering dynamic couplings. Firstly, the concept of dynamic couplings generated by polymorphism is introduced, and a strategy for measuring the stubbing complexity of simulating dynamic dependencies is proposed. Then, we combine this new stubbing complexity with a reinforcement learning method to generate class integration test orders and achieve the optimal result with minimal overall stubbing complexity. Comprehensive experiments show that our proposed approach outperforms other methods in measuring the cost of generating class integration test orders.

Keywords: Class integration test order · Stubbing complexity · Dynamic coupling · Reinforcement learning

1 Introduction

In recent years, with the rapid development of artificial intelligence, Reinforcement Learning (RL) has made significant progress in various fields, such as games [1], robotics, computer vision [2], and software testing optimization [3], especially in solving sequential decision-making problems [4]. The Class Integration Test Order (CITO) generation problem is a key decision-making problem in integration testing, which is an important link in software engineering. Researchers have successfully applied RL to solve this problem [5,6], but still, there are some challenges that need to be addressed.

B. Luo et al. (Eds.): ICONIP 2023, LNCS 14448, pp. 95–107, 2024.
https://doi.org/10.1007/978-981-99-8082-6_8

When generating CITOs, stubs are constructed to emulate the dependencies between non-integrated and integrated classes [7]. Constructing stubs requires an understanding of related classes, encompassing member functions and call relations. If a stub is too simplistic to replicate required call behavior, it postpones failure detection, leading to errors [8]. Minimizing the cost of constructing stubs demands rational class integration test order generation.

Initially, the number of stubs was used to evaluate stubbing cost. Yet, distinct stubs simulate different methods or attributes, and their cost varies. Relying on inter-class dependencies, Briand et al. [9] adopted stubbing complexity to evaluate the stubbing cost. However, prior stubbing complexity only modeled static dependencies [5,6,9,10], neglecting dynamic dependencies induced by polymorphism, an inherent feature of object-oriented programs [10,11]. Both static and dynamic dependencies should be leveraged as coupling information.

Diverse studies [10–16] have explored dynamic dependencies' effect on CITO generation, yet possess limitations. Zhang et al. [14] treated associations and dynamic dependencies' stubbing complexity as equal; [10–15] solely minimized the number of stubs. These methods can be inaccurate, as dynamic binding between classes may not merely entail association but aggregation too [11]. Meng et al. [16] differentiated dynamic couplings-inter-class call activities-from intra-class calls. None of the previous methods considers the stubbing cost of dynamic dependencies, causing the stubbing complexity is not accurate.

In this study, we propose a new RL-based method to generate CITOs along with a redesigned stubbing complexity measurement strategy, which can evaluate the impact of dynamic couplings on CITOs. We evaluate our method on five programs and find that dynamic couplings significantly affect CITO generation. Specifically, the contributions of this paper are as follows:

- We propose a stubbing cost measurement that considers dynamic couplings. This measurement is a novel contribution to the CITO generation problem.
- We apply the new stubbing cost measurement to the RL algorithm, enabling the generation of CITOs with a minimum overall stubbing cost.
- Experimental results show that the proposed method can enhance the completeness of stubbing cost evaluation and minimize it as much as possible.

The rest of this paper is organized as follows. Section 2 introduces the related work. Our method is presented in Sect. 3. Experiments and result analysis follow in Sect. 4. Section 5 summarizes the full text.

2   Related Work

According to techniques, the existing CITO generation methods can be categorized into three types: graph-based, search-based, and RL-based methods.

The graph-based method proposed by Kung et al. [17] was the first method that identified the CITO problem during integration testing. They solved it by a graph-based method, which uses the Object Relationship Diagram (ORD) to depict the inter-class dependencies of the program under test. By identifying and breaking strongly connected components, this method obtains CITOs. To minimize the number of stubs constructed, researchers assigned values to the dependencies and removed the ones with lower costs to reduce the stubbing cost [18,19]. Briand et al. [20] demonstrated that it is inaccurate to rely solely on quantity but ignore the internal couplings. Instead, they proposed an indicator called stubbing complexity, which accounted for the cost of simulating different dependencies by analyzing and distinguishing method calls and attributes of inter-class relationships.

The search-based method introduced by Briand et al. [9] solved the CITO generation problem by regarding it as a multi-objective optimization problem (MOP). Making use of the stubbing complexity as guidance information, Briand et al. [9] developed a search-based strategy to generate CITOs based on Genetic Algorithm (GA). Encoding the class first and then designing a population evolution strategy around the fitness function, this method finds the optimal particle that corresponds to the optimal CITO. Other search-based methods, such as Random Interaction Algorithm (RIA) [21] and Particle Swarm Optimization (PSO) [22], have also achieved good results by combining the stubbing complexity with fitness to guide the population's evolution.

Recently, with the rise of artificial intelligence, RL-based methods have emerged and have been found to yield better results when applied to solving the CITO problem. Czibula et al. [5] were the first to apply the RL algorithm to address this problem. They trained an agent using the Q-learning algorithm and recorded the agent's behavioral paths as CITOs, resulting in a reduction in the number of required stubs. Building upon this work, an RL-based method for generating CITOs was proposed with the aim of reducing overall stubbing complexity [6]. This method decreases the stubbing cost by constructing less expensive stubs and confirms the applicability of applying RL to the CITO generation problem.

3 Method

In this section, we introduce a new method for measuring dynamic couplings, which helps calculate the stubbing complexity in simulating dynamic dependencies. We then incorporate this strategy into an RL algorithm to generate CITOs.

3.1 Dynamic Coupling

Object-oriented programming employs polymorphism, wherein a reference variable's method invocation and bound content can vary based on different static dependencies. As an object-oriented programming language, Java formats polymorphism by interface implementation, base class inheritance, and method overriding, resulting in dynamic dependencies [23]. Specifically, dynamic dependencies can be categorized into the following cases:

(1) Interface implementation (Fig. 1(a)): When class A implements interface I, it must implement all the abstract methods of the interface. If class B calls

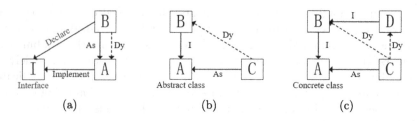

Fig. 1. The formation of dynamic dependencies

class A, a dynamic dependency is formed between classes B and A when class B declares an I interface variable and instantiates it with class A.

(2) Abstract class inheritance (Fig. 1(b)): If class A is an abstract class and class B inherits from class A while implementing the abstract methods, a dynamic dependency is formed between classes C and B when class C declares a class A variable and instantiates it with class B, and class C calls the abstract method overridden by class B.

(3) Invocation of virtual methods (Fig. 1(c)): If class A is a concrete class and class B inherits from class A while overriding methods, a dynamic dependency is formed between classes C and B when class C declares an A class variable and instantiates it with class B, and class C calls the overridden method.

Multiple inheritance can also raise dynamic dependencies. In Fig. 1(c), when class D inherits from class B and overrides a method from the ancestor class A, we consider that classes C, B, and D all form dynamic dependencies. Considering the potential redundancy and recommended limit of three levels in multi-level inheritance code, we solely focused on the impact of two-level inheritance.

3.2 Stubbing Complexity

The stubbing complexity, which is represented by $SCplx(i,j)$, is used to measure the cost of constructing a stub for class i to simulate the behavior of class j that i called. The original $SCplx(i,j)$ [9] is calculated by $A(i,j)$ and $M(i,j)$, where $A(i,j)$ is the number of attributes accessed by i from j, $M(i,j)$ is determined by the number of distinct parameters transferred from j to i.

This method only considers static coupling. We represent the complexity of the dynamic dependency as $D(i,j)$, which is determined by the number of methods that are implemented through interface implementation, abstract class inheritance, and invocation of virtual methods between classes i and j. We add dynamic coupling to improve the calculation of $SCplx(i,j)$, as shown in Eq. (1), where $\overline{A(i,j)}$, $\overline{M(i,j)}$, and $\overline{D(i,j)}$ represent the normalization results of $A(i,j)$, $M(i,j)$, and $D(i,j)$, and W_A, W_M, and W_D indicate their weights.

$$SCplx(i,j) = [W_A \times \overline{A(i,j)}^2 + W_M \times \overline{M(i,j)}^2 + W_D \times \overline{D(i,j)}^2]^{\frac{1}{2}} \quad (1)$$

Previous weights $W_f = 1/num(f)$, where $f \in (A, M, D)$ and $num(f)$ equals the number of indicators, which is 3. Considering the differences between different indicators, we use the entropy weight method [24] to measure their weights. The process to calculate the weight of different indicators is as follows:

To begin, a $m*3$ evaluation matrix is established to represent inter-class relationships of a program, where m represents the program containing m classes. The matrix $\mathbf{R}$ is calculated by Eq. (2), where r_{ij} represents the attribute, method, and dynamic coupling complexity of class i when $j = 1, 2, 3$.

$$R = (r_{ij})_{m \times 3}(i = 1, 2, ..., m; j = 1, 2, 3) \tag{2}$$

Next, the proportion of the evaluation value for each indicator, represented by p_{ij}, is calculated by Eq. (3), which is then used to calculate the information entropy e_j for the jth indicator as shown in Eq. (4). The constant K is determined by Eq. (5), where m is the number of classes.

$$p_{ij} = \frac{r_{ij}}{\sum_{i=1}^{m}}(i = 1, 2, ..., m; j = 1, 2, 3) \tag{3}$$

$$e_j = -K\sum_{i=1}^{m} p_{ij} \ln p_{ij}(i = 1, 2, ..., m; j = 1, 2, 3) \tag{4}$$

$$K = \frac{1}{\ln m} \tag{5}$$

Finally, the weight of each indicator is denoted by W_j, where j corresponds to the weight of attribute, method, and dynamic coupling complexity. The calculation method is shown in Eq. (6).

$$W_j = \frac{1 - e_j}{\sum_{j=1}^{2}(1 - e_j)}(j = 1, 2, 3) \tag{6}$$

When we use o to represent a CITO, $OCplx(o)$ to represent the overall stubbing complexity, and N to represent the set of stubs to be constructed, the overall stubbing complexity for generating CITOs can be obtained as follows:

$$OCplx(o) = \sum_{(i,j) \in N} SCplx(i, j) \tag{7}$$

3.3   CITO Generation Based on Reinforcement Learning

In RL, the agent interacts with the environment to obtain a maximize reward [4]. When the agent picks action a_t from state s_t at time t, it will get a reward r that reflects action outcomes, boosting or reducing trends. Then, the environment shifts to s_{t+1}, and the agent' learn persists until obtains a maximal gain.

Assuming a program has n classes, each state offers n actions. Actions lead to distinct states, yielding n options per state. This creates n potential successor states, and max possible states are $(n^{n+1}-1)/(n-1)$. If the agent traverses $n+1$ states sequentially, from initial to fth (f being final), and takes n actions, s_f is treated as the terminal state. In this procedure, the state transition function is shown in Eq. (8) [5], where $\delta(s_i)$ represents the union of all possible states in the ith group, and $s' \in \delta(s)$ is called a neighbor of s. Using $\sigma = (\sigma_0, \sigma_1, ..., \sigma_n)$ for state path initial-final s_f, where $\sigma_0 = s_1$ and $\sigma_{k+1} \in \delta(\sigma_k) \; \forall \{k \in 1, ..., n-1\}$. Agent's n actions on this path are a_σ, forming action history. If path has no repeated actions, a_σ becomes an alternative class order for integration and testing.

$$\delta(s_i) = \bigcup_{k=1}^{n} \{s_{n*i-n+1+k}\} \; \forall i \in \left\{1, ..., \frac{n^n - 1}{n-1}\right\} \; \forall k \in \{1, ..., n-1\} \qquad (8)$$

In RL, rewards drive agent exploration [4]. Unfavorable learning yields minimum reward as a penalty. For instance, duplicate action class prompts reward $-\infty$, denoting avoidance. Upon non-repeated final state arrival, an alternate action path is found. If its cost beats prior orders, CITO gets a higher reward. Merging this history curbs overall stubbing complexity, reducing expenses. To reduce CITOs' total stubbing cost, the reward function in RL is remodeled, coupled with stubbing complexity calculation (Eq. (9)), where c is constant, $OCplx(a_\sigma)$ is stubbing complexity of actions by agent a_σ via state path σ:

$$r(\sigma) = \left\{ \begin{array}{l} -\infty \\ Max - c \times OCplx(a_\sigma) \end{array} \right\} \qquad (9)$$

In this section, CITOs are generated via Q-learning, aiming to find optimal Q-values through Bellman equation resolution. It selects the highest Q-value action for the optimal state, iteratively uplifting policy by maintaining new value superiority. Q-value follows Eq. (10) [4], with α (learning rate), r (action a reward in state s), s' (next state), a' (next action), and γ (discount factor).

$$Q(s, a) = Q(s, a) + \alpha(r + \gamma maxQ(s', a') - Q(s, a)) \qquad (10)$$

To escape local optima, ϵ-Greedy mechanism [5] adapts as:

* Action with highest Q-value chosen with $1 - \epsilon$ probability.
* Action with smallest n_1/n_2 ratio selected with ϵ^2 probability, where n_1 is stubs needed to add action, and n_2 is avoidable stubs.
* Random action selection.

As the agent explores infinite state-action pairs, Q-values converge to optimum [25]. Accordingly, the action history associated with the path learned by the agent will converge to an extent, and the optimal CITO will be obtained.

4 Experiments

4.1 Experimental Settings

Parameters. We configure Q-learning algorithm parameters as follows: Learning rate α initiates at 0.9, decrementing to 0.01 in training for future rewards focus. Discount rate γ, influencing future rewards on the present state, is set at 0.9, prioritizing future rewards. Exploration rate ϵ at 0.8 balances exploration and exploitation. Constant c set to 100 as a balanced value within the stubbing complexity range. When the agent's explored CITO has minimal stubbing cost, reward value Max (set at 1000) is assigned. After 2 million training iterations, enabling better comparison with previous methods [5], optimal CITO is determined via action sequence yielding the highest overall reward value.

Datasets. We select five commonly used open-source programs from the Software artifact Infrastructure Repository (SIR)[1]. To obtain information on coupling, we employ Soot, a Java program analysis framework that identifies and analyzes inter-class dependencies in bytecode files. Table 1 presents program details, including the program name, description, number of classes (*Cla.*), dependencies (*Dep.*), number of cycles (*Cir.*), and lines of codes (*LOC*).

Table 1. Description of subject programs

Programs	Description	Cla	Dep	Cir	LOC
ANT	A Java based build tool	25	83	654	4093
email_spl	Email tool	39	63	38	2276
DNS	Domain Name System	61	266	16	6710
notepad_spl	Source code editor	65	142	227	2419
Jboss	Application server	91	83	1	5252

We conduct experiments on an H3C server equipped with an Intel(R) Xeon (R) Gold 5120 CPU @ 2.20GHz, 24 cores, and 48 GB RAM, using JDK 1.8.

4.2 Results and Analysis

Effect of Considering Dynamic Couplings on the Generated CITOs. We employ the algorithms described in Seciton 3.3 to generate CITOs, both with and without considering dynamic couplings, and then compare these orders to assess the impact of dynamic couplings on CITOs.

Firstly, we perform a static analysis on five programs and distinguish static and dynamic dependencies. The analysis results are presented in Table 2, where

---

[1] http://sir.unl.edu/portal/index.html.

"Dep."' and "Cir."' represent the dependencies and circles after considering dynamic couplings. The information in this table indicates that, except for *notepad_spl*, the consideration of dynamic couplings leads to increased inter-class dependencies or cycles. *Notepad_spl* does not have dynamic dependencies, whereas *DNS* and *jboss* have fewer and more dynamic dependencies, respectively, that do not form cycles. Thus, we categorize the tested programs into three groups: *ANT* and *email_spl*, which have multiple dynamic dependencies causing cycles, *DNS* and *jboss* with dynamic dependencies that do not cause cycles, and *notepad_spl*, which has no dynamic dependencies.

Table 2. Changes in programs after considering dynamic couplings

Programs	Cla	Dep	Dep.'	Cir	Cir.'
ANT	25	83	112	654	10651
email_spl	39	50	57	5	38
DNS	61	266	461	16	16
jboss	91	106	108	1	1
notepad_spl	65	98	98	227	227

Then, we employ the RL algorithm to generate CITOs and compare the results with and without considering dynamic couplings. The generated CITOs and constructed stubs for ANT are presented in Tables 3, where "Class Order" indicates the CITO of ANT excluding dynamic couplings and "Class Order" indicates the CITO considering dynamic couplings. "St" indicates the number of stubs simulating static dependencies, "Dy" indicates the number of stubs simulating dynamic dependencies, and "Sum" indicates the total number of stubs. Without considering dynamic couplings, 11 stubs are constructed, while 15 stubs required for dynamic dependencies are disregarded, totaling 26 stubs are needed. When considering dynamic couplings, 14 stubs are constructed for static dependencies and 6 for dynamic dependencies, totaling 20 stubs are needed.

Table 3. ANT without and with the consideration of dynamic couplings.

Cass Order	11	2	3	22	4	16	8	14	15	18	20	21	1	9	0	23	5	19	17	6	7	10	12	13	24	Sum
St	0	0	1	1	0	1	0	0	4	1	1	0	1	0	0	0	0	1	0	0	0	0	0	0	0	11
Dy	0	0	0	2	0	0	0	0	7	0	0	0	1	2	0	1	0	0	0	0	0	0	2	0	0	15

Cass Order'	2	3	4	11	21	8	14	0	16	15	18	20	23	22	19	1	5	13	9	24	17	6	7	10	12	Sum
St	1	1	0	0	2	0	0	1	1	3	1	1	2	0	1	0	0	0	0	0	0	0	0	0	0	14
Dy	0	0	0	0	0	0	0	0	0	5	0	0	1	0	0	0	0	0	0	0	0	0	0	0	0	6

For *DNS*, both methods construct 11 stubs, but without considering dynamic couplings, the actual number of stubs to break 97 dynamic dependencies is neglected. For *email_spl*, five static stubs are required without considering dynamic stubs, and four static and one dynamic stubs are needed while considering dynamic couplings. For *jboss*, without considering dynamic couplings requires four stubs, while considering them requires three. However, the former neglects one dynamic dependency's stubbing cost. Lastly, *notepad_spl* requires 47 and 46 stubs for both methods, respectively, to address static dependencies.

In conclusion, not accounting for dynamic couplings significantly impacts generated CITOs, especially for programs with extensive dynamic couplings.

Effect of Considering Dynamic Couplings on the Stubbing Cost of CITOs. We conducted experiments using "RL" without considering dynamic couplings, and using "RL" while considering dynamic couplings. Each of the five test programs was executed ten times, and the averages for the following metrics are shown in Table 4: the number of constructed stubs (*"Stubs"*), attribute coupling complexity (*"ACost"*), method coupling complexity (*"MCost"*), dynamic coupling complexity (*"DCost"*), and overall stubbing complexity (*"OCplx"*) of the generated CITOs. The last column, labeled as *"T/T"*, represents the ratio of the average execution time without to that with dynamic dependencies.

Firstly, regarding *Stubs*, we observe that "RL" generates more stubs than the other systems, except for *notepad_spl*, which lacks dynamic dependencies. This disparity arises because these systems possess varying degrees of dynamic dependencies, whereas "RL" disregards the statistics on the number of stubs required for dynamic dependencies.

Accounting for dynamic dependencies complements the necessary stubbing cost, while minimizing the total CITOs cost by constructing fewer stubs. For instance, in the case of *ANT*, "RL" incurs higher *ACost*, *MCost*, and *OCplx* than the conventional method "RL", which disregards dynamic dependencies. In fact, on average, *ANT* requires 15 stubs for simulating dynamic dependencies, which involves 63 dynamic dependencies. This simulation necessitates a dynamic coupling complexity of 8.99. Using the weight value considering dynamic dependency for calculation, the overall stubbing complexity becomes 7.73, much higher than the proposed overall stubbing complexity of 5.60, which belongs to "RL" considering dynamic couplings. This indicates that the proposed method is more efficient in reducing overall CITOs' costs.

In terms of runtime, we have observed that the time difference between the methods that consider dynamic couplings and those that only consider static coupling is insignificant.

Performance of Our RL Method Combined with Dynamic Couplings. To evaluate our RL method's performance, we contrast it with two dynamic-coupling-aware graph-based techniques (test-level method [14] and Class-HITS method [15]), and three search-based methods that do not consider dynamic couplings. (GA [9], RIA [21], and PSO methods [22]). Notably, as per our knowledge, all existing search-based methods didn't account for dynamic couplings.

Table 4. Comparison of results without and with the consideration of dynamic couplings.

Programs	RL				RL'					T/T'
	Stubs	ACost	MCost	OCplx	Stubs	ACost	MCost	DCost	OCplx	
ANT	11	2.58	1.5	2.4	19	4.46	3.24	4.33	5.6	0.96
Email-spl	5	0.2	0.39	0.38	6	0.18	0.4	1	1.01	0.97
DNS	10	1.92	1.47	2.02	10	2.5	1.81	0	1.02	1.01
jboss	4	1.5	0.44	0.86	4	1.5	0.44	0	0.35	0.99
notepad_spl	46	5.35	2.86	5.19	46	5.15	2.87	0	5.05	0.98

Test-level method [14] and Class-HITS method [15] conducted experiments on *ANT* and *DNS*. A comparison between these methods and ours for *ANT* and *DNS* is shown in Table 5. *"Stubs"* indicates the number of stubs, *"OCplx"* depicts the overall stubbing complexity of CITOs, "Num_a" represents attributes need to be simulated, and "Num_a" signifies methods need to be simulated.

Table 5. Comparison between graph-based methods and our proposed method

Programs	Methods	Stubs	OCplx	Num_a	Num_m
ANT	Method [14]	23	3.84	87	305
	Method [15]	30	3.65	65	187
	Ours	19	5.52	142	45
DNS	Method [14]	6	1.47	11	19
	Method [15]	6	1.47	11	19
	Ours	11	1.01	30	18

For *ANT*, our method yields fewer stubs but higher construction complexity. This arises as prior graph-based methods overlooked varying stubbing costs for dynamic dependencies. They mainly addressed dynamic dependencies from associations and inheritance, neglecting polymorphic invocation coupling measurement. Programs like *ANT*, with multiple dynamic dependencies forming cycles, require breaking more dynamic dependencies to eliminate all cycles. Thus, ignoring dynamic coupling calculation results in substantial cost disparity.

For *DNS*, we constructed a large number of stubs with lower overall complexity. Compared to *ANT*, *DNS* requires simulating fewer attributes and methods, resulting in a smaller difference between *ANT*s. Moreover, *DNS* does not result in redundant cycles due to dynamic dependencies. Although our approach considers the cost of constructing stubs for dynamic couplings in *DNS*, it still incurs less overall stubbing costs compared to previous methods.

Search-based methods are prevalent for CITOs but frequently overlook dynamic dependencies. We examined three notable search-based methods: GA

[9], PSO [22], and RIA [21]. These methods employ enhanced fitness functions guiding population evolution toward lower-cost CITOs. Our study improved fitness function calculation, incorporating dynamic dependency measurements for evolution guidance. Results combining static and dynamic dependencies are in Table 6. The left term *"St"* indicates static dependencies consideration, while right *"Dy"* signifies dynamic dependencies inclusion.

Table 6. Comparison between search-based methods and our proposed method

Programs	St						Dy						
	Methods	Runtime	Stubs	ACost	MCost	OCplx	Methods	Runtime	Stubs	ACost	MCost	DCost	OCplx
ANT	GA	**137**	17	2.24	2.91	2.79	GA'	**175**	31	**2.05**	**2.61**	10.63	10.22
	PSO	2777	14	**1.46**	2.35	**2.04**	PSO'	4262	30	6.94	4.48	**1.01**	**4.33**
	RIA	1447	33	9.65	6.56	9.38	RIA'	1891	45	10.16	6.43	8.26	12.03
	RL	29204	**11**	2.58	**1.5**	2.4	RL'	30518	**19**	4.46	3.24	4.33	5.6
DNS	GA	**3264**	42	5.97	5.17	6.58	GA'	**4333**	106	9.03	6.45	21.12	22.82
	PSO	9763	37	5.2	5.56	6.23	PSO'	18185	75	20.18	9.62	7.1	13.15
	RIA	92381	114	31.23	18.03	28.82	RIA'	129038	199	29.42	16.53	40.02	46.6
	RL	181266	**10**	**1.92**	**1.47**	**2.02**	RL'	186586	**10**	**2.5**	**1.81**	0	**1.02**
Email-spl	GA	**384**	16	0.4	0.76	0.73	GA'	**423**	16	0.36	0.92	0.9	1.2
	PSO	4121	17	0.37	0.69	0.68	PSO'	3939	18	0.32	1.25	0	**0.62**
	RIA	8657	32	1.57	2.9	2.63	RIA'	12289	33	1.29	2.72	0.6	1.96
	RL	97994	**5**	**0.2**	**0.39**	**0.38**	RL'	96967	**6**	**0.18**	**0.4**	1	1.01
jboss	GA	**2745**	19	8.7	1.83	4.3	GA'	**3254**	20	8.9	1.89	0.5	2.26
	PSO	7761	16	7.55	1.56	4	PSO'	7732	14	6.85	1.21	0	1.47
	RIA	368927	52	23.6	6.73	12.97	RIA'	458631	57	26.2	6.77	0.8	6.39
	RL	278292	**4**	**1.5**	**0.44**	**0.86**	RL'	280250	**4**	**1.5**	**0.44**	0	**0.35**
notepad_spl	GA	**2309**	55	3.65	2.06	3.82	GA'	**2556**	55	3.55	**1.92**	0	3.7
	PSO	6751	57	**2.65**	**2.04**	**3.19**	PSO'	6787	55	**2.75**	2.05	0	**3.26**
	RIA	108441	70	21.55	2.66	16.6	RIA'	122292	75	20.6	2.49	0	15.93
	RL	181847	**46**	5.05	2.83	4.99	RL'	178516	**46**	5.15	2.87	0	5.05

Considering dynamic couplings, GA's runtime is only 1.06–1.32 times longer than without them. RL's longer runtime stems from a large number of iterations (2 million) and extensive Q-value storage, though memory clears in later iterations, slightly lengthening runtime. In contrast, GA performs 100 iterations, using an old-new switch to trim memory usage and cleanup time. In the future, we aim to cut runtime while minimizing stubbing costs.

Regarding *Stubs*, "RL" and "RL" require fewer than other methods. For *ACost* and *MCost*, "RL" generates CITOs with the lowest *ACost* for three out of five programs and the lowest *MCost* for four programs. The overall stubbing complexities of three programs are the lowest, followed by the PSO method. Hence, "RL" is most effective in simulating static dependencies. Considering dynamic couplings, "RL" outperforms others in three programs' *ACost* and *MCost*, except *notepad_spl*. Others' results vary, but RL's remain consistent.

For *ANT* and *email_spl*, the PSO method performs best. Despite this fact that for *ANT*, PSO and PSO' calculate 16 and 12 additional stubs compared to our methods "RL" and "RL", respectively, increasing the overall stubbing complexity by 2.29. On the other hand, for *email_spl*, the overall stubbing complexity is reduced by 0.06 since the method assigns different weights to dynamic dependencies based on their complexity.

For *DNS* and *jboss*, our method "RL" significantly outperforms other methods in terms of results, especially for programs with multiple dependencies that do not construct redundant cycles. In these cases, our method can avoid constructing redundant stubs that simulate dynamic dependencies, ensuring a balanced simulation of static and dynamic dependencies.

Based on these results, our approach, which incorporates the stubbing cost measurement of dynamic couplings, exhibits several advantages when compared to other methods. Specifically, it effectively balances the various types of dependencies and generates CITOs with minimal stubbing cost.

5   Conclusions

This paper presents a novel RL method to generate CITOs combined with a new measurement of the complexity involved in constructing stubs for dynamic couplings. To achieve this, we first redefine dynamic dependencies to provide a more comprehensive description of inter-class relationships and measure dynamic couplings. Then, we integrate this measurement strategy of dynamic and static coupling to propose a new method for calculating the overall stubbing complexity. Finally, we apply this new overall stubbing complexity calculation method to the RL algorithm to generate CITOs. The experimental results demonstrate that our method can measure the stubbing complexity of CITOs more accurately and minimize the overall stubbing cost of CITOs as much as possible. In the future, we will extend our method to more real programs to better explore the impact of dynamic couplings on the generation of CITOs.

Acknowledgements. This work is supported by "the Fundamental Research Funds for the Central Universities" under grant No. 2022XSCX18.

References

1. Vinyals, O., et al.: Grandmaster level in starcraft ii using multi-agent reinforcement learning. Nature **575**(7782), 350–354 (2019)
2. Luong, N.C., et al.: Applications of deep reinforcement learning in communications and networking: a survey. IEEE Commun. Surv. Tutor. **21**(4), 3133–3174 (2019)
3. Spieker, H., Gotlieb, A., Marijan, D., Mossige, M.: Reinforcement learning for automatic test case prioritization and selection in continuous integration. In: ACM SIGSOFT International Symposium on Software Testing and Analysis, pp. 12–22 (2017)
4. Sutton, R.S., Barto, A.G.: Reinforcement Learning: An Introduction. MIT press, Massachusetts (2018)
5. Czibula, G., Czibula, I.G., Marian, Z.: An effective approach for determining the class integration test order using reinforcement learning. Appl. Soft Comput. **65**, 517–530 (2018)
6. Ding, Y., Zhang, Y., Shujuan, J., Yuan, G., Wang, R., Qian, J.: Generation method of class integration test order based on reinforcement learning (in Chinese). J. Softw. **33**(5), 1674–1698 (2022)

7. Stopford, B.: Test-oriented languages: is it time for a new era? In: International Conference on Software Testing, Verification and Validation Workshops, pp. 444–449 (2011)
8. Hashim, N.L., Schmidt, H.W., Ramakrishnan, S.: Test order for class-based integration testing of java applications. In: International Conference on Quality Software, pp. 11–18 (2005)
9. Briand, L.C., Feng, J., Labiche, Y.: Using genetic algorithms and coupling measures to devise optimal integration test orders. In: International Conference on Software Engineering and Knowledge Engineering, pp. 43–50 (2002)
10. Labiche, Y., Thévenod-Fosse, P., Waeselynck, H., Durand, M.H.: Testing levels for object-oriented software. In: International Conference on Software Engineering, pp. 136–145 (2000)
11. Malloy, B.A., Clarke, P.J., Lloyd, E.L.: A parameterized cost model to order classes for class-based testing of c++ applications. In: International Symposium on Software Reliability Engineering, pp. 353–364 (2003)
12. Kraft, N.A., Lloyd, E.L., Malloy, B.A., Clarke, P.J.: The implementation of an extensible system for comparison and visualization of class ordering methodologies. J. Syst. Softw. **79**(8), 1092–1109 (2006)
13. Bansal, P., Sabharwal, S., Sidhu, P.: An investigation of strategies for finding test order during integration testing of object oriented applications. In: International Conference on Methods and Models in Computer Science, pp. 1–8 (2009)
14. Zhang, Y., Jiang, S., Yuan, G., Ju, X., Zhang, H.: An approach of class integration test order determination based on test levels. Software: Pract. Exper. **45**(5), 657–687 (2015)
15. Wang, Y., Zhu, Z., Yu, H., Yang, B.: Risk analysis on multi-granular flow network for software integration testing (in Chinese). IEEE Trans. Circuits-II **65**(8), 1059–1063 (2017)
16. Meng, F., Wang, Y., Yu, H., Zhu, Z.: Devising optimal integration test orders using cost-benefit analysis. Front. Inform. Tech. El. **23**(5), 692–714 (2022)
17. Kung, D.C., Gao, J., Hsia, P., Lin, J., Toyoshima, Y.: Class firewall, test order, and regression testing of object-oriented programs. J. Object-Oriented Prog. **8**(2), 51–65 (1995)
18. Tai, K.C., Daniels, F.J.: Test order for inter-class integration testing of object-oriented software. In: International Computer Software and Applications Conference, pp. 602–607 (1997)
19. Le Traon, Y., Jéron, T., Jézéquel, J.M., Morel, P.: Efficient object-oriented integration and regression testing. IEEE Trans. Reliab. **49**(1), 12–25 (2000)
20. Briand, L.C., Labiche, Y., Wang, Y.: Revisiting strategies for ordering class integration testing in the presence of dependency cycles. In: International Symposium on Software Reliability Engineering, pp. 287–296 (2001)
21. Wang, Z., Li, B., Wang, L., Li, Q.: An effective approach for automatic generation of class integration test order. In: Computer Software and Applications Conference, pp. 680–681 (2011)
22. Zhang, Y., Jiang, S., Ding, Y., Yuan, G., Liu, J., Lu, D., Qian, J.: Generating optimal class integration test orders using genetic algorithms. Int. J. Softw. Eng. Know. **32**(06), 871–892 (2022)
23. Ogihara, M.: Interfaces, inheritance, and polymorphism, pp. 427–455. Springer International Publishing, Cham (2018)
24. He, D., Xu, J., Chen, X.: Information-theoretic-entropy based weight aggregation method in multiple-attribute group decision-making. Entropy **18**(6), 171 (2016)
25. Watkins, C.J.C.H., Dayan, P.: Q-learning. Mach. Learn. **8**(3–4), 279–292 (1992)

A Novel Machine Learning Model Using CNN-LSTM Parallel Networks for Predicting Ship Fuel Consumption

Xinyu Li[1], Yi Zuo[1(✉)], Tieshan Li[2], and C. L. Philip Chen[3]

[1] Dalian Maritime University, Dalian 116000, China
{lxy503,zuo}@dlmu.edu.cn
[2] University of Electronic Science and Technology of China, Chengdu 610000, China
[3] South China University of Technology, Guangzhou 510000, China
philip.chen@ieee.org

Abstract. With continuous increasing of carbon emission, prediction of ship fuel consumption is gaining significance in reduction of energy consumption and emissions for ships. This paper proposes a novel model of parallel network by combining convolutional neural network and long short-term memory (CNN-LSTM). The proposed model integrates three advantages. The CNN part of proposed model can extract spatial features, the LSTM part of proposed model can capture temporal relationships, and the parallel structure of proposed model can obtain feature fusion from both of CNN and LSTM based on multi-source data. Experimental outcomes reveal that CNN-LSTM parallel networks can obtain best results of MAE and RMSE, which outperformed single LSTM, single CNN and other neural networks with decreasing of 48.06%, 64.06% and 48.56% in MAE, and 35.71%, 58.25% and 37.85% in RMSE.

Keywords: Ship Fuel Consumption · Neural Network · CNN · LSTM · Parallel Networks · Prediction

1 Introduction

Shipping is one of the most significant sectors to promoting the growth of oversea trade and the development of global transportation [1]. With increasing of maritime transportation, the accompanying carbon emissions and environmental pollution should be also brought to attention [2]. Shipping urgently needs to operate with improved cost efficiency due to the rising oil prices and the cost of ship fuel accounting for a significant portion of running expenses [3]. It is required to achieve reduction of fuel consumption and environmental impact, and then promote the development of shipping sector in the direction of low carbon and low pollution. Prediction of ship fuel consumption (SFC) has been widely accepted as an important research topic [4].

One of the most popular methods is to learn prediction models based on artificial neural networks (ANN) [5]. Farag et al. proposed an ANN-based model to predict SFC

© The Author(s), under exclusive license to Springer Nature Singapore Pte Ltd. 2024
B. Luo et al. (Eds.): ICONIP 2023, LNCS 14448, pp. 108–118, 2024.
https://doi.org/10.1007/978-981-99-8082-6_9

[6]. Then, a novel machine learning model of combining ANNs was proposed SFC prediction [7]. Here, long short-term memory (LSTM) [8] and convolutional neural network (CNN) [9] were used under background of liquefied petroleum gas (LPG) carrier as a case ship. Based on the comprehensive consideration of the time series and nonlinear characteristics of SFC data, this paper utilizes different ANNs and mechanism of parallel learning to achieve multi-source data fusion [10]. The proposed model integrates three advantages. The CNN part of proposed model is used as first layer to extract spatial features from input data, the LSTM part of proposed model is used as second layer to capture temporal relationships from the same input data, and the parallel structure of proposed model can obtain feature fusion from both layers of CNN and LSTM based on multi-source data. We also design three modules as data acquisition, data processing and SFC prediction. In data acquisition, multi-source SFC data are gathered from four sources. In data processing, we perform operations such as data cleaning, feature selection and data normalization on the collected data. In SFC prediction, the proposed CNN-LSTM parallel networks is compared with single LSTM, single CNN and other ANNs. The results indicate that our model can obtain the best MAE and RMSE for SFC prediction in the actual voyage than other baseline methods. Therefore, the proposed model can significantly enhance energy efficiency of the ship and reduce operating expenses and emissions.

The remaining part of the article can be structured in following way. Section 2 gives a concise summary of relevant research of SFC prediction methods. Section 3 introduces the data processing and modeling techniques used for SFC prediction. Section 4 presents a comparative experiment and a subsequent discussion of the results. Finally, Sect. 5 provides the conclusion.

2 Related Study

Early studies have been carried out on prediction of SFC using statistical methods. Medina et al. provided semi-empirical formulas to predict fuel efficiency of the case ship considering the influence of the wind and wave correlation on Beaufort scale [11]. Bialystocki et al. offered a solution to obtain an accurate SFC and speed curve by statistically analyzing the data related to SFC [12]. Then, many machine learning techniques were also included to enhance prediction accuracy of SFC. ANN is widely used in SFC prediction [13]. Peng et al. developed five ANN methods to predict SFC [14]. Panapakidis et al. combined traditional ANN and deep neural networks to predict SFC [15].

Recently, LSTM and CNN have also attracted much attention from academia and industry [16, 17]. Nevertheless, both CNN and LSTM have their limitations of accurately predicting issues on multi-source and time-series SFC data [18]. Therefore, this article developed a novel machine learning model with CNN-LSTM parallel networks by combining ANN, LSTM and CNN models and conducts a comparative study of SFC prediction with the basic three models.

3 Methodology

3.1 Overview of SFC Prediction

SFC prediction of this paper is illustrated in Fig. 1. Firstly, to consider both complexity of the own engine power system and maritime and meteorological conditions during navigation, multi-source data collected in this paper that affect SFC cover cabin log data, sensor data, as well as available oceanographic and meteorological data [19]. Due to factors such as sensor failure, human error, etc., there may be outliers in the collected data. Moreover, the raw data is a complex multivariate time series, this article implements data analysis and processing through three procedures, namely data cleaning, features selection, and data normalization.

Fig. 1. Overview of SFC prediction process.

3.2 Modeling of SFC Prediction

The constructed CNN-LSTM parallel network for SFC prediction is obtained by combining three models of CNN, LSTM and ANN through the concatenate function, depicted in Fig. 2.

The blue dashed box indicates CNN component of the model, which is composed of trainable multi-layer architectures such as convolutional layer, pooling layer, and full connected layer.

The convolution layer extracts local features through convolution operation. The procedure can be represented by Eq. (1). Y_c denotes output of convolution layer, f indicates activation function, W_c denotes weights, $\otimes$ indicates convolution operator, X_c denotes the input multi-source data of SFC, and b denotes the bias values.

$$Y_c = f(W_c X_c + b_c) \tag{1}$$

After the convolution layer, a maximum pooling layer is connected to realize the dimensionality reduction and further feature extraction by summarizing the features acquired through the process of convolution. The procedure is expressed by Eq. (2), where Y_P indicates the pooling layer output, and $Pool_{max}$ denotes the maximum pooling function. The characteristics of input SFC data acquired after pooling procedure

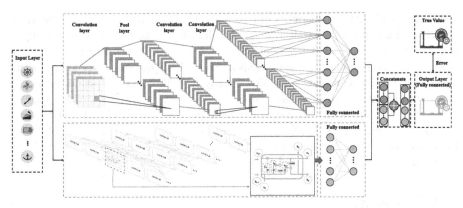

Fig. 2. Modeling of SFC based on CNN-LSTM parallel networks.

are ultimately aggregated by the full connection layer to complete the output of CNN component.

$$Y_p = Pool_{max}(Y_c) \tag{2}$$

The orange dashed box in Fig. 2 indicates the LSTM component in the SFC prediction model. Compared with CNN, LSTM considers the correlation between SFC-related data in different time series, which help to improve the reliability of model prediction. From Fig. 2 it is evident that specific composition of the LSTM cell consists of input gate i_k, forget gate f_k, and output gate o_k. The specific calculation process is shown in Eqs. (3)–(5). For input gate i_k, the activation value h_{k-1} of previous moment hidden layer and the current input value x_k are used to identify additional information that can be incorporated into cell state. The forget gate f_k decides which information need to be eliminated from the cell state c_{k-1}. In addition, the output gate o_k controls what information from h_{k-1}, x_k, c_k is passed to the hidden layer state h_k. Eqs. (6), (7) represent the solution process of c_k and h_k, respectively [8].

$$i_k = \sigma\left(W_i\left[h_{k-1}, x_k\right] + b_i\right) \tag{3}$$

$$f_k = \sigma\left(W_f\left[h_{k-1}, x_k\right] + b_f\right) \tag{4}$$

$$o_k = \sigma\left(W_o\left[h_{k-1}, x_k\right] + b_o\right) \tag{5}$$

$$c_k = f_k\, c_{k-1} + i_k\, tanh\left(W_e\left[h_{k-1}, x_k\right] + b_e\right) \tag{6}$$

$$h_k = o_k\, tanh(c_k) \tag{7}$$

where x_k denotes the input data at moment k, σ is the activation function, W_i, W_f, W_o and W_e indicate the weight matrices, b_i, b_f, b_o and b_e denote the bias values.

The structure of CNN-LSTM parallel networks can achieve effective fusion, which not only takes advantage of CNN ability to capture spatial characteristics of SFC input data, but also inherits the superiority of LSTM ability to capture the correlation between a series and a sequence, as well as long- and short-term characteristics of SFC input data. As shown in the dotted green line in Fig. 2, the fully connected neural network, a type of ANN, is connected behind the CNN and LSTM, respectively. Then the extracted features are fused by concatenate function. Finally, the fully connected neural network is employed to process data after concatenate function fusion to acquire forecast outputs. The learning ability of network can be improved by adjusting layer count of CNN and LSTM. Furthermore, every CNN layer is connected behind a maximum pooling layer to limit parameter count. To avoid overfitting, a dropout layer is added after every LSTM layer [10].

4 Experiments

4.1 Data Processing

Data Source. The data utilized in this article come from an LPG carrier with capacity of 54340 DWT. And its Length Overall and Molded Width are 225 m and 37 m, respectively [20]. The raw dataset was collected continuously for a period of 7 months, including 28 features and 18499 data points. By calculating the number of count, mean, standard deviation, extreme value and quartile of each feature variable in the dataset, the basic statistical analysis is obtained as shown in Table 1.

Table 1. Statistical analysis of source data.

Feature	Mean	Std	Min	25%	50%	75%	Max
RPM (r/min)	78.0	17.4	−1.0	74.9	85.3	86.5	90.0
SFC (mt)	1399.3	426.5	−1.0	1135.7	1563.4	1697.4	2000.0
SOG (knots)	14.2	3.4	0.0	13.5	15.2	16.1	19.2
AT (°C)	34.9	7.7	2.9	33.5	35.7	37.2	150.0
Trim (m)	2.0	1.2	0.5	0.7	2.6	3.0	4.8

Data Cleaning. First remove the obvious outliers in the original data according to the actual sailing conditions, such as the minimum values of RPM, SFC and SOG in Table 1. Taking into account the practical situation of navigation, a method of setting thresholds on SOG and RPM was employed to clean the data. From the perspective of data analysis, two reasonable threshold ranges are set corresponding to 60–90 r/min and 7–20 kn respectively. Data points for RPM and SOG outside the threshold ranges are then removed. Simultaneously, the data points related to other features corresponding to the same time outside this range are also filtered out.

Feature Selection. In terms of data type, certain features that correspond to data that cannot be precisely determined within a specific range will be removed [21], such as propeller slip, wind sea waves, swell and wind. In addition, the features related to the following two cases also need to be screened out. (1) The same features collected from different sources, such as course over ground provided by electronic chart display and information system and measured by ship automation system. (2) Features that can be converted to each other according to the formula, such as wind speed can be calculated from Uwind, Vwind and wind direction. After the above selection of features, 15 feature variables are retained as input and SFC as output, as shown in Table 2.

Table 2. The abbreviations of SFC feature variables.

No	Feature variables	Abbr	No	Feature variables	Abbr
1	Ship ME total FC	SFC	9	Uwind	Uwind
2	Revolution per minute	RPM	10	Vwind	Vwind
3	Speed over ground	SOG	11	Significant wave height	SWH
4	Course over ground	COG	12	Wave direction	WD
5	Ambient air temperature	AT	13	Wave period	WP
6	Sea water temperature	SWT	14	Sea direction	SD
7	Trim	Trim	15	Sea current speed	SCS
8	List	List			

Data Normalization. The Min-Max normalization is applied for scaling data. Let $x'_{i_{Min}}$ and $x'_{i_{Max}}$ denote the lower and upper bounds of the feature variable. To acquire the value x_i, the approach transforms the raw value x'_i of feature variable to the range between 0 and 1, as demonstrated by the following formula.

$$x_i = \frac{x'_i - x'_{i_{Min}}}{x'_{i_{Max}} - x'_{i_{Min}}} \tag{8}$$

After fine filtering of data, data points count is decreased to 17287. Finally, the normalized data is segmented into 70% training, 10% validation and 20% test set employing a randomly sampling approach.

4.2 Numerical Experiments

Evaluation Indicator. This study requires an evaluation of the prognostic performance of SFC model, therefore the mean absolute error (MAE), root mean square error (RMSE) and R^2 are applied as depicted by Eqs. (9), (10) and (11). Where y'_i is predicted SFC value, y_i denotes actual SFC value, $\bar{y}_i$ indicate average value and m signifies test data points count.

$$MAE = \frac{1}{m} \sum_{i=1}^{m} |y_i' - y_i| \tag{9}$$

$$RMSE = \sqrt{\frac{1}{m} \sum_{i=1}^{m} (y_i' - y_i)^2} \tag{10}$$

$$R^2 = 1 - \frac{\sum_i^m (y_i' - y_i)^2}{\sum_i^m (\bar{y}_i - y_i)^2} \tag{11}$$

Comparison with SFC Methods. To objectively evaluate the predictive capabilities of various SFC models, simulation training using default parameters in the Python programming language. The parameter settings for the CNN, ANN, LSTM, and CNN-LSTM parallel networks are shown in Table 3. Additionally, the time step for all models were set to 4 based on our experience. The R^2 values for different models are discussed as shown in Fig. 3.

Table 3. Parameter settings of models.

Prediction model	Parameter setting
CNN	Quantity of convolution layers = 2; Quantity of convolution kernel = (4, 8); Activation function = Sigmoid
ANN	Quantity of hidden layers = 2; Quantity of neurons = (50, 40); Activation function = Relu
LSTM	Quantity of hidden layers = 2; Quantity of neurons = (50, 40); Activation function = Tanh
CNN-LSTM parallel networks	Synthesize the parameters of all the above three single models

Figure 3 compares the test results of predicted SFC values for each model with actual SFC values. Consequently, the prediction result of CNN-LSTM parallel networks is the closest to reality, with the highest R^2 reaching 0.9702. Therefore, the proposed model can predict SFC more accurately.

4.3 Parameter Tuning and Optimization

For above four SFC prediction models, the optimal setting for every parameter is established by evaluating the MAE and RMSE of test set. The outcomes of each model are averaged over 100 runs to avoid overtraining and validate generalizations. Early stopping approach was adopted to terminate training when the training loss was less than 50 epochs to avoid overfitting.

For CNN prediction models, the impact of the quantity of convolution layers and kernel on results were discussed. The range of quantity of convolution layers between

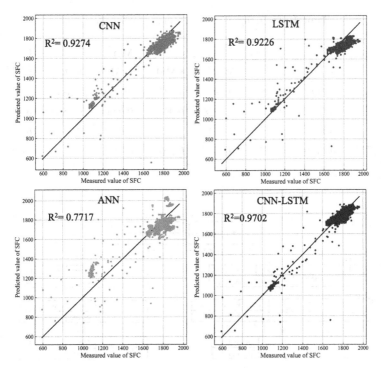

Fig. 3. Prediction accuracy of different models.

1 and 3 was compared, and the number of convolution kernel in every layer was set to 4, 8, 16 respectively; For ANN and LSTM prediction models, time step, the quantity of hidden layers and neurons were discussed. The quantity of hidden layers is adjusted from 1 to 5, and the quantity of neurons contained in every layer is 50, 40, 40, 20, 20, in order. For CNN-LSTM parallel networks prediction model, all parameters involved in the above models need to be adjusted. Specifically, the number of convolution layers and kernel was first explored, and parameter optimization method adopted was the same as that in CNN model. Multiple experiments show that when the number of convolutional kernels is 16, the effect is the best with MAE and RMSE reaching 0.0161 and 0.0243. Then the corresponding parameters of LSTM part are discussed under the optimal parameters of CNN component. In addition, for all of the above models, the optimization process involves setting time step parameters within the range of 1–10. Table 4 presents a summary of the experimental results.

Table 4 statistics the maximum, average and minimum RMSE values of every model. The optimal parameter combination (number of convolution layers, quantity of hidden layers, time step) corresponding to the minimum RMSE is also attached. It can be seen more clearly that the predictive capability of CNN-LSTM parallel networks is superior to the other three single models and has a significant improvement in prediction performance. The minimum MAE of CNN-LSTM parallel networks can reach 0.0161, which is 48.06%, 64.06% and 48.56% higher than that of CNN, ANN and LSTM. The minimum RMSE of CNN-LSTM parallel networks can reach 0.0243, which is 35.71%,

Table 4. The result of parameter tuning and optimization.

Model	Max		Average		Min		Optimal parameter
	MAE	RMSE	MAE	RMSE	MAE	RMSE	
CNN	0.1817	0.1996	0.1311	0.1471	0.0310	0.0378	(1, 0, 4)
ANN	0.1162	0.1453	0.0778	0.0918	0.0448	0.0582	(0, 2, 3)
LSTM	0.1374	0.1522	0.0595	0.0686	0.0313	0.0391	(0, 4, 6)
CNN-LSTM parallel networks	0.1166	0.1243	0.0558	0.0672	0.0161	0.0243	(3, 5, 4)

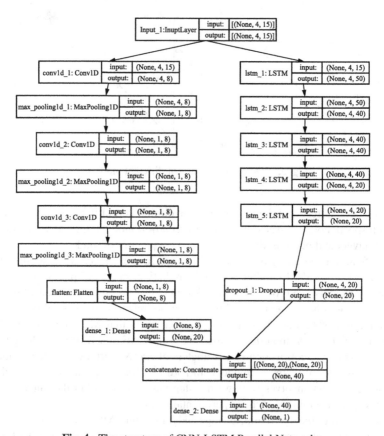

Fig. 4. The structure of CNN-LSTM Parallel Networks.

58.25% and 37.85% higher than that of CNN, ANN and LSTM. The optimal parameter in this case is (3, 5, 4) and the network architecture as depicted in Fig. 4. It can be found from Fig. 4 that the input and output dimensions of each part of the proposed model. The path on the left is CNN and the path on the right is LSTM. The data processed by the two paths are represented in the form of vectors, and the final vector representation is obtained by concatenation. Finally output through the full connection layer of ANN. It turns out that the variant outperforms the single CNN, ANN and LSTM. During the parameter tuning process, the minimum MAE and RMSE of each model and its corresponding variance are shown in Fig. 5. The observation shows that the optimal parameter combination has the highest accuracy and the overall prediction is relatively stable. Therefore, CNN-LSTM parallel networks model is the best choice for SFC prediction in actual sailing.

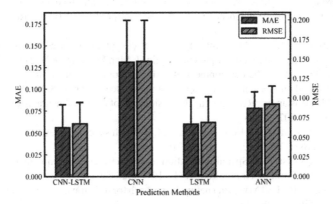

Fig. 5. The result of parameter tuning and optimization.

5 Conclusions

To strengthen the study of capturing both short and long term characteristics and time series properties of SFC data, and improve prediction accuracy, this paper aims to identify a suitable combination of single CNN, ANN, and LSTM models to enhance the current SFC prediction models. Consequently, a novel machine learning model with CNN-LSTM parallel networks for SFC prediction is developed. The proposed model integrates advantages of three single models, such as based on the comprehensive consideration of the time series and nonlinear characteristics of SFC data through CNN and LSTM, and utilizing ANN and mechanism of parallel learning to achieve multi-source data fusion. Additionally, based on the multi-source SFC data of LPG, the validity of CNN-LSTM parallel networks model is verified. In comparison to single CNN, ANN and LSTM, CNN-LSTM parallel networks has the smallest RMSE, which decreased by 5.81%, 58.25% and 37.85%, respectively. CNN-LSTM parallel networks not only realizes the optimal fitting effect of real SFC, but further evidences the combined model exhibits superior performance. In future work, we plan to further refine the proposed model by incorporating attention mechanisms to obtain better results.

Acknowledgments. Partial support for this work was provided by the National Natural Science Foundation of China (grant nos. 52131101, 51939001), the LiaoNing Revitalization Talents Program (grant no. XLYC1807046) and the Science and Technology Fund for Distinguished Young Scholars of Dalian (grant no. 2021RJ08).

References

1. Ailong, F., Jian, Y., Liu, Y.: A review of ship fuel consumption models. Ocean Eng. **264**, 112405 (2022)
2. Wang, S., Psaraftis, H.N., Qi, J.: Paradox of international maritime organization's carbon intensity indicator. Commun. Transp. Res. **1**, 100005 (2021)
3. Wan, Z., El Makhloufi, A., Chen, Y.: Decarbonizing the international shipping industry: solutions and policy recommendations. Mar. Pollut. Bull. **126**, 428–435 (2018)
4. Ran, Y., Wang, S., Harilaos, N.: Data analytics for fuel consumption management in maritime transportation: status and perspectives. Transp. Res. Part E: Logist. Transp. Rev. **155**, 102489 (2021)
5. Zhu, Y., Zuo, Y., Li, T.: Modeling of ship fuel consumption based on multisource and heterogeneous data: case study of passenger ship. J. Mar. Sci. Eng. **9**(3), 273 (2021)
6. Farag, Y., Ölçer, A.I.: The development of a ship performance model in varying operating conditions based on ANN and regression techniques. Ocean Eng. **198**, 106972 (2020)
7. Hinton, G.E., Ruslan, R.: Reducing the dimensionality of data with neural networks. Science **313**(5786), 504–507 (2006)
8. Greff, K., Srivastava, R.K., Koutník, J.: LSTM: a search space odyssey. IEEE Trans. Neural Netw. Learn. Syst. **28**(10), 2222–2232 (2016)
9. Shomron, G., Weiser, U.: Spatial correlation and value prediction in convolutional neural networks. IEEE Comput. Archit. Lett. **18**(1), 10–13 (2019)
10. Guo, J.: A CNN-Bi_LSTM parallel network approach for train travel time prediction. Knowl.-Based Syst. **256**, 109796 (2022)
11. Medina, J.R., Molines, J., González-Escrivá, J.A.: Bunker consumption of containerships considering sailing speed and wind conditions. Transp. Res. Part D: Transp. Environ. **87**, 102494 (2020)
12. Bialystocki, N., Konovessis, D.: On the estimation of ship's fuel consumption and speed curve: a statistical approach. J. Ocean Eng. Sci. **1**(2), 157–166 (2016)
13. Li, X., Zhu, Y., Zuo, Y.: Prediction of ship fuel consumption based on broad learning system. In: International Conference on Security, Pattern Analysis, and Cybernetics, pp. 54–58. IEEE, Guangzhou, China (2019)
14. Peng, Y., Liu, H., Li, X.: Machine learning method for energy consumption prediction of ships in port considering green ports. J. Clean. Prod. **264**, 121564 (2020)
15. Panapakidis, I., Sourtzi, V.M., Dagoumas, A.: Forecasting the fuel consumption of passenger ships with a combination of shallow and deep learning. Electronics **9**(5), 776 (2020)
16. Kim, J., Oh, S., Kim, H.: Learning Traffic as Images: tutorial on time series prediction using 1D-CNN and BiLSTM: a case example of peak electricity demand and system marginal price prediction. Eng. Appl. Artif. Intell. **126**, 106817 (2023)
17. Zhu, Y., Zuo, Y., Li, T.: Predicting ship fuel consumption based on LSTM neural network. In: 7th International Conference on Information, Cybernetics, and Computational Social Systems, pp. 310–313. IEEE, Guangzhou, China (2020)
18. Ke, J., Zheng, H., Yang, H.: Short-term forecasting of passenger demand under on-demand ride services: a spatio-temporal deep learning approach. Transp. Res. Part C: Emerg. Technol. **85**, 591–608 (2017)
19. Li, X., Zuo, Y., Jiang, J.: Application of regression analysis using broad learning system for time-series forecast of ship fuel consumption. Sustainability **15**(1), 380 (2023)
20. Vorkapić, A., Radonja, R., Martinčić-Ipšić, S.: Predicting seagoing ship energy efficiency from the operational data. Sensors **21**(8), 2832 (2021)
21. Jiang, J., Zuo, Y.: Prediction of ship trajectory in nearby port waters based on attention mechanism model. Sustainability **15**(9), 7435 (2023)

Two-Stage Attention Model to Solve Large-Scale Traveling Salesman Problems

Qi He[1], Feng Wang[1(✉)], and Jingge Song[2]

[1] School of Computer Science, Wuhan University, Wuhan, China
{heqi_049,fengwang}@whu.edu.cn
[2] Hongyi Honor College, Wuhan University, Wuhan, China
song_jg@whu.edu.cn

Abstract. The Traveling Salesman Problem (TSP) widely exists in real-world scenarios. Various methods, such as exact methods, heuristic methods, and deep learning-based methods, can solve TSPs efficiently. However, as the size of the problems increases, these methods become increasingly time-consuming due to the high complexity of large-scale TSPs. This paper proposes a two-stage attention model (TSAM) that incorporates the divide-and-conquer strategy and attention model to solve large-scale TSPs efficiently. Experimental results demonstrate that TSAM can rapidly produce promising solutions for TSP instances ranging from 500 to 10,000 nodes.

Keywords: Large-Scale TSP · Reinforcement Learning · Attention Mechanism

1 Introduction

The Traveling Salesman Problem (TSP) is a classical problem in the field of Combinatorial Optimization (CO), and it has practical applications in various domains, including genetic sequencing, route planning, and electronic device layout. Given a set of cities with their positions, a salesman aims to visit each city exactly once and return to the depot. The goal of TSP is to minimize the total distance of the visited tour.

Since TSPs widely exist in the real world, researchers are dedicated to designing effective methods to solve them. In the past few decades, exact methods, such as Concorde solver [1,2], and heuristic methods, such as the Lin-Kernighan heuristic (LKH) [3,4], have been proposed for TSPs. Although these methods achieve significant performance, they still face several challenges. Since they require specifically crafted rules and heavily depend on expert knowledge, they are not flexible to be generalized to different problems. Moreover, due to the demand for numerous iterations, they consume a lot of time to find satisfactory solutions, especially when dealing with large-scale problems.

B. Luo et al. (Eds.): ICONIP 2023, LNCS 14448, pp. 119–130, 2024.
https://doi.org/10.1007/978-981-99-8082-6_10

In the past ten years, deep learning has achieved significant success in many fields, such as computer vision and natural language processing. Researchers have also proposed many deep learning-based (DL-based) methods to solve CO problems. These methods can partially address the challenges faced by traditional methods. They can be categorized into two main classes: learn-to-construct methods and learn-to-improve methods. The former train a deep neural network (DNN) to construct a solution from scratch step by step, while the latter train the DNN to iteratively improve the current solution. Due to their offline training nature, DL-based methods can save a lot of time during the inference process. Additionally, models trained by these methods can be easily modified to solve various types of CO problems, including TSP, CVRP, and Knapsack Problems.

However, DL-based methods require training different models for problems of various sizes, and the training process can be very time-consuming as the problem scale increases. Additionally, extracting features from long sequence inputs with existing techniques is challenging, which leads to a drastic decrease in terms of solution quality when the number of cities exceeds 500. Furthermore, since DNNs are primarily trained on GPUs and GPU memory is often limited, machines may run out of memory when training DNNs for large-scale problems.

To address these challenges, we propose a two-stage attention model (TSAM) that combines two different attention models using a divide-and-conquer strategy to solve large-scale TSPs. The primary contributions of our paper are as follows.

Firstly, two distinct end-to-end models are designed and trained to solve the large-scale TSP hierarchically. One model follows the architecture proposed by Kool et al. [5] to solve TSPs, while the other model is specifically modified to address open-loop TSPs with fixed source and target nodes.

Additionally, an efficient approach is developed to connect the sub-paths generated from the second stage that minimizes the distance between each sub-path. As a result, this strategy generates a complete solution for large-scale TSPs.

Furthermore, a novel strategy is introduced to enhance the quality of the solution. This approach randomly samples one solution into multiple sub-tours and subsequently reconstruct these sub-tours to build a new solution.

It is worthwhile to mention that the traditional POPMUSIC algorithm [6] shares a similar concept with ours, which leverages a divide-and-conquer method to solve large-scale TSPs.

2   Related Works

Existing methods for solving TSPs can be classified as exact methods, heuristic methods, and DL-based methods. In this section, we provide a concise overview of DL-based methods, which can be further categorized as learn-to-construct methods and learn-to-improve methods.

2.1 Learn-to-Construct

Vinyals et al. [7] introduced the Pointer Network (PN) to solve TSP for the first time. The PN utilizes an encoder-decoder architecture that consists of several LSTM layers. The encoder computes the embedding of each node, and the decoder generates the probability distribution of the nodes to be visited at each time step.

Some researchers have discovered the efficiency of reinforcement learning in solving the TSP. Bello et al. [8] suggested training the PN with the asynchronous advantage actor-critic (A3C) [9] without supervised solution labels. Additionally, they designed an active search strategy to enhance the solution quality further.

The success of the Transformer architecture inspires some researchers to employ it to solve TSP. Deudon et al. [10] and Kool et al. [5] concurrently proposed an attention model (AM) that leverages the Transformer architecture introduced by Vaswani et al. [11]. Kool et al. trained the network using the REINFORCE algorithm with the rollout baseline instead of the actor-critic to decrease the number of parameters and reduce gradient variance.

Other researchers have attempted to leverage some intrinsic characteristics to improve the performance of AM. Kwon et al. [12] proposed the POMO model with the shared baseline that exploits the solution symmetry in TSP. The model first samples trajectories with multiple nodes as starting nodes and utilizes the average of these trajectories as the baseline. In addition, they leveraged an instance augmentation [13] to further enhance the solution quality. Subsequently, Kim et al. [14] designed a novel loss function to exploit the symmetry in TSP.

Combining a Graph Neural Network (GNN) with supervised learning enables generating probabilities for each edge to be included in the optimal solution. Joshi et al. [15] leveraged a Graph Convolutional Network (GCN) to represent the feature of the graph. This GCN-based network generates a heatmap indicating the probabilities of edges being part of the optimal tour. Then a beam search strategy is employed to construct a promising solution based on the heatmap.

2.2 Learn-to-Improve

The process of local search can be learned using the PN. Wu et al. [16], Costa et al. [17] and Ma et al. [18] leveraged the Transformer [11] architecture to rewrite the solution through iterative improvement. The network learns to generate the probability distribution of node pairs to be rewritten by a specific local operator. Besides, Chen et al. [19], Gao et al. [20], and Lu et al. [21] designed the PN-based models to solve CVRP and achieve satisfactory solution quality.

The GNN has the potential to significantly improve the effectiveness of heuristic methods. Some researchers have explored the integration of GNNs into existing methods, such as the LKH algorithm, by replacing the traditional α-value with the output of the GNN. Xin et al. [22] introduced the NeuroLKH model that trains a Sparse Graph Network (SGN) to enhance the α-value employed in the LKH algorithm. By training the SGN, the model generates edge scores and node penalties that provide valuable guidance to the LKH algorithm, ultimately improving its performance. Besides, Kool et al. [23] employed the GNN to guide the dynamic programming algorithm.

Notably, Fu et al. [24] leveraged the GCN-based model introduced by Joshi et al. [15] and Monte Carlo Tree Search (MCTS) [25] to solve large-scale TSPs and achieved promising performance on 10,000-scale TSP. They designed a sampling-and-merging strategy that leverages the concept of divide-and-conquer to construct a heatmap. The MCTS algorithm learns to improve a feasible solution based on the k-opt algorithm and the heatmap.

3   Methods

In this section, we will introduce the TSAM method in three subsections. Firstly, as presented in Subsect. 3.1, we train the two-stage attention model to solve TSP and open-loop TSP separately. Secondly, as shown in Subsect. 3.2, we devise a strategy for constructing complete solutions. Thirdly, as shown in Subsect. 3.3, we develop a sampling strategy to improve the quality of the solution. The pipeline of our method is illustrated in Fig. 1.

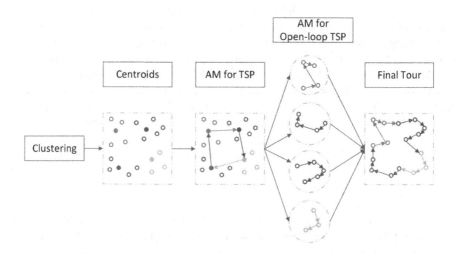

Fig. 1. The pipeline of two-stage attention model.

3.1   Two-Stage Attention Model

The large-scale TSP is divided into two sub-problems: TSP and open-loop TSP. The models for solving TSP are introduced by Kool et al. [5] and Kwon et al. [12]. In this section, we pay attention to the method for solving open-loop TSP and give a detailed description of the model.

Open-Loop TSP. The open-loop TSP is a variant of TSP where the salesman is required to visit a set of cities without returning to the starting city. In this problem, the salesman begins the tour at a designated city, travels to each city exactly once, and concludes the tour at a different specified city.

Attention Model. The AM contains an encoder and a decoder architecture. The encoder can be regarded as a Graph Attention Network (GAT) [26] that contains several Multi-head Attention (MHA) layers and Feed-forward (FF) layers to calculate the embeddings of all nodes. The node embeddings represent information about each node and its relevance with all neighbors. The self-attention mechanism, introduced by Vaswani et al. [11], is the essential element in the MHA. The self-attention is calculated as follows.

$$Q_i^m = W_Q^m X_i, K_i^m = W_K^m X_i, V_i^m = W_V^m X_i \tag{1}$$

$$MHA(Q, K, V) = concat(softmax\left(\frac{Q^m K^{mT}}{\sqrt{d_k}}\right) V^m) W_O \tag{2}$$

where $m \in \{1, 2, \ldots, M\}$ denotes the number of heads in the MHA.

The skip-connection [27] and batch normalization [28] are added to alleviate gradient vanishing and explosion.

$$\hat{h}_i = BN(X_i + MHA(Q_i, K_i, V_i)), h_i = BN(\hat{h}_i + FF(\hat{h}_i)) \tag{3}$$

The decoder uses the attention mechanism to compute the attention weights between the node embeddings and the context embedding, then generates the solution sequence using a greedy strategy. In the decoder, the origin context embedding for solving TSP contains the embedding of the graph, the first visited node π_1 and the last (previous) visited node π_{t-1}, v^1 and v^f are used as placeholders for input when $t = 1$:

$$h_{context} = \begin{cases} [h_{graph}, h_{\pi_1}, h_{\pi_{t-1}}], & t > 1 \\ [h_{graph}, v^1, v^f], & t = 1 \end{cases} \tag{4}$$

where $h_{graph} = \bar{h} = \frac{1}{N}(\sum_{i=1}^N h_i)$.

The context embedding in the decoder is modified for solving open-loop TSP by adding the embedding of the source and target node:

$$h_{context} = \begin{cases} [h_{graph}, h_{source}, h_{\pi_1}, h_{\pi_{t-1}}, h_{target}], & t > 1 \\ [h_{graph}, h_{source}, v^1, v^f, h_{target}], & t = 1 \end{cases} \tag{5}$$

Training Scheme. Our method follows the REINFORCE algorithm with the shared baseline, which is introduced by Kwon et al. [12]. First, the algorithm samples $N - 2$ trajectories with distinct starting nodes after visiting the source node. Then, the algorithm calculates the average reward across all these trajectories and employs it as the shared baseline $\bar{b}(s)$. Give the trajectories $\{\tau^1, \tau^2, \ldots, \tau^{N-2}\}$, The loss J is optimized using gradient ascent:

$$\nabla_\theta J(\theta) = E_{p_\theta(\tau^i|s)}(R(\tau^i) - \bar{b}(s))\nabla_\theta logp_\theta(\tau^i|s) \tag{6}$$

The shared baseline is calculated as $\bar{b}(s) = \frac{1}{N-2}\sum_{i=1}^{N-2} R(\tau^i)$.

3.2 Solution Construction

A hierarchical architecture is devised to construct the final solution for the large-scale TSP. This architecture consists of clustering, graph normalization, endpoint selection, and tour construction. Algorithm 1 depicts the whole process of solution construction.

Algorithm 1. Solution Construction

    Input: coordinates of cities $V = \{v_1, v_2, \ldots, v_N\}$, number of clusters K.
    Output: solution $\pi = \{\pi_1, \pi_2, \ldots, \pi_{N-2}\}$.
1: $Centroids \leftarrow Clustering(V, K)$.
2: $Tour \leftarrow AM1(Centroids)$.
3: $Subsets \leftarrow EndpointsSelection(Centroids, Tour, K)$.
4: $\pi \leftarrow \{\}$.
5: **for** $i = 1$ to K **do**
6:    **if** $len(Subset_i) \leq 2$ **then**
7:       add $Subset_i$ to π.
8:    **else**
9:       $Subpath \leftarrow AM2(Subsets_i)$
10:      add $Subpath$ to π.
11:   **end if**
12: **end for**
13: **return** π

Clustering. The k-means algorithm is employed to partition the nodes into multiple clusters. The algorithm begins by randomly initializing K centroids and then computes the distance between each node and the centroids. Subsequently, it assigns the nodes to their nearest clusters. The centroids are updated by calculating the average coordinates of nodes assigned to each cluster until convergence is reached. To avoid empty sub-paths, a random point will be reassigned to each empty cluster.

Graph Normalization. Since the model is trained using randomly generated instances with a uniform distribution ranging from 0 to 1, normalization is crucial for each cluster to generate promising sub-solutions. The coordinates of cities in each cluster can be transformed from (x_i, y_i) to (x_i^{new}, y_i^{new}) through Min-Max Normalization.

$$x_i^{new} = \frac{x_i - x_{min}}{x_{max} - x_{min}}, y_i^{new} = \frac{y_i - y_{min}}{y_{max} - y_{min}} \tag{7}$$

where $x_{min} = \min_{i \in C} x_i$, $x_{max} = \max_{i \in C} x_i$, $y_{min} = \min_{i \in C} y_i$, $y_{max} = \max_{i \in C} y_i$.

Endpoints Selection. To construct the complete solution, the decision on which two endpoints to connect between consecutive clusters is crucial. For endpoint selection, our method starts by calculating the distances between each point in the two clusters. Subsequently, the two points that are closest to each other are chosen as the endpoints for the two clusters. Algorithm 2 describe the method of endpoints selection.

Algorithm 2. Endpoints Selection

    Input: initial subsets of nodes *Subsets*, tour of centroids *Tour*, number of clusters K.

    Output: subsets with selected endpoints $Subsets = \{Subset_1, \dots, Subset_K\}$

1: sort *Subsets* by *Tour*.
2: **for** $i = 1$ to K **do**
3:     $j \leftarrow (i+1) \bmod K$.
4:     calculate the distance matrix D_{ij} of nodes between $Subset_i$ and $Subset_j$.
5:     find the node $\{a, b\}$ with minimum value of D_{ij}.
6:     move a to the end of $Subset_i$.
7:     move b to the begin of $Subset_j$.
8: **end for**
9: **return** *Subsets*

Tour Construction. After obtaining the tour of the centroids, the endpoints of each cluster, and the solutions for open-loop TSPs, our method connects the subpaths to construct a complete solution for the large-scale TSP. It is worthwhile to note that the AM for open-loop TSP is applicable only to input instances with more than two nodes, as the source and target nodes must be included. For clusters containing no more than two nodes, the nodes are directly added to the final solution.

3.3 Improving Strategy

Initially, a complete tour is divided into sub-tours of equal size by randomly selecting a starting index. Subsequently, these sub-tours are reconstructed using the model trained for open-loop TSP. Finally, the worse sub-tours from the original solution are replaced with the reconstructed sub-tours. This process can be repeated for a specified number of iterations. By employing this approach, the incorrect ordering within the sub-tours is effectively rectified, resulting in a shorter final tour.

4 Experiments

Our experiments were conducted on TSP instances with 500, 1000, 2000, 5000, and 10,000 nodes. The results were compared with several learning-based and non-learning methods. To ensure fair comparisons, all methods were executed on a single system comprising an Intel Xeon E5-2640v4 2.4 GHz processor with 20 cores and an Nvidia Tesla V100 GPU with 16 GB of memory.

4.1 Hyper-parameters and Baselines

Comparative experiments were conducted to evaluate the effectiveness of our TSAM method and various baselines, including Concorde solver [1], LKH3 solver [4], AM [5], POMO [12], and Att-GCN+MCTS [24]. For the learning-based methods, the hyper-parameters were set according to their respective papers, and the pre-trained models available online is directly utilized to verify their performance. Since the learn-to-improve method requires multiple iterations which will slower the inference process, to ensure fair comparisons, Att-GCN-MCTS was run in parallel with a batch size of 8, while the other learning-based methods were run sequentially. It should be noted that when $n = 10000$, the number of iterations of the LKH algorithm is set to 1 in order to reduce runtime. Additionally, the beam search width of the AM algorithm is set to 10 in order to prevent out-of-memory issues.

The hyper-parameters of TSAM followed those of POMO. Our TSAM model was trained solely on TSP and open-loop TSP instances with 20, 50, and 100 nodes. Furthermore, a customized combination of parameters was developed to efficiently solve problems of different scales. In addition, to balance solution quality and time consumption, the number of iterations for k-means in TSAM is set to 100.

4.2 Datasets Setup

The training sets contains batches of 2D-Euclidean TSP instances randomly generated under a uniform distribution ranging from 0 to 1. To ensure fair comparisons, the testing sets were customized to evaluate the performance of all methods. Specifically, we generated 128 instances for the TSP with $n = 500, 1000$, and 16 instances for the TSP with $n = 2000, 5000, 10000$, respectively.

4.3 Comparative Study

Table 1 shows the absolute results and the relative optimality gap compared to the Concorde solver, for TSAM and the baselines on 128 TSP instances with $n = 500, 1000$. Table 2 presents the absolute results and the relative optimality gap compared to the LKH3 solver on 16 TSP instances with $n = 2000, 5000, 10000$. Len. denotes the length of the tour, Gap is defined as $(Len. - Opt.)/Opt. \times 100\%$, and $Time$ indicates the total runtime of all instances. To prevent out-of-memory issues, we excluded the POMO with ×8 augmentation and AM with sampling in Table 2, and reduced the beam search width to 10 in AM when $n = 10000$.

As illustrated in Table 1, the Concorde solver generates optimal solutions for TSP with 500 and 1000 nodes, while Att-GCN+MCTS achieves the shortest tours compared to other DL-based methods. Among all the DL-based methods, TSAM exhibits slightly lower solution quality than Att-GCN+MCTS, but it significantly reduces the runtime, amounting to only one-fifth of the runtime of Att-GCN+MCTS. Table 2 reveals that the LKH3 solver produces optimal solutions for TSP with 2000, 5000, and 10,000 nodes, while Att-GCN+MCTS generates the best solutions among all DL-based methods. TSAM achieves the shortest runtime among all methods, taking only 1% of the runtime compared to LKH3, while maintaining an optimality gap of approximately 12%. Additionally, the improving strategy reduces the gap to about 8% without significant increases in runtime.

It is worthwhile to note that this paper does not aim to directly surpass exact and heuristic solvers in terms of solution quality. Instead, our focus is on efficiently obtaining promising solutions. Although Att-GCN+MCTS, LKH3, and Concorde outperform TSAM in terms of solution quality, TSAM is tens to hundreds of times faster than the former. The experimental results show that TSAM can achieves solutions with a small optimality gap while maintaining a minute-level runtime.

Table 1. Our results w.r.t. the baselines, tested on 128 instances respectively with $n = 500, 1000$. The gap % is w.r.t. the Concorde across all methods.

Method	Type	500			1000		
		Len.	Gap	Time	Len.	Gap	Time
Concorde	Exact	**16.0298**	**0.0000%**	45.87 min	**22.0078**	**0.0000%**	14.74h
LKH3 (10 runs)	Heuristic	16.4891	2.8656%	21.64 min	23.1205	5.0560%	2.02 h
AM (sampling)	Construct	22.5755	40.8351%	3.23 min	42.7149	94.0896%	11.17 min
AM (beam 100)	Construct	19.4722	21.4754%	1.68 min	29.9425	36.0542%	4.85 min
POMO (×8 aug.)	Construct	20.1439	25.6656%	**1.60 min**	32.5202	47.7666%	10.43 min
Att-GCN+MCTS	Improve	16.8917	5.3770%	11.73 min	23.8957	8.5781%	24.11 min
TSAM	Construct	18.6336	16.2435%	2.41 min	26.2303	19.1862%	**4.71 min**
TSAM(impr.20)	Construct	17.9521	11.9920%	3.14 min	25.3531	15.2005%	5.63 min

Table 2. Our results w.r.t. the baselines, tested on 16 instances respectively with $n = 2000, 5000, 10000$. The gap % is w.r.t. the LKH3 across all methods.

Method	Type	2000			5000			10000		
		Len.	Gap	Time	Len.	Gap	Time	Len.	Gap	Time
LKH3 (10 runs)	Heuristic	**32.3674**	**0.0000%**	1.35h	**50.9608**	**0.0000%**	10.85h	**71.7799**	**0.0000%**	8.47h*
AM (beam 100)	Construct	45.9695	42.0241%	2.30 min	82.2582	61.4149%	20.97 min	130.8284	82.2632%	16.87 min*
POMO (no aug.)	Construct	50.9726	57.4811%	1.47 min	88.3271	73.3238%	19.50 min	–		
Att-GCN+MCTS	Improve	33.5982	3.8024%	6.44 min	53.0587	4.1168%	21.46 min	74.6946	4.0605%	1.53h
TSAM	Construct	36.5634	12.9636%	**49.18 s**	57.4219	12.6788%	**1.47 min**	80.4318	12.0533%	**2.59 min**
TSAM(impr.50)	Construct	34.9274	7.9092%	1.50 min	55.0023	7.9306%	3.63 min	78.2354	8.9935%	13.05 min

* For n=10000, the LKH3's number of iterations is set to 1, and the beam size of AM is set to 10.

4.4 Ablation Study

The improving strategy can further enhance the solution quality of TSAM. However, as the number of iterations increases, the runtime will also increase.

Therefore, we conduct experiments to compare the effects of improving strategy with no iteration, 20 iterations, and 50 iterations. According to the results in Table 3, we can select the appropriate number of iterations for TSAM. When $n = 500, 1000$, Impr.20 provides promising solutions with a shorter runtime compared to Impr.50. When $n = 2000, 5000, 10000$, Impr.50 achieves smaller gaps than Impr.20, and the runtime is acceptable.

Table 3. The gaps when using the improving strategy with the different number of iterations.

Size	No impr.	Impr.20	Impr.50
500	16.24%	11.99%	11.97%
1000	19.19%	15.20%	15.07%
2000	12.96%	8.14%	7.91%
5000	12.68%	8.48%	7.93%
10000	12.05%	9.88%	8.99%

5   Conclusions

In this paper, we present a two-stage attention model designed to address large-scale TSPs. The model contains two stages where the first stage focuses on solving a smaller-scale TSP, while the second stage tackles a group of open-loop TSPs. Then we devise an effective approach to construct the final solution. Finally, we apply an improving strategy to enhance the solution quality. Experimental results demonstrate that TSAM can generate promising solutions without a significant increase in runtime as the problem scale increases. The method is well-suited for real-time applications in real-world scenarios.

In the future, we will work on enhancing solution quality when solving large-scale TSPs. Furthermore, we will dedicate more effort to exploring how to adapt TSAM to address various CO problems.

Acknowledgment. This work is supported by the National Nature Science Foundation of China [Grant Nos. 62173258, 61773296].

References

1. Applegate, D., Bixby, R., Chvatal, V., Cook, W.: Concorde tsp solver (2006)
2. Applegate, D.L., et al.: Certification of an optimal tsp tour through 85,900 cities. In: Operations Research Letters, pp. 11–15 (2009)
3. Helsgaun, K.: General k-opt submoves for the lin-kernighan tsp heuristic. Math. Program. Comput. 1, 119–163 (2009). https://doi.org/10.1007/s12532-009-0004-6

4. Helsgaun, K.: An Extension of the lin-kernighan-helsgaun TSP Solver for Constrained Traveling Salesman and Vehicle Routing Problems, Roskilde University, Roskilde, pp. 1–60 (2017)
5. Kool, W., van Hoof, H., Welling, M.: Attention, learn to solve routing problems! In: Proceedings of the 7th International Conference on Learning Representations, pp. 1–25 (2019)
6. Taillard, É. D., Helsgaun, K.: Popmusic for the travelling salesman problem. Eur. J. Oper. Res. **272**, 420–429 (2019)
7. Vinyals, O., Fortunato, M., Jaitly, N.: Pointer networks. In: Proceedings of the 28th International Conference on Neural Information Processing Systems, pp. 2692–2700 (2015)
8. Bello, I., Pham, H., Le, Q.V., Norouzi, M., Bengio, S.: Neural combinatorial optimization with reinforcement learning, arXiv preprint arXiv:1611.09940 (2016)
9. Mnih, V., et al.: Asynchronous methods for deep reinforcement learning. In: International Conference on Machine Learning, pp. 1928–1937 (2016)
10. Deudon, Michel, Cournut, Pierre, Lacoste, Alexandre, Adulyasak, Yossiri, Rousseau, Louis-Martin.: Learning heuristics for the TSP by policy gradient. In: van Hoeve, Willem-Jan. (ed.) CPAIOR 2018. LNCS, vol. 10848, pp. 170–181. Springer, Cham (2018). https://doi.org/10.1007/978-3-319-93031-2_12
11. Vaswani, A., et al.: Attention is all you need. In: Proceedings of the 31st International Conference on Neural Information Processing Systems, pp. 6000–6010 (2017)
12. Kwon, Y.D., Choo, J., Kim, B., Yoon, I., Gwon, Y., Min, S.: Pomo: policy optimization with multiple optima for reinforcement learning. In: Proceedings of the 34th Conference on Neural Information Processing Systems, pp. 21 188–21 198 (2020)
13. Gidaris, S., Singh, P., Komodakis, N.: Unsupervised representation learning by predicting image rotations, arXiv preprint arXiv:1803.07728 (2018)
14. Kim, M., Park, J., Park, J.: Sym-NCO: leveraging symmetricity for neural combinatorial optimization. In: Proceedings of the 36th Conference on Neural Information Processing Systems, pp. 1936–1949 (2022)
15. Joshi, C.K., Laurent, T., Bresson, X.: An efficient graph convolutional network technique for the travelling salesman problem, arXiv preprint arXiv:1906.01227 (2019)
16. Wu, Y., Song, W., Cao, Z., Zhang, J., Lim, A.: Learning improvement heuristics for solving routing problems. IEEE Trans. Neural Netw. Learn. Syst. **33**(9), 5057–5069 (2021)
17. d O Costa, P.R., Rhuggenaath, J., Zhang, Y., Akcay, A.: Learning 2-opt heuristics for the traveling salesman problem via deep reinforcement learning. In: Asian Conference on Machine Learning, pp. 465–480 (2020)
18. Ma, Y., Li, J., Cao, Z., Song, W., Zhang, L., Chen, Z., Tang, J.: Learning to iteratively solve routing problems with dual-aspect collaborative transformer. In: Proceedings of the 35th International Conference on Neural Information Processing Systems, pp. 11 096–11 107 (2021)
19. Chen, X., Tian, Y.: Learning to perform local rewriting for combinatorial optimization. In: Proceedings of the 33rd International Conference on Neural Information Processing Systems, pp. 6281–6292 (2019)
20. Gao, L., Chen, M., Chen, Q., Luo, G., Zhu, N., Liu, Z.: Learn to design the heuristics for vehicle routing problem, arXiv preprint arXiv:2002.08539 (2020)

21. Lu, H., Zhang, X., Yang, S.: A learning-based iterative method for solving vehicle routing problems. In: Proceedings of the 8th International Conference on Learning Representations, pp. 1–15 (2020)
22. Xin, L., Song, W., Cao, Z., Zhang, J.: Neurolkh: combining deep learning model with lin-kernighan-helsgaun heuristic for solving the traveling salesman problem. In: Proceedings of the 35th International Conference on Neural Information Processing Systems, pp. 7472–7483 (2021)
23. Kool, W., van Hoof, H., Gromicho, J., Welling, M.: Deep policy dynamic programming for vehicle routing problems. In: Proceedings of Integration of Constraint Programming, Artificial Intelligence, and Operations Research: 19th International Conference, pp. 190–213 (2022)
24. Fu, Z.-H., Qiu, K.-B., Zha, H.: Generalize a small pre-trained model to arbitrarily large tsp instances. In: Proceedings of the AAAI Conference on Artificial Intelligence, pp. 7474–7482 (2021)
25. Chaslot, G. M. J.-B. C.: Monte-carlo tree search (2010)
26. Veličković, P., Cucurull, G., Casanova, A., Romero, A., Liò, P., Bengio, Y.: Graph attention networks. In: Proceedings of the 6th International Conference on Learning Representations, pp. 1–12 (2018)
27. He, K., Zhang, X., Ren, S., Sun, J.: Deep residual learning for image recognition. In: Proceedings of the IEEE Conference on Computer Vision and Pattern Recognition, pp. 770–778 (2016)
28. Ioffe, S., Szegedy, C.: Batch normalization: accelerating deep network training by reducing internal covariate shift. In: Proceedings of the International Conference on Machine Learning, pp. 448–456 (2015)

Learning Primitive-Aware Discriminative Representations for Few-Shot Learning

Jianpeng Yang[1], Yuhang Niu[1], Xuemei Xie[1,2], and Guangming Shi[1(✉)]

[1] School of Artificial Intelligence, Xidian University, Xian, China
yangjp@stu.xidian.edu.cn, gmshi@xidian.edu.cn
[2] Guangzhou institute of technology, Xidian University, Xian, China
xmxie@mail.xidian.edu.cn

Abstract. Few-shot Learning (FSL) aims to learn a classifier that can be easily adapted to recognize novel classes with only a few labeled examples. Recently, some works about FSL have yielded promising classification performance, where the image-level feature is used to calculate the similarity among samples for classification. However, the image-level feature ignores abundant fine-grained and structural information of objects that could be transferable and consistent between seen and unseen classes. How can humans easily identify novel classes with several samples? Some studies from cognitive science argue that humans recognize novel categories based on primitives. Although base and novel categories are non-overlapping, they share some primitives in common. Inspired by above research, we propose a Primitive Mining and Reasoning Network (PMRN) to learn primitive-aware representations based on metric-based FSL model. Concretely, we first add Self-supervision Jigsaw task (SSJ) for feature extractor parallelly, guiding the model encoding visual pattern corresponding to object parts into feature channels. Moreover, to mine discriminative representations, an Adaptive Channel Grouping (ACG) method is applied to cluster and weight spatially and semantically related visual patterns to generate a set of visual primitives. To further enhance the discriminability and transferability of primitives, we propose a visual primitive Correlation Reasoning Network (CRN) based on Graph Convolutional network to learn abundant structural information and internal correlation among primitives. Finally, a primitive-level metric is conducted for classification in a meta-task based on episodic training strategy. Extensive experiments show that our method achieves state-of-the-art results on miniImageNet and Caltech-UCSD Birds.

Keywords: Few-shot Learning · Visual Primitive · Graph Convolution · Episodic Training

---

(a) The preprint of this paper: Yang J, Niu Y, Xie X, Shi G. Learning Primitive-aware Discriminative Representations for Few-shot Learning[J]. arXiv preprint arXiv:2208.09717, 2022. https://arxiv.org/abs/2208.09717.

1 Introduction

In recent years, deep learning (DL) has achieved tremendous success in various recognition tasks with abundant labeled data. To train supervised model efficiently, we need lots of annotated samples that are expensive and time-consuming to obtain. Therefore, how to recognize novel classes with few labeled samples has attracted more attention. To reduce the reliance on human annotation, Few-shot Learning (FSL) has been proposed and studied widely [1–3], which aims to learn a classifier that can be rapidly adapted to novel classes given just several labeled images per class.

FSL attempts to transfer knowledge acquired from base classes with sufficient labeled data to novel classes with only several examples. Humans can rapidly classify an object into one of several novel classes by recognizing the differences between them. Inspired by this ability of humans, a series of Few-shot Learning methods adopt metric-based algorithm [2, 3], which learns a global image-level representation in an appropriate feature space and directly calculates the distances between the query and support images for classification. Nevertheless, most of them measure similarity on image-level feature or feature after the pooling operation for classification, which destroys the structure of objects and ignore local clues.

Revisiting the process that humans recognize new concepts or objects, humans can first learn primitives from plenty of known classes and then apply them to identify novel ones. In practice, primitives are viewed as object parts, or regions capturing the compositional structure of the examples [9], whose the boundary is vague and unclear. Some researches on interpretability of deep networks show that CNN feature channels often correspond to some visual patterns. Inspired by above studies, CPDE [9] selects a single channel of image feature as a primitive by soft composition mechanism. However, this method learns primitives by directly feeding them to general classifier, which is composed of fully connected layers (FC) and softmax, which makes primitive lack of generalization in few-shot scenario.

Compared to these methods, we thereby compose primitives via selecting feature channels related to object parts rather than a single channel, which aggregates visual pattern that is coherent on semantic and spatial relation. Moreover, to the best of our knowledge, the local representations based methods [5–8, 10] lose sight of internal correlation among them and structural information of objects. Therefore, we propose to learn and encode internal correlation among visual primitives to improve discriminability of primitive-aware representation.

In this paper, we propose a Primitive Mining and Reasoning Network (PMRN) to learn discriminative and transferable primitive-aware representations for metric-based FSL model based on episodic training mechanism. The main contributions of our work are summarized as follows:

- We develop a Primitive Mining Network (PMN) to learn visual primitives. Firstly, this network guides feature extractor to encode visual patterns related to object parts into feature channels by adding a special Self-supervision Jigsaw (SSJ) task parallelly. Then, it weights and clusters feature channels that are consistent on semantics and spatial location to generate a set of visual primitives through an Adaptive Channel Grouping (ACG) module.

- To capture structural information and inter correlation among primitives, we propose a Correlation Reasoning Network (CRN) to jointly learn the semantic and spatial dependency among primitives by constructing a graph on visual primitives and conduct convolution reasoning operation.
- We design a Task-specific Weight method aiming at measuring the importance of primitives in a metric-based few-shot task, which adaptively generates task-specific weight for weighting primitive-level metric-based classification. To enhance algorithm's generalization across task, all the modules are embedded into meta-task to simulate few-shot scenario by episodic training mechanism.
- We conduct experiments on miniImageNet and Caltech-UCSD Birds and the results reveal the effectiveness of our methodology.

2   Related Work

Few-Shot Learning (FSL). Most recent literature about few-shot learning mainly involve two types of research methods, metric-based and meta-learning based methods.

The meta-learning based methods [1, 26] optimize a meta-learner utilizing learning-to-learn paradigm, which can rapidly adapt to novel classes with just few samples for FSL. An external memory module is employed by [26] to communicate with an LSTM-based meta learner to update weights. MAML [1] and some variants aim to learn a better parameter initialization that can be quickly adapted to a novel task.

The metric-based methods [2–4] learn the feature representations for input samples in an appropriate embedding space, where the similarity between images is calculated among different classes through diverse distance metrics. The application of Metric-based method to few-shot learning is firstly proposed by Koch [27], which aims to generalize representations to novel classes through a Siamese Neural Network. MatchingNet [2] utilizes episodic training strategy and selects cosine similarity as metric to solve FSL problem. ProtoNet [3] regards the mean value of each class's embedding as prototype and calculates Euclidean distance between support and query samples for classification. The proposed PMRN belongs to metric-based methods, but our method adaptively mines and exploits potential local representations related to object parts, which is more discriminative and transferable among the base and novel categories than the image-level representations utilized by above methods.

Dense Local Feature Based FSL. In contrast to previous methods, some FSL work [5–8] focus on local representations and try to exploit the discriminative ability of local patches.

Specifically, the local patch is considered as each spatial grid in the feature map and all the patch-level distance is aggregated as result. DN4 [5] introduces Naive-Bayes Nearest Neighbor into FSL and computes image-level similarity via a k-nearest neighbor search over local patches. ATL-Net [7] proposes to adaptively select important semantic patches by an episodic attention mechanism. DeepEMD [6] conducts a many-to-many matching method among local patches via the earth mover's distance. MCL [8] consider the mutual affiliations between the query and support features to thoroughly affiliate two disjoint sets of dense local features.

Self-Supervised Learning (SSL). Manual labels are expensive and time-consuming to collect in practice. SSL aims at learning representations from structural information in the object itself without label. Recently, some work [9, 11] introduce SSL methods such as predicting the rotation and the relative position to FSL. In this paper, we propose to encourage the backbone to learn visual patterns related to object parts by using SSL loss as a regularizer.

3 Method

In this section, we first introduce the general settings of few-shot learning method. Then we introduce our proposed Primitive Mining and Reasoning Network (PMRN) concretely, and it is composed of Primitive Mining Network (PMN) in Sect. 3.2 and visual primitive Correlation Reasoning Network (CRN) in Sect. 3.3.

3.1 Description and Formulation of Few-Shot Learning

In few-shot learning scenario, three sets of data are provided, support set S. , query set Q, and an auxiliary set B. The support set S contains N novel classes, and each class has K samples. FSL aims at recognizing an unlabeled query sample $q \in Q$ into one of the N novel classes of S, and we call such a task as an N-way K-shot task. However, the support set S only has several labeled samples per class, and an auxiliary set B is employed to train the model thorough episodic training mechanism for learning transferable knowledge. In episodic training, the auxiliary set B is divided into many N-way K-shot tasks T randomly (also called episodes), where each T contains an auxiary support set B_S and an auxiliary query set B_Q. Each B_Q contains M samples per class. During the training process, the model is forced to conduct hundreds of tasks T to simulate the test scenario, in order to obtain the generalization ability across tasks and learn transferable knowledge that can be used in new N-way K-shot tasks. Note that auxiliary the set B contains abundant labeled classes and samples, but its label space is disjoint with the set S and Q.

3.2 Primitive Mining Network

As introduced in Sect. 1, our proposed visual primitive shows its discriminability and transferability in few-shot learning. So, we replace general local representation with visual primitive in metric-based classification and design a Primitive Mining Network (PMN) to learn visual primitive. The detail of model is shown in Fig. 1.

Self-Supervision Jigsaw Task. To assist the primitive mining operation, we adopt a special self-supervision jigsaw task loss to FSL model as a regularizer, which encourage model to recognize related location of image patches by learning visual pattern corresponding to object parts. Note that self-supervision auxiliary task does not need any extra part annotation.

Concretely. we first divide an input image $x \in B = \{(x_i, y_i)\}_i^n$ into $h \cdot w$ patches along rows and columns. Afterwards, these patches are permuted randomly as input x^p and we get index of the permutation as the target label y^p. The train goal is to predict each

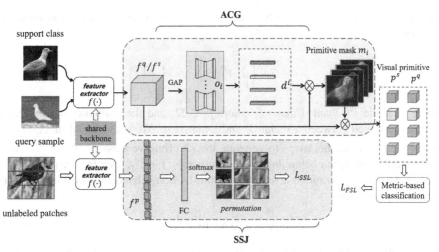

Fig. 1. The architecture of Primitive Mining Network (PMN). It is composed of a Self-supervision Jigsaw (SSJ) task and an Adaptive Channel Grouping (ACG) module.

index of permutation for patches of an image. Then, all permuted patches are fed into feature extractor $f(\cdot)$ to obtain $h \cdot w$ features and we concatenate the permuted features. Finally, a FC layer for classification with a cross-entropy loss can be utilized to train the model. Let's denote the concatenated features as f^p, and the classification loss can be formulated as the negative log-probability:

$$L_{ssl} = - \sum_{x \in B} \log p(y^p | f^p) \qquad (1)$$

where $f^p \in R^{h \cdot w \cdot d}$, $h \cdot w$ is the number of patches and d is the dimension of feature map. In practice, $h \cdot w$ is set as 3×3 and the number of index reach $9!$.

Adaptive Channel Grouping Module. ACG synthesizes a set of visual primitives from these feature maps adaptively by clustering and weighting a set of spatially and semantically consistent visual patterns encoded in feature channels.

In each episode, all the images from support and query set $B_S = \left\{ \left(x_j^s, y_j^s \right) \right\}, j \in [1, NK]$, $B_Q = \left\{ \left(x_j^q, y_j^q \right) \right\}, j \in [1, NM]$ are fed into feature extractor $f(\cdot)$ to obtain a collection of features $\Omega = \{f_i^s\}_i^{NK} \cup \{f_i^q\}_i^{NM}$, and the dimension of these feature is $H \times W \times C$, where H, W, C indicate height, width, and the number of feature channels. Afterwards, a set of parallel channel grouping operations are employed to weight and cluster feature into k groups of feature channels to generate k primitives $P = \{p_1, p_2, \ldots p_k\}$ for each sample in Ω, and these operations are denoted as $O = \{o_1, o_2, \ldots, o_k\}$ Therefore, each channel grouping operation o_i is responsible for the generation of a set of weights for each primitive:

$$D_i = o_i(f), i \in [1, k] \qquad (2)$$

where $f \in \Omega$, and D_i is a set of weights $\left[d_1^i, d_2^i, \cdots d_c^i \right]$ for the generation of ith primitive p_i. After that, we separately employ each set of weights to cluster all the channels into

k groups spatially and semantically consistent channels and obtain k primitive masks as follows:

$$m_i = \sigma \left(\sum_{j=1}^{c} d_j^i . a_j \right), i \in [1, k] \tag{3}$$

where a_j is the jth channel map of feature $f \in \Omega$, and the multiplication operation here is conducted between a scalar d_j^i and a matrix a_j. We use sigmoid function $\sigma(\cdot)$ on the weighted channel maps to generate the ith primitive mask $m_i \in R^{H \cdot W}$, which covers activation region belonging to i^{th} primitive p_i. Finally, each feature $f \in R^{H \cdot W \cdot C}$ is filtered through k primitive-level attention masks along channel dimension to acquire the initial primitives:

$$p_i = f \otimes m_i, i \in [1, k] \tag{4}$$

where $\otimes$ denotes the element wise multiplication between m_i and each channel of f.

However, the diversity of primitive can further provide rich information and contribute to robust recognition, especially for occlusion cases. To avoid the duplication of learned primitives, we utilize a channel grouping diversity loss inspired by [10], and it is formulated as:

$$L_{div} = \sum_{i=1, i \neq j}^{k} \sum_{j=1}^{k} \delta \big(g(p_i), g(p_j) \big) \tag{5}$$

where $\delta(\cdot)$ denotes the cosine similarity function between primitives from a sample, and $g(\cdot)$ is the global average pooling operation in the spatial dimension.

3.3 Correlation Reasoning Network

To further improve the discriminability and transferability across task of visual primitive, we propose a visual primitive Correlation Reasoning Network (CRN), which constructs a graph structure on visual primitives and designs a special graph convolutional network for reasoning internal correlation among primitives and structural information. In addition, a Task-specific Weight (TSW) method is applied to measure the importance of primitives in primitive-level metric-based classification.

Visual Primitive Graph Structure. Given initial visual primitives $P = \{p_1, p_2, \ldots p_k\}$ have been obtained through PMN. we construct a graph on visual primitives at first. The illustration of visual primitive graph structure is shown as Fig. 2.

In detail, global average pooling operation is applied to primitives $P = \{p_1, p_2, \ldots, p_k\}$ to get k primitive feature vectors as node embedding $N = \{n_1, n_2, \ldots, n_k\}, n_i \in R^C$. Then, the similarity matrix of node embeddings should be calculated as adjacent matrix of graph, which encodes semantic and spatial correlation between visual primitives. Moreover, the normalized embedded Gaussian function is adopted to calculate the similarity of the two nodes as follows:

$$S(n_i, n_j) = \frac{e^{\theta(n_i)\omega(n_j)^T}}{\sum_{j=1}^{k} e^{\theta(n_i)\omega(n_j)^T}} \tag{6}$$

where N is the total number of the nodes corresponding to primitives. The dot product is used to measure the similarity of the two nodes in an embedding space. To sum up, visual primitive graph can be formulated as $\xi(N, \vartheta)$, in which ϑ denotes connection between nodes.

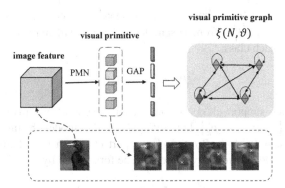

Fig. 2. The illustration of visual primitive graph structure. The primitive vectors are embedded into nodes of graph and internal correlation among primitives are regarded as adjacent matrix.

Adaptive Graph Convolution Layer. According to the characteristics of visual primitive graph, a special adaptive graph convolutional network is designed to update the visual primitive graph and reason semantic and spatial correlation between visual primitives.

Specifically, given input vector $v \in R^{k \cdot c}$, k is the number of nodes and c is the dimension of each node embedding, we first project it into $k \times c_e$ with two projection functions, i.e., $\theta(\cdot)$ and $\varphi(\cdot)$, which are implemented with 1×1 convolutional layer. Then two feature vectors are rearranged and reshaped to an $k \times c_e$ matrix and a $c_e \times k$ matrix separately, which are multiplied to obtain a $K \times K$ similarity matrix C. The element C_{ij} of C represents the similarity of node n_i and n_j, and we conduct softmax operation along the row of matrix for normalization. Based on Eq. (6), the adjacent matrix C can be calculated as follows:

$$C = \sigma\left(\left(vw_\theta^T\right) \otimes \left(w_\varphi v^T\right)\right) \tag{7}$$

where σ is softmax operation, $\otimes$ is matrix computation, w_φ and w_θ are parameters of $\varphi(\cdot)$ and $\theta(\cdot)$ respectively.

The adaptive graph convolutional layer employs node features $v \in R^{k \cdot c}$ and adjacent matrix $C \in R^{k \cdot k}$ as inputs. A single graph convolution layer can be formulated as:

$$v^* = G(v, C) = \varepsilon(Cvw_\omega) \tag{8}$$

where $w_\omega \in R^{c \cdot c^*}$ is the learned weight parameters of 1×1 convolution operation, and $\varepsilon(\cdot)$ is ReLU function. The complete adaptive graph convolutional network based on visual primitives is constructed by stacking multiple graph convolution layers.

Task-Specific Weight Method. For each sample in a task T, a set of transferable and discriminative visual primitives $P = \{p_1, p_2, \ldots, p_k\}$ can be produced through PMN and

CRN. Intuitively speaking, the importance of different primitives in a task is different. So that we propose a task-specific weight (TSW) method to measure importance of visual primitives in a task and calculate primitive-level similarity of support and query set.

For each query sample $x^q \in B_Q$, we can extract a set of query primitives $P^q = \{p_1^q, p_2^q, \ldots, p_k^q\}$, and classify it into one of N support classes by calculating similarity between them. For n^{th} support class, we can get M set of primitives $P^{s,j} = \{p_1^{s,j}, p_2^{s,j}, \ldots, p_k^{s,j}\}_{j=1}^M$ from M samples and average them to get a primitive-level representation for each class:

$$P^s = \frac{1}{M} \sum_{j=1}^M P^{s,j} \tag{9}$$

After that, each support class also have a set of support primitives, which can be denoted as $P^s = \{p_1^s, p_2^s, \ldots, p_k^s\}$. In traditional few-shot model, the final similarity between query sample $x^q \in B_Q$ and c^{th} support class can be calculated simply by summing primitive-level similarity, which can be formulated by:

$$I(x^q, c) = \sum_i^k \varphi\big(g\big(p_i^q\big), g\big(p_i^s\big)\big) \tag{10}$$

where $\varphi(\cdot)$ is metric function, and it is implemented as cosine similarity in this paper. $g(\cdot)$ is global average pooling operation.

As above analysis that contribution of different primitives is discrepant, equally aggregating the primitive-level similarity makes no sense. Therefore, we design a task-specific weight module to adaptively assign appropriate weight for each pair primitive $\{p_i^q, p_i^s\}_{i=1}^k$. Concretely, each primitive in a pair primitive $\{p_i^q, p_i^s\}$ should be compressed into a map to obtain a pair of maps $\{m_i^q, m_i^s\}$. Then we concatenate them along channel dimension as follows:

$$m_i = m_i^q \oplus m_i^s, m_i \in R^{H \cdot W \cdot 2} \tag{11}$$

After that, a task-specific attention feature $F_a \in R^{A \cdot H \cdot W}$ is reconstructed by further concatenating all pairs of maps $\{m_i^q, m_i^s\}_{i=1}^k$. The number of channels in the F_a is A, and $A = 2k$. In contrast to general image-level representation, channels of F_a corresponds to primitive pairs. Thus, we use a weight generator $G(\cdot)$ to measure the importance among different pairs of primitives by learning the importance of different channels of F_a. In this way, a set of task-specific weight $W \in R^k$ for each pair of primitives can be generated as follows:

$$W = sigmoid(\gamma(g(F_a), \gamma(m(F_a))) \tag{12}$$

where $g(\cdot)$ and $m(\cdot)$ are global average pooling and max pooling separately. $\gamma(\cdot)$ consists of two consistent 1×1 convolution layers and a sigmoid function.

A large number of task-specific attention feature F_a is applied to train weight generator $G(\cdot)$ across tasks and it adaptively assign higher weight to significant primitives. Therefore, the final similarity should be adjusted to the weighted sum of primitive-level similarity, and Eq. (10) can be changed as follows:

$$I(x^q, c) = \sum_{i=1}^k W_i \varphi\big(g\big(p_i^q\big), g\big(p_i^s\big)\big) \tag{13}$$

3.4 Loss and Train

For few-shot learning, the final similarity $I(\cdot)$ between query sample x^q and class c can be calculated by the weighted sum of primitive-level similarity. Hence, the probability that each query sample $x^q \in Q = \left\{ \left(x_j^q, y_j^q \right) \right\}, j = 1, 2, \cdots NM$ is classified in class c should be formulated as:

$$p(y_c|x^q) = \frac{\exp(I(x^q,c))}{\sum_{c^s=1}^{N} \exp(I(x^q,c^s))} \tag{14}$$

We compute the classification probability by using a softmax operation and then cross-entropy loss is selected as the few-shot learning loss:

$$L_{cls} = \sum_{x^q \in Q} -\log(p(y_q|x^q)) \tag{15}$$

consisting of L_{cls}, L_{ssl} and L_{div}, the loss of our proposed PMN can be defined as:

$$L = L_{cls} + \lambda L_{div} + \alpha L_{ssl} \tag{16}$$

where hyper-parameters λ and α control the importance of the self-supervision loss and diversity loss respectively.

4 Experiments

In this section, we first introduce datasets involved in our experiments and then present some key implementation details. Afterwards, we compare our methods with the state-of-the-art methods on general few-shot learning datasets and fine-grained few-shot learning datasets respectively. Finally, we conduct qualitative analysis and show some ablation experiments to validate each module in our network.

4.1 Dataset Description

To evaluate the performance of our proposed PMRN, we conduct extensive experiments on a widely used few-shot learning dataset and five fine-grained datasets:

miniImageNet [28] consists of 100 classes with 600 images selected from the ILSVRC-2012 [30]. We take all the classes into 64, 16 and 20 classes as train set, validation set, and test set separately. **Caltech-UCSDBirds-200-2011** [29] contain 11, 788 images from 200 bird classes. Following the splits in [12], we divide them into 100/50/50 classes for train/val/test and each image is first cropped to a human-annotated bounding box.

4.2 Implementation Details

PMRN adopts the widely used ResNet-18 [13] as the backbone of our feature extractor $f(\cdot)$, and remove the last pooling layer of it. PMRN is learned by episodic training mechanism and each episode consists of an N-way K-shot task and 16 query samples are provided for each class. Specifically, there are 5 support images and 80 query images for 5-way 1-shot setting while 25 support images and 80 query images for 5-way 5-shot setting in a single episode. We train the network by ADAM [14] with a learning rate of 0.001. Note that the number of episodes is 100,000 for 5-way and 5-shot setting and 300,000 for 5-way and 1-shot setting. Our model is implemented with the Pytorch based on the codebase for few-shot learning denoted in [11].

In the testing stage, we randomly sample 1000 episodes from the test set and use the top-1 mean accuracy as the evaluation criterion. We report the final mean accuracy with the 95% confidence intervals. All the modules of PMRN are trained from scratch in an end-to-end manner and do not need fine-tuning in the test stage.

For self-supervision task, we first randomly crop the original images to get a 255×255 region with random scaling between [0.5, 1.0]. Then we split it into 3×3 regions, which contains nine random patches of size 64×64. The number of primitives k is set as 4 on all the datasets and the default value of hyper-parameter λ and α is set as 0.4 and 1.0 separately. All experiments were performed on two TiTan RTX GPU.

4.3 General Few-Shot Classification Results

We show experimental result of PMRN for general few-shot learning task on miniImageNet. Table 1 shows the comparison of our proposed PMRN with general few-shot learning methods, including local feature-based methods and the state-of-the-arts.

Comparison with Local Feature Based Methods. Because our method belongs metric-based few-shot learning branch based on local representations, we first compare our method with some popular metric-based methods that exploits local representations. The detailed results shown in Table 1.

Comparison with the State-of-the-Arts. We compare our PMRN with some state-of-the-art methods on miniImageNet.

As Table 1 shows, our proposed PMRN achieves the new state-of-the-art performance on all settings (5-way 1-shot and 5-way 5-shot). Compared with the best method HCT [20], we achieve a remarkable 5.92% performance gain in 5-way 1-shot setting and 2.84% performance gain in 5-way 5-shot setting. The better results indicate improvement of PMRN in general few-shot learning. In contrast to HCT and EASY3-R [21], PMRN systematically design special network and mechanism to learn discriminative visual primitives as local representations for few-shot learning, rather than simply stacks networks or increases the depth of backbone.

4.4 Fine-Grained Few-Shot Classification Results

To further demonstrate the effectiveness of PMRN, we conduct extensive experiments on various fine-grained datasets for fine-grained few-shot learning task.

As shown in Table 2, PMRN also achieves new state-of-the-art performance. Compared with best methods on Caltech-UCSD Birds-200-2011, PMRN has 3.89% accuracy gain under 5-way 1-shot setting and 1.64% accuracy gain under the 5-way 5-shot setting.

Table 1. Comparison of our method with the state-of-the-art methods. Few-shot classification (%) results with 95% confidence intervals on miniImageNet.

Method	Backbone	miniImageNet	
		5-way 1-shot	5-way 5-shot
MatchingNet [2, 6, 15]	ResNet-12	65.64	78.72
RelationNet [4, 8]	ResNet-12	60.97	75.32
ProtoNet [3, 8]	ResNet-12	62.67	77.88
CAN [16]	ResNet-12	63.85	79.44
DN4 [5]	ResNet-12	65.35	81.10
DeepEMD [6]	ResNet-12	65.91	82.41
ATL-Net [7]	ConvNet	54.30	73.22
DSN [18]	ResNet-12	62.64	78.83
FRN [17]	ResNet-12	66.45	82.83
CPDE [9]	ResNet-18	65.55	80.66
TPMN [10]	ResNet-12	67.64	83.44
CTM [19]	ResNet-18	64.12	80.51
MCL [8]	ResNet-12	67.85	84.47
EASY3-R [21]	3xResNet-12	71.75	87.15
HCT [20]	3xTransformers	74.62	89.19
PMRN(ours)	ResNet-18	**80.54**	**92.03**

It is obvious that our proposed PMRN achieves larger performance gain on fine-grained datasets than general datasets. For fine-grained few-shot learning task, fine-grained information and clues could be more competitive due to small inter-class differences and large intra-class differences.

4.5 Qualitative Analysis

To further evaluate the effectiveness of PMRN and illustrate its mechanism, we conduct special and ample qualitative analysis.

Local Representations Related to Structural Clues. As shown in Fig. 3, the activation regions of visual primitive roughly correspond to certain structural parts or semantic regions of object, such as legs and abdomen of birds.

For the same class, such as support sample and query sample (a), the activation regions of visual primitive are consistent in object structure to a certain extent, which shows that the same visual primitives tend to cover the same structural parts or semantic regions. The above analysis illustrates that PMN can adaptively mine and generate visual primitives corresponding to parts or structure of object, which could be more discriminative local representations, and it possesses certain interpretability and transferability on semantic or spatial location.

Table 2. Comparison of our method with the state-of-the-art few-shot learning methods on fine-grained dataset. The results with 95% confidence intervals on **Caltech-UCSD Birds**.

Method	Backbone	Caltech-UCSD Birds	
		5-way 1-shot	5-way 5-shot
MatchNet [2, 11]	ResNet-18	73.49	84.45
MatchNet [2, 6, 12]	ResNet-12	71.87	85.08
RelationNet [4, 11]	ResNet-18	68.58	84.05
RelationNet [4, 11]	ResNet-34	66.2	82.30
ProtoNet [3, 11]	ResNet-18	72.99	86.64
MAML [1, 13]	ResNet-18	68.42	83.47
SCA+MAML++ [22]	DenseNet	70.33	85.47
Baseline [11]	ResNet-18	65.51	82.85
Baseline++ [11]	ResNet-18	67.02	83.58
S2M2 [23]	ResNet-18	71.43	85.55
DeepEMD [6]	ResNet-12	75.65	88.69
DEML [24]	ResNet-50	67.28	83.47
ATL-Net [7]	ConvNet	60.91	77.05
Cosine classifier [11]	ResNet-18	72.22	86.41
DSN [18]	ResNet-12	80.80	91.19
FRN [17]	ResNet-12	83.16	92.59
CPDE [10]	ResNet-18	80.11	89.28
CFA [25]	ResNet-18	73.90	86.80
MCL [8]	ResNet-12	85.63	93.18
PMRN (ours)	ResNet-18	**89.25**	**94.82**

For different classes, such as query sample (b) and support sample in Fig. 3, although they have huge differences in structure, visual primitives still roughly cover similar structural parts or semantic regions, for example, visual primitive activating neck and head of a bird also covers the neck and head of a dog. Above analysis reflects the transferability and generalization across tasks of visual primitives. It also shows that visual primitives based on episodic training mechanism can adapt to few-shot scenario well.

Inter Correlation Among Visual Primitives. For several visual primitives with poor interpretability and discriminability produced by PMN, CRN makes them more natural in structural parts and distinguishable in spatial regions. Because visual primitives are related to object parts, strong inter correlation on structure and semantics should be mined necessarily. It can be concluded that spatial and structural information among visual primitives is constrained and optimized by constructing visual primitive graph

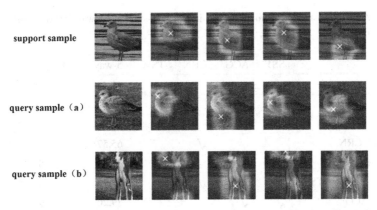

support sample

query sample（a）

query sample（b）

Fig. 3. The visual primitive visualization of a support sample and two query samples, where query sample (a) and support sample are of the same class. Note that visual primitives are produced by PMN.

and reasoning by graph convolution network, so that some confusing visual primitives can be optimized.

4.6 Ablation Study

To assess the effectiveness of each module in PMRN, we conduct detailed ablation studies on the miniImageNet dataset.

We first introduce our baseline as basis reference for validation study of other modules. Specifically, we use the ResNet-18 as the backbone, then the episodic training mechanism is used to classify each query sample into one of the N support classes, called N-way K-shot task as the ProtoNet [3] shows. The main difference is that we change similarity function from euclidean distance to cosine similarity. As shown in Table 3, we add various modules on the baseline respectively to verify the effectiveness of each module.

Compared to baseline method, our proposed ACG module improve the accuracy by 2.51% in 1-shot setting and 4.06% in 5-shot setting. This large improvement proves that primitive-level representations possess better discriminative power. Based on primitive-level representations produced by ACG module, we utilize CRN and task-specific weight module (TSW) separately to enhance the model: With the application of CRN module, further improvements by 2.87% in 5-shot setting indicates that considering the internal correlation among primitives is necessary. Meanwhile, TSW achieves a large accuracy gain of 5.25% and 7.52% in 5-shot and 1-shot setting respectively on the top of ACG module, which reveals our proposed task-specific weight generation mechanism can select both transferable and discriminative primitives across tasks indeed. It is worth noting that the combination of CRN and TSW acquire remarkable performance improvement than any single addition of them. The 9.86% and 13.09% performance gain in 5-shot and 1-shot setting can adequately demonstrate they can mutually be optimized and enhanced in the training stage.

Table 3. Ablation results on **miniImageNet** in 5-way 1-shot and 5-way 5-shot settings for the proposed PMRN.

Method	SSJ	ACG	CRN	TSW	5-way 1-shot	5-way 5-shot
baseline					61.51	75.32
+ SSJ	✓				62.04	76.12
+ ACG		✓			64.02	79.38
+ ACG + CRN		✓	✓		65.57	82.25
+ ACG + TSW		✓		✓	71.54	85.21
+ ACG + CRN + TSW		✓	✓	✓	77.13	89.24
+ SSJ + ACG	✓	✓			79.26	90.02
+ SSJ + ACG + CRN + TSW	✓	✓	✓	✓	80.54	92.03

5 Conclusion

Inspired by rapid recognition ability of humans, we research visual primitive as local representations for metric-based few-shot learning. In this paper, we propose a Primitive Mining and Reasoning Network (PMRN) for few-shot learning, and it is composed of Primitive Ming Network (PMN) and visual primitive Correlation Reasoning Network (CRN). PMN adaptively generates primitive-aware representations by mining and clustering visual patterns related object parts based on episodic training strategy. Then visual primitives are applied to conduct primitive-level metric for classification. CRN constructs special visual primitive graph structure and graph convolution network to reason internal correlation among visual primitives. Based qualitative and quantitative analysis, visual primitives in PMRN show remarkable transferable and discriminative power across tasks and domains. Extensive experiments indicate the effectiveness of our method.

Acknowledgements. This research was financially and technically supported by Guangzhou Key Research and Development Program (202206030003) and the Guangzhou Key Laboratory of Scene Understanding and Intelligent Interaction (No. 202201000001).

References

1. Finn, C., Abbeel, P., Levine, S.: Model-agnostic meta-learning for fast adaptation of deep net-works. In: International Conference on Machine Learning. PMLR, pp. 1126–1135 (2017)
2. Vinyals, O., Blundell, C., Lillicrap, T., et al.: Matching networks for one shot learning. In: Advances in Neural Information Processing Systems, vol. 29 (2016)
3. Snell, J., Swersky, K., Zemel, R.: Prototypical networks for few-shot learning. In: Advances in Neural Information Processing Systems, vol. 30 (2017)
4. Sung, F., Yang, Y., Zhang, L., et al.: Learning to compare: relation network for few-shot learning. In: Proceedings of the IEEE Conference on Computer Vision and Pattern Recognition, pp. 1199–1208 (2018)

5. Li, W., Wang, L., Xu, J., et al.: Revisiting local descriptor based image-to-class measure for few-shot learning. In: Proceedings of the IEEE/CVF Conference on Computer Vision and Pattern Recognition, pp. 7260–7268 (2019)
6. Zhang, C., Cai, Y., Lin, G., et al.: DeepEMD: differentiable earth mover's distance for few-shot learning. IEEE Trans. Pattern Anal. Mach. Intell. (2022)
7. Dong, C., Li, W., Huo, J., et al.: Learning task-aware local representations for few-shot learning. In: Proceedings of the Twenty-Ninth International Conference on International Joint Conferences on Artificial Intelligence, pp. 716–722 (2021)
8. Liu, Y., Zhang, W., Xiang, C., et al.: Learning to affiliate: mutual centralized learning for few-shot classification. In: Proceedings of the IEEE/CVF Conference on Computer Vision and Pattern Recognition, pp. 14411–14420 (2022)
9. Zhou, B., Khosla, A., Lapedriza, A., et al.: Learning deep features for discriminative localization. In: Proceedings of the IEEE Conference on Computer Vision and Pattern Recognition, pp. 2921–2929 (2016)
10. Wu, J., Zhang, T., Zhang, Y., et al.: Task-aware part mining network for few-shot learning. In: Proceedings of the IEEE/CVF International Conference on Computer Vision, pp. 8433–8442 (2021)
11. Ashok, A., Aekula, H.: When does self-supervision improve few-shot learning?-A reproducibility report. In: ML Reproducibility Challenge (Fall Edition) (2021)
12. Zheng, H., Fu, J., Mei, T., et al.: Learning multi-attention convolutional neural network for fine-grained image recognition. In: Proceedings of the IEEE International Conference on Computer Vision, pp. 5209–5217 (2017)
13. He, K., Zhang, X., Ren, S., et al.: Deep residual learning for image recognition. In: Proceedings of the IEEE Conference on Computer Vision and Pattern Recognition, pp. 770–778 (2016)
14. Kingma, D.P., Ba, J.: Adam: a method for stochastic optimization. arXiv pre-print arXiv: 1412.6980 (2014)
15. Ye, H.J., Hu, H., Zhan, D.C., et al.: Few-shot learning via embedding adaptation with set-to-set functions. In: Proceedings of the IEEE/CVF Conference on Computer Vision and Pattern Recognition, pp. 8808–8817 (2020)
16. Hou, R., Chang, H., Ma, B., et al.: Cross attention network for few-shot classification. In: Advances in Neural Information Processing Systems, 32 (2019)
17. Wertheimer, D., Tang, L., Hariharan, B.: Few-shot classification with feature map reconstruction networks. In: Proceedings of the IEEE/CVF Conference on Computer Vision and Pattern Recognition, pp. 8012–8021 (2021)
18. Simon, C., Koniusz, P., Nock, R., et al.: Adaptive subspaces for few-shot learn-ing. In: Proceedings of the IEEE/CVF Conference on Computer Vision and Pattern Recognition, pp. 4136–4145 (2020)
19. Li, H., Eigen, D., Dodge, S., et al.: Finding task-relevant features for few-shot learning by cate-gory traversal. In: Proceedings of the IEEE/CVF Conference on Computer Vision and Pattern Recognition, pp. 1–10 (2019)
20. He, Y., Liang, W., Zhao, D., et al.: Attribute surrogates learning and spectral tokens pooling in transformers for few-shot learning. In: Proceedings of the IEEE/CVF Conference on Computer Vision and Pattern Recognition, pp. 9119–9129 (2022)
21. Bendou, Y., Hu, Y., Lafargue, R., et al.: Easy—ensemble augmented-shot-Y-shaped learning: state-of-the-art few-shot classification with simple components. J. Imaging **8**(7), 179 (2022)
22. Antoniou, A., Storkey, A.J.: Learning to learn by self-critique. In: Advances in Neural Information Processing Systems, 32 (2019)
23. Mangla, P., Kumari, N., Sinha, A., et al.: Charting the right manifold: manifold mixup for few-shot learning. In: Proceedings of the IEEE/CVF Winter Conference on Applications of Computer Vision, pp. 2218–2227 (2020)

24. Zhou, F., Wu, B., Li, Z.: Deep meta learning: learning to learn in the concept space. arXiv preprint arXiv:1802.03596 (2018)
25. Hu, P., Sun, X., Saenko, K., et al.: Weakly-supervised compositional featureaggregation for few-shot recognition. arXiv preprint arXiv:1906.04833 (2019)
26. Santoro, A., Bartunov, S., Botvinick, M., et al.: Meta-learning with memory-augmented neural networks. In: International Conference on Machine Learning. PMLR, pp. 1842–1850 (2016)
27. Koch, G., Zemel, R., Salakhutdinov, R.: Siamese neural networks for one-shot image recognition. In: ICML Deep Learning Workshop, vol. 2(1) (2015)
28. Krizhevsky, A., Sutskever, I., Hinton, G.E.: ImageNet classification with deep convolutional neural networks. Commun. ACM **60**(6), 84–90 (2017)
29. Welinder, P., et al.: Caltechucsd birds 200 (2010)
30. Russakovsky, O., et al.: ImageNet large scale visual recognition challenge. Int. J. Comput. Vision **115**(3), 211–252 (2015). https://doi.org/10.1007/s11263-015-0816-y

Time-Series Forecasting Through Contrastive Learning with a Two-Dimensional Self-attention Mechanism

Linling Jiang[1], Fan Zhang[1,4]($\boxtimes$) (iD), Mingli Zhang[1,2,4], and Caiming Zhang[3]

[1] School of Computer Science and Technology, Shandong Technology and Business University, Yantai 264005, Shandong, China
`zhangfan@sdtbu.edu.cn`
[2] McGill Centre for Integrative Neuroscience, Montreal Neurological Institute, McGill University, Montreal, QC H3A 2B4, Canada
[3] Shandong University, Jinan 250100, Shandong, China
[4] Shandong Future Intelligent Financial Engineering Laboratory, Yantai 264005, China

Abstract. Contrastive learning methods have impressive capabilities in time-series representation; however, challenges in capturing contextual consistency and extracting features that meet the requirements of representation learning remain. To address these problems, this study proposed a time-series prediction contrastive learning model based on a two-dimensional self-attention mechanism. The main innovations of this model were as follows: First, long short-term memory (LSTM) adaptive pruning was used to form two subsequences with overlapping parts to provide robust context representation for each timestamp. Second, the model extracted sequence data features in both global and local dimensions. In the channel dimension, the model encoded sequence data using a combination of a self-attention mechanism and dilated convolution to extract key features for capturing long-term trends and periodic changes in data. In the spatial dimension, the model adopted a sliding-window self-attention mechanism to encode sequence data, thereby improving its perceptual ability for local features. Finally, the model introduced a self-correlation attention mechanism that converted the similarity calculation from the real domain to the frequency domain through a Fourier transform, better capturing the periodicity and trends in the data. The experimental results showed that the proposed model outperformed existing models in multiple time-series prediction tasks, demonstrating its effectiveness and feasibility in time-series prediction tasks.

Keywords: Time-series prediction · Contrastive learning · Adaptive pruning · Self-attention mechanism · Representation learning

B. Luo et al. (Eds.): ICONIP 2023, LNCS 14448, pp. 147–165, 2024.
https://doi.org/10.1007/978-981-99-8082-6_12

1   Introduction

Long-term forecasting involves the prediction of time-series data with long-term dependencies. It plays a crucial role in the fields of finance [19], economics [3], meteorology [2], healthcare [34], and transportation [22]. Long-term forecasting can help people forecast future trends, formulate reasonable plans and policies, and assist in decision-making. At the same time, long-term time-series forecasting can help discover patterns and regularities in historical data, thereby providing a better understanding of the nature of time-series data. Moreover, long-term time-series forecasting based on contrastive learning [29] has recently become an emerging research trend [25]. Unlike traditional supervised learning methods [14], contrastive learning attempts to learn the similarities and differences from two or more related sequences [26], using this information to make predictions, thereby improving a model's generalization ability and robustness. This method can improve the model's understanding of data and achieve better results.

Feature extraction is a key step in contrastive learning, and considerable progress has been made in the development of deformable convolutional network (DCN) [4]. DCN is a classic time-series prediction model based on convolutional neural network (CNN) [1] that uses dilated convolution [27] to expand the receptive field of CNN, thereby capturing long-term temporal dependencies. However, limitations using this method remain. DCN learns local information in a sequence through convolutional operations, making it difficult to handle global information. Additionally, modeling limitations exist for complex nonlinear sequences. Therefore, this study synchronously modeled the time-series in the channel and spatial dimensions, allowing the model to simultaneously focus on global and local information, thereby providing stronger feature extraction capabilities.

In long-term forecasting in fields such as power [10], transportation, and economics, relevant networks-including recurrent neural network (RNN) [30] and LSTM [28]-typically require time-series to be sampled at regular intervals. However, in practical applications, time-series are often irregular, limiting their use in real-world applications. Additionally, the recurrent structure used by the RNN and LSTM model can make it difficult to capture long-term dependencies when the sequence length is very long, leading to a decrease in prediction accuracy. The Transformer [21], however, is a model constructed based on self-attention mechanisms, and its powerful global modeling ability shows great potential for time-series forecasting, attracting increasing attention [33]. Based on the existing research, this study directly modeled a time-series using self-attention mechanisms in the channel dimension, thereby guiding the model to focus on global features. Additionally, a sliding self-attention mechanism was constructed in the spatial dimension by setting a fixed-size window to enhance local information. The sliding attention mechanism considers only the correlation between the positions within the window and other positions, which ensures that local information is unaffected by the entire sequence. This makes it easier for the model to capture local features and patterns. Moreover, the sliding window's stride can be adjusted, while the sliding attention mechanism controls the degree of overlap between each window, affecting the continuity and smoothness of the calculation, thereby enhancing the model's robustness and generalization performance.

Contrastive learning improves the generalization ability of a model through similarity comparisons between data. Typically, two samples are used as inputs-that is, one as an anchor and the other as a positive sample-a negative sample that is dissimilar to the anchor being selected from the dataset. The key is to map similar samples to nearby spatial positions and dissimilar samples to distant spatial positions [31]. Consequently, the selection of the positive and negative sample pairs is critical. For example, U-representation [16] can generate positive samples from different time series with the same timestamp, the representation of the time-series being as close as possible to the subsequence from which it was selected; TNEN [5] can generate positive samples through temporal proximity to enhance the local smoothness of the representation; and URL-MTS can generate positive samples through data augmentation with the expectation that the model can learn representations that are invariant to changes. However, these strategies are not suitable for time-series analysis. TS2vec uses random pruning to obtain two segments of the time-series with overlapping parts as positive and negative samples, but it cannot guarantee that the time span of the positive samples is within a reasonable range. To address these problems, this study proposed the use of an LSTM model to adaptively learn two reasonable segments of a time-series as positive and negative samples. The LSTM model can adaptively learn long-term dependencies in sequence data and relationships between multiple variables, enabling it to better capture the differences between positive and negative samples [11]. With the adaptive feature-learning ability of the LSTM model, the uncertainty of feature selection can be avoided, providing strong support for the subsequent training of the model. The contributions of this study are as follows:

- A dual-dimensional feature extraction module that could learn representations of time-series in both the channel and spatial dimensions was proposed. The module can improve the model's understanding and predictive ability of time-series data, thereby enhancing its effectiveness and robustness.
- The LSTM model was used to adaptively prune the input sequence to obtain the optimal lengths of the positive and negative samples. The LSTM model could adaptively learn long-term dependencies in a sequence, thereby avoiding the uncertainty and randomness caused by traditional random pruning.
- A self-correlation attention mechanism was introduced into the model and a Fourier transform was used to convert the similarity calculation from the real domain to the frequency domain, thus better capturing the periodicity and trends in the data.
- Through extensive experiments, it was shown that the proposed model exhibited excellent performance in long-term time-series prediction, outperforming current state-of-the-art methods. Moreover, under different evaluation settings, the proposed model exhibited a more stable predictive performance, further validating its practicality and reliability.

2   Related Works

2.1   Contrastive Learning

Contrastive learning is an unsupervised learning method aimed at obtaining effective feature representations by learning similarities and differences between samples. Based on this concept, a discriminative representation learning framework can be constructed by determining similar and dissimilar positive and negative samples using the designed model structure and a contrastive loss function to maximize the similarity of the positive samples and minimize the similarity of the negative samples to optimize the model parameters. This method can promote better differentiation of sample feature representations, thereby achieving clustering [23].

With the rapid development of deep learning technology, researchers have proposed contrasting deep-learning based methods. For example, the Siamese CNN model [6] can simultaneously process image and text inputs and learn their similarities through contrastive learning. The deep ranking [24] model uses contrastive learning to learn the relative order and similarity between images. Contrastive structured world models [12] can learn the structural representation of objects from visual inputs and understand the similarities between objects through contrastive learning, thereby learning the object structure representation. These methods typically use deep learning models-such as CNNs or RNNs-to learn representations of time-series data and have achieved good results in tasks such as time-series prediction, anomaly detection, and classification. This study applied contrastive learning methods to long-term time-series prediction and proposed a new model structure for representation learning of time-series data. The model used different dimensions to synchronously model the data, learn the similarities between time-series data, and further increase the complexity of the model, which helped the model better fit the data and improve its prediction accuracy.

This study applied contrastive learning methods to long-term time-series prediction and proposed a new model structure for representation learning of time-series data. The model used different dimensions to synchronously model the data, learn the similarities between time-series data, and further increase the complexity of the model, which helped the model better fit the data and improve its prediction accuracy.

2.2   Self-attention Mechanism

The self-attention mechanism is a widely used attention mechanism in deep learning that can be used to model correlations between sequential data. It allows the model to dynamically calculate the similarity between representations of different positions in the sequence and uses these similarities as weights for information-weighted combinations to better capture long-term sequence dependencies.

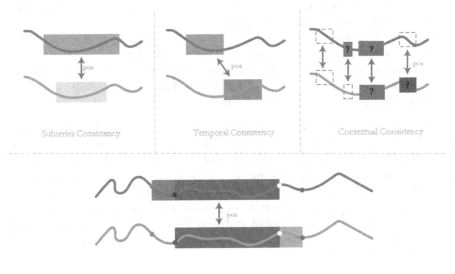

Fig. 1. Three traditional time-series data pruning methods and the LSTM adaptive pruning.

For sequential data, the self-attention mechanism exhibits powerful capabilities. For example, in the transformer model, the multi-head self-attention mechanism can divide the input into several groups and perform self-attention calculations for each group separately, which can capture the dependency relationships between different focus points in the input and improve the model's expressive power [32]. The Swin transformer [15] uses a local self-attention mechanism to reduce complexity and constructs hierarchical feature maps by merging image patches. The Conformer [18] uses a self-attention mechanism to learn global interactions and convolutional modules to capture the local features of relative offsets by combining these two modules. Consequently, these models have achieved advanced performance in various tasks, inspiring our study.

In the process of time-series modeling, the modeling of both local and global relationships is particularly important and can help the model better capture information in the sequence and improve its predictive performance. This study considered both local and global contexts and modelled them from different perspectives through a self-attention mechanism to effectively use the input information.

2.3   Data Pruning

The construction of positive and negative sample pairs is critical in contrastive learning. As illustrated in Fig. 1, the traditional construction methods mainly include the following:

(1) Subsequence Consistency [9]: Encourages the representation of the time-series to be closer to its sampled subsequence.
(2) Temporal Consistency [20]: Enforces the local smoothness of the representation by selecting adjacent segments as positive samples.
(3) Contextual Consistency: For any input time-series, contextual representations are formed by randomly sampling two overlapping time intervals, where the overlapping portion serves as a positive sample pair.

These strategies are based on strong assumptions regarding data distribution and are not suitable for time-series data. Subsequence consistency assumes that the sampled subsequences represent the features of the entire sequence. However, in some cases, the sampled subsequence may not fully represent the characteristics of the sequence, resulting in lost or inaccurate representation. Similarly, temporal consistency may excessively emphasize local smoothness, resulting in inadequate representation of global features. To address these problems, this paper proposes a method that used LSTM adaptive pruning to generate positive and negative sample pairs.

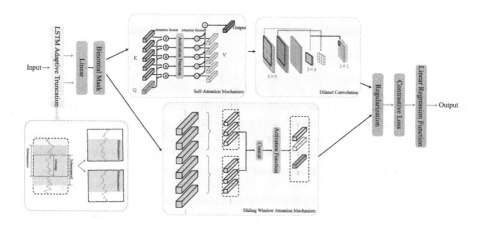

Fig. 2. The time-series data is first processed through LSTM adaptive pruning, before being parallelized across the channel dimensionincluding self-attention mechanism and dilated convolution, andacross the spatial dimension including sliding window attentionmechanism. The contrastive loss is then calculated, and thepredicted values are obtained by the trained regression function.

3   Methodology

Here, we discuss the problem of predicting long time-series. Given n observations of a time-series, the goal is to predict the next m observations. Compared to traditional supervised learning methods, contrastive learning methods can learn better feature representations from unlabeled data, thereby improving prediction accuracy and robustness. However, in long time-series data, there are often complex temporal relationships and contextual information that make it difficult to capture and extract contextual consistency and meet the requirements of representation learning. To address this, we proposed a dual-dimensional feature extraction model based on contrastive learning, as shown in Fig. 2. The time-series data are first subjected to LSTM adaptive pruning to obtain two subsequences. These two subsequences are then fed in parallel into the channel and spatial dimensions of the model for representation learning and feature extraction. Finally, a linear regression model is trained using the L2 norm penalty, which takes the last timestamp of the input time-series as the starting point and directly predicts m future observations. Section 3.1 introduces the LSTM adaptive pruning module, which cuts out two overlapping subsequences and Sect. 3.2 details the dual-dimensional feature extraction module, which extracts features in different dimensions to more accurately characterize the features of the time-series and improve prediction accuracy.

3.1 LSTM Adaptive Pruning Method

The LSTM model has a memory unit and gate mechanism that can store and update past information and control the flow of information, respectively, to avoid information loss during subsequent selection. During data pruning, the LSTM model is used to model the sequence data, and the optimal segmentation point is selected based on the output of the model. The sequence data are then divided into multiple subsequences, each containing the key information required by the model. This reduces the bias between the subsequence and the global representation, making it more representative. To achieve this, we designed a subsequence-generation module based on an LSTM model.

Specifically, the subsequence-generation module first maps the input sequence to a hidden state sequence using the LSTM model before mapping the hidden state sequence to an output sequence through a fully connected layer. To generate two subsequences, the model compresses each element of the output sequence between 0 and 1 using the sigmoid function and selects a position to divide the sequence into two subsequences. The LSTM model then computes the probability of each position as a splitting point and sets positions with probability values below a threshold value as invalid. Next, the model selects the average value of all valid positions as the splitting point and divides the input sequence into first and second halves. The model also has a hyperparameter called overlap_size, which controls the size of the overlapping area between two subsequences.

The input sequence is denoted as $x \epsilon \mathbf{R}^{B \times T \times D}$, where B denotes the batch size, T denotes the sequence length, and D denotes the feature dimensions at each time step. The forward computation process for the model can be expressed as follows:

$$\begin{cases} p_t = FC(LSTM(x_t)) \\ \quad p_t[: t/2 - k] \\ \quad p_t[t/2 + k :] \end{cases} \tag{1}$$

$$i = argmax(mean(p_t, -1)) \qquad (2)$$

$$\begin{cases} s_1 = x_{1:i+1} \\ s_2 = x_{i-k:} \end{cases} \qquad (3)$$

where p_t denotes the corresponding segmentation probability, i denotes the position of the segmentation point, and s_1 and s_2 denote the subsequences resulting from the segmentation, *mean* refers to an average function, $[: t/2 - k]$ encompassing all the elements from the beginning of the sequence to position $t/2 - k$, and $[t/2 + k :]$ encompassing all the elements from position $t/2 + k$ to the end of the sequence.

3.2 Two-Dimensional Feature Extraction Module

By contrast, feature extraction is the process of transforming an input time-series into a fixed-dimensional vector representation that can measure the similarity between different time-series [36]. As a key step in contrastive learning, feature extraction directly affects subsequent similarity measurements and comparisons. The goal is to convert the original time-series into a vector representation containing important features. This vector should have a certain degree of invariance to enable the same feature representation in different time-series. By comparing the feature representations of different time-series, their similarity can be determined and used for contrastive learning. Contrastive predictive coding (CPC) [17] is a contrastive learning method based on autoencoders that learns the representation of time-series by predicting future subsequences in the time-series. However, CPC has obvious shortcomings-such as the need to segment the time-series and performance degradation over a long time-series. By contrast, a multiscale convolutional siamese network (MC-Siamese) [7] is a contrastive learning method designed for multiscale time-series. It uses multiple convolutional layers to extract features from the input time-series, and multiple similarity measures to compare feature representations at different scales. However, it requires adjusting multiple hyperparameters to optimize model performance. To address these problems, we introduced the self-attention mechanism in transformers into contrastive learning, leveraging its powerful global perception ability to model long time-series more effectively, thereby avoiding the segmentation problem in CPC. This method combines the autocorrelation attention mechanism and multi-head mechanism of the transformer model to capture the overall structure of the time-series.

In recent years, transformer-based methods have achieved remarkable results in long-term sequence prediction tasks; however, they do not allow the targeted modeling of local features—such as CNN structures. This study focused on applying the transformer model to time-series prediction tasks to improve their prediction accuracy and efficiency. To address the problems of local features and global correlations in time-series data, we proposed a new method that used small receptive fields of convolution to mitigate the information redundancy problem in the global modeling of the transformer, thereby improving the prediction performance and generalization ability of the model. This method combined the sliding-window approach with a self-attention mechanism to capture local features. The new method was extensively validated for multiple time-series prediction tasks, and the results showed that it achieved excellent performance in terms of prediction accuracy and computational efficiency.

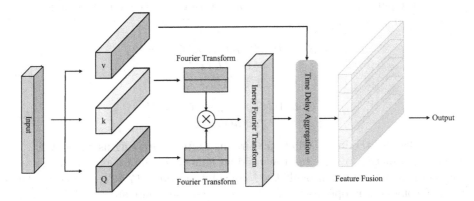

Fig. 3. This figure demonstrates structural diagram of the self-correlation attention mechanism.

The proposed method encoded sequence data in the channel dimension using a combination of a self-attention mechanism and dilated convolution to extract key features of the sequence. The self-attention mechanism modelled the global dependency of the data—that is, the correlation between different time steps—capturing long-term trends and periodic changes in the time-series data. Dilated convolution expanded the receptive field to extract more contextual information. The combination of both could better learn global features and improved the long-term prediction performance. In the self-attention mechanism, similarity calculations between query and key vectors are usually performed in the real domain using a dot product or other similarity metrics. However, this method cannot capture complex relationships in data—such as periodicity and trends in time-series data. The Fourier transform is a method that decomposes a signal into a series of frequency components represented by sine and cosine functions. By comparing the spectral differences of the signals at different frequencies in the frequency domain, signal similarity can be better assessed. To achieve this, the Fourier transform was introduced to convert the similarity calculation from the time domain to the frequency domain using the autocorrelation attention mechanism to capture the periodicity and trend characteristics in the data more effectively. Figure 3 shows how the subsequences obtained through LSTM adaptive pruning are transformed into the frequency domain using a Fourier transform to capture data features. Subsequently, an inverse Fourier transform is applied to convert the signal back to the time domain and merge the frequency components into the original signal. The experimental results show that replacing the self-attention mechanism with an autocorrelation attention mechanism can greatly improve a model's performance in long-term prediction. Additionally, an extended CNN structure with residual connections and dilated convolution can be used to extract context representations for each timestamp. Each residual block contains two 1-D convolution layers—the dilation factor of the lth block being $2^\wedge l$—providing a larger receptive field for different layers to further enhance the performance. The overall computational process for the channel can be expressed as follows:

$$\begin{cases} z = SamePadConv(x, C_{in}, C_{out}, k, d) \\ \quad x = GELU(z) \\ \quad r = P(x) \end{cases} \tag{4}$$

$$P = \begin{cases} Conv1d(C_{in}, C_{out}, 1) \ C_{in} \neq C_{out} \\ \quad\quad otherwise \end{cases} \tag{5}$$

where x denotes the intermediate feature map after processing using the GELU activation function, r denotes the residual connection input after undergoing channel transformation through 1×1 convolution, $SamePadConv$ denotes the dilated convolution operation on the input feature map after padding, and $Conv1d$ denotes the 1×1 convolution operation.

In the spatial dimension, the module uses a sliding-window self-attention mechanism to encode sequence data. The sliding-window autocorrelation attention mechanism helps the model learn local features in the sequence data, thereby improving its generalization ability and accuracy. The algorithm for the sliding-window self-attention mechanism is introduced in the pseudocode to illustrate the computation process in the spatial dimension.

Assuming there is a time-series x of length n, and each time step has a feature vector x_i of dimension d, then using a sliding window of size k to process x, we can calculate the self-attention representation for each time step. The pseudocode for the sliding-window self-attention mechanism algorithm is shown in Table 1.

The formula for calculating the attention scores can be expressed as follows:

$$score = q_i^T k_j \tag{6}$$

The formula for calculating the attention weights can be expressed as follows:

$$\partial_{i,j} = \frac{exp(score)}{\sum_{j=1}^{k} exp(score)} \tag{7}$$

where $q_i = w_q x_i$ is the query vector of the input sequence x_i, $k_j = w_k x_i$ is the key vector of the input sequence x_i, and $v_j = w_v x_i$ is the value vector of the input sequence x_i.

Table 1. The sliding-window self-attention mechanism algorithm.

Sliding-Window Self-attention Mechanism Algorithm	
1	Input: time-series x, sliding-window size k
2	for i = 1 to n do
3	// compute attention scores
4	scores = []
5	for j = max(1, i-k+1) to i do
6	score = dot_product(Wq * x_i, Wk * x_j)
7	scores.append(score)
8	// compute attention weights
9	attention = softmax(scores)
10	// compute attention representation
11	self_attention = 0
12	for j = max(1, i-k+1) to i do
13	self_attention += attention[j-i+k-1] * (Wv * x_j)
14	output[i] = self_attention

Finally, the encoding results of the channel and spatial dimensions can be added based on the different weights to obtain the final representation of the sequence data. Overall, the model combines the autocorrelation attention mechanism, dilated convolution, and sliding-window self-attention mechanism, fully utilizing the global and local information in the sequence data and improving the model's performance and generalization ability.

Table 2. Evaluation metrics.

Metric	Method
Mean Square Error (MSE)	$MSE = \frac{1}{n} \sum_{i=1}^{n} (y_i - \hat{y}_i)^2$
Master of Aerospace Engineering (MAE)	$MAE = \frac{1}{n} \sum_{i=1}^{n} |(y_i - \hat{y}_i)|$

4 Experiments

4.1 Experimental Setup

In this section, the experimental results are presented to demonstrate the performance of the proposed model. The experiments were conducted on five datasets, of which the first four were real-world industrial datasets and the fifth was a commonly used benchmark dataset. Single-variate time-series long-term predictions were performed for each dataset. The evaluation metrics are listed in Table 2.

4.2   The Detail of Dataset

Electricity[1]: The Electricity dataset is a widely used benchmark dataset in the time-series domain for evaluating the performance of time-series prediction algorithms. It recorded hourly electricity consumption data from 2004 to 2005—including 321 days of data, with 24 data points per day. Electricity datasets have become standard practice for evaluating the performance of time-series prediction algorithms and have been widely used in various time-series prediction tasks.

Table 3. Comparison with other unsupervised representation learning methods.

Method		Ours		Unsupervised representation learning							
				TS2Vec		SimTS		CoST		TNC	
Evaluation metric		MSE	MAE	MSE	MAE	MSE	MAE	MSE	MAE	MSE	MAE
ETTh1	96	**0.086**	**0.223**	0.101	0.254	0.089	0.231	0.092	0.237	0.143	0.299
	192	**0.093**	**0.233**	0.123	0.295	0.094	0.235	0.104	0.245	0.159	0.321
	336	0.101	0.245	0.160	0.316	**0.100**	**0.239**	0.112	0.258	0.179	0.345
	720	**0.125**	0.278	0.179	0.345	0.126	**0.277**	0.148	0.306	0.235	0.408
ETTh2	96	**0.148**	**0.299**	0.164	0.300	0.149	0.301	0.168	0.313	0.111	0.259
	192	**0.172**	**0.328**	0.198	0.355	0.193	0.342	0.190	0.340	0.199	0.355
	336	**0.180**	**0.338**	0.205	0.364	0.199	0.354	0.206	0.360	0.207	0.366
	720	**0.204**	**0.369**	0.208	0.371	0.212	0.370	0.214	0.371	0.207	0.370
ETTm1	96	**0.038**	0.148	0.045	0.162	0.041	**0.143**	0.038	0.147	0.054	0.178
	192	**0.054**	**0.185**	0.077	0.224	0.085	0.190	0.070	0.191	0.086	0.225
	288	**0.076**	0.212	0.095	0.235	0.098	**0.207**	0.077	0.209	0.142	0.290
	672	**0.102**	**0.241**	0.142	0.290	0.117	0.242	0.113	0.257	0.136	0.290
ETTm2	96	0.082	0.222	0.089	0.225	**0.068**	**0.189**	0.072	0.196	0.094	0.229
	192	**0.056**	**0.176**	0.091	0.241	0.071	0.240	0.069	0.239	0.083	0.274
	288	**0.126**	**0.276**	0.161	0.306	0.160	0.272	0.153	0.307	0.155	0.09
	672	**0.163**	**0.318**	0.201	0.351	0.249	0.334	0.183	0.329	0.197	0.352
Electricity	96	**0.269**	0.386	0.332	0.346	0.284	**0.299**	0.396	0.382	0.454	0.501
	192	**0.312**	0.407	0.379	**0.401**	0.325	0.370	0.423	0.494	0.562	0.572
	336	**0.329**	**0.427**	0.565	0.535	0.523	0.522	0.674	0.655	0.875	0.835
	720	**0.417**	**0.486**	0.861	0.902	0.847	0.832	0.975	0.982	1.132	0.987

Electric Power Transformer Temperature (ETT)[2]: The ETT dataset provides two consecutive years of power data from two different regions in the same province in China. The dataset includes two variants with different granularities—that is, one with hourly data, including ETTh1 and ETTh2, and

---

[1] https://archive.ics.uci.edu/ml/datasets/ElectricityLoadDiagrams20112014.
[2] https://github.com/liyaguang/ETDataset.

the other with 15-min data, including ETTm1 and ETTm2. Each data point contains eight features, including the recording date, the predicted "oil temperature," and six different types of external load values. The dataset had training, validation, and testing times of 12, 4, and 4 months, respectively. This dataset can be used for the research and evaluation of power-load forecasting tasks and has become an important benchmark dataset in this field.

4.3   Experimental Results

Baseline: The proposed model was compared with several advanced methods developed in recent years, divided into two categories—that is, time-series prediction using contrastive learning (including SimTS [35], TS2Vec, TNC [20], and CoST [25]) and end-to-end time-series prediction methods (including FEDformer [38], Autoformer [8], Informer [37], and LogTrans [13]). Each method treats the prediction as a single variable in a time-series. Tables 3 and 4 present the results of several time-series prediction methods for the five datasets.

Table 4. Comparison with other end-to-end time-series prediction methods.

Method		Ours		End-to-end prediction							
				FEDformer		Autoformer		Informer		LogTrans	
Evaluation metric		MSE	MAE	MSE	MAE	MSE	MAE	MSE	MAE	MSE	MAE
ETTh1	96	0.086	0.223	0.079	0.215	**0.071**	**0.206**	0.193	0.377	0.283	0.468
	192	**0.093**	**0.233**	0.104	0.245	0.114	0.262	0.217	0.395	0.234	0.409
	336	**0.101**	**0.245**	0.119	0.270	0.107	0.258	0.202	0.381	0.386	0.546
	720	**0.125**	**0.278**	0.142	0.299	0.126	0.283	0.183	0.355	0.475	0.628
ETTh2	96	0.148	0.299	**0.128**	**0.271**	0.153	0.306	0.213	0.373	0.217	0.379
	192	**0.172**	**0.328**	0.182	0.330	0.204	0.351	0.227	0.387	0.281	0.429
	336	**0.180**	**0.338**	0.231	0.378	0.246	0.389	0.242	0.401	0.293	0.437
	720	**0.204**	**0.369**	0.278	0.420	0.268	0.409	0.291	0.439	0.218	0.387
ETTm1	96	0.038	0.148	**0.033**	**0.140**	0.056	0.183	0.109	0.277	0.049	0.171
	192	**0.054**	**0.185**	0.058	0.186	0.081	0.216	0.151	0.310	0.157	0.317
	288	**0.076**	**0.212**	0.084	0.232	0.081	0.225	0.089	0.237	–	–
	672	**0.102**	**0.241**	0.113	0.287	0.109	0.250	0.117	0.293	–	–
ETTm2	96	0.082	0.222	0.067	0.198	**0.065**	**0.189**	0.088	0.225	0.075	0.208
	192	**0.056**	**0.176**	0.110	0.245	0.118	0.256	0.132	0.283	0.129	0.275
	288	**0.126**	**0.276**	0.134	0.293	0.131	0.289	0.151	0.304	–	–
	672	**0.163**	**0.318**	0.194	0.357	0.179	1.337	0.204	0.385	–	–
Electricity	96	0.269	0.386	0.262	0.378	0.341	0.438	**0.258**	**0.368**	0.288	0.393
	192	0.312	0.407	0.316	0.410	0.345	0.428	**0.285**	**0.388**	0.432	0.483
	336	**0.329**	**0.427**	0.361	0.445	0.406	0.470	0.336	0.423	0.430	0.483
	720	**0.417**	**0.486**	0.448	0.501	0.545	0.581	0.607	0.599	0.491	0.531

Table 3 presents a comparison of the results of the proposed method with advanced unsupervised learning models developed in recent years. From the data, it is evident that the proposed method exhibits advanced performance in most cases, particularly in long-term prediction. For example, when the prediction length is 672, the proposed method achieves improvements of 18.9%, 34.5%, 10.9%, and 17.3% compared with TS2Vec, SimTS, CoST, and TNC, respectively, on the ETTm2 dataset. For the other datasets, the proposed method achieves considerable performance improvements at longer prediction lengths. For shorter prediction lengths, the proposed method shows smaller improvements than the other methods. This is because the proposed self-attention mechanism models the global dependencies, thereby predicting long-term trends more accurately. Table 4 compares the proposed method with end-to-end time-series prediction models commonly used in recent years, the results showing that the proposed model achieves competitive prediction accuracy. Especially when predicting time-series with length of 672 and 720, the effect is significantly improved. The dual-dimensional module can better capture the correlation between features through the embedding and interaction in the feature dimension, thus predicting the future trend of the time-series more accurately.

The proposed method also conducts a sensitivity analysis of multiple adjustable parameters. (1) When using the LSTM model for subsequence generation, the length of the common subsequence must be specified in advance to ensure that the two subsequences have overlapping parts. We studied and analyzed the value of this length, the results of which are shown in Fig. 4, which indicate that when the overlap window is too short, the model cannot obtain rich neighboring sample information, limiting its ability to learn the representation of the time-series. Moreover, when the overlap window is too long, the model focuses too much on positive samples, leading to overfitting or weak generalization ability. The experiments show that the prediction ability of the model is optimal when the overlap window is approximately 200. (2) When fusing the features extracted from the channel and spatial dimensions, different weights are assigned to the two dimensions, with the weight of the channel dimension set to alpha and the weight of the spatial dimension set to 1-alpha. Sensitivity analysis of the alpha range is conducted using the ETTh1 and ETTh2 datasets as examples. The experimental results are shown in Fig. 5. It is evident that when either the dimension weight is too large or too small, the effect is not ideal, the best performance being achieved when the weight of the channel dimension is set to approximately 0.6. In long-term prediction tasks, global information obtained from the channel dimension is more important for trend predictions.

4.4 Ablation Study

Numerous ablation experiments were conducted to verify the effectiveness of each component of the proposed model. Specifically, using the proposed model as the baseline, various components were removed and the remaining model was compared with the baseline to analyze the importance of each module based on the experimental results.

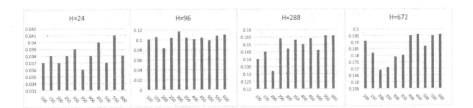

Fig. 4. Effects of different overlapping lengths in the LSTM subsequence consistency pruning method. When the overlapping window is around 200, the predictive ability of the model reaches its optimal level.

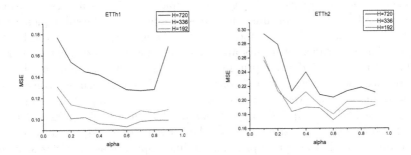

Fig. 5. Effects of different overlapping lengths in the LSTM subsequence consistency pruning method. The best effect is achieved when the weight of the channel dimension is set to around 0.6.

Table 5. Ablation experiments of the LSTM subsequence generation module.

Prediction length		96	192	336	720
Subsequence consistency	MSE	0.102	0.121	0.135	0.154
	MAE	0.275	0.298	0.299	0.341
Temporal consistency	MSE	0.098	0.117	0.128	0.150
	MAE	0.271	0.288	0.290	0.339
Contextual consistency	MSE	0.092	0.105	0.112	0.142
	MAE	0.246	0.269	0.272	0.307
LSTM adaptive trunc (ours)	MSE	**0.086**	**0.093**	**0.101**	**0.128**
	MAE	**0.223**	**0.233**	**0.245**	**0.284**

Table 6. Ablation experiments of the dual-dimensional feature extraction module.

Prediction length		96	192	336	720
Channel dimension	MSE	0.108	0.115	0.126	0.142
	MAE	0.267	0.286	0.298	0.315
Spatial dimension	MSE	0.104	0.107	0.119	0.136
	MAE	0.271	0.285	0.293	0.303
Ours	MSE	**0.086**	**0.093**	**0.101**	**0.128**
	MAE	**0.223**	**0.233**	**0.245**	**0.284**

Analysis of the effectiveness of the LSTM adaptive pruning module: The LSTM adaptive pruning module is a sequence data pruning method based on the LSTM model. This method adaptively divides a long sequence into two subsequences to form positive and negative sample pairs for contrastive learning. Here, the LSTM adaptive pruning module is compared with the subsequence consistency, temporal consistency, and random pruning consistency on the ETTh1 dataset. The experimental results are listed in Table 5. It is evident that using the LSTM model for adaptive pruning, compared with the pruning methods of subsequence consistency, temporal consistency, and random pruning consistency, the performance improves by approximately 18%, 20%, and 10%, respectively. Overall, the LSTM adaptive pruning module has unique advantages for processing sequence data, and can provide better subsequence-generation results. In the future, LSTM adaptive pruning module applications—such as sequence classification, sequence generation, sequence completion, and other tasks—can be further explored. Additionally, more advanced adaptive pruning methods—such as attention-based adaptive pruning—can be studied to improve the efficiency and accuracy of sequence data processing.

Analysis of dual-dimensional feature extraction module effectiveness: The dual-dimensional feature extraction module is a new feature extraction method that aims to extract feature information simultaneously from multiple dimensions for better data representation. We conducted an effectiveness analysis by comparing it with other feature extraction methods to validate the effectiveness of the module.

In this section, three scenarios are presented for comparison. The first scenario used only the autocorrelation attention mechanism and dilated convolution to extract features in the channel dimension. This method only considered feature extraction in the channel dimension and did not consider information in the spatial dimension. The second scenario used only the sliding-window attention mechanism to extract features in the spatial dimension. This method considered only feature extraction in the spatial dimension and did not consider information in the channel dimension. The third scenario synchronously modeled the channel and spatial dimensions and added their results, together with the different weights assigned to each dimension. This method considered feature extraction in both the channel and spatial dimensions and combined them with a specific weight allocation strategy. The experimental results for the ETTh1 and ETTh2 datasets are presented in Table 6. Overall, the dual-dimensional feature extraction module demonstrates good feature extraction ability for multiple datasets. This method simultaneously considers feature extraction in both the channel and spatial dimensions and comprehensively describes the data. Through a specific weight allocation strategy, it then combines feature information from both dimensions to improve the efficiency and accuracy of feature extraction.

5   Conclusions

This study proposed a dual-dimensional feature extraction framework based on contrastive learning that extracted features from both the channel and spatial dimensions of time-series data. In the channel dimension, the self-attention mechanism was used to perform a global correlation on the data, and convolution was then used to further extract sequence features, thus improving the accuracy of the model's learning in the channel dimension. In the spatial dimension, the sliding-window attention mechanism was used to adaptively weigh the data in each window, complementing the local features ignored by global representation learning in the channel dimension, thereby learning more comprehensive time-series data features. Dual-dimensional synchronous modeling enabled the model to achieve a good balance between global and local information, thereby improving the feature extraction ability and achieving better performance. To better construct positive and negative pairs in contrastive learning, we designed an LSTM-based data-pruning strategy that used LSTM's traversing learning ability on time-series data to adaptively select subsequence pairs that were more beneficial for subsequent model training. Numerous experiments demonstrated the effectiveness of the proposed modeling method for long-term prediction tasks. In the future, we will further investigate the application of the self-attention mechanism and convolution in contrastive learning, seeking more effective fusion methods to promote the development of time-series prediction tasks.

Acknowledgements. This work is supported by the National Natural Science Foundation of China (62272281), the Special Funds for Taishan Scholars Project(tsqn202306274), and the Youth Innovation Technology Project of Higher School in Shandong Province (2019KJN042).

References

1. Albawi, S., Mohammed, T.A., Al-Zawi, S.: Understanding of a convolutional neural network. In: International Conference on Engineering and Technology (ICET), pp. 1–6. IEEE (2017)
2. Angryk, R.A., et al.: Multivariate time series dataset for space weather data analytics. Sci. Data **7**(1), 227 (2020)
3. Beck, N., Katz, J.N.: Modeling dynamics in time-series-cross-section political economy data. Ann. Rev. Polit. Sci. **14**, 331–352 (2011)

4. Börjesson, L., Singull, M.: Forecasting financial time series through causal and dilated convolutional neural networks. Entropy **22**(10), 1094 (2020)
5. Bose, A.J., Ling, H., Cao, Y.: Adversarial contrastive estimation. In: International Conference on Machine Learning (2018)
6. Bromley, J., Guyon, I., LeCun, Y., Säckinger, E., Shah, R.: Signature verification using a "siamese" time delay neural network. Adv. Neural Inf. Process. Syst. **6** (1993)
7. Chen, H., Wu, C., Du, B., Zhang, L.: Deep siamese multi-scale convolutional network for change detection in multi-temporal VHR images. In: 2019 10th International Workshop on the Analysis of Multitemporal Remote Sensing Images (MultiTemp), pp. 1–4. IEEE (2019)
8. Chen, M., Peng, H., Fu, J., Ling, H.: Autoformer: searching transformers for visual recognition. In: Proceedings of the IEEE/CVF International Conference on Computer Vision, pp. 12270–12280 (2021)
9. Franceschi, J.Y., Dieuleveut, A., Jaggi, M.: Unsupervised scalable representation learning for multivariate time series. Adv. Neural Inf. Process. Syst. **32**, 1–11 (2019)
10. Gasparin, A., Lukovic, S., Alippi, C.: Deep learning for time series forecasting: the electric load case. CAAI Trans. Intell. Technol. **7**(1), 1–25 (2022)
11. Graves, A., Graves, A.: Supervised Sequence Labelling. Springer, Heidelberg (2012). https://doi.org/10.1007/978-3-642-24797-2_2
12. Kipf, T., Van der Pol, E., Welling, M.: Contrastive learning of structured world models. In: International Conference on Learning Representations (2019)
13. Li, S., et al.: Enhancing the locality and breaking the memory bottleneck of transformer on time series forecasting. Adv. Neural Inf. Process. Syst. **32**, 1–11 (2019)
14. Liu, X., Guo, J., Wang, H., Zhang, F.: Prediction of stock market index based on ISSA-BP neural network. Expert Syst. Appl. **204**, 117604 (2022)
15. Liu, Z., et al.: Swin transformer: hierarchical vision transformer using shifted windows. In: Proceedings of the IEEE/CVF International Conference on Computer Vision, pp. 10012–10022 (2021)
16. Ma, C., Wen, J., Bengio, Y.: Universal successor representations for transfer reinforcement learning. In: International Conference on Learning Representations (2018)
17. Oord, A.v.d., Li, Y., Vinyals, O.: Representation learning with contrastive predictive coding. In: Conference on Neural Information Processing Systems (2018)
18. Peng, Z., et al.: Conformer: local features coupling global representations for visual recognition. In: Proceedings of the IEEE/CVF International Conference on Computer Vision, pp. 367–376 (2021)
19. Sezer, O.B., Gudelek, M.U., Ozbayoglu, A.M.: Financial time series forecasting with deep learning: a systematic literature review: 2005–2019. Appl. Soft Comput. **90**, 106181 (2020)
20. Tonekaboni, S., Eytan, D., Goldenberg, A.: Unsupervised representation learning for time series with temporal neighborhood coding. In: International Conference on Learning Representations (2021)
21. Vaswani, A., et al.: Attention is all you need. Adv. Neural Inf. Process. Syst. **30**, 1–11 (2017)
22. Vlahogianni, E.I., Karlaftis, M.G.: Testing and comparing neural network and statistical approaches for predicting transportation time series. Transp. Res. Rec. **2399**(1), 9–22 (2013)
23. Wang, J., Zhou, F., Wen, S., Liu, X., Lin, Y.: Deep metric learning with angular loss. In: Proceedings of the IEEE International Conference on Computer Vision, pp. 2593–2601 (2017)

24. Wang, J., et al.: Learning fine-grained image similarity with deep ranking. In: Proceedings of the IEEE Conference on Computer Vision and Pattern Recognition, pp. 1386–1393 (2014)
25. Woo, G., Liu, C., Sahoo, D., Kumar, A., Hoi, S.: Cost: contrastive learning of disentangled seasonal-trend representations for time series forecasting. In: International Conference on Learning Representations (2022)
26. Wu, Z., Xiong, Y., Yu, S.X., Lin, D.: Unsupervised feature learning via nonparametric instance discrimination. In: Proceedings of the IEEE Conference on Computer Vision and Pattern Recognition, pp. 3733–3742 (2018)
27. Yu, F., Koltun, V.: Multi-scale context aggregation by dilated convolutions. In: International Conference on Learning Representations (2016)
28. Yu, Y., Si, X., Hu, C., Zhang, J.: A review of recurrent neural networks: LSTM cells and network architectures. Neural Comput. **31**(7), 1235–1270 (2019)
29. Yue, Z., et al.: Ts2vec: towards universal representation of time series. In: Proceedings of the AAAI Conference on Artificial Intelligence, vol. 36, pp. 8980–8987 (2022)
30. Zaremba, W., Sutskever, I., Vinyals, O.: Recurrent neural network regularization. In: International Conference on Learning Representations (2014)
31. Zhang, D., Han, J., Zhang, Y.: Supervision by fusion: towards unsupervised learning of deep salient object detector. In: Proceedings of the IEEE International Conference on Computer Vision, pp. 4048–4056 (2017)
32. Zhang, F., Chen, G., Wang, H., Li, J., Zhang, C.: Multi-scale video super-resolution transformer with polynomial approximation. IEEE Trans. Circ. Syst. Video Technol. **33**, 4496–4506 (2023). https://doi.org/10.1109/TCSVT.2023.3278131
33. Zhang, F., Guo, T., Wang, H.: DFNET: decomposition fusion model for long sequence time-series forecasting. Knowl.-Based Syst. **277**, 110794 (2023)
34. Zhang, J., Nawata, K.: Multi-step prediction for influenza outbreak by an adjusted long short-term memory. Epidemiol. Infect. **146**(7), 809–816 (2018)
35. Zheng, X., Chen, X., Schürch, M., Mollaysa, A., Allam, A., Krauthammer, M.: SimTS: rethinking contrastive representation learning for time series forecasting. In: International Joint Conference on Artificial Intelligence (2023)
36. Zhou, B., Khosla, A., Lapedriza, A., Oliva, A., Torralba, A.: Learning deep features for discriminative localization. In: Proceedings of the IEEE Conference on Computer Vision and Pattern Recognition, pp. 2921–2929 (2016)
37. Zhou, H., et al.: Informer: beyond efficient transformer for long sequence time-series forecasting. In: Proceedings of the AAAI Conference on Artificial Intelligence, vol. 35, pp. 11106–11115 (2021)
38. Zhou, T., Ma, Z., Wen, Q., Wang, X., Sun, L., Jin, R.: Fedformer: frequency enhanced decomposed transformer for long-term series forecasting. In: International Conference on Machine Learning, pp. 27268–27286 (2022)

Task Scheduling with Multi-strategy Improved Sparrow Search Algorithm in Cloud Datacenters

Yao Liu, Wenlong Ni[✉], Yang Bi, Lingyue Lai, Xinyu Zhou, and Hua Chen

School of Computer and Information Engineering, Jiangxi Normal University,
Nanchang 330022, China
{liuy,wni,byang,laily,xyzhou,huachen}@jxnu.edu.cn

Abstract. How to efficiently schedule tasks is the focus of cloud computing. Combining the task scheduling characteristics of the cloud computing environment, a multi-strategy improved sparrow search algorithm (MISSA) that takes into account task completion time, task completion cost and load balancing is proposed. First, the initialization of the population using piecewise linear chaotic map (PWLCM) enhances the degree of individual dispersion. After that, the global search phase in the marine predator algorithm (MPA) is incorporated to increase the scope of the search space. The introduction of dynamic adjustment factors in the joiner part strengthens the search ability of the algorithm in the early stage and the convergence ability in the late stage. Finally, the greedy strategy is used to update the joiner's position so that the information of the optimal solution and the worst solution can be uesd to guide the next generation of position updates. Using CloudSim for simulation, the experimental results show that the proposed algorithm has a shorter task completion time and a more balanced system load. Compared with the ant colony optimization (ACO), MPA, and sparrow search algorithm (SSA), the MISSA improves the integrated fitness function values by 20%, 22%, and 17%, confirming the feasibility of the proposed algorithm.

Keywords: Sparrow Search Algorithm · Marine Predator Algorithm · Greedy Strategy · Cloud Computing · Multi-Objective Task Scheduling

1 Introduction

Modern society has entered the era of big data, the traditional computing model can no longer meet the processing needs of big data, cloud computing has risen as internetbased model. It provides access to a configurable pool of assets on demand, which can be immediately provided and discharged with very little administration or cloud provider cooperation [1].

The task scheduling problem has been a hot issue in cloud computing research, and its core lies in how to reasonably manage and schedule user tasks

B. Luo et al. (Eds.): ICONIP 2023, LNCS 14448, pp. 166–177, 2024.
https://doi.org/10.1007/978-981-99-8082-6_13

and computing resources. In recent years, numerous studies have been conducted using heuristic algorithms to solve the task scheduling problem. The load balancing ant colony optimization algorithm (LB-ACO) proposed in the literature [2] provides some improvement load balancing than the other algorithm, but does not address the problem of pheromone scarcity and slow convergence at the beginning of the iteration. The literature [3] incorporates cuckoo search (CS) and particle swarm optimization (PSO) to reduce the task completion time, cost and violation rate, but does not consider the load balancing problem of virtual machines(VMs).

The SSA is a new population optimization method proposed by Xue in 2020 [4]. Currently, SSA are widely used to solve problems in different fields. Abdulhammed [5] used SSA to improve the cloud computing IoT load balancing problem. Qiu [6] used improved SSA applied in data compression for edge computing. The results of the study revealed that these optimization algorithms led to a significant improvement in the accuracy and efficiency of the research objectives.

In this paper, MISSA is proposed. To address the problem of uneven population initialization in the original algorithm, PWLCM is introduced to increase the individual dispersion of the initial population. For the defect that the discoverer is easy to converge to zero, the global search of the marine predator algorithm is utilized to extend the search space of the discoverer. Secondly, for the defect of slow convergence, a dynamic adjustment factor is introduced to enhance the convergence of individuals in the late iteration. A greedy learning strategy is added to retain better adapted individuals. Compared with the SSA, the main contributions of this paper are as follows:

- The PWLCM is introduced to increase the dispersion of individuals in the initial population.
- The global search method of marine predator algorithm is introduced in the update mechanism of discoverer to expand the search space.
- Greedy strategy is added to obtain the relative optimal position, and the addition of dynamic adjustment factor improves the convergence speed.

This experiment simulates the task scheduling problem in cloud computing by CloudSim software, and compares the MISSA with ACO, MPA and SSA, and finds out that MISSA outperforms other algorithms in terms of solution time and solution quality.

The rest of this paper is organized as follows. Section 1 describes the concept of cloud computing task scheduling, the problem encoding, and the fitness function. Section 2 describes the optimization process of the sparrow search algorithm. Section 3 puts the MISSA and other algorithms through CloudSim for simulation and analyzes the experimental results. Finally, in Sect. 4, conclusions and future studies are presented.

2   Problem Description

Distributed computing is the authorized access to a public pool of assets through an Internet service. With the increasing number of cloud clients repeating daily, task scheduling becomes a necessary issue to manage [7]. Task scheduling in cloud computing is to improve the quality of service to users through a reasonable scheme of task assignment to nodes so that the total task completion cost is minimized.

The way task scheduling done in cloud computing can be described as follows: assign n tasks to be executed on m VMs (where m is the number of VM). The cloud task list is represented as CloudletList $= \{t_1, t_2, ..., t_n\}$, t_i ($i = 1, 2, ..., n$) denotes the task i. The VM is represented as VMList $= \{vm_1, vm_2, ..., vm_m\}$, vm_j ($j = 1, 2, ..., m$), denotes the vm_j, a task can only be assigned to run on one VM. The scheduling algorithm generates an optimal resource allocation matrix T at the end of the iteration as the scheduling solution, each element in the matrix T is represented as $t_{i,j}$ ($i = 1, 2, ..., n$, $j = 1, 2, ..., m$), the value of $t_{i,j}$ is taken as:

$$t_{i,j} = \begin{cases} 0, & \text{task } i \text{ is not assigned to } vm_j, \\ 1, & \text{task } i \text{ is assigned to } vm_j, \end{cases} \tag{1}$$

2.1 Encoder and Decoder

In an experiment simulating a cloud environment, the population consisting of d sparrows can be expressed in the following form:

$$X = \begin{bmatrix} x_1^1 & x_1^2 & \cdots & x_1^n \\ x_2^1 & x_2^2 & \cdots & x_2^n \\ \cdots & \cdots & \cdots & \cdots \\ x_d^1 & x_d^2 & \cdots & x_d^n \end{bmatrix}, \tag{2}$$

the location of each individual sparrow is used as an allocation policy.

The specific treatment is that if the position information of individual i in the process of iteration is $x_i = \{x_i^1, x_i^2, ..., x_i^n\}$, the deployment scheme of the corresponding task is: $d(x_i) = \{\lfloor|(x_i^1)|\rfloor, \lfloor|(x_i^2)|\rfloor, ..., \lfloor|(x_i^n)|\rfloor\}$. Suppose we need to assign 5 tasks to 3 VMs with the numbers $\{0,1,2\}$, and the position information of the i individual in the population at a certain time is $\{0.334, 1.75, 1.63, 0.83, 2.67\}$, using the SVP from the literature [8], which is encoded as $\{0,1,1,0,2\}$. This assignment scheme indicates that the first task is assigned to the first VM for execution, the second task is assigned to the sec ond VM for execution, ..., and the fifth task is assigned to the third VM for execution.

2.2 Objective Function

The optimization objectives in this paper are to minimize (1) task completion time and (2) execution cost as well as to maximize the (3) average load factor. Therefore, task completion time, average load factor and total cost of execution need to be defined, the details are as follows.

(1) Task Completion Time

   In the cloud task scheduling environment, the ETC (Expect Time to Complete) matrix is used to calculate the time required for all tasks to run and complete on different VMs, and the task completion time is the task completion time of the VM with the longest execution time:

$$T_{\max} = max \left(\sum_{i=1}^{k} ETC_{ij} \right), j \in [1, m], \tag{3}$$

where k denotes the number of tasks executed on vm_j.

(2) Total Cost of Task Execution

The total cost of task execution is defined as the sum of the costs of all tasks completed, expressed as:

$$C_s = \sum_{i=1}^{n} ETC_{ij} \cdot Cast_j, \tag{4}$$

where $Cast_j$ denotes the cost of executing tasks per unit time of vm_j.

(3) Average System Load Ratio

The higher the average load factor of the system, the more fully utilized the resources are. The average load factor of the system is the average value of the load factor of each VM, expressed as:

$$L_s = \frac{1}{m} \sum_{j=1}^{m} \frac{T_j}{T_{\max}}, \tag{5}$$

where T_j denotes the total time that vm_j performs the task and m is the total number of VMs.

2.3   Fitness Function

In this paper search algorithm, the lower the value of the fitness function, the better the optimization result of the algorithm. We use the linear weighting method to transform the multi-objective problem in Sect. 2.2 into a single-objective problem, which is described as follows.

The time, cost and load factor are first constrained using the method in the literature [9]. The task completion time and the total cost of execution are treated as shown in Eqs. (6):

$$f_T = \frac{T_{\max} - T_{Min}}{T_{Max} - T_{Min}}, f_C = \frac{C_s - C_{Min}}{C_{Max} - C_{Min}}, f_L = \frac{l_r - l_{Min}}{l_{Max} - l_{Min}}, \tag{6}$$

where f_T, f_C and f_L are the average load factor metrics for time, cost and after constraint processing. T_{Min}, T_{Max} denote the minimum and maximum total cost of performing the task. C_{Min}, C_{Max} indicates the minimum and maximum total cost of performing the task. l_r, l_{Max} and l_{Min} are the values after taking the inverse of L_s, maximum and minimum load factors. As a result, the fitness optimization function of this paper is set as shown in Eq. (7):

$$f = \omega_1 \cdot f_T + \omega_2 \cdot f_C + \omega_3 \cdot f_L. \tag{7}$$

3 Proposed Algorithm

SSA was used in many scenarios like [10–12]. For cloud computing task scheduling scenarios, SSA is improved as follows.

3.1 Strategy 1: PWLCM Mapping

The PWLCM mapping is used in the search algorithm to initialize the population, which has a simple mathematical form, with the characteristics of randomness, convenience, overall stability and local instability, and the generated sequence has good statistical properties [13], which can satisfy the initial assignment of cloud computing tasks.
The PWLCM as follows:

$$x(t+1) = \begin{cases} m \cdot \left(\dfrac{x(t)}{p} \right), & 0 \le x(t) < p, \\[2mm] m \cdot \left(\dfrac{x(t) - p}{\frac{m}{2} - p} \right), & p \le x(t) < \dfrac{m}{2}, \\[2mm] m \cdot \left(\dfrac{m - p - x(t)}{\frac{m}{2} - p} \right), & \dfrac{m}{2} \le x(t) < 1 - p, \\[2mm] m \cdot \left(\dfrac{m - x(t)}{p} \right), & m - p \le x(t) < m, \end{cases} \tag{8}$$

where $x(t)$ denotes the t-dimensional position of the sparrow, m is the number of virtual nodes and p is the control parameter used to determine the non-overlapping part of the distribution, p is taken as $0.4m$ in this experiment.

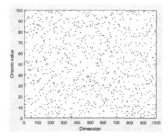

Fig. 1. Initialization population

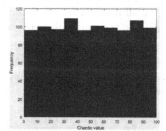

Fig. 2. Frequency of distribution

Figure 1 shows the population distribution after chaotic mapping with PWLCM, and Fig. 2 shows the frequencies of the distribution, where m is taken as 100. It can be seen that the population distribution is random, the distributions do not overlap, and the frequencies of the distributions are similar in each interval, proving that the initial population distribution is uniform.

3.2 Strategy 2: Discoverer Strategy

The search formula of the first stage in MPA [14] is introduced to improve the location update strategy of the discoverer in SSA, as shown in Eq. (9):

$$X_{i,j}(t+1) = \begin{cases} X_{i,j}(t) + P \cdot R \cdot (X_{best}(t) - \beta \cdot X_{i,j}(t)), & R_2 < ST, \\ X_{i,j}(t) + Q \cdot L, & R_2 \geq ST, \end{cases} \quad (9)$$

where $X_{i,j}(t)$ denotes the j-dimensional position information of sparrow i in t iterations, and P is a constant equal to 0.5. R is a vector of uniformly distributed random numbers between 0 and 1 with dimension d, $X_{best}(t)$ is the position of the best individual in the population, β is a random number obeying normal distribution. Q is a random number obeying a normal distribution, and L is an all-1 $1 \times d$ matrix, ST is a constant of [0.5, 1] indicating the safety value and R_2 is a random number of [0, 1] indicating the warning value.

The addition of the random number and the position of the best individual solves the problem of over-convergence point of the original formula at the zero point, and gives the discoverer more search space in the iterative process, and the communication with the optimal solution also makes the position of the optimal solution fully utilized.

3.3 Strategy 3: Joiner Strategy

The strategy of the joiner exists to converge to the optimal solution characteristics, so it is update using the following two methods.

(1) Dynamic Adjustment Factor

As we can see from the joiner's position update, when $i > n/2$, if the number of individuals in the population is too large, i will be too large to make the joiner's position converge to a minimal value, changing i^2 to i expands the search range of the population. When $i \leq n/2$ the sparrow position will be randomly updated into the neighborhood of the best individual, when the number of tasks to be arranged at one time is too large, it will lead to A^+ too small, and the variance from the best individual will keep decreasing after each dimensional update, so that the population will fall into the local optimal solution.

The update formula for the accessions is adjusted as in Eq. (10):

$$X_{i,j}(t+1) = \begin{cases} Q \cdot \exp\left(\dfrac{X_{worst}(t) - X_{i,j}(t)}{i}\right), & if \quad i > \dfrac{n}{2}, \\ X_p(t+1) + |X_{i,j}(t) - X_p(t+1)| \cdot A^+ \cdot L \cdot w, & otherwise, \end{cases}$$
$$(10)$$

where X_{worst} is the position of the worst individual, $A^+ = A^T(AA^T)^{-1}$, A is a $1 \times d$ matrix with elements randomly assigned to 1 or –1, and X_p is the position of the current best finder.

An adjustment factor w is introduced, where w is defined as:

$$w = \frac{(T-t) \cdot n}{4T}, \quad (11)$$

where t and T are the current and total number of iterations. The adjustment factor is large at the beginning of the iteration to facilitate global search, and as the number of iterations increases, the local search capability of the algorithm is enhanced to improve the convergence accuracy.

(2) Greedy Strategy

Based on the study of literature [15,16], this paper suggests that a greedy strategy. The greedy idea is to process a best solution every time. When $i>n/2$, the sparrows in the population that have found food and the hungriest sparrows will direct the movements of the current generation of sparrows. The equation for updating the position of this class of sparrows is shown in (12):

$$X_{i,j}(t+1) = X_{i,j}(t) + a \cdot |X_{best}(t) - X_{worst}(t)|. \tag{12}$$

The greedy strategy is shown in Eq. (13):

$$X_{i,j}(t+1) = \begin{cases} Q \cdot \exp\left(\dfrac{X_{\text{worst}}(t) - X_{i,j}(t)}{i}\right), & if \ f_e \leq f_s, \\ X_{i,j}(t) + a \cdot |X_{\text{best}}(t) - X_{\text{worst}}(t)|, & if \ f_e > f_s, \end{cases} \tag{13}$$

where a is a random number of $[0, 1]$, f_e is the adaptation obtained by the original joiner under the position update strategy at $i>n/2$ and f_s is the adaptation of the individual after the learning strategy.

4 Simulation Results

In this paper, we use CloudSim [17] platform to simulate the cloud environment. The performance of MISSA is verified by comparing ACO, MPA, SSA, and MISSA. We set up 100 tasks and 1000 tasks to simulate both small task size and large task size cases.

4.1 Parameter Setting

In order to combine task completion time, cost and average system load factor, in Eq. (7) we set the $\omega_1 = \omega_2 = \omega_3 = 1/3$. Table 1 shows the experimental parameters of the cloud environment, Table 2 shows the computational cost of the VM, and Table 3 shows the parameters of the algorithm, which are set as follows:

4.2 Numerical Analysis

In the simulation experiments, the task size and the performance of the VM are generated randomly, and each algorithm is run multiple times and its average value is taken for analysis.

(1) The experimental results of the 100 tasks are shown in Fig. 3 and Fig. 4: From the local view of Fig. 3, the starting point of iteration of MISSA is lower compared with other algorithms, which is the result of using chaotic mapping

Table 1. Experimental parameter settings

Parameter	Values
Task size/kb	500–5000
Number of tasks	100, 1000
Number of VMs	20
VM Processing Speed/(MB/s)	500–2000
VN Memory (RAM)/MB	512
VM Bandwidth/(MB/s)	1000

Table 2. VM computing cost settings

Performance	1000MIPS Price
$[500, 1000)$	0.03
$[1000, 1500]$	0.027
$(500, 2000]$	0.025

Table 3. Algorithm parameter settings

Algorithm	Parameters
ACO	$\alpha = 2.5$, $\beta = 3.2$, $\rho = 0.5$, $\gamma = 1.33$, r $= 0.2$
MPA	FADS $= 0.2$, P $= 0.5$
SSA	ST $= 0.7$, PD $= 0.3$*pop, SD $= 0.2$*pop
MISSA	ST $= 0.7$, PD $= 0.3$*pop, SD $= 0.2$*pop

initialization by enhance the degree of individual dispersion. The iteration curves of ACO and MPA are easy to smooth at the beginning of iteration, while MISSA is more capable of jumping out of the local optimum. In the middle of iteration, it can be seen that MISSA can find better scheduling results than other algorithms within a relatively small number of iterations. In the late iteration compared with ACO, MPA, and SSA, MISSA converges faster and finds the global optimal solution easily. From an overall perspective, MISSA converges slightly faster than other algorithms in the case of small-scale tasks, and the iterative results are better.

Figure 4 shows the optimization of each algorithm for total task completion cost, task completion time, and average system load factor under small-scale tasks. From the figure, we can see that ACO is better than SSA in terms of time, but the task cost is slightly higher, the average system load ratio is lower, and the resources are not fully utilized. Compared with ACO, MPA, and SSA, MISSA achieves better results, which proves that the MISSA algorithm has significant advantages in small-scale tasks.

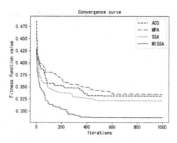

Fig. 3. Convergence curve of the algorithm at the scale of 100 tasks

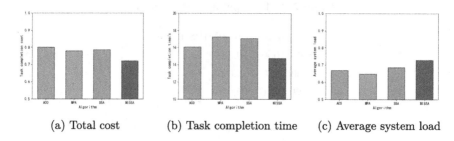

(a) Total cost   (b) Task completion time   (c) Average system load

Fig. 4. Optimization of each objective for a task size of 100

(2) The experimental results for the case of 1000 tasks are shown in Fig. 5 and Fig. 6:

From Fig. 5, it can be seen that MISSA performs better and better in terms of convergence as the number of tasks increases, MISSA also outperforms other algorithms in terms of convergence accuracy.

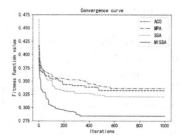

Fig. 5. Convergence curve of the algorithm at the scale of 1000 tasks

Figure 6 shows the optimization of each algorithm under the large-scale task, and MISSA outperforms the ACO, MPA, and SSA algorithms in all three metrics. Combining the convergence curves of the fitness functions for the two task

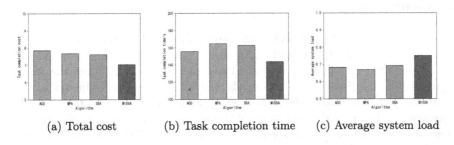

(a) Total cost   (b) Task completion time   (c) Average system load

Fig. 6. Optimization of each objective for a task size of 1000

scales, MISSA improves by 20%, 22%, and 17% compared to the ACO, MPA, and SSA. This shows that the MISSA outperforms the other algorithms for both different task scales.

(3) Comparison of the effects of algorithms with different virtual machine sizes
In order to compare the effects of different VM counts on the stability of the ACO, MPA, and SSA and the MISSA. Compare the fitness function values of these algorithms for 10 and 50 virtual machines sizes and for different numbers of tasks. All parameters were the same as Table 1 except for the modification of the number of VMs, yielding the experimental results shown in Fig. 7.

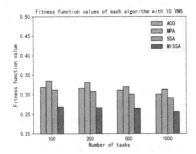
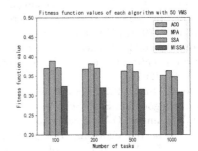

Fig. 7. Value of fitness function for different task sets

From Fig. 7, it can be seen that the fitness value of each algorithm decreases with the number of tasks and increases with the number of VMs. As the number of tasks and the number of VMs rise, the capability of the MISSA algorithm does not diminish, and the values of the fitness function of MISSA decrease by 18% to 20%, 19% to 22%, and 15% to 17% compared to the ACO, MPA, and SSA algorithms, respectively. It proves that the algorithm still has some stability of the algorithm in different scale VM operation environment.

5   Conclusion

In this paper, we improved SSA with multi-strategy. The initialization of the population with PWLCM chaotic mapping enhances the dispersion of individuals. The global search phase of the MPA is introduced to enhance the search capability of the discoverer. The introduction of the dynamic adjustment factor in the joiner enhances the global search ability in the early stage and the convergence ability in the late stage of the algorithm, and the greedy strategy preserves the better solution. From the comparison of the results of simulation experiments, the MISSA proposed in this paper converges faster and with higher convergence accuracy than the standard SSA, and has better ability to jump out of the local optimum, and the scheduling strategy is also better than the ACO, MPA, and SSA. Therefore, the MISSA proposed in this paper can better solve the task scheduling problem in cloud computing.

In this paper, we consider the task completion time, the cost and the average system load. In future studies, the energy consumption of the physical machine as well as the memory of the VM will be considered in the task scheduling process.

References

1. Keshanchi, B., Souri, A., Navimipour, N.J.: An improved genetic algorithm for task scheduling in the cloud environments using the priority queues: formal verification, simulation, and statistical testing. J. Syst. Softw. **124**, 1–21 (2017)
2. Gupta, A., Garg, R.: Load balancing based task scheduling with ACO in cloud computing. In: 2017 International Conference on Computer and Applications (ICCA), pp. 174–179. IEEE (2017)
3. Prem Jacob, T., Pradeep, K.: A multi-objective optimal task scheduling in cloud environment using cuckoo particle swarm optimization. Wirel. Pers. Commun. **109**, 315–331 (2019)
4. Xue, J., Shen, B.: A novel swarm intelligence optimization approach: sparrow search algorithm. Syst. Sci. Control Eng. **8**(1), 22–34 (2020)
5. Abdulhammed, O.Y.: Load balancing of IoT tasks in the cloud computing by using sparrow search algorithm. J. Supercomput. **78**(3), 3266–3287 (2022)
6. Qiu, S., Li, A.: Application of chaos mutation adaptive sparrow search algorithm in edge data compression. Sensors **22**(14), 5425 (2022)
7. Arunarani, A.R., Manjula, D., Sugumaran, V.: Task scheduling techniques in cloud computing: a literature survey. Futur. Gener. Comput. Syst. **91**, 407–415 (2019)
8. Alguliyev, R.M., Imamverdiyev, Y.N., Abdullayeva, F.J.: PSO-based load balancing method in cloud computing. Autom. Control. Comput. Sci. **53**, 45–55 (2019)
9. Woldesenbet, Y.G., Yen, G.G., Tessema, B.G.: Constraint handling in multiobjective evolutionary optimization. IEEE Trans. Evol. Comput. **13**(3), 514–525 (2009)
10. Zhang, Z., He, R., Yang, K.: A bioinspired path planning approach for mobile robots based on improved sparrow search algorithm. Adv. Manuf. **10**(1), 114–130 (2022)

11. Tuerxun, W., Chang, X., Hongyu, G., Zhijie, J., Huajian, Z.: Fault diagnosis of wind turbines based on a support vector machine optimized by the sparrow search algorithm. IEEE Access **9**, 69307–69315 (2021)
12. Liu, T., Yuan, Z., Wu, L., Badami, B.: Optimal brain tumor diagnosis based on deep learning and balanced sparrow search algorithm. Int. J. Imaging Syst. Technol. **31**(4), 1921–1935 (2021)
13. Luo, Y., Zhou, R., Liu, J., Cao, Y., Ding, X.: A parallel image encryption algorithm based on the piecewise linear chaotic map and hyper-chaotic map. Nonlinear Dyn. **93**, 1165–1181 (2018)
14. Faramarzi, A., Heidarinejad, M., Mirjalili, S., Gandomi, A.H.: Marine predators algorithm: a nature-inspired metaheuristic. Expert Syst. Appl. **152**, 113377 (2020)
15. Yan, S., Yang, P., Zhu, D., Zheng, W., Wu, F.: Improved sparrow search algorithm based on iterative local search. Comput. Intell. Neurosci. **2021** (2021)
16. Wang, Z., Huang, X., Zhu, D.: A multistrategy-integrated learning sparrow search algorithm and optimization of engineering problems. Comput. Intell. Neurosci. **2022** (2022)
17. Calheiros, R.N., Ranjan, R., Beloglazov, A., De Rose, C.A., Buyya, R.: CloudSim: a toolkit for modeling and simulation of cloud computing environments and evaluation of resource provisioning algorithms. Softw. Pract. Exp. **41**(1), 23–50 (2011)

Advanced State-Aware Traffic Light Optimization Control with Deep Q-Network

Wenlong Ni[1,2(✉)], Zehong Li[2], Peng Wang[2], and Chuanzhaung Li[2]

[1] School of Computer Information Engineering, JiangXi Normal University,
NanChang, China
wni@jxnu.edu.cn
[2] School of Digital Industry, JiangXi Normal University, ShangRao, China
{Zehong.Li,peng.wang,lichuanzhuang}@jxnu.edu.cn

Abstract. The former traffic light control (TLC) system cannot effectively regulate the traffic conditions dynamically in real time due to urban growth. The Dueling Double Deep Recurrent Q-Network with Attention Mechanism (3DRQN-AM) method for TLC is proposed in this study. The proposed method is based on Deep Q-Network and employs target network, double learning method and dueling network to boost its learning efficiency. In order to integrate the past state of the vehicle's motion trajectory with the current state of the vehicle for the best decision-making, the Long-Short Term Memory (LSTM) is introduced. While this is going on, an Attention Mechanism is introduced to help the neural network automatically focus on crucial state components and improve its capacity to represent state. According to experimental findings, the Dueling Double Deep Q-Network with Attention Mechanism (3DQN-AM), Dueling Double Deep Recurrent Q-Network (3DRQN), Dueling Double Deep Q-Network (3DQN), Fixed-Time-3DRQN-AM (FT-3DRQN-AM) signal management methods are compared. The techniques presented in this work lower the average waiting time under typical traffic flow by about $46.2\%, 53.3\%, 85.1\%,$ and 30.0% respectively, and the average queue length by about $41.9\%, 44.6\%, 76.0\%,$ and 21.7% respectively. Under peak traffic conditions, the average waiting time is decreased by around $20.8\%, 32.1\%,$ $36.7\%,$ and 38.7% respectively, while the average queue is decreased by roughly $2.8\%, 2.8\%, 21.3\%,$ and $44.9\%.$

Keywords: Traffic Signal Control · Attention Mechanism · Deep Reinforcement Learning · Variable Time Interval · Multi-indicator Optimization

1 Introduction

The construction of current roads in cities cannot meet the needs of travel in modern society, resulting in increased travel time, unnecessary fuel consumption, and environmental pollution [1]. There is an urgent need to find a way to alleviate the current traffic congestion. Literature [2,3] proposes a predefined

B. Luo et al. (Eds.): ICONIP 2023, LNCS 14448, pp. 178–190, 2024.
https://doi.org/10.1007/978-981-99-8082-6_14

fixed-duration scheme, which does not efficiently handle dynamic traffic flows in real time, leading to inefficient traffic signal control systems in some cases, to greater traffic congestion. With the wave of artificial intelligence, Deep Learning (DL) and Reinforcement Learning (RL) are being applied to solve various engineering problems. Traffic signal optimization is also one of the hotspots of current research. Deep RL can simulate traffic police to direct traffic. Advanced sensors and internet technology are utilized to obtain real-time data such as vehicle speed and position. The brain of the traffic police is simulated through RL to observe the current traffic situation and learn how to adjust the control strategy of traffic lights according to the rewards given by the environment, as shown in Fig. 1. Where the choice of action will directly determine whether the traffic is congested or not, but most of the studies set the phase duration to a fixed value, which may cause unnecessary delays to vehicles [4,5,9,12]. In this paper, we propose a variable phase duration. That is, each phase corresponds to multiple phase durations. The intelligent body can choose the most appropriate phase action and phase duration according to the traffic state.

Literature [6,7] assumes that traffic states are totally visible and uses Convolutional Neural Network (CNN) to obtain the main features of traffic states. However, this assumption is not applicable in practice. Therefore, it is necessary to consider the case where sensor errors lead to incomplete observed states. In this paper, we take advantage of the fact that vehicle trajectories have the characteristics of continuity and persistence and process the traffic state through recurrent network that it enables the LSTM to effectively capture the long-term dependencies in the sequence of vehicle states and compensate for the missing states caused by the inevitable sensor noise and errors in the real environment [8]. In addition, an attention mechanism is introduced to improve the focus on key state components in the sequence and suppress unimportant features [11].

In summary, the main contributions of this paper are as follows:

- A deep reinforcement learning signal control algorithm based on attention mechanism (3DRQN-AM) is proposed. The neural network can give additional attention to analyzing the sequential changes of vehicle trajectory and compensate the incomplete state acquisition due to sensor fault.
- By using a variable time interval to define the phase duration, the intelligent agent can choose the most suitable action and phase duration according to the traffic status.
- We will simulate normal traffic flow and peak traffic flow on the synthetic dataset for experiments to compare various benchmark algorithms and verify the effectiveness and rationality of the proposed algorithm in this paper.

2   Related Works

In early research, dynamic control of adaptive traffic lights mainly used fuzzy logic and linear programming [8] to solve the problem due to limitations in computational power and data collection and transmission. These methods generally use limited information when modeling road traffic and therefore have limitations in large-scale applications. In order to solve this problem, real-time traffic signal control has become an important topic in this field. Long et al. [9] proposed a self-organizing control model with road capacity as the constraint and minimum expected traffic flow delay as the goal to achieve real-time adaptive control of traffic signal phase timing, and experiments showed that the control effect is better than Sydney Coordinated Adaptive Traffic Control System in the same situation. With the application of DL and RL to traffic signals, many traffic signal optimization methods based on deep learning have emerged. Although such a design takes into account the waste of phase duration, it does not have substantial flexibility in choosing a fixed time interval for the same traffic light phase, which tends to cause moderate convergence of the algorithm.

Yu et al. [11] proposed the principle of minimum pressure difference that analogizing the traffic flow at an intersection to a fluid and traffic congestion is analogous to the resistance of the fluid. By adjusting the control time of the signal, the drag force on the traffic fluid is reduced. Klein et al. [12] used discrete traffic state encode (DTSE) to transform the specific vehicle position and speed in an intersection into a two-dimensional matrix of position-velocity, which effectively reduces the amount of input information and thus the computational complexity of the algorithm. Huang et al. [14] proposed a model based on a 3DQN. The model is based on DQN combining dueling network, Double Q-learning methods and target network to mitigate overestimation and improve data usage efficiency, and verified that the 3DQN model outperforms the DQN model. Kodama et al. [15] proposed a traffic signal control system based on deep reinforcement learning, emphasizing the reinforcement success experience approach, which combines successful experiences in a multi-intelligence environment without information transfer between intersections and uses a dual objective algorithm to reduce the impact of the simultaneous learning problem, and the results show that compared to a conventional traffic light control system using deep RL traffic light waiting time is significantly reduced.

Based on most of the above studies, the instability of real traffic conditions (e.g., sensor failures) is not taken into account. In this paper, we propose a new technical direction, i.e., replacing CNN with LSTM-Attention, so that the intelligent body can also select actions based on historical experience information. In addition, the introduction of the attention mechanism can help the intelligent system to focus on the key traffic flow information such as vehicle density near the intersection, vehicle trajectory, and so on. Thereby providing a more accurate understanding of the current traffic situation.

3   Algorithm Model

In this section, we will define the state, action, and reward used in the 3DRQN-AM algorithm. We provide a detailed explanation of these three crucial factors in reinforcement learning.

3.1 State

In this paper, the traffic state s is represented by the vehicle position, speed, and current traffic light phase and duration. In order to represent the intersection state more accurately, this paper adopts the DTSE method to grid the inbound road in each direction, as shown in Fig. 2. Specifically, each inbound lane is divided into a number of grids, and the width of each grid is the same as the lane width, and the length is the same as the sum of the vehicle length and minimum clearance. If the center of the vehicle is in the grid, 1 is assigned to the grid. Otherwise, 0 is assigned to the grid. Thus, all grids form a vehicle position matrix where the number of rows is equal to the number of lanes and the number of columns is the number of grids in each lane. Similarly, a speed matrix is obtained by assigning the speed of the vehicle to the corresponding cell and normalizing the speed of the vehicle to calculate it. In addition, the current signal state is also an essential factor that affects the intersection traffic efficiency [16]. Setting the traffic light phase encoded as a vector of 1 (green) and 0 (red), if the current signal state is [1, 0, 0, 0, 15] means that the north-south direction is green and the current phase signal duration is 15 s.

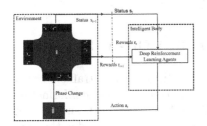

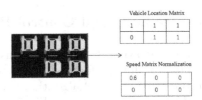

Fig. 1. Intersection signal control model.

Fig. 2. Traffic status information setting.

3.2 Action

Inspired by [10], this paper considers traffic light phase with variable time interval. In this paper, four signal phases are set up: the first phase is north-south straight, right and left turn, the second phase is north-south left turn, the third phase is east-west straight, right and left turn, and the fourth phase is east-west left turn. Three optional time, 10, 15, and 20 s, are set for each phase, constituting the action space set A = [0, 1, 2, 3, 4, 5, 6, 7, 8, 9, 10, 11]. To prevent long waiting time at different intersections, the same action selection can not be made more than three times. And a 3-s yellow light time is set between red and green lights. To verify that the variable duration used in this paper's algorithm works better than the predefined fixed duration. The algorithm of this paper is compared with the FT-3DRQN-AM algorithm. In this case, there are only 4 phases in FT-3DRQN-AM, and each phase is set with a fixed duration of 15 s, otherwise the situation is same as 3DRQN-AM algorithm.

3.3   Reward

In the paper, the number of vehicle stops and the average speed are used to jointly define the reward function, as shown in Eq. 1.

$$r_t = k_1(h_{t-1} - h_t) + k_2 s_t \tag{1}$$

where: k is the weight coefficient, and after several experiments it is known that k_1 is taken as -1 and k_2 is taken as -0.25. h_t is the number of all vehicles stopped at the intersection at moment t, h_{t-1} is the number of all vehicles stopped at the intersection at moment t-1, and s_t is the average speed of all vehicles at moment t. If $h_{t-1} - h_t > 0$, then it means that after executing the action at moment t, it makes the traffic become smoother, and vice versa, it is more congested. The use of multiple indicators to design the reward and punishment function is beneficial for the intelligences to make better decisions and improve the convergence speed of the algorithm.

4   Traffic Signal Control Based on 3DRQN-AM

Due to the existence of sensor transmission data errors in real traffic environments, it is difficult for intelligent bodies to obtain complete traffic information. In this paper, a novel algorithm named 3DRQN-AM is designed, and the network structure is shown in Fig. 5. In this paper, we compensate for the state error by using the feature that the vehicle has continuity and continuity in the direction of traffic road movement, as shown in Fig. 3, at $t-2$ and $t-1$ moments of the vehicle position and speed exercised by the lane and other features at time t is the same, that is, the state of the vehicle has a certain degree of temporal order. We use recurrent network to process the states related to temporality. The attention mechanism is also introduced, in which the vehicle movement direction changes with the vehicle close to the intersection should be given greater weight, so that the intelligent body pays automatic attention to the vital state components and suppresses the unimportant features, thus improving the expression ability of the network.

4.1   Network Model and Algorithm for 3DRQN-AM

In the paper, the deep neural network model of the algorithm and the associated parameters are shown in Fig. 5. The model mainly includes SENet model, LSTM model, and dueling neural network. Among them, LSTM [17] is able to capture the vehicle history state information and its calculation process is as follows: let x_t, h_t and C_t be the inputs at time step t, the control state and the unit state.

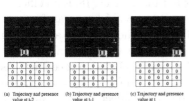

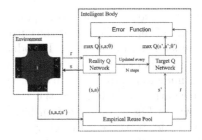

Fig. 3. Presence values of vehicle trajectories at intersections.

Fig. 4. Algorithm flow of 3DRQN-AM.

Given an input sequence $(x_1, x_2,... x_n)$, LSTM computes h-sequence $(h_1, h_2,... h_t)$ and c-sequence $(C_1, C_2,... C_n)$ as follows:

$$f_t = \sigma(W_f \cdot [h_{t-1}, x_t] + b_f) \tag{2}$$

$$i_t = \sigma(W_t \cdot [h_{t-1}, x] + b_t) \tag{3}$$

$$c_t = \tanh(W_C \cdot [h_{t-1}, x] + b_C) \tag{4}$$

$$C_t = f_t * C_{t-1} + i_t * c_t \tag{5}$$

$$o_t = \sigma(W_o \cdot [h_{t-1}, x_t] + b_o) \tag{6}$$

$$h_t = o_t * \tanh(C_t) \tag{7}$$

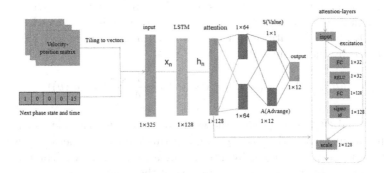

Fig. 5. The architecture of 3DRQN-AM.

Where σ is a logistic sigmoid function, * is an element-wise multiplication, and C_t is the state of the cell that needs to be updated. W and b denote the weight matrix and bias, respectively, and (W_f, b_f), (W_t, b_t), (W_C, b_C), and (W_o, b_o) correspond to the weight matrix and bias of the forgetting gate, the input gate, the state unit, and the output gate, respectively. f_t, i_t and o_t are the activation functions corresponding to the output of the gate. h_t denotes the output of the LSTM cell at the current moment in time. SENet [18] is an efficient channel attention network. Introducing SENet on the LSTM layer can focus on the information features that are more significant to the current task among the numerous input information, suppress the attention to other unimportant information, and improve the efficiency and accuracy of task processing. As shown in Fig. 5, the attention layer mainly contains two operations, excitation and scale. In the excitation operation, the two fully-connected layers are first dimensionally reduced and then dimensionally increased to explicitly model the correlation between states. The first fully-connected layer compresses the 128 channels into 32 channels to reduce the computational effort (ReLU activation function), and the second fully-connected layer restores the 128 channels. The second fully-connected layer restores 128 channels (sigmoid activation function) to obtain normalized (0–1) state weights w:

$$vec(w) = \sigma(F_2 \times \delta(F_1 \times vec(h))) \tag{8}$$

where F_1 denotes the fully connected layer parameter for downscaling; F_2 denotes the fully connected layer parameter of the ascending dimension; σ is the sigmoid function; δ is the ReLU function; vec is the vectorization operator. The scale operation updates the original state matrix h with the state weight matrix w obtained from excitation to obtain the attention input state s_{am}.

$$s_{am} = w \odot h \tag{9}$$

In the traditional DQN, different state behaviors correspond to different value functions $Q(s, a)$, but in fact, in some states, the magnitude of the value function is independent of the action (such as the absence of a car in the traffic environment) and is mainly related to the state value. In order to better express the value function [14], a competitive network architecture is introduced to decompose the Q-value into state value $V(s; \theta)$ and action advantage $A(s, a; \theta)$, which is calculated by the formula 10.

$$Q(s, a'; \theta) = V(s; \theta) + (A(s, a; \theta) - \frac{1}{|A|} \sum_{a'} A(s, a'; \theta)) \tag{10}$$

Traditional DQN use the same network to update the parameters each time, resulting in unstable estimated Q-value. In this paper, two networks with the same structure are established by introducing a target network. The target network is updated by copying the parameters of the reality network to the target network at each step. This ensures that the parameters of the target network are stable over time. If the same network characteristics are used for selecting and evaluating actions for calculating the Q-value, this may lead to an overestimation of the target value [13]. Therefore, the reality network is used to select actions and the target network is used for action evaluation. The formula for this is as follows.

$$Q_{\text{target}} = r + \gamma Q(s', \arg\max(s', a; \theta); \theta') \tag{11}$$

In this paper, the algorithm uses two neural networks for updating the parameters of the neural network. As shown in Fig. 4. In this paper, the algorithm, by observing the state s, the after taking action a through the behavioral strategy $\pi = p(a|s)$, at each time step t with γ as the decay coefficient to obtain the maximum gain r, and use a neural network to approximate the true Q-value, where the reality network is denoted by and the goal network is denoted by, where θ_i and θ'_i to denote the network parameters at the ith iteration, and through a loss function that reduce the error between Q reality and Q target to train the network, the error function is:

$$L_i(\theta_i) = E_{s,a}[(r + \Upsilon Q(s'_i, \arg\max Q(s'_i, a'_i; \theta_i); \theta'_i)) - Q(s_i, a_i; \theta_i)]^2. \quad (12)$$

In the above equation, s represents the current state and s' represents the next state of the current state. θ denotes the realistic network parameters and θ' denotes the target network parameters.

5 Experiment

In this section, we introduce the experimental setup, hyperparameter settings, experimental evaluation, and experimental results. We then compare four benchmark algorithms in simulation experiments that simulate normal traffic flow and peak traffic flow to assess the effectiveness of the algorithms in this paper.

5.1 Simulation Environment and Hyperparameter Setting

The simulation scenario in this paper is a single intersection as shown in Fig. 1. The intersection is connected to four 500-m-long road sections, each with two incoming lanes and two outgoing lanes, respectively. The dataset used in this paper is a synthetic dataset where vehicle generation is simulated according to the Weber distribution. The vehicle length is $5\,m$, acceleration is $1\,m/s^2$ and deceleration is $4.5\,m/s^2$.

Two iterative training methods are used on the model. The first method converges iterations by using the same traffic file as the algorithm's training traffic file. The second method creates a distinct traffic file for each round by setting a random number seed and tying it to the number of iterative rounds. The hyperparameters were derived according to references [14–16] and combined with extensive experiments: The learning rate is 0.001, the batch size is 32, the experience pool size is 3200, the discount factor is 0.75, the number of training rounds is 300, and the training iterations per round are 3600.

5.2 Experimental Evaluation and Analysis of Results

In order to verify the effectiveness of the proposed 3DRQN-AM algorithm. The paper is designed to simulate normal traffic flow in a single intersection, with a vehicle arrival rate of 500 veh/h (veh/h denotes the number of vehicles passing per hour) and peak traffic flow, with a vehicle arrival rate of 1500 veh/h for each road. The algorithm is compared with four benchmark algorithms to analyze the simulation results in terms of three metrics: average cumulative reward, average queue length, and average waiting time.

5.2.1   Training Results

Figure 6 and Fig. 7 show the effect of the algorithm in this paper and the shift of loss value. As can be seen from the graphs, in the pre-training phase, the algorithm of this paper uses ε-greedy strategy for random exploration, which results in large fluctuation of reward value. With the gradual increase of the number of events, the ε-greedy strategy for random exploration decreases, the number of choices for the current basic estimated Q-value increases, and the reward value and loss value gradually stabilize.

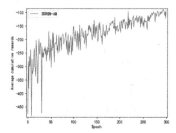
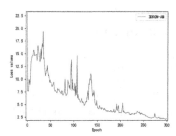

Fig. 6. Convergence of 3DRQN-AM algorithm.

Fig. 7. Change of loss value for 3DRQN-AM.

5.2.2   Performance Comparison of Traffic Signal Control Strategies

(1) Average Cumulative Rewards: From Fig. 8 and Table 1, in normal traffic flow this paper's algorithm converges faster and fluctuates smaller compared to 3DQN-AM, 3DRQN, but the reward value does not modify significantly after

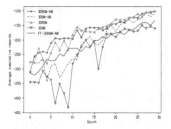
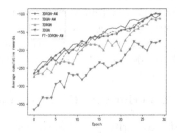

(a) Average cumulative reward variation under normal traffic flow

(b) Change in average cumulative rewards under peak traffic flow

Fig. 8. Variation of average cumulative rewards under different traffic flows

final stabilization. Compared with 3DQN, FT-3DRQN-AM, the reward value of this paper's algorithm is significantly improved. In peak traffic flow this paper's algorithm has no significant change in convergence compared to 3DQN-AM, 3DRQN, and FT-3DRQN-AM but has significant improvement over 3DQN.

(2) Average Vehicle Queue Length: According to Fig. 9a and Table 1, under typical traffic conditions, 3DRQN-AM shortens the average queue by around 41.9%, 44.6%, 76.0% and 21.7% when compared to 3DQN-AM, 3DRQN, 3DQN and FT-3DRQN-AM respectively. The findings for average queue length under peak traffic flow from Fig. 9b and Table 1 demonstrate that 3DRQN-AM lowers around 2.8%, 2.8%, 21.3% and 44.9% compared to 3DQN-AM, 3DRQN, 3DQN and FT-3DRQN-AM respectively.

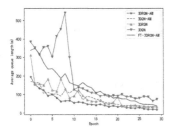 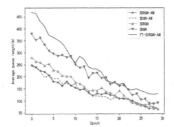

(a) Average queue length variation under normal traffic flow

(b) Average queue length variation under peak traffic flow

Fig. 9. Variation of the average queue length under different traffic flows

(3) Average Waiting Time: Fig. 10 illustrates how the addition of an attention mechanism or LSTM based on the 3DQN algorithm might shorten the average waiting period. In comparison to the four benchmark algorithms, the method in this study has the lowest average vehicle waiting time and the smallest variation in both regular and peak traffic flows. The average waiting time of 3DRQN-AM is decreased by around 46.2%, 53.3%, 85.1% and 30.0% when compared to 3DQN-AM, DRQN, 3DQN and FT-3DRQN-AM respectively, in the situation of typical traffic flow. As can be seen from Fig. 10a and Table 1. As can be shown from Fig. 10b and Table 1, 3DRQN-AM has a lower average waiting time during peak traffic flow than 3DQN-AM, DRQN, and 3DQN, respectively, by 20.8%, 32.1%, 36.7% and 38.7%.

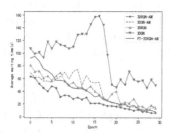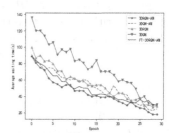

(a) Average waiting time variation under normal traffic flow

(b) Change in average waiting time under peak traffic flow

Fig. 10. Variation of the average queue length under different traffic flows

Table 1. Average value of each algorithm index under different traffic volumes.

Algorithm	Normal traffic flow			Peak traffic flow		
	rewards	queue length/m	waiting time/s	rewards	queue length/m	waiting time/s
3DRQN-AM	−108	36	7	−104	70	19
3DQN-AM	−112	62	13	−104	72	24
3DRQN	−110	65	15	−119	72	28
3DQN	−158	150	47	−139	89	30
FT-3DRQN-AM	−141	46	10	−106	127	31

6 Summary

(1) We suggest the use of LSTM network to improve the problem of incomplete acquisition of traffic status caused by sensor fault by utilizing the continuous and persistent characteristics of vehicle motion direction in order to address the issue of incomplete or inaccurate acquisition of traffic status by deep reinforcement learning control traffic signals. In addition, the attention mechanism is included to enhance the intelligent body's automatic attention to significant state components (such as changes in the direction of travel of the vehicle).
(2) Comparative experiments were conducted at a single intersection establish with normal and peak traffic flows. The experiments demonstrate that 3DRQN-AM is applicable to different traffic environments and can effectively improve the efficiency of traffic intersections, and verify that the algorithm outperforms 3DQN-AM, 3DRQN, 3DQN and FT-3DRQN-AM.
(3) The next research will focus on timing methods for multiple intersection traffic signals, joint timing mechanisms between intersections, and methods for handling complex traffic state information.

References

1. Xu, L., Kun, W., Pengfei, L., Miaoyu, X.: Deep learning short-term traffic flow prediction based on lane changing behavior recognition. In: 2020 2nd International Conference on Information Technology and Computer Application (ITCA), pp. 760–763 (2020). https://doi.org/10.1109/ITCA52113.2020.00163

2. Tran-Van, N.Y., Nguyerr, X.H., Le, K.H.: Towards smart traffic lights based on deep learning and traffic flow information. In: 2022 9th NAFOSTED Conference on Information and Computer Science (NICS), pp. 1–6 (2022). https://doi.org/10.1109/NICS56915.2022.10013375

3. Maduka, N.C., Ajibade, I.I., Bello, M.: Modelling and optimization of smart traffic light control system. In: 2022 IEEE Nigeria 4th International Conference on Disruptive Technologies for Sustainable Development (NIGERCON), pp. 1–5 (2022). https://doi.org/10.1109/NIGERCON54645.2022.9803073

4. Abhishek, A., Nayak, P., Hegde, K.P., Lakshmi Prasad, A., Nagegowda, K.S.: Smart traffic light controller using deep reinforcement learning. In: 2022 3rd International Conference for Emerging Technology (INCET), pp. 1–5 (2022). https://doi.org/10.1109/INCET54531.2022.9824501

5. Raeisi, M., Mahboob, A.S.: Intelligent control of urban intersection traffic light based on reinforcement learning algorithm. In: 2021 26th International Computer Conference, Computer Society of Iran (CSICC), pp. 1–5 (2021). https://doi.org/10.1109/CSICC52343.2021.9420622

6. Wu, T., Kong, F., Peng, P., Fan, Z.: Road intersection model based reward function design in deep q-learning network for traffic light control. In: 2020 5th International Conference on Robotics and Automation Engineering (ICRAE), pp. 182–186 (2020). https://doi.org/10.1109/ICRAE50850.2020.9310858

7. Yang, D., Seo, S.W.: Traffic light detection using attention-guided continuous conditional random fields. In: 2022 International Conference on Electronics, Information, and Communication (ICEIC), pp. 1–3 (2022). https://doi.org/10.1109/ICEIC54506.2022.9748468

8. Huang, L., Huang, H., Xiao, J., Lv, X.: Research on adaptive adjustment method of intelligent traffic light based on real-time traffic flow detection. In: 2022 5th International Conference on Artificial Intelligence and Big Data (ICAIBD), pp. 644–647 (2022). https://doi.org/10.1109/ICAIBD55127.2022.9820545

9. Long, G., Wang, A., Jiang, T.: Traffic signal self-organizing control with road capacity constraints. IEEE Trans. Intell. Transp. Syst. **23**(10), 18502–18511 (2022). https://doi.org/10.1109/TITS.2022.3152060

10. Zhu, Y., Cai, M., Schwarz, C., Junchao, L., Shaoping, X.: Intelligent traffic light via policy-based deep reinforcement learning. Int. J. Intell. Transport. Syst. Res. **20**, 734–744 (2022). https://doi.org/10.1007/s13177-022-00321-5

11. Yu, P., Luo, J.: Minimize pressure difference traffic signal control based on deep reinforcement learning. In: 2022 41st Chinese Control Conference (CCC), pp. 5493–5498 (2022). https://doi.org/10.23919/CCC55666.2022.9901790

12. Klein, N., Prünte, J.: A new deep reinforcement learning algorithm for the online stochastic profitable tour problem. In: 2022 IEEE International Conference on Industrial Engineering and Engineering Management (IEEM), pp. 0635–0639 (2022). https://doi.org/10.1109/IEEM55944.2022.9989933

13. Gu, J., Fang, Y., Sheng, Z., Wen, P.: Double deep q-network with a dual-agent for traffic signal control. Appl. Sci. **10**, 1622 (2020). https://doi.org/10.3390/app10051622

14. Huang, J., et al.: Application of deep reinforcement learning in optimization of traffic signal control. In: 2021 IEEE 23rd International Conference on High Performance Computing & Communications, pp. 1958–1964 (2021). https://doi.org/10.1109/HPCC-DSS-SmartCity-DependSys53884.2021.00293
15. Kodama, N., Harada, T., Miyazaki, K.: Traffic signal control system using deep reinforcement learning with emphasis on reinforcing successful experiences. IEEE Access **10**, 128943–128950 (2022). https://doi.org/10.1109/ACCESS.2022.3225431
16. Xu, J., Li, L., Zhu, R., Lv, P.: Communication information fusion based multi-agent reinforcement learning for adaptive traffic light control. In: 2023 3rd International Conference on Neural Networks, Information and Communication Engineering (NNICE), pp. 488–492 (2023). https://doi.org/10.1109/NNICE58320.2023.10105755
17. Choe, C.J., Baek, S., Woon, B., Kong, S.H.: Deep q learning with LSTM for traffic light control. In: 2018 24th Asia-Pacific Conference on Communications (APCC), pp. 331–336 (2018). https://doi.org/10.1109/APCC.2018.8633520
18. Zhang, X., Liang, X., Zhiyuli, A., Zhang, S., Xu, R., Wu, B.: At-lstm: an attention-based lstm model for financial time series prediction. In: IOP Conference Series: Materials Science and Engineering, vol. 569, no. 5, pp. 052037 (2019). https://doi.org/10.1088/1757-899X/569/5/052037

Impulsive Accelerated Reinforcement Learning for H_∞ Control

Yan Wu, Shixian Luo$^{(\boxtimes)}$, and Yan Jiang

School of Electrical Engineering, Guangxi University, Nanning 530004,
People's Republic of China
shixian_luo@gxu.edu.cn, shixianluo@126.com

Abstract. This paper revisits reinforcement learning for H_∞ control of affine nonlinear systems with partially unknown dynamics. By incorporating an impulsive momentum-based control into the conventional critic neural network, an impulsive accelerated reinforcement learning algorithm with a restart mechanism is proposed to improve the convergence speed and transient performance compared to traditional gradient descent-based techniques or continuously accelerated gradient methods. Moreover, by utilizing the quasi-periodic Lyapunov function method, sufficient condition for input-to-state stability with respect to approximation errors of the closed-loop system is established. A numerical example with comparisons is provided to illustrate the theoretical results.

Keywords: Reinforcement Learning · Impulsive systems · H_∞ control

1 Introduction

H_∞ control has attracted extensive research attention due to its ability to handle external disturbances. It achieves robust stability and performance by optimizing controller design to minimize the sensitivity of the system to uncertainties and disturbances. Many important works about H_∞ control have been reported in [1–5].

Remarkably, most existing results on H_∞ control require complete system parameter information. Accurate dynamic models of practical systems are difficult to obtain, which is the key challenge to solving model-based control problems. Recently, the so-called model-free reinforcement learning algorithms based on data-assisted feedback control techniques have been proposed and successfully applied to various real-life scenarios, such as autonomous driving [6], cyber-physical systems [7], and industrial robots [8]. However, compared to model-based H_∞ control, there have been fewer achievements in model-free H_∞ control, perhaps because striking a balance between the performance and robustness of the system in the model-free case is a challenging problem. Some scholars attempted to solve this challenge. For instance, the works

This work is supported by the National Natural Science Foundation of China under Grant 62003104, the Guangxi Natural Science Foundation under Grant 2022GXNSFBA035649, the Guangxi Science and Technology Planning Project under Grant AD23026217, and the Guangxi University Natural Science and Technological Innovation Development Multiplication Plan Project under Grant 2023BZRC018.

B. Luo et al. (Eds.): ICONIP 2023, LNCS 14448, pp. 191–203, 2024.
https://doi.org/10.1007/978-981-99-8082-6_15

[9,10] proposed model-free algorithms to solve H_∞ control problems for a class of discrete-time and continuous-time systems, respectively. In addition, the complexity of H_∞ control algorithms usually leads to slow convergence, which is a drawback of reinforcement learning. To overcome the limitation of the convergence, some accelerated reinforcement learning algorithms have been proposed. For instance, Wilson et al. [11] introduced auxiliary variables to accelerate reinforcement learning using a continuous-time system; Pertsch et al. [12] proposed an accelerated reinforcement learning algorithm using an acquired skill prior; the works of [13,14] proposed a hybrid acceleration algorithm based on hybrid systems framework. It should be mentioned that these accelerated algorithms lack robustness when applied to systems with disturbance. Moreover, to the best of our knowledge, there are currently no accelerated reinforcement learning algorithms for solving H_∞ control problems.

Motivated by the abovementioned analysis, we revisit the H_∞ control problems in the reinforcement learning framework. Following the works [2,16], we use the critic neural network to approximate the optimal value function, leveraging sufficiently rich recorded data and current output data to find the solution to the Hamilton-Jacobi-Bellman (HJB) equation in real time. The main contributions are as follows:

i) An accelerated reinforcement learning algorithm based on the impulsive control framework is proposed for H_∞ control of affine nonlinear systems. The accelerated reinforcement learning algorithm introduces an impulsive system restarting the gradient at the impulses to regulate and (or) improve the stability and transient properties of the accelerated time-varying optimal dynamics. In addition, the impulsive system includes several adjustable variables, which makes the closed-loop system more flexible and optimizable.

ii) Utilizing the quasi-periodic Lyapunov function method, sufficient condition for input-to-state stability with respect to approximation errors of the closed-loop system in the impulsive control framework is established.

The remainder of this paper is organized as follows. In Sect. 2, the notations and the dynamic model are introduced. Section 3 introduces the impulsive momentum-based critic dynamics and analyzes its stability. Section 4 presents the simulation results. Finally, the conclusion is drawn in Sect. 5.

2   System Description and Preliminaries

Notation. $\mathbb{R}^n$ and $\mathbb{R}^{m \times n}$ represent the set of real n-vector and $m \times n$ matrices, respectively. The notation $P \succ 0 (\succeq 0)$ means that the matrix P is positive (semi) definite. $\|\cdot\|_p = [\sum_{i=1}^{n} |x_i|^p]^{1/p}$, $1 \le p < \infty$, denotes the p-norm of a vector. Given $A \in \mathbb{R}^{n \times n}$, we use $\lambda_{\max}(A)(\lambda_{\min}(A))$ to represent maximun eigenvalue (minimum eigenvalue). For any $M \succ 0$, $\|x\|_M$ defines as $x^T M x$. The gradient of a scalar-valued function V with respect to a vector-valued variable x is defined as a row vector and is denoted by ∇V. A function $\gamma : \mathbb{R}_{\ge 0} \to \mathbb{R}_{\ge 0}$ is said to be of class-$\mathcal{K}_\infty$, if it is continuous, zero at zero, and strictly increasing. Given a compact set $\mathcal{A} \subset \mathbb{R}^n$ and a vector $z \in \mathbb{R}^n$ one uses $|z|_{\mathcal{A}} = \min_{s \in \mathcal{A}} \|z - s\|_2$ to represent the minimum distance of z to $\mathcal{A}$.

Consider the following affine nonlinear systems

$$\begin{cases} \dot{x}(t) = f(x) + g(x)u + h(x)v \\ z(t) = k(x) \end{cases}, \tag{1}$$

where $x \in \mathbb{R}^n$, $u \in \mathbb{R}^{m_u}$ and $v \in \mathbb{R}^{m_v}$ represent the system state, control input, and the disturbance, respectively; $f : \mathbb{R}^n \to \mathbb{R}^n$ with $f(0) = 0$. $g(x)$, $h(x)$, and $k(x)$ are known continuous vector or matrix functions of appropriate dimensions.

This paper has two main objectives: 1) use a neural network to find the state feedback control law u^* such that the system (1) is asymptotically stable, and has L_2-gain less than or equal to $\bar{\gamma}$, that is,

$$\int_0^\infty (\|z(t)\|_2^2 + \|u(t)\|_{R_u}^2) dt \le \bar{\gamma}^2 \int_0^\infty \|v(t)\|_2^2 dt, \tag{2}$$

where $0 \prec R_u \in \mathbb{R}^{m_u}$, and $\bar{\gamma} > 0$ is some prescribed level of disturbance attenuation; 2) propose an impulsive momentum-based control strategy to accelerate the convergence speed of critic weights and improve transient performance of the closed-loop system.

According to the objective function (2), the value function is given by

$$V(x_0) \triangleq \min_u \max_v \int_0^\infty (\|z(t)\|_2^2 + \|u(t)\|_{R_u}^2 - \bar{\gamma}^2 \|v(t)\|_2^2) dt.$$

Define the Hamiltonian function as

$$H(x, u, v, \nabla V^T(x)) \triangleq \|z\|_2^2 + \|u\|_{R_u}^2 - \bar{\gamma}^2 \|v\|_2^2 + \nabla V(x) F(x, u, v), \tag{3}$$

where $F(x, u, v) = f(x) + g(x)u + h(x)v$.

Applying the stationarity conditions to the Eq. (3), one has

$$\begin{cases} \frac{\partial H(x,u,v,\nabla V^T(x))}{\partial u} = 0 \Rightarrow u^*(x) = -\frac{1}{2} R_u^{-1} g^T(x) \nabla V^T(x), \\ \frac{\partial H(x,u,v,\nabla V^T(x))}{\partial v} = 0 \Rightarrow v^*(x) = \frac{1}{2} \bar{\gamma}^{-2} h^T(x) \nabla V^T(x). \end{cases}$$

On the other hand, the $V(x)$ satisfies the HJB equation [18]

$$\frac{\partial V(x)}{\partial t} = -H(x, u^*, v^*, \nabla V^T(x)), \forall x \in \mathbb{R}^n.$$

Since $\frac{\partial V(x)}{\partial t} = 0$, one obtains

$$\nabla V(x) F(x, u^*, v^*) + \|z\|_2^2 + \|u\|_{R_u}^2 - \bar{\gamma}^2 \|v\|_2^2 = 0. \tag{4}$$

Unfortunately, obtaining an explicit expression for $V(x)$ using Eq. (4) and subsequently solving the optimal control u^* and the worst disturbance v^* is often a challenging problem. The next section will utilize a critic neural network and Eq. (4) to estimate u^* and v^*.

3   Critic Dynamics

To utilize Eq. (4) and according to the Weierstrass higher order approximation theorem
[19], one can use a neural network to approximate $V(x)$ on a compact set $\mathcal{D} \subset \mathbb{R}^n$, and
then obtain $\nabla V(x)$.

$$
\begin{cases}
V(x) = W^{*T}\phi(x) + \epsilon(x), \ x \in \mathcal{D} \\
\nabla V(x) = W^{*T}\nabla\phi(x) + \nabla\epsilon(x) ,
\end{cases}
\tag{5}
$$

where $W^* \in \mathbb{R}^N$ is an ideal constant weight vector satisfying $\|W^*\|_2 \le \bar{W}, \phi : \mathcal{D} \to$
$\mathbb{R}^N$ is a vector of basis functions, and $\epsilon(x)$ is an approximation error.

The optimal control law and the worst disturbance can be approximated as

$$
\begin{cases}
u^*(x) = -\dfrac{1}{2}R_u^{-1}g^T(x)[\nabla\phi^T(x)W^* + \nabla\epsilon^T(x)], & \text{(6a)} \\[2mm]
v^*(x) = \dfrac{1}{2}\bar{\gamma}^{-2}h^T(x)[\nabla\phi^T(x)W^* + \nabla\epsilon^T(x)]. \ x \in \mathcal{D} & \text{(6b)}
\end{cases}
$$

Substituting (5) into (3), one gets the approximate HJB equation

$$
\begin{aligned}
H(x, u^*, v^*, \nabla\phi^T(x)W^*) &= W^{*T}\nabla\phi(x)F(x, u^*, v^*) + \|z\|_2^2 + \|u^*\|_{R_u}^2 - \bar{\gamma}^2\|v^*\|_2^2 \\
&= \epsilon_{HJB} , \ x \in \mathcal{D}
\end{aligned}
\tag{7}
$$

where $\epsilon_{HJB} \triangleq -\nabla\epsilon(x)F(x, u^*, v^*)$ is the residual error.

Due to the ideal weights W^* being unknown, one defines a critic neural network
with estimates $\hat{W}$ as

$$
\hat{V}(x) = \hat{W}^T\phi(x), \ x \in \mathcal{D}
\tag{8}
$$

and the approximate control and disturbance policies become

$$
\begin{cases}
\hat{u}(x) = -\dfrac{1}{2}R_u^{-1}g^T(x)\nabla\phi^T(x)\hat{W}, & \text{(9a)} \\[2mm]
\hat{v}(x) = \dfrac{1}{2}\bar{\gamma}^{-2}h^T(x)\nabla\phi^T(x)\hat{W}. \ x \in \mathcal{D} & \text{(9b)}
\end{cases}
$$

Substituting (8), (9a), and (9b) into (3), one can get the approximate HJB equation

$$
\begin{aligned}
\hat{H}(x, \hat{u}, \hat{v}, \nabla\phi^T(x)\hat{W}) &= \hat{W}^T\nabla\phi(x)F(x, \hat{u}, \hat{v}) + \|z\|_2^2 \\
&\quad + \|\hat{u}\|_{R_u}^2 - \bar{\gamma}^2\|\hat{v}\|_2^2, \ x \in \mathcal{D}
\end{aligned}
\tag{10}
$$

We will employ Eq. (10) to the updating dynamics of the critic parameter $\hat{W}$. To further
analysis, one provides an assumption that is accompanied by a sufficient "richness" in
recorded data.

Assumption 1. *[14]. Let $\{\varphi(x_i, \hat{u}(x_i), \hat{v}(x_i))\}_{i=1}^k$ be a sequence of recorded data, and define*

$$
\begin{cases}
\Lambda = \sum_{i=1}^k \bar{\varphi}(x_i, \hat{u}(x_i), \hat{v}(x_i)) \bar{\varphi}^T(x_i, \hat{u}(x_i), \hat{v}(x_i)), \\
\bar{\varphi}(x, u, v) = \frac{\varphi(x,u,v)}{1 + \varphi^T(x,u,v)\varphi(x,u,v)}, \\
\varphi(x, u, v) = \nabla \phi(x) F(x, u, v).
\end{cases}
$$

Then there exists $\underline{\lambda} > 0$ such that $\Lambda \succeq \underline{\lambda} I$.

Remark 1. Acquiring exploration signals that meet the requirements of the standard persistence of excitation attributes can be challenging in practical scenarios. However, Assumption 1 is relatively more lenient, which is comparatively easier to satisfy. One may refer [20] for more details.

Assumption 2. *There exists positive constants $\bar{g}$, $\bar{h}$, $\bar{\phi}$, $\overline{\epsilon_{HJB}}$, $\overline{d\phi}$, $\bar{\epsilon}$, and $\overline{d\epsilon}$ such that*

$$\|g(x)\|_2 \le \bar{g}, \ \|h(x)\|_2 \le \bar{h}, \ \|\phi(x)\|_2 \le \bar{\phi}, \ \|\epsilon_{HJB}\|_2 \le \overline{\epsilon_{HJB}},$$

$$\|\nabla \phi(x)\|_2 \le \overline{d\phi}, \ \|\epsilon(x)\|_2 \le \bar{\epsilon}, \ \|\nabla \epsilon(x)\|_2 \le \overline{d\epsilon},$$

hold for all $x \in \mathcal{D}$.

Now, we will define the Hamiltonian estimation error for the current time and for the recorded data at $0 \le t_1 < \cdots < t_i < t$, respectively.

$$
\begin{aligned}
e(t) =& \hat{H}(x(t), \hat{u}(t), \hat{v}(t), \nabla^T \phi(x)\hat{W}(t)) - H(x, u^*(x), v^*(x), \nabla V) \\
=& \hat{W}^T(t)\varphi(x(t), \hat{u}(t), \hat{v}(t)) + \|z(t)\|_2^2 + \|\hat{u}(t)\|_{R_u}^2 - \bar{\gamma}^2 \|\hat{v}(t)\|_2^2,
\end{aligned}
$$

$$
\begin{aligned}
e(t_i, t) =& \hat{H}(x(t_i), \hat{u}(t_i), \hat{v}(t_i), \nabla^T \phi(x_i)\hat{W}(t)) - H(x_i, u^*(x_i), v^*(x_i), \nabla V) \\
=& \hat{W}^T(t)\varphi(x(t_i), \hat{u}(t_i), \hat{v}(t_i)) + \|z(t_i)\|_2^2 + \|\hat{u}(t_i)\|_{R_u}^2 - \bar{\gamma}^2 \|\hat{v}(t_i)\|_2^2,
\end{aligned}
$$

Next, the total error for Hamiltonian estimates is defined as follows

$$
\begin{aligned}
E(\hat{W}(t)) =& \frac{1}{2} \frac{(e(t))^2}{(1 + \varphi^T(x(t), \hat{u}(t), \hat{v}(t))\varphi(x(t), \hat{u}(t), \hat{v}(t)))^2} \\
&+ \frac{1}{2} \sum_{i=1}^k \frac{(e(t_i))^2}{(1 + \varphi^T(x_i, \hat{u}(t_i), \hat{v}(t_i))\varphi(x_i, \hat{u}(t_i), \hat{v}(t_i)))^2},
\end{aligned}
\tag{11}
$$

then, one computes the gradient of (11) with respect to $\hat{W}(t)$ as follows

$$
\begin{aligned}
\nabla E(\hat{W}(t)) =& \bar{\varphi}(x(t), \hat{u}(t), \hat{v}(t)) \bar{\varphi}^T(x(t), \hat{u}(t), \hat{v}(t))\hat{W}(t) \\
&+ \frac{\varphi(x(t), \hat{u}(t), \hat{v}(t))[\|z(t)\|_2^2 + \|\hat{u}(t)\|_{R_u}^2 - \bar{\gamma}^2 \|\hat{v}(t)\|_2^2]}{\Upsilon^2} \\
&+ \Lambda \hat{W}(t) + \sum_{i=1}^k \frac{\varphi(x_i, \hat{u}(t_i), \hat{v}(t_i))[\|z(t_i)\|_2^2 + \|\hat{u}(t_i)\|_{R_u}^2 - \bar{\gamma}^2 \|\hat{v}(t_i)\|_2^2]}{\Upsilon_i^2},
\end{aligned}
\tag{12}
$$

where $\Upsilon = 1 + \varphi^T(x, \hat{u}, \hat{v})\varphi(x, \hat{u}, \hat{v})$, $\Upsilon_i = 1 + \varphi^T(x_i, \hat{u}(t_i), \hat{v}(t_i))\varphi(x_i, \hat{u}(t_i), \hat{v}(t_i))$, Λ and $\bar{\varphi}$ are defined in Assumption 1.

3.1   Critic Dynamic via Impulsive Momentum-Based Control

To increase the speed of convergence of $\hat{W}$ to W^*, we propose an impulsive momentum-based control strategy inspired by the literature [13–15, 17]. Specifically, we consider the following impulsive dynamics system

$$
\begin{cases}
\dot{y}(t) = \begin{bmatrix} \frac{1}{t - t_k + T_0}(\xi - \hat{W}(t)) \\ -\alpha \nabla E(\hat{W}(t)) \end{bmatrix}, & t \in (t_k, t_{k+1}) \\
y(t_k^+) = \begin{bmatrix} \hat{W}(t_k^-) \\ \hat{W}(t_k^-) \end{bmatrix}, & k \in \mathbb{N}_0,
\end{cases}
\tag{13}
$$

with initial condition $y_0 = [\hat{W}_0^T, \xi_0^T]^T$, where $y = [\hat{W}^T, \xi^T]^T$, ξ is auxiliary variables; $\alpha > 0$ is the learning rate; $t_k = kT$, $k \in \mathbb{N}_0$, are known as impulsive instants; $T > 0$ is the impulsive period; $T_0 > 0$ is a tunable parameter; $y(t_k^+) = \lim_{s \to 0^+} y(t_k + s)$ and $y(t_k^-) = \lim_{s \to 0^+} y(t_k - s)$. It is clear that the auxiliary variable ξ is reset to $\hat{W}$ at impulsive instants, while $\hat{W}$ is unaffected.

The following theorem provides the sufficient condition for stability and convergence of impulsive system (13).

Theorem 1. *Given a critic neural network with N-dimensional basis functions $\phi(x)$ and a compact set $\mathcal{D} \subset \mathbb{R}^n$. Suppose Assumptions 1 and 2 are satisfied. If use the controller $\hat{u}$ (defined in (9a)) with disturbance $\hat{v}$ (defined in (9b)) on the plant (1), and the following conditions hold*

$$
T > \frac{1}{2\alpha \underline{\lambda}} \text{ and } T + T_0 < \frac{1}{\alpha \bar{\varphi}^T(x, \hat{u}, \hat{v})\bar{\varphi}(x, \hat{u}, \hat{v})},
\tag{14}
$$

then there exist positive constants ρ_1, μ_1, and class-$\mathcal{K}_\infty$ functions γ_{1i}, $i \in \overline{1,2}$ such that

$$
\|\hat{W}(t) - W^*\|_2 \leq \rho_1 e^{-\mu_1 t} |y_0|_{\mathcal{A}} + \gamma_{11}(\overline{\epsilon_{HJB}}) + \gamma_{12}(\overline{d\epsilon}), \ t \geq 0.
\tag{15}
$$

where $\mathcal{A} := \{[\hat{W}^T, \xi^T]^T \in \mathbb{R}^{2N} \mid \hat{W} = W^*, \xi = \hat{W}\}$.

Proof. According to (7), we have

$$
\varphi^T(x, u^*(x), v^*(x))\hat{W} + \|z\|_2^2 + \|u^*\|_{R_u}^2 - \bar{\gamma}^2\|v^*\|_2^2 = \epsilon_{HJB}.
\tag{16}
$$

Then, substituting (16) into (12), the equation (12) can be rewritten as

$$
\begin{aligned}
\nabla E(\hat{W}(t)) =&\Theta(\hat{W} - W^*) + \frac{\varphi(x,\hat{u},\hat{v})\epsilon_{HJB}(x)}{\Upsilon^2} + \sum_{i=1}^{k}\frac{\varphi(x_i,\hat{u}_i,\hat{v}_i)\epsilon_{HJB}(x_i)}{\Upsilon_i^2}\\
&+ \frac{\varphi(x,\hat{u},\hat{v})[\nabla\phi(x)(g(x)(\hat{u}-u^*)+h(x)(\hat{v}-v^*))]^T W^*}{\Upsilon^2}\\
&+ \frac{\varphi(x,\hat{u},\hat{v})[\|\hat{u}\|_{R_u}^2 - \|u^*\|_{R_u}^2 - \bar{\gamma}^2\|\hat{v}\|_2^2 + \bar{\gamma}^2\|v^*\|_2^2]}{\Upsilon^2}\\
&+ \sum_{i=1}^{k}\frac{\varphi(x_i,\hat{u}_i,\hat{v}_i)[\|\hat{u}_i\|_{R_u}^2 - \|u_i^*\|_{R_u}^2 - \bar{\gamma}^2\|\hat{v}_i\|_2^2 + \bar{\gamma}^2\|v_i^*\|_2^2]}{\Upsilon_i^2}\\
&+ \sum_{i=1}^{k}\frac{\varphi(x_i,\hat{u}_i,\hat{v}_i)[\nabla\phi(x_i)(g(x_i)(\hat{u}_i-u_i^*)+h(x_i)(\hat{v}_i-v_i^*))]^T W^*}{\Upsilon_i^2}\\
=&\Theta(\hat{W}-W^*) + \mu_\epsilon(x,\hat{u},\hat{v}) + \mu_{\epsilon_i}(x_i,\hat{u}_i,\hat{v}_i) + \zeta^i(x,\hat{u},\hat{v}), \qquad (17)
\end{aligned}
$$

where $\Theta = \bar{\varphi}(x,\hat{u},\hat{v})\bar{\varphi}^T(x,\hat{u},\hat{v}) + \Lambda$, $\mu_\epsilon(x,\hat{u},\hat{v}) = \frac{\varphi(x,\hat{u},\hat{v})}{\Upsilon^2}[W^{*T}\nabla\phi(x)(g(x)(\hat{u}-u^*) + h(x)(\hat{v}-v^*)) + \|\hat{u}\|_{R_u}^2 - \|u^*\|_{R_u}^2 - \bar{\gamma}^2\|\hat{v}\|_2^2 + \bar{\gamma}^2\|v^*\|_2^2]$, $\mu_{\epsilon_i}(x_i,\hat{u}_i,\hat{v}_i) = \sum_{i=1}^{k}\{\frac{\varphi(x_i,\hat{u}_i,\hat{v}_i)}{\Upsilon_i^2}[W^{*T}\nabla\phi(x_i)(g(x_i)(\hat{u}_i-u_i^*)+h(x_i)(\hat{v}_i-v_i^*)) + \|\hat{u}_i\|_{R_u}^2 - \|u_i^*\|_{R_u}^2 - \bar{\gamma}^2\|\hat{v}_i\|_2^2 + \bar{\gamma}^2\|v_i^*\|_2^2]\}$, $\zeta^i(x,\hat{u},\hat{v}) = \frac{\varphi(x,\hat{u},\hat{v})\epsilon_{HJB}(x)}{\Upsilon^2} + \sum_{i=1}^{k}\frac{\varphi(x_i,\hat{u}_i,\hat{v}_i)\epsilon_{HJB}(x_i)}{\Upsilon_i^2}$.

According to Assumption 2, combining $\|\frac{\varphi(x,u,v)}{\Upsilon^2}\|_2 \le \delta$, (6a) and (6b), we obtain

$$
\begin{aligned}
\mu_\epsilon(x,\hat{u},\hat{v}) =&\frac{\varphi(x,\hat{u},\hat{v})}{\Upsilon^2}[(\hat{u}-u^*)^T R_u(\hat{u}-u^*) - (\hat{v}-v^*)^T \bar{\gamma}^2 I(\hat{v}-v^*)\\
&- \nabla\epsilon(x)g(x)(\hat{u}-u^*) - \nabla\epsilon(x)h(x)(\hat{v}-v^*)]\\
\Rightarrow \|\mu_\epsilon(x,\hat{u},\hat{v})\|_2 \le&\delta(\lambda_{\max}(R_u)\|\hat{u}-u^*\|_2^2 + \bar{\gamma}^2\|\hat{v}-v^*\|_2^2\\
&+ \overline{d\epsilon g}\|\hat{u}-u^*\|_2 + \overline{d\epsilon h}\|\hat{v}-v^*\|_2). \qquad (18)
\end{aligned}
$$

Similarly, we derive

$$
\begin{cases}
\|\mu_{\epsilon_i}(x_i,\hat{u}_i,\hat{v}_i)\|_2 \le \delta\sum_{i=1}^{k}\{(\lambda_{\max}(R_u)\|\hat{u}_i-u_i^*\|_2^2\\
\quad + \bar{\gamma}^2\|\hat{v}_i-v_i^*\|_2^2 + \overline{d\epsilon g}\|\hat{u}_i-u_i^*\|_2 + \overline{d\epsilon h}\|\hat{v}_i-v_i^*\|_2)\}, & (19a)\\
\|\zeta^i(x,\hat{u},\hat{v})\|_2 \le \delta(1+k)\overline{\epsilon_{HJB}}. & (19b)
\end{cases}
$$

Consider the following candidate Lyapunov function

$$
\begin{aligned}
V(t,y) =&\frac{\|\xi - \hat{W}\|_2^2}{2} + \frac{\|\xi - W^*\|_2^2}{2}\\
&+ \alpha(t - t_k + T_0)(\hat{W} - W^*)^T \Lambda(\hat{W} - W^*). \qquad (20)
\end{aligned}
$$

Then, compute the derivative of (20) and then substitute the result into Eqs. (13) and (17). One has

$$
\begin{aligned}
\dot{V}(t,y) = & -\frac{1}{t - t_k + T_0}\|\xi - \hat{W}\|_2^2 + 2\alpha(\hat{W} - W^*)^T \Lambda(\xi - W^*) - \alpha(\xi - \hat{W})^T \nabla E(\hat{W}) \\
& - \alpha(\xi - W^*)^T \nabla E(\hat{W}) + \alpha(\hat{W} - W^*)^T \Lambda(\hat{W} - W^*) \\
= & -\frac{1}{t - t_k + T_0}\|\xi - \hat{W}\|_2^2 + (\xi - \hat{W})^T (2\alpha\Lambda - 2\alpha\Theta)(\hat{W} - W^*) \\
& - (\hat{W} - W^*)^T (\alpha\Theta - \alpha\Lambda)(\hat{W} - W^*) - 2\alpha(\xi - \hat{W})^T (\mu_\epsilon + \mu_{\epsilon_i} + \zeta^i) \\
& - \alpha(\hat{W} - W^*)^T (\mu_\epsilon + \mu_{\epsilon_i} + \zeta^i) \\
= & -[(\xi - \hat{W})^T \ (\hat{W} - W^*)^T]\Delta \begin{bmatrix} (\xi - \hat{W}) \\ (\hat{W} - W^*) \end{bmatrix} - 2\alpha(\xi - \hat{W})^T (\mu_\epsilon + \mu_{\epsilon_i} + \zeta^i) \\
& - \alpha(\hat{W} - W^*)^T (\mu_\epsilon + \mu_{\epsilon_i} + \zeta^i),
\end{aligned}
$$

where $\Delta = \begin{bmatrix} \frac{1}{t - t_k + T_0}I & \alpha\bar{\varphi}\bar{\varphi}^T \\ \alpha\bar{\varphi}\bar{\varphi}^T & \alpha\bar{\varphi}\bar{\varphi}^T \end{bmatrix}$.

According to Schur's Complement Lemma, $\Delta \succ 0$ if and only if

$$
\alpha\bar{\varphi}\bar{\varphi}^T - \alpha\bar{\varphi}\bar{\varphi}^T (t - t_k + T_0)\alpha\bar{\varphi}\bar{\varphi}^T \succ 0 \Leftrightarrow (\alpha - \alpha^2(t - t_k + T_0)\bar{\varphi}^T \bar{\varphi})\bar{\varphi}\bar{\varphi}^T \succ 0
$$

Moreover, if (14) holds, one can guarantee $\Delta \succ 0$.

Based on $a + b \leq \sqrt{2(a^2 + b^2)}$, $\theta \in (0, 1)$, (18), (19a) and (19b), we have

$$
\begin{aligned}
\dot{V}(t,y) \leq & -\lambda_{\min}(\Delta)[(\xi - \hat{W})^T \ (\hat{W} - W^*)^T] \begin{bmatrix} (\xi - \hat{W}) \\ (\hat{W} - W^*) \end{bmatrix} \\
& + 2\alpha\|\xi - \hat{W}\|_2\|\mu_\epsilon + \mu_{\epsilon_i} + \zeta^i\| + 2\alpha\|\hat{W} - W^*\|_2\|\mu_\epsilon + \mu_{\epsilon_i} + \zeta^i\|_2 \\
\leq & -(1 - \theta)\lambda_{\min}(\Delta)|y|_\mathcal{A}^2 - \theta\lambda_{\min}(\Delta)|y|_\mathcal{A}^2 + 2\sqrt{2}\alpha\|\mu_\epsilon + \mu_{\epsilon_i} + \zeta^i\|_2|y|_\mathcal{A} \\
\leq & -(1 - \theta)\lambda_{\min}(\Delta)|y|_\mathcal{A}^2, \ \forall |y|_\mathcal{A} \geq \frac{2\sqrt{2}\alpha\|\mu_\epsilon + \mu_{\epsilon_i} + \zeta^i\|_2}{\theta\lambda_{\min}(\Delta)}. \quad (21)
\end{aligned}
$$

On the other hand, according to Assumption 1 and (20), the relationship between the candidate Lyapunov function $V(t, y)$ at impulse instants is given as follows

$$
\begin{aligned}
V(t^+, y^+) - V(t^-, y^-) = & \frac{\|\hat{W} - W^*\|_2^2}{2} - \alpha T(\hat{W} - W^*)^T \Lambda(\hat{W} - W^*) \\
& - \frac{\|\xi - \hat{W}\|_2^2 + \|\xi - W^*\|_2^2}{2} \\
\leq & \frac{1 - 2\alpha T\underline{\lambda}}{2}\|\hat{W} - W^*\|_2^2 - \frac{\|\xi - \hat{W}\|_2^2 + \|\xi - W^*\|_2^2}{2}.
\end{aligned}
$$

If (14) holds, one obtains $V(t^+, y^+) - V(t^-, y^-) \leq 0$. Moreover, using (6a), (6b), (9a) and (9b), one obtains

$$
\begin{cases}
\|\hat{u} - u^*\|_2 \leq \frac{1}{2}\lambda_{\max}(R_u^{-1})\bar{g}[\overline{d\phi}\|\hat{W} - W^*\|_2 + \overline{d\epsilon}], \\
\|\hat{v} - v^*\|_2 = \frac{1}{2\bar{\gamma}^2}\bar{h}[\overline{d\phi}\|\hat{W} - W^*\|_2 + \overline{d\epsilon}].
\end{cases} \quad (22)
$$

According to [14, Th.1], and deflate using (22), we derive

$$
\begin{aligned}
||\hat{W}(t) - W^*||_2 &\le \rho e^{-\mu t}|y_0|_{\mathcal{A}} + \gamma_1(||\hat{u}(t) - u^*(t)||_2) \\
&\quad + \gamma_2(||\hat{v}(t) - v^*||_2) + \gamma_3(\overline{\epsilon_{HJB}}) + \gamma_4(\overline{d\epsilon}) \\
&\le \rho_1 e^{-\mu_1 t}|y_0|_{\mathcal{A}} + \gamma_{11}(\overline{\epsilon_{HJB}}) + \gamma_{12}(\overline{d\epsilon}),
\end{aligned}
$$

where ρ, ρ_1, μ and μ_1 are positive constants, $\gamma_i, i \in \overline{1,4}$ and $\gamma_{1i}, i \in \overline{1,2}$ are class-$\mathcal{K}_\infty$ functions. The proof of Theorem 1 is completed.

Remark 2. The traditional gradient descent method in neural networks experiences a reduced rate of gradient change in the mid and late stages, resulting in a slower update speed for critic weights. However, impulsive momentum-based accelerated critic architecture restarts the gradients at impulsive instants, inducing an acceleration flow that effectively enhances the update speed of critic weights. Specifically, the main source of acceleration is achieved by incorporating momentum into the gradient-based dynamics, these dynamics to be viewed as the continuous-time counterpart of Nesterov's accelerated optimization algorithm, which minimizes smooth convex functions at a rate of $\mathcal{O}(1/t^2)$ (see [15,17] for a detailed description).

According to system (1) and impulsive system (13), define $\mathcal{Z} = [x^T, \hat{W}^T, \xi^T]^T$, $\mathcal{Z} \in \mathcal{D}_\mathcal{Z} \subset \mathbb{R}^{n+2N}$, $\mathcal{A}_\mathcal{Z} = \{[x, \hat{W}^T, \xi^T]^T \in \mathbb{R}^{n+2N} \mid x = 0, \hat{W} = W^*, \xi = \hat{W}\}$, the closed-loop system is given as

$$
\begin{cases}
\dot{\mathcal{Z}}(t) = \begin{bmatrix} f(x) + g(x)\hat{u}(t) + h(x)\hat{v}(t) \\ \frac{1}{t-t_k+T_0}(\xi - \hat{W}(t)) \\ -\alpha \nabla E(\hat{W}(t)) \end{bmatrix}, \quad t \in (t_k, t_{k+1}) \\[3em]
\mathcal{Z}(t_k^+) = \begin{bmatrix} x(t_k^-) \\ \hat{W}(t_k^-) \\ \hat{W}(t_k^-) \end{bmatrix}, \quad k \in \mathbb{N}_0
\end{cases}
\tag{23}
$$

with initial condition $\mathcal{Z}_0 = [x_0^T, \hat{W}_0^T, \xi_0^T]^T$.

Next, the stability analysis of the closed-loop system will be carried out.

Theorem 2. *Given a critic neural network with N-dimensional basis functions $\phi(x)$ and a compact set $\mathcal{Z} \in \mathcal{D}_\mathcal{Z} \subset \mathbb{R}^{n+2N}$. Suppose Assumptions 1 and 2 are satisfied. If the tunable parameters α, T_0, T satisfy (14) and using the approximate optimal controller $\hat{u}$ (defined in (9a)) with worst-case disturbance $\hat{v}$ (defined in (9b)) on the plant (1), then there exist positive constants $\rho_2, \mu_2,$ and class-$\mathcal{K}_\infty$ functions $\gamma_{2i}, i \in \overline{1,2}$ such that*

$$
|\mathcal{Z}(t)|_{\mathcal{A}_\mathcal{Z}} \le \rho_2 e^{-\mu_2 t}|\mathcal{Z}_0|_{\mathcal{A}_\mathcal{Z}} + \gamma_{21}(\overline{\epsilon_{HJB}}) + \gamma_{22}(\overline{d\epsilon}), \quad t \ge 0.
\tag{24}
$$

Proof. Consider the following candidate Lyapunov function

$$
V(t, \mathcal{Z}) = V(x) + V(t, y).
$$

According to (22) and [21, Th. 1], then consider the derivative of $V(x)$, we have

$$
\begin{aligned}
\dot{V}(x) &= \nabla V(x)[f(x) + g(x)\hat{u} + h(x)\hat{v}] \\
&= \nabla V(x)[f(x) + g(x)u^* + h(x)v^* + g(x)(\hat{u} - u^*) + h(x)\hat{v} - v^*] \\
&\leq \frac{1}{2}(\bar{W}\overline{d\phi} + \overline{de})(\bar{g}^2 \lambda_{\max}(R_u^{-1}) + \bar{h}^2\bar{\gamma}^{-2})(\overline{d\phi}\|\hat{W} - W^*\|_2 + \overline{de}) - cV(x),
\end{aligned}
$$

where $c > 0$, there exists a number ε, such that $\frac{1}{2}(\bar{W}\overline{d\phi} + \overline{de})(\bar{g}^2\lambda_{\max}(R_u^{-1}) + \bar{h}^2\bar{\gamma}^{-2})(\overline{d\phi}\|\hat{W} - W^*\|_2 + \overline{de}) \leq \varepsilon\|\hat{W} - W^*\|_2$. Hence

$$
\dot{V}(x) \leq -cV(x) + \varepsilon\|\hat{W} - W^*\|_2 \leq -cV(x) + \varepsilon\|y\|_{\mathcal{A}}. \tag{25}
$$

Using (21) and (25), we obtain

$$
\begin{aligned}
V(t, \mathcal{Z}) &= V(x) + V(t, y) \\
&\leq -(1 - \theta)\lambda_{\min}(\Delta)|y|_{\mathcal{A}}^2 - cV(x) - \theta\lambda_{\min}(\Delta)|y|_{\mathcal{A}}^2 \\
&\quad + (\varepsilon + 2\sqrt{2}\alpha\|\mu_\epsilon + \mu_{\epsilon_i} + \zeta^i\|_2)|y|_{\mathcal{A}} \\
&\leq -(1 - \theta)\lambda_{\min}(\Delta)|y|_{\mathcal{A}}^2 - cV(x), \ \forall\, |y|_{\mathcal{A}} \geq \frac{\varepsilon + 2\sqrt{2}\alpha\|\mu_\epsilon + \mu_{\epsilon_i} + \zeta^i\|_2}{\theta\lambda_{\min}(\Delta)}.
\end{aligned}
$$

On the other hand, due to x do not changing at impulse instants, we have $V(t^+, \mathcal{Z}^+) - V(t^-, \mathcal{Z}^-) = V(t^+, y^+) - V(t^-, y^-) \leq 0$. Furthermore, according to Theorem 1 and [14, Th.1], one has

$$
\begin{aligned}
|\mathcal{Z}|_{\mathcal{A}_{\mathcal{Z}}} &\leq \bar{\rho}e^{-\bar{\mu}t}|\mathcal{Z}_0|_{\mathcal{A}_{\mathcal{Z}}} + \gamma_1(\|\hat{W} - W^*\|_2) + \gamma_2(\overline{\epsilon_{HJB}}) + \gamma_3(\overline{de}) \\
&\leq \rho_2 e^{-\mu_2 t}|\mathcal{Z}_0|_{\mathcal{A}_{\mathcal{Z}}} + \gamma_{21}(\overline{\epsilon_{HJB}}) + \gamma_{22}(\overline{de}),
\end{aligned}
$$

where $\bar{\rho}, \rho_2, \bar{\mu}$ and μ_2 are positive constants, $\gamma_i, i \in \overline{1,3}$ and $\gamma_{2i}, i \in \overline{1,2}$ are class-$\mathcal{K}_\infty$ functions. The proof of Theorem 2 is completed.

4   Simulation Results

Consider the following affine nonlinear system [21],

$$
\begin{cases}
\dot{x}(t) = f(x) + g(x)u + h(x)v \\
z(t) = x
\end{cases},
$$

where $f(x) = [\tanh(x_1 + x_2)x_2, \ -\tanh(x_1 + x_2)x_2]^T$, $g(x) = I_2$, $h(x) = -I_2$, and $v(t) = 2e^{-0.1t}\cos(t)$.

Select the impulsive period $T = 0.5$s, learning rate $\alpha = 60$, and $T_0 = 0.01$s, such that (14) hold. Then, other parameters are chosen as $R_u = \frac{1}{2}I_2$, $\bar{\gamma} = 5$ and $x_0 = [0, 0]^T$. The initial weights $\hat{W}_0$ are randomly initialized within the interval $[-1, 1]$, the basis function $\phi(x) = [x_1^2, \ x_1 x_2, \ x_2^2]^T$. To ensure that the collected data satisfies Assumption 1, a detection signal lasting 0.5 s is added to (9a).

Figure 1(a) illustrates the state response process of the affine nonlinear system (1), indicating its convergence to the origin at approximately 60 s. Moreover, Fig. 1(b) depicts the state trajectory of the impulsive system (13). Figure 2 presents a comparison among the impulsive-based critic neural network, the continuous accelerated critic neural network [11], and the traditional critic neural network. Results showcase the impulsive-

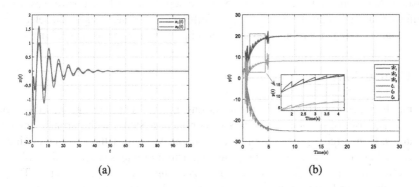

(a)   (b)

Fig. 1. (a) State response of system (1); (b) The trajectory of the impulsive period system (13).

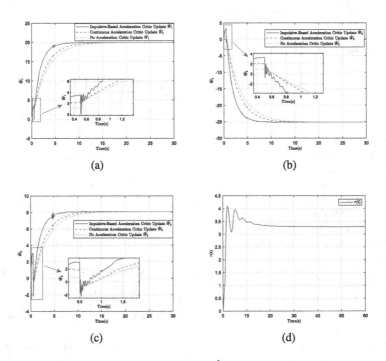

(a)   (b)

(c)   (d)

Fig. 2. (a) Convergence process of critics' weight $\hat{W}_1$ for the three methods, (b) Convergence process of critics' weight $\hat{W}_2$ for the three methods, (c) Convergence process of critics' weight $\hat{W}_3$ for the three methods, (d) The curve of $r(t)$.

based critic neural network achieving the swiftest convergence to steady-state, followed by the continuous accelerated variant, and lastly, the traditional one. To demonstrate the correlation between L_2-gain and time, define the following ratio of disturbance attenuation as $r(t) = \left(\frac{\int_0^t (\|z(\tau)\|_2^2 + \|u(\tau)\|_{R_u}^2) \mathrm{d}\tau}{\int_0^t \|v(\tau)\|_2^2 \mathrm{d}\tau} \right)^{\frac{1}{2}}$. Figure 2(d) illustrates the curve of $r(t)$, where it converges to 3.293 ($< \bar{\gamma} = 5$) as time increases, which implies that (2) is satisfied. Additionally, Eqs. (9a)–(10) reveal the dependency of the control law $\hat{u}$ on $\bar{\gamma}$, leading to varying control performance for different $\bar{\gamma}$. Further insights on selecting an appropriate $\bar{\gamma}$ to balance robustness and control performance are discussed in [5].

5 Conclusion

An impulsive accelerated reinforcement learning algorithm has been proposed for solving the H_∞ control problem of a class of nonlinear systems with partially unknown dynamics. Compared with the traditional gradient descent method or the continuous accelerated gradient methods, the proposed algorithm provides several adjustable variables, effectively improving the convergence speed and transient performance. The stability of the closed-loop system has been analyzed by the quasi-periodic Lyapunov function method. Potential extensions include exploring similar accelerated training techniques for the actor neural network.

References

1. Yang, X., Xu, M., Wei, Q.: Adaptive dynamic programming for nonlinear-constrained H_∞ control. IEEE Trans. Syst. Man Cybern. Syst. **53**(7), 4393–4403 (2023)
2. Luo, B., Wu, H.N., Huang, T.: Off-policy reinforcement learning for H_∞ control design. IEEE Trans. Cybern. **45**(1), 65–76 (2014)
3. Wang, D., Mu, C.X., Liu, D.R., Ma, H.W.: On mixed data and event driven design for adaptive-critic-based nonlinear H_∞ control. IEEE Trans. Neural Networks Learn. Syst. **29**(4), 993–1005 (2017)
4. Kiumarsi, B., Vamvoudakis, K.G., Modares, H., Lewis, F.L.: Optimal and autonomous control using reinforcement learning: a survey. IEEE Trans. Neural Networks Learn. Syst. **29**(6), 2042–2062 (2017)
5. Başar, T., Bernhard, P.: H_∞ Optimal Control and Related Minimax Design Problems: A Dynamic Game Approach. Springer, Boston (2008). https://doi.org/10.1007/978-0-8176-4757-5
6. Liang, X., Liu, Y., Chen, T., Liu, M., Yang, Q.: Federated transfer reinforcement learning for autonomous driving. In: IFederated and Transfer Learning, pp. 357–371 (2022)
7. Stanly, J.J., Priyadarsini, M.J.P., Parameshachari, B.D., Karimi, H.R., Gurumoorthy, S.: Deep Q-network with reinforcement learning for fault detection in cyber-physical systems. J. Circuits, Syst. Comput. **31**(9), 2250158 (2022)
8. Liu, Q., Liu, Z., Xiong, B., Xu, W., Liu, Y.: Deep reinforcement learning-based safe interaction for industrial human-robot collaboration using intrinsic reward function. Adv. Eng. Inform. **49**, 101360 (2021)
9. Yang, Y., Wan, Y., Zhu, J., Lewis, F.L.: H_∞ tracking control for linear discrete-time systems: model-free Q-learning designs. IEEE Control Syst. Lett. **5**(1), 175–180 (2020)

10. Vamvoudakis, K.G., Ferraz, H.: Event-triggered H_∞ control for unknown continuous-time linear systems using Q-learning. In: 2016 IEEE 55th Conference on Decision and Control (CDC), pp. 1376–1381 (2016)
11. Wilson, A.C., Recht, B., Jordan, M.I.: A Lyapunov analysis of accelerated methods in optimization. J. Mach. Learn. Res. **22**(1), 5040–5073 (2021)
12. Pertsch, K., Lee, Y., Lim, J.: Accelerating reinforcement learning with learned skill priors. In: Conference on Robot Learning, pp. 188–204 (2021)
13. Poveda, J.I., Li, N.: Robust hybrid zero-order optimization algorithms with acceleration via averaging in time. Automatica, pp. 109361 (2021)
14. Ochoa, D.E., Poveda, J.I.: Accelerated continuous-time approximate dynamic programming via data-assisted hybrid control. IFAC-PapersOnLine **55**(12), 561–566 (2022)
15. Su, W., Boyd, S., Candes, E.: A differential equation for modeling Nesterov's accelerated gradient method: theory and insights. J. Mach. Learn. Res. **17**(153), 1–43 (2016)
16. Wang, D., He, H.B., Liu, D.R.: Adaptive critic nonlinear robust control: a survey. IEEE Trans. Cybern. **47**(10), 3429–3451 (2017)
17. Brendan, O.D., Candes, E.: Adaptive restart for accelerated gradient schemes. Found. Comput. Math. **15**, 715–732 (2015)
18. Kamalapurkar, R., Walters, P., Rosenfeld, J., Dixon, W.: Reinforcement Learning for Optimal Feedback Control. Springer, Cham (2018). https://doi.org/10.1007/978-3-319-78384-0
19. Hornik, K., Stinchcombe, M., White, H.: Universal approximation of an unknown mapping and its derivatives using multilayer feedforward networks. Neural Networks **3**(5), 551–560 (1990)
20. Chowdhary, G., Johnson, E.: Concurrent learning for convergence in adaptive control without persistency of excitation. In: 49th IEEE Conference on Decision and Control (CDC), pp. 3674–3679 (2010)
21. Kokolakis, N.M.T., Vamvoudakis, K.G.: Safety-aware pursuit-evasion games in unknown environments using gaussian processes and finite-time convergent reinforcement learning. IEEE Trans. Neural Networks Learn. Syst. (2022). https://doi.org/10.1109/TNNLS.2022.3203977

MRRC: Multi-agent Reinforcement Learning with Rectification Capability in Cooperative Tasks

Sheng Yu, Wei Zhu$^{(\boxtimes)}$, Shuhong Liu, Zhengwen Gong, and Haoran Chen

School of Information and Communication, National University of Defense Technology, Wuhan 430014, China
yusheng17@nudt.edu.cn, zhuwei929@hotmail.com

Abstract. Motivated by the centralised training with decentralised execution (CTDE) paradigm, multi-agent reinforcement learning (MARL) algorithms have made significant strides in addressing cooperative tasks. However, the challenges of sparse environmental rewards and limited scalability have impeded further advancements in MARL. In response, MRRC, a novel actor-critic-based approach is proposed. MRRC tackles the sparse reward problem by equipping each agent with both an individual policy and a cooperative policy, harnessing the benefits of the individual policy's rapid convergence and the cooperative policy's global optimality. To enhance scalability, MRRC employs a monotonic mix network to rectify the state-action value function Q for each agent, yielding the joint value function Q_{tot} to facilitate global updates of the entire critic network. Additionally, the Gumbel-Softmax technique is introduced to rectify discrete actions, enabling MRRC to handle discrete tasks effectively. By comparing MRRC with advanced baseline algorithms in the "Predator-Prey" and challenging "SMAC" environments, as well as conducting ablation experiments, the superior performance of MRRC is demonstrated in this study. The experimental results reveal the efficacy of MRRC in reward-sparse environments and its ability to scale well with increasing numbers of agents.

Keywords: Multi-agent reinforcement learning · Cooperative task · Individual reward rectification · Monotonic mix function

1 Introduction

In recent years, artificial intelligence (AI) technologies have been extensively applied in various domains such as medical services [3,15], multiplayer games [13,34], and image processing [4,25]. Within these domains, effectively handling multi-agent cooperative tasks has emerged as a crucial challenge. The advent of the centralised training with decentralised execution (CTDE) [11] paradigm has significantly advanced the field of multi-agent reinforcement learning (MARL) in addressing such tasks through centralized agent training. However, challenges arise due to the conditions under which the environment provides rewards to agents

[19]. For instance, in games like Go, it is difficult to determine the winner until the game's completes, making it arduous for agents to obtain rewards during gameplay. Consequently, this hampers the convergence of the algorithm, increases the cost of sample acquisition, and diminishes the utilization of sample [1].

To tackle the sparse reward problem, researchers have proposed various methods, including reward function reconstruction [10,20] and multi-objective learning [7,23]. While reconstructing the reward function can partially alleviate the slow update issue caused by sparse rewards, it typically applies exclusively to specific tasks. Moreover, migrating to a new environment necessitates designing a novel reward function. Additionally, reconstructed reward functions occasionally misguide agents and require constant debugging to provide meaningful guidance, which reduces efficiency [5]. On the other hand, multi-objective learning involves augmenting the original learning goal, enabling agents to receive rewards upon reaching certain milestones. This approach expedites training in the initial stages, and the agent's policy improves through accumulated successful experiences. However, these additional rewards introduce significant errors that accumulate over time, ultimately impeding the attainment of a globally optimal solution [16].

As the number of agents participating in cooperative tasks grows, several MARL algorithms, including FOP [35], QPLEX [28], and MADDPG [12], encounter challenges related to dimensionality. These challenges, commonly referred to as dimensional catastrophes, hinder the efficient operation of the algorithms. To enhance the scalability of algorithms within the CTDE framework, researchers have proposed various methodologies, including value-based approaches [21,26,31] and actor-critic methods [32,36]. The value-based approach typically revolves around Q-values updated by agents. It often employs a mix network that combines the Q-values of each agent resulting in a joint value function Q_{tot} with global nature. This approach improves scalability to some extent. Nonetheless, in this approach, the agent needs to recalculate the value function and select the next action at each moment t, leading to a significant reduction in operational efficiency. In contrast, algorithms based on the actor-critic framework utilize policy gradients for updates. By incorporating a global state value function, these algorithms can achieve greater scalability. However, policy gradients are normally employed in tasks with continuous action spaces, which imposes certain limitations [2].

A novel method, Multi-agent Reinforcement learning with Rectification Capability in cooperative tasks (MRRC), is proposed in this study to address the challenges of sparse reward and scalability in multi-agent cooperative tasks. The MRRC method incorporates both individual and cooperative policies for each agent, with the individual policy aiming for individual rewards and the cooperative policy targeting cooperative rewards. By employing an actor network, the individual reward rectifies the policy and produces the action a, while the critic network generates Q-values based on this action. These Q-values are further adjusted by the mix network to form a global value function Q_{tot}. The loss value is calculated using Q_{tot}, and the MRRC's network is updated accordingly.

MRRC incorporates a discrete action corrector to handle discrete actions and estimate the policy gradient. The key contributions of this study are as follows:

- A novel MARL method based on the actor-critic framework is proposed, specifically designed to tackle multi-agent cooperative tasks.
- An individual reward rectification module is introduced which equips each agent with two policies: a individual policy and a cooperative policy. By combining the benefits of rapid convergence from the individual policy and the global optimal solution from the cooperative policy, the actor network efficiently outputs the optimal next action, effectively addressing the challenges posed by sparse rewards.
- The mix network rectification module and the discrete action rectification module are introduced in this study, which significantly enhance the scalability of the algorithm. The linear combination of the value functions Q from all agents forms the global joint value function Q_{tot}, which proves to be effective, even in situations with changing environments.

2   Background

2.1   Dec-POMDP

In the pursuit of solving fully cooperative multi-agent tasks, it is often necessary to decompose the task and model it as a decentralized partially observable Markov decision process (Dec-POMDP) [17]. This modeling approach is normally represented by a one-tuple, denoted as $G \ =< \ S, A, P, r, \gamma, N, \Omega, O >$. Among the tuple $N \equiv \{1, 2, ..., n\}$ represents the set of all agents, S represents the set of environment states, and the observation $o_i \in \Omega$ is obtained from the observation kernel $O(s, i)$. The discount factor is denoted as $\gamma \in [0, 1)$. At time step t, each agent i selects an action $a_i \in A$ based on its observation o_i and the collective actions of all agents form a joint action denoted as a. This joint action influences the environment, leading to a change in the state and the generation of a reward value $r = R(s, a)$. Each agent then transitions to the next state s' through the state transition function $P(s'|s, a) : S \times A \rightarrow [0, 1]$. Based on their respective observation values, each agent maintains its own action-observation history, denoted as $\tau_i \in T \equiv \{\Omega \times A\}$. The overarching objective throughout this process is to discover a joint policy that maximizes the joint action-value function $Q(s_t, a_t) = \mathbb{E}[R_t|s_t, a_t]$. In tasks where actions are continuous, agents utilize a continuous policy μ, while discrete actions correspond to a discrete policy π.

2.2   MERL and IRAT

MERL [14] and IRAT [30] are both approaches aimed at addressing the sparse reward problem in multi-agent tasks. They utilize gradient-based optimizers to train agents with higher reward potential, ultimately maximizing the overall reward.

MERL takes a hierarchical approach and employs two optimization processes to handle different objectives. Initially, an evolutionary algorithm is utilized to neuroevolve the entire population of agents, focusing on maximizing the collective reward in scenarios with sparse rewards. Subsequently, policy gradients are incorporated into the agent population for training. This involves information transfer between agents and the population, and culminates in the optimization of the overall policy using the acquired rewards.

In contrast, IRAT utilizes a team policy to guide the optimization of agent-specific policies. For a specific agent i, the optimization goal is to maximize its individual rewards, denoted as $J(\theta_i) = \mathbb{E}\left[\sum_{t=0}^{\infty} \gamma^t r_i^t\right]$. Since the team policy in IRAT represents a joint policy, the actions outputted by each agent i are executed using the respective individual policies π_i. The objective for each agent i is to maximize the following function:

$$J(\theta_i) = \begin{cases} \mathbb{E}[\max(J^{CLIP}(\theta_i), J^{IRAT}(\theta_i))], & \sigma_i^t \leq 1 \\ \mathbb{E}[\min(J^{CLIP}(\theta_i), J^{IRAT}(\theta_i))], & \sigma_i^t > 1 \end{cases}, \tag{1}$$

where σ_i^t is the coefficient of similarity. $J^{IRAT}(\theta_i) = \mathbb{E}\left[\text{clip}\left(\sigma_i^t(\theta_i), 1 - \xi, 1 + \xi\right) A_i^t\right]$ is a cooperation-oriented goal and $J^{CLIP}(\theta_i) = \mathbb{E}[\min(\eta_i^t A_i^t, \text{clip}(\eta_i^t, 1 - \epsilon, 1 + \epsilon) A_i^t)]$ is a goal reward of a single agent.

2.3 MADDPG, QMIX and FACMAC

MADDPG [12], QMIX [21], and FACMAC [18] are all algorithms employed in the CTDE paradigm to address cooperative tasks involving multiple agents. QMIX is a value-based algorithm, while MADDPG and FACMAC are actor-critic framework-based algorithms. In MADDPG, each agent possesses its own independent actor and critic, enabling autonomous learning. For agent i, a joint action value function $Q_i^\mu(s, a_1, ..., a_n; \phi_i)$ is maintained, and the critic network is updated by minimizing the following loss function:

$$LOSS(\phi_i) = \mathbb{E}_D[(y^i - Q_i^\mu(s, a_1, ..., a_n; \phi_i))^2], \tag{2}$$

where $y^i = r_i + \gamma Q_i^\mu(s', a_1', ..., a_n'; \phi_i^-), a_i' = \mu_i(\tau_i'; \theta_i^-)$ and r_i represent the reward for agent i, $\{a_1', ..., a_n'\}$ denotes the set of actions taken by other agents obtained from the replay buffer D, and ϕ_i^- represents the parameter of the target critic for agent i. In contrast, the QMIX algorithm builds upon Q-learning and utilizes a linear, monotonic mix function to compose the global value function Q_{tot}, expressed as:

$$Q_{tot}(\boldsymbol{\tau}, \boldsymbol{a}, s; \boldsymbol{\phi}, \omega) = f_\omega(s, Q_1(\tau_1, a_1; \phi_1), ..., Q_n(\tau_n, a_n; \phi_n)), \tag{3}$$

where ω represents the parameter of the monotonic mixture function f.

Taking advantage of both MADDPG and QMIX, the FACMAC algorithm is introduced. It incorporates mix functions within the actor-critic framework. Specifically, the Q-values of each agent is combined with nonlinear functions to obtain the joint action value function $Q_{tot}^\mu(\boldsymbol{\tau}, \boldsymbol{a}, s; \boldsymbol{\phi}, \omega) =$

$g_\omega(s, \{Q_i^{\mu_i}(\tau_i, a_i; \phi_i)\}_{i=1}^n)$, parameters ϕ and ϕ_i represent the parameters of the joint value function Q and $Q_i^{\mu_i}$ for agent i, respectively. The parameter of the nonlinear function is represented by ω. The critic network is updated by minimizing the following loss function:

$$LOSS(\phi, \omega) = \mathbb{E}_D[(y^{tot} - Q_{tot}^\mu(\tau, a, s; \phi, \omega))^2], \tag{4}$$

where $y^{tot} = r + \gamma Q_{tot}^\mu(\tau', \mu(\tau'; \theta^-), s'; \phi^-, \omega^-)$, ω^-, θ^-, and ϕ^- denote the parameters of the mix function, the target actor network, and the critic network, respectively.

3   Method

This section describes the components and processes of MRRC, and the overall framework of MRRC is shown in Fig. 1.

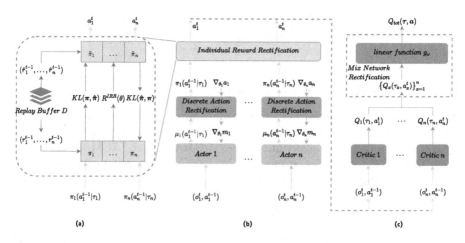

Fig. 1. Overview of the proposed method MRRC. **(a)** The overall process of individual reward rectification. **(b)** Sampling of continuous policy and policy gradients using discrete action rectification. **(c)** Computation of the joint value function Q_{tot} using mix network

3.1   Individual Reward Rectification

By leveraging the widespread implementation of the CTDE paradigm in MARL algorithms, researchers have achieved remarkable advancements in addressing collaborative multi-agent tasks, including QMIX [21] and MADDPG [12]. Nevertheless, such tasks often provide cooperative rewards exclusively upon accomplishing specific objectives. As the complexity of the environment escalates, agents encounter difficulties in attaining rewards within a designated timeframe, leading to sparse reward predicaments which yield sluggish policy convergence [33]. To tackle this issue, the integration of individual reward rectification into the actor-critic framework is proposed in this study. Individual reward rectification involves utilizing each agent's personal reward to rectify cooperative reward, thereby mitigating the sparsity inherent in the reward environment. Unlike IRAT, this approach necessitates the acquisition of two policies per agent-an individual policy and a cooperative policy-rendering it more compatible with the actor-critic framework and significantly enhancing the agent's learning efficiency. The exact process of Individual Reward Rectification is shown in Fig. 2.

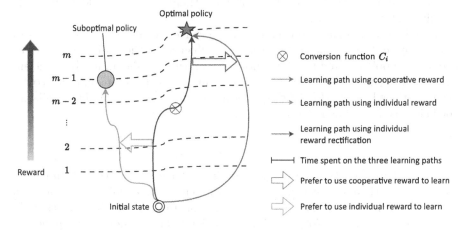

Fig. 2. Example of individual reward rectification

In Fig. 2, the far-right blue line represents the cooperative policy, which relies on sparse cooperative rewards for learning, resulting in a slow learning process. Conversely, the far-left orange line represents the individual policy, which utilizes dense individual rewards for learning, leading to a faster learning process but with a tendency to converge towards local optima. In order to leverage the strengths of both approaches while mitigating their drawbacks, we introduce individual reward rectification. This approach combines the benefits of dense individual rewards for rapid learning in the initial stage, and subsequently relies on the cooperative policy to attain the optimal reward position in the later stage.

Regarding the concept of "two policies per agent", specifically, agent i is required to acquire an individual policy $\hat{\pi}_i$ with parameter $\hat{\theta}_i$ and a cooperative policy π_i with parameter θ_i. Their objectives encompass maximizing the cumulative individual reward $\hat{R}(\hat{\theta}_i) = \mathbb{E}\left[\sum_{t=0}^{\infty} \gamma^t \hat{r}_i^t\right]$ and the cumulative cooperative reward $R(\theta_i) = \mathbb{E}\left[\sum_{t=0}^{\infty} \gamma^t r_i^t\right]$, respectively. It is noteworthy that these two policies have mutually influencing effects. The process of individual reward rectification employs cumulative increasing KL regularization [27] and cumulative decreasing KL regularization to strike a balance between the two policies, ultimately leading to the optimal action produced by the individual policy. The cumulative decreasing KL regularization is employed to learn the actions performed by the individual policy, while the cumulative increasing KL regularization refines and samples the actions of the cooperative policy.

The cumulative increasing KL regularization does not exert its influence during the initial stages of learning for agent i. Consequently, the actions acquired by agent i tend to learn individual policies. This accelerates the learning efficiency and assists in escaping the sparse reward state in the early stages of learning. Subsequently, the role of the regularizer employed to learn cooperative policies gains prominence in the later stages of learning. Agent i becomes inclined to acquire cooperative policies associated with substantial reward values, thus eliminating the drawback of suboptimal solutions yielded by solely learning individual policies.

Based on the aforementioned analysis, the current challenge focuses on determining the optimal balance between the influence of the individual and the cooperative policies, and the difference between the two should not be sufficient to generate conflicts between the policies. Similar to IRAT, a similarity coefficient is utilized to ascertain the equilibrium point, denoted as δ. The similarity coefficient is defined as follows:

$$\delta_i^t(\theta_i, \hat{\theta}_i) = \frac{\hat{\pi}(a_i^t|\tau_i^t, \hat{\theta}_i)}{\pi(a_i^t|\tau_i^t, \theta_i)}, \tag{5}$$

where δ_i^t represents the similarity coefficient of agent i at time t and $\delta_i^t \in [1-x, 1+x]$, therein x serving as the limiting factor. Examining the expression of the similarity coefficient reveals a distinction from IRAT. In this research, each agent corresponds to an individual policy $\hat{\pi}$ and a cooperative policy π, which facilitates the computation of the similarity coefficient, enhances operational efficiency, and adapts more effectively to the actor-critic framework within the CTDE paradigm.

In the initial execution of individual reward rectification, significant emphasis is placed on the individual reward, necessitating the utilization of a decreasing KL regularizer to amplify the effectiveness of the individual policy. The target reward for the individual policy $\hat{R}(\hat{\theta}_i)$ is represented as follows:

$$\begin{aligned}
\hat{R}(\hat{\theta}_i) = \mathbb{E}[&\mathbb{I}_{\delta \geq 1} \max(\delta_i^t(\theta_i, \hat{\theta}_i), \hat{\delta}_i^t(\hat{\theta}_i, \theta_i))\hat{A}_i \\
&+ \mathbb{I}_{\delta < 1} \min(\delta_i^t(\theta_i, \hat{\theta}_i), \hat{\delta}_i^t(\hat{\theta}_i, \theta_i))\hat{A}_i \\
&+ \alpha KL(\pi_i, \hat{\pi}_i)],
\end{aligned} \tag{6}$$

where $\mathbb{I}$ denotes the conditional selection function, α indicates the decreasing coefficient, and $\hat{A}_i$ denotes the set of actions taken by agent i using the individual policy. Subsequently, a progressively increasing KL regularizer is introduced to enhance its influence when a cooperative policy is required, and the target reward for the cooperative policy $R(\theta_i)$ is expressed as:

$$\begin{aligned}
R(\theta_i) = \mathbb{E}[&\mathbb{I}_{\delta \geq 1} \min(\delta_i^t(\theta_i, \hat{\theta}_i), \hat{\delta}_i^t(\hat{\theta}_i, \theta_i))A_i \\
&+ \mathbb{I}_{\delta < 1} \max(\delta_i^t(\theta_i, \hat{\theta}_i), \hat{\delta}_i^t(\hat{\theta}_i, \theta_i))A_i \\
&+ \beta KL(\hat{\pi}_i, \pi_i)],
\end{aligned} \tag{7}$$

where β represents the incremental coefficient, and $\boldsymbol{A}_i$ denotes the set of actions taken by agent i using the cooperative policy. Following the computation of the individual and cooperative rewards, which necessarily translate into the target reward for individual reward rectification $R^{IRR}(\theta)$, which can be described as:

$$R^{IRR}(\theta_i, \hat{\theta}_i) = \begin{cases} C_i^t \hat{R}(\hat{\theta}_i) - (C_i^t - C_i^{t-1})R(\theta_i), & C_i^t \geq 0 \\ -C_i^t R(\theta_i) + (C_i^{t-1} - C_i^t)\hat{R}(\hat{\theta}_i), & C_i^t < 0 \end{cases}, \qquad (8)$$

where C_i^t denotes the conversion function and $C_i^t = r_i^t + \gamma V(s_t) - V(s_{t-1})$, $V(s_t) = \mathbb{E}_{s_{t+1}:\infty, a_t:\infty}[\sum_{l=0}^{\infty} r_{t+l}]$ [22]. When $C_i^t \geq 0$, the individual reward holds dominates, whereas when $C_i^t < 0$, the cooperative reward takes precedence. The output action set $\boldsymbol{A}^{IRR}$ is calculated from the rectified individual policy and $\boldsymbol{A}^{IRR} = \{a_1, a_2, ..., a_n\}$.

3.2 Mix Network Rectification

As a prominent representative of the CTDE paradigm, MADDPG excels at swiftly attaining the global optimal solution through centralized critic training. However, as the number of agents increases, obtaining a single global state-value function through centralized training for all agents becomes challenging [24]. To address this issue, mix network rectification is proposed, wherein a mix network [6] is employed to factorize the Q-value of each agent and derive the global value, denoted as Q_{tot}. Distinct from FACMAC, the factorization function employed is linear, which promotes scalability.

Specifically, in mix network rectification, the critic undergoes centralized training using the value functions of all agents to acquire Q_{tot}, as depicted by:

$$\begin{aligned} Q_{tot}^{\pi}(\boldsymbol{\tau}, \boldsymbol{a}, s; \boldsymbol{\phi}, \omega) =& g_\omega(s, Q_1^{\pi_1}(\tau_1, a_1; \phi_1), \\ & ..., Q_n^{\pi_n}(\tau_n, a_n; \phi_n)), \end{aligned} \qquad (9)$$

where Q_{tot}, the joint action value function, depends on the parameters $\boldsymbol{\phi}$. On the other hand, the value function for a single agent, Q_n, depends on ϕ_n. The mix network parameter is denoted as ψ, and the mixing function, g_ψ, is a linear function. During the training process, the critic network is updated by minimizing the loss function, which is given by:

$$LOSS(\boldsymbol{\phi}, \omega) = \mathbb{E}_D[(y^{tot} - Q_{tot}^{\pi}(\boldsymbol{\tau}, \boldsymbol{a}, s; \boldsymbol{\phi}, \omega))^2], \qquad (10)$$

where $y^{tot} = r + \gamma Q_{tot}^{\pi}(\boldsymbol{\tau}', \boldsymbol{\pi}(\boldsymbol{a}, \boldsymbol{\tau}'; \boldsymbol{\psi}^-), s'; \boldsymbol{\phi}^-, \omega^-)$, and D represents the replay buffer. The parameters $\boldsymbol{\psi}^-$, $\boldsymbol{\phi}^-$, and ω^- represent the target actor, target critic, and target mix network, respectively.

Once the global state value function Q_{tot} is computed, the actor network requires updating. Taking MADDPG as an example, each agent's actor parameters are updated through its own policy gradient, where the policy gradient of each agent is affected by the actions of all agents, typically obtained from the replay buffer D. Nevertheless, this approach typically leads to reduced computational efficiency. To overcome this challenge, a mix network is adopted to introduce a novel policy gradient calculator. This innovation effectively accelerates the efficiency of global policy updates for actor networks, thereby enhancing the algorithm's scalability.

Similar to FACMAC, during actor network updates, the action set is denoted as $A = \{a_1, a_2, ..., a_n\}$, is extracted from the current policies of all agents rather than solely relying on the replay buffer D. However, unlike FACMAC, the mix network in this study utilizes linearity, which facilitates the implementation of policy gradient estimation and reduces computational complexity. The actor network's policy gradient can be expressed as follows:

$$\nabla_\psi J(\pi) = \mathbb{E}_D[\nabla_\psi \pi \nabla_\pi Q_{tot}^\pi(\tau, \pi_1(\tau_1, a_1), ..., \pi_n(\tau_n, a_n), s)], \qquad (11)$$

where $\pi = \{\pi_1(\tau_1, a_1|\psi_1), ..., \pi_n(\tau_n, a_n|\psi_n)\}$ denotes the set of current policies for all agents, and ψ indicates the actor network's parameter.

3.3 Discrete Action Rectification

Conventionally, policy gradients are computed based on continuous policies, denoted as μ. However, many cooperative tasks inherently involve discrete actions. This poses a challenge for algorithms like MADDPG, as they struggle to handle tasks with discrete actions effectively. The discrete action rectification is introduced to settle this issue. Specifically, the Gumbel-Softmax [9] technique is utilized to sample from the continuous policy, resulting in the output of the sampled agent being $\pi(a^{t-1}, \tau)$. Moreover, Gumbel-Softmax enables the computation of gradients for discrete samples, thereby approximating the policy gradient. The process of gradient approximation is described as follows:

$$\begin{aligned} \nabla_\psi J(\mu) &= \mathbb{E}_D[\nabla_\psi \mu \nabla_\mu Q_{tot}^\mu(\tau, \mu_1(\tau_1), ..., \mu_n(\tau_n), s)] \\ &\approx \mathbb{E}_D[\nabla_\psi v \nabla_v Q_{tot}^v(\tau, v_1, ..., v_n, s)] = \nabla_\psi J(\pi), \end{aligned} \qquad (12)$$

where $v = \pi(a|\tau)$, and $v = \{v_1, ..., v_n\}$ demonstrates the actions performed during the continuous sampling process.

4   Experiments and Results

This section presents the experimental results of the MRRC algorithm conducted in two cooperative environments: "Predator-Prey" and "StarCraft Multi-Agent Challenge (SMAC)". To assess the effectiveness of MRRC, it was compared against several state-of-the-art actor-critic algorithms, namely FACMAC [18] and MASAC [8], as well as the value-based algorithms QMIX [21] and QTRAN [26], and the policy-based algorithm IRAT [30]. Additionally, ablation experiments were performed to evaluate the impact of the "Individual Reward Rectification" and "Mix Network Rectification" modules in MRRC. Specifically, two new algorithms were created in this study, "MRRC-IRR" by removing individual reward rectification, and "MRRC-MNR" by excluding mix network rectification. In the "Predator-Prey" environment, all experiments were conducted with a training duration of 1 million steps, while for the more challenging "SMAC" environment, the training duration was set to 2 million steps. Each experiment was executed with seven random seeds, and the 95% confidence interval is shown shaded.

4.1 Predator-Prey

Experiments were conducted in the "Predator-Prey" environment, utilizing a 12×12 grid world. The preys were controlled by built-in AI, and the objective of the MARL algorithm was to manipulate the predators to capture these preys. The observation range of the predators was limited to a 2×2 grid. The aim was to verify the effectiveness of the MRRC algorithm in environments with sparse rewards and scalability when the number of agents increases. Two experimental scenarios were established for this purpose: Predator-Prey with decreasing preys and Predator-Prey with increasing agents. Three experiments were conducted in the Predator-Prey with decreasing preys. The number of predators was fixed at 6, while the number of preys varied, specifically 3, 2, and 1. The experimental results are presented in Fig. 3. Similarly, three experiments were conducted in the Predator-Prey with increasing agents. The number of predators was set to 3, 6, and 9, respectively. The corresponding results are illustrated in Fig. 4.

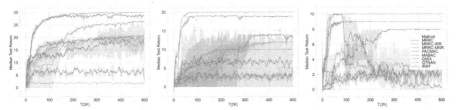

(a) 6 predators and 3 preys (b) 6 predators and 2 preys (c) 6 predators and 1 prey

Fig. 3. Median test returns of Predator-Prey with decreasing preys

As depicted in Fig. 3, the MRRC algorithm exhibits superior performance compared to all other baseline algorithms in Predator-Prey with decreasing preys. When the number of preys is high, as illustrated in Fig. 3(a), most algorithms manage to maintain relatively better performance. However, as the number of preys decreases, resulting in increasingly sparse rewards in the environment, only MRRC, MRRC-MNR, and IRAT demonstrate significant performance, particularly when only one prey remaining, as shown in Fig. 3(c). Notably, the reward value of QMIX algorithm decline over time, indicating its poor performance in an environment with extremely sparse rewards. Among all the ablation experiments conducted in the sparse reward scenario, MRRC-MNR achieves the best performance, suggesting that individual reward rectification effectively addresses the challenges posed by sparse rewards in the environment.

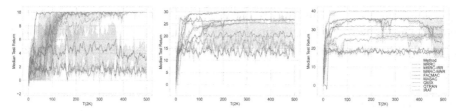

(a) 3 predators and 3 preys (b) 6 predators and 6 preys (c) 9 predators and 9 preys

Fig. 4. Median test returns of Predator-Prey with increasing agents

Remarkably, the performance of both MRRC and MRRC-IRR is lower than that of MRRC-MNR, suggesting that mix network rectification may introduce additional complexity to the algorithm, which potentially leads to negative effects.

Figure 4 demonstrates that MRRC surpasses all other baseline algorithms in the Predator-Prey with increasing agents. When the number of agents is small, as illustrated in Fig. 4(a) and 4(b), most algorithms exhibit better performance. Nevertheless, as the number of agents increases, as shown in Fig. 4(c), the performance of the baseline algorithms QMIX and FACMAC declines significantly, highlighting their poor scalability. In contrast, the MRRC algorithm which incorporates the monotonic mix function performs better, emphasizing the crucial role of mix network rectification in enhancing scalability. The performance of MRRC is noticeably lower than that of MRRC-IRR in Fig. 4(c), indicating that individual reward rectification has a negative impact on scalability.

4.2   More Challenging SMAC

To assess the adaptability of the MRRC algorithm in complex tasks, six distinct experiments were conducted in the more challenging SMAC environment. The details of the experimental maps and parameters can be found in Table 1, and accroding to the research of Wang et al. [29] the other parameters were set. The results of these experiments were illustrated in Fig. 5.

Table 1. The maps and parameters of the SMAC in experiment

Map name	Difficulty	Ally	Opponent
2s3z	Easy	2 Stalkers & 3 Zealots	—
3s5z	Easy	3 Stalkers & 5 Zealots	—
3 s_vs_5z	Hard	3 Stalkers	5 Zealots
5m_vs_6m	Hard	5 Marines	6 Marines
3s5z_vs_3s6z	Super hard	3 Stalkers & 5 Zealots	3 Stalkers & 6 Zealots
27m_vs_30m	Super hard	27 Marines	30 Marines

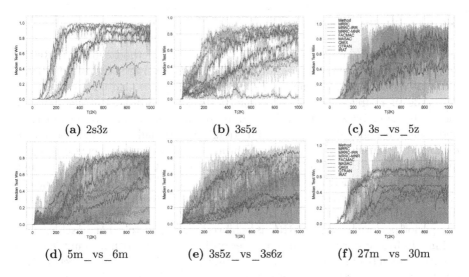

(a) 2s3z   (b) 3s5z   (c) 3s_vs_5z

(d) 5m_vs_6m   (e) 3s5z_vs_3s6z   (f) 27m_vs_30m

Fig. 5. Median test win for easy (a) (b), hard (c) (d) and super-hard (e) (f) maps of SMAC

As presented in Fig. 5, the MRRC algorithm demonstrates strong performance across SMAC environments with varying maps and difficulty levels. In simpler tasks, as shown in Fig. 5(a) and 5(b), MRRC, MRRC-IRR, MRRC-MNR, FACMAC, and QMIX all exhibit excellent performance. As task complexity increases, as indicated in Fig. 5(c), 5(d), 5(e) and 5(f), the MRRC algorithm consistently outperforms the other baseline algorithms, demonstrating its superiority and adaptability in diverse cooperative environments. Notably, the QTRAN and MASAC algorithms consistently underperform across all tasks, indicating their limited adaptability in different environments.

The results were analyzed in detail in the ablation experiments. In simpler tasks, as shown in Fig. 5(a)and5(b), both the MRRC-IRR and MRRC-MNR algorithms outperform the MRRC algorithm. This suggests that in simple tasks, excessive optimization may be unnecessary, as solving such tasks does not require the full range of modules. Alternatively, incorporating an excessive number of modules increases computational burden and degrades algorithm performance. As task complexity increases, the MRRC algorithm outperforms the two module-removed algorithms, indicating the significance of both individual reward rectification and mix network rectification in challenging environments. Figure 5(e) shows that MRRC-MNR outperforms MRRC-IRR in the "3s5z_vs_3s6z" task, which features sparse rewards. This highlights the effectiveness of individual reward rectification in addressing the sparse reward problem. Notably, in an environment with a large number of agents, as depicted in Fig. 5(f), MRRC-IRR outperforms MRRC-MNR, indicating that mix network rectification excels at enhancing algorithm scalability.

5   Conclusion

In this study, we propose a novel actor-critic based method called MRRC for addressing multi-agent cooperation tasks. MRRC leverages the rapid convergence of the individual policy to rectify agent behaviors and effectively tackle the issue of sparse rewards. To enhance the scalability of the algorithm, a monotonic mix network is introduced to rectify the agents' value functions and constructs global state value functions. Additionally, Gumbel-Softmax is combined to handle discrete actions. Through extensive experiments conducted in the "Predator-Prey" and "StarCraft Multi-Agent Challenge (SMAC)" environments, we demonstrate that MRRC effectively guides the actions of agents, enabling them to overcome the slow convergence caused by sparse rewards and improving the scalability of the algorithm. Moreover, the results indicate that a superior performance of MRRC is achieved compared to other baseline algorithms. In future work, we aim to further optimize the algorithm to mitigate the negative impacts of individual reward rectification and mix network rectification on the overall algorithm's performance.

Acknowledgements. This work is sponsored by Equipment Advance Research Fund (NO. 61406190118).

References

1. Aleardi, M., Vinciguerra, A., Stucchi, E., Hojat, A.: Machine learning-accelerated gradient-based Markov chain monte Carlo inversion applied to electrical resistivity tomography. Near Surface Geophys. **20**(4), 440–461 (2022)
2. Barth-Maron, G., et al.: Distributed distributional deterministic policy gradients. arXiv preprint arXiv:1804.08617 (2018)
3. Castiglioni, I., et al.: AI applications to medical images: from machine learning to deep learning. Physica Med. **83**, 9–24 (2021)
4. Cetinic, E., She, J.: Understanding and creating art with AI: review and outlook. ACM Trans. Multimedia Comput. Commun. Appl. (TOMM) **18**(2), 1–22 (2022)
5. Dalal, G., Hallak, A., Dalton, S., Mannor, S., Chechik, G., et al.: Improve agents without retraining: parallel tree search with off-policy correction. Adv. Neural. Inf. Process. Syst. **34**, 5518–5530 (2021)
6. Ha, D., Dai, A., Le, Q.V.: Hypernetworks. arXiv preprint arXiv:1609.09106 (2016)
7. Huang, S., et al.: A constrained multi-objective reinforcement learning framework. In: Conference on Robot Learning, pp. 883–893. PMLR (2022)
8. Iqbal, S., Sha, F.: Actor-attention-critic for multi-agent reinforcement learning. In: International Conference on Machine Learning, pp. 2961–2970. PMLR (2019)
9. Jang, E., Gu, S., Poole, B.: Categorical reparameterization with gumbel-softmax. arXiv preprint arXiv:1611.01144 (2016)
10. Jin, L., Qian, S., Owens, A., Fouhey, D.F.: Planar surface reconstruction from sparse views. In: Proceedings of the IEEE/CVF International Conference on Computer Vision, pp. 12991–13000 (2021)
11. Kraemer, L., Banerjee, B.: Multi-agent reinforcement learning as a rehearsal for decentralized planning. Neurocomputing **190**, 82–94 (2016)

12. Lowe, R., Wu, Y.I., Tamar, A., Harb, J., Pieter Abbeel, O., Mordatch, I.: Multi-agent actor-critic for mixed cooperative-competitive environments. In: Advances in Neural Information Processing Systems, vol. 30 (2017)
13. Lu, Y., Li, W.: Techniques and paradigms in modern game AI systems. Algorithms **15**(8), 282 (2022)
14. Majumdar, S., Khadka, S., Miret, S., McAleer, S., Tumer, K.: Evolutionary reinforcement learning for sample-efficient multiagent coordination. In: International Conference on Machine Learning, pp. 6651–6660. PMLR (2020)
15. Mansour, R.F., El Amraoui, A., Nouaouri, I., Díaz, V.G., Gupta, D., Kumar, S.: Artificial intelligence and internet of things enabled disease diagnosis model for smart healthcare systems. IEEE Access **9**, 45137–45146 (2021)
16. Nian, R., Liu, J., Huang, B.: A review on reinforcement learning: introduction and applications in industrial process control. Comput. Chem. Eng. **139**, 106886 (2020)
17. Oliehoek, F.A., Amato, C.: A concise introduction to decentralized pomdps (2015)
18. Peng, B., et al.: FACMAC: factored multi-agent centralised policy gradients. In: Advances in Neural Information Processing Systems, vol. 34, pp. 12208–12221 (2021)
19. Qin, Z., Zhang, K., Chen, Y., Chen, J., Fan, C.: Learning safe multi-agent control with decentralized neural barrier certificates. arXiv preprint arXiv:2101.05436 (2021)
20. Rajeswar, S., et al.: Haptics-based curiosity for sparse-reward tasks. In: Conference on Robot Learning, pp. 395–405. PMLR (2022)
21. Rashid, T., Samvelyan, M., De Witt, C.S., Farquhar, G., Foerster, J., Whiteson, S.: Monotonic value function factorisation for deep multi-agent reinforcement learning. J. Mach. Learn. Res. **21**(1), 7234–7284 (2020)
22. Schulman, J., Moritz, P., Levine, S., Jordan, M., Abbeel, P.: High-dimensional continuous control using generalized advantage estimation. arXiv preprint arXiv:1506.02438 (2015)
23. Shao, Y., et al.: Multi-objective neural evolutionary algorithm for combinatorial optimization problems. IEEE Trans. Neural Networks Learn. Syst. **34**, 2133–2143 (2021)
24. Sharma, P.K., Fernandez, R., Zaroukian, E., Dorothy, M., Basak, A., Asher, D.E.: Survey of recent multi-agent reinforcement learning algorithms utilizing centralized training. In: Artificial Intelligence and Machine Learning for Multi-domain Operations Applications III, vol. 11746, pp. 665–676. SPIE (2021)
25. Shen, Y., Song, K., Tan, X., Li, D., Lu, W., Zhuang, Y.: Hugginggpt: solving AI tasks with chatgpt and its friends in huggingface. arXiv preprint arXiv:2303.17580 (2023)
26. Son, K., Kim, D., Kang, W.J., Hostallero, D.E., Yi, Y.: Qtran: learning to factorize with transformation for cooperative multi-agent reinforcement learning. In: International Conference on Machine Learning, pp. 5887–5896. PMLR (2019)
27. Vieillard, N., Kozuno, T., Scherrer, B., Pietquin, O., Munos, R., Geist, M.: Leverage the average: an analysis of kl regularization in reinforcement learning. Adv. Neural. Inf. Process. Syst. **33**, 12163–12174 (2020)
28. Wang, J., Ren, Z., Liu, T., Yu, Y., Zhang, C.: Qplex: duplex dueling multi-agent q-learning. arXiv preprint arXiv:2008.01062 (2020)
29. Wang, J., Zhang, Y., Gu, Y., Kim, T.K.: Shaq: Incorporating shapley value theory into multi-agent q-learning. Adv. Neural. Inf. Process. Syst. **35**, 5941–5954 (2022)
30. Wang, L., et al.: Individual reward assisted multi-agent reinforcement learning. In: International Conference on Machine Learning, pp. 23417–23432. PMLR (2022)

31. Wang, T., Wang, J., Zheng, C., Zhang, C.: Learning nearly decomposable value functions via communication minimization. arXiv preprint arXiv:1910.05366 (2019)
32. Wang, Y., Han, B., Wang, T., Dong, H., Zhang, C.: Off-policy multi-agent decomposed policy gradients. arXiv preprint arXiv:2007.12322 (2020)
33. Yan, Y., Chow, A.H., Ho, C.P., Kuo, Y.H., Wu, Q., Ying, C.: Reinforcement learning for logistics and supply chain management: methodologies, state of the art, and future opportunities. Transp. Res. Part E Logist. Transp. Rev. **162**, 102712 (2022)
34. Zhang, R., McNeese, N.J., Freeman, G., Musick, G.: "An ideal human" expectations of AI teammates in human-AI teaming. Proc. ACM Hum.-Comput. Inter. **4**(CSCW3), 1–25 (2021)
35. Zhang, T., Li, Y., Wang, C., Xie, G., Lu, Z.: Fop: factorizing optimal joint policy of maximum-entropy multi-agent reinforcement learning. In: International Conference on Machine Learning, pp. 12491–12500. PMLR (2021)
36. Zhou, H., Lan, T., Aggarwal, V.: PAC: assisted value factorisation with counterfactual predictions in multi-agent reinforcement learning. arXiv preprint arXiv:2206.11420 (2022)

Latent Causal Dynamics Model
for Model-Based Reinforcement Learning

Zhifeng Hao[1,2], Haipeng Zhu[1], Wei Chen[1(✉)], and Ruichu Cai[1(✉)]

[1] School of Computer Science,
Guangdong University of Technology, Guangzhou, China
zfhao@gdut.edu.cn, {chenweiDelight,cairuichu}@gmail.com
[2] College of Engineering, Shantou University, Shantou, China

Abstract. Learning an accurate dynamics model is the key task for model-based reinforcement learning (MBRL). Most existing MBRL methods learn the dynamics model over states. But in most cases, the relationships among states are complex because the states are affected by the interaction of various factors in the environment. Recently some works are proposed to learn the dynamics model on latent representations space. But the learned model is dense and may contain spurious associations between latent representations. To deal with these problems, we introduce a latent causal dynamics model over latent representations and provide a learning method for MBRL. Specifically, we first learn the latent representations from the observed state space. Second, we learn a latent causal dynamics model among latent representations by a causal discovery method. Finally, the latent causal dynamics model is used to aid policy learning. The above steps are iterative to update the unified loss function until convergence. Experimental results on four tasks show that the performance of our proposed method benefits from the causality and the learned latent representations.

Keywords: Reinforcement learning · Causal discovery · Latent representation · Dynamics model

1 Introduction

Reinforcement learning has achieved impressive success in many challenging real-world domains, including video games [14,23], robot control [4,16] and autonomous driving [10]. One of the typical methods for reinforcement learning is Model-Based Reinforcement Learning (MBRL), which aims to learn an environment dynamics model through the interaction between the agent and the environment. MBRL enables an agent to predict the outcomes of its actions and learn from an imagined future. Therefore, the performance of the MBRL method is sensitive to the learned dynamics models.

Many approaches are proposed to learn an accurate dynamics model [3,12,15]. They directly learn the dynamics model in the original observed state

B. Luo et al. (Eds.): ICONIP 2023, LNCS 14448, pp. 219–230, 2024.
https://doi.org/10.1007/978-981-99-8082-6_17

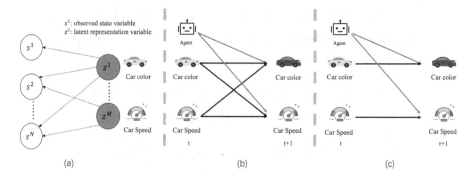

Fig. 1. Two latent dynamics models on latent representations (car's speed and color) that the agent can intervene. (a) The illustration of the data generating process, where $\{z_1, z_2, \ldots, z_m\}$ is denoted a set of latent representation variables, and $\{s_1, s_2, \ldots, s_n\}$ is denoted a set of observed state variables. (b) A latent dynamics model learned by existing methods. (c) A latent dynamics model built upon the causal structure among latent representation variables by our proposed method.

space. But recently, some works have argued that modeling all observed state variables is expensive and unnecessary, such as PlaNet [8] and ALM [6]. Because in most real-world situations, the state variables are very high-dimensional, which leads to hard training. In addition, some observed state variables are not related to the task, without semantic information. Thus, they focus on modeling the task-related information in the environment from states, which contain semantic information. Those are referred to as latent *representations*.

The dynamics model learned by existing methods tends to be dense, where all latent representations are interconnected. This implies that when predicting the value of each latent representation at a given timestamp, the model relies on the current action and all latent representations from the previous timestamp. However, not all latent representations are dependent on each other. For example, in Fig. 1 (a), the car's color is independent of the car's speed. In this case, the correlation between the color and the speed is spurious in the dynamics model. If using this model to predict the car's speed by taking the car's color at the previous time into consideration, it may lead to incorrect predictions.

To learn a good dynamics model, two main challenges need to be tackled: 1) learning task-relevant latent representations from states and 2) removing spurious relationships between latent representations to recover the true dynamics model. Based on the above analysis, we find that the environmental information received by an agent can be categorized into two kinds: 1) task-relevant information; 2) task-irrelevant or weakly relevant information. Therefore, a good latent representation should only capture task-relevant information. As for the second challenge, spurious relationships between latent representations arise when they are not truly dependent. These spurious relationships can be detected through causal discovery methods. Thus, to overcome these challenges, we propose a Latent Causal dynamics model learning method. First, we obtain task-relevant latent representations by constraining them to be associated with state changes and received rewards. Second, to remove the spurious correlations between the latent representations, we use causal discovery methods for recovering the causal structure between the latent representations, which is used to obtain the correct dynamics model as well as the reward model. These steps are iteratively performed to update the latent representation learning, latent causal dynamics model learning, and policy learning until convergence is achieved.

2 Preliminary

2.1 Reinforcement Learning

In this paper, we consider infinite-horizon Markov Decision Processes (MDP), which is described by a tuple $(\mathcal{S}, \mathcal{A}, p, \mathcal{R}, \gamma, p_0)$. Here, $\mathcal{S} \in R^n$ and $\mathcal{A} \in R^m$ represent continuous state and action spaces respectively, $p(s_{t+1}|s_t, a_t)$ denotes the transition distribution, $\mathcal{R} : \mathcal{S} \times \mathcal{A} \to \mathbb{R}$ or $\mathcal{S} \to \mathbb{R}$ is a reward function, γ is the discount factor which lies in $(0,1)$, and p_0 represents the initial state distribution. Model-based reinforcement learning is one of the typical methods of reinforcement learning. It aims to construct a model of the transition distribution, denoted as $p_\theta(s_{t+1}|s_t, a_t)$. This model is learned using data collected from interacting with the environment. Additionally, a reward model is learned through supervised learning to approximate the true reward function $\mathcal{R}$. In MBRL, the dynamics and reward functions are usually assumed to be unknown.

2.2 Structural Causal Model

The *Structural Causal Model* (SCM) represents causal relationships between two variables as their functional dependencies [20]. The definition is as follows:

Definition 1. *(Structural Causal Model). A Structural Causal Model (SCM) is a tuple $(V, G, F, \mathcal{N})$, where*

- *V is a set of random variables involved in the model;*
- *$\mathcal{G}$ is a directed acyclic graph (DAG), in which the node corresponds to variables in V and the edges indicate the causal connections between variables.*
- *F is a set of structural functions that describe the functional relationships between the variables;*
- *$\mathcal{N}$ is a set of noise terms associated with each variable.*

Furthermore, for each variable $v_i \in V$, v_i is associated with a function that describes how it is influenced by its directed causes, which is formulated as: $v_i = f_i((v_{pa_i}), n_i)$, where $v_i \in V, f_i \in F, n_i \in \mathcal{N}$, and v_{pa_i} represents parents of v_i.

3 Latent Causal Dynamics Model Learning

In this section, we introduce a method for learning a latent causal dynamics model (LCDM), and then derive a policy through planning based on the model.

3.1 Latent Causal Dynamics Learning

Given a fully observed Markov Process with $V = \{s_t^N, a_t, s_{t+1}^N, r_{t+1}\}$, where s_t^N is denoted as the state variables s^N at timestamp t and N represents the dimensional of states. We aim to learn a good dynamics model for MBRL. In MBRL, if an agent observes all the information in the environment, then we can directly extract all the task-relevant information from the information observed by the agent. In the real world, however, we can not observe all information. Hence, our focus is on recovering or extracting the task-relevant information from what we observe. This kind of information is implied in the observed state variables, which may represent a combination of multiple state variables. Take Fig. 1 as an example, the speed of the car is a task-relevant representation that is a combination of multiple sensor data.

Specifically, the extracted information is referred to *latent representations*, which contains the task-related latent representations defined as Z and task-independent latent representations defined as U. Then, we only need to care about Z ($|Z| = M$). Let z_t denote the latent representation z that is extra from S at time step t. The generation process of representations can be represented by a causal graph $\mathcal{G} = \{Z, E\}$, where E are causal edges that describe the causal relationships between z_t^i and z_{t+1}^i, $i \in \{1, 2, \ldots, M\}$.

Therefore, the aim of latent causal dynamics learning includes 1) learning the latent representations and 2) discovering the causal structure among the latent representations. Regarding the first task, we can extract the representations from the observed states. Inspired by [6,9], we use an encoder g_θ to model the state-to-latent representation mapping, then use a set of neural network to model the latent representations transition.

To ensure the stability of the learning process of latent representations, we use the L2 distance between z_{t+1} and the $g_\theta(s_{t+1})$ as a consistency loss. This loss encourages the latent representations to remain consistent and prevents them from diverging during the learning process. The reason why we do not use a decoder to reconstruct the state here is that we do not require the latent representations to encapsulate all the information present in the environment. Thus, the learning function is formalized as follows:

$$\mathcal{L}_\theta = \sum_{t=0}^{T} ||f_\theta(z_t, a_t, \mathcal{G}), g_\theta(s_{t+1})||_2, \tag{1}$$

where $\mathcal{G}$ is a causal graph that contains the relationships among Z.

Recently, some studies [6,9] consider that the representations are correlated with each other. But actually, these correlations may be spurious and not truly indicative of causal relationships. For example, in Fig. 1, the speed and color of the car are actually independent of each other. This kind of independence can be detected by causality since causality describes the mechanism of data generation. Hence, we incorporate causal structure as a constraint in the latent representation learning procedure. This ensures that the learned latent representation transition function is more reasonable. Additionally, our reward model is built upon latent representations, which enables the latent representations to capture and incorporate as much task-relevant information as possible.

Learning a causal structure among representations involves determining whether a causal edge exists between two latent representations [2,17]. That is, determining whether $z_t^i \to z_{t+1}^j$ or $z_t^j \to z_{t+1}^i$ is corresponding to the truth causal relationship. To make this inference, certain assumptions are required.

Assumptions A1. [17,20] Causal Markov and Faithfulness assumptions in the underlying dynamics.

Assumptions A2. The latent dynamics is Markovian and stationarity.

Assumptions A3. The edge $z_t^i \to z_{t+1}^i$ exists for all variables z^i.

Assumptions A4. No simultaneous or backward edges in time, i.e., for all i, j, $z_t^i \nrightarrow s_t^j$ and $z_t^i \nrightarrow z_{t-1}^j$.

Based on the above assumptions, we can identify the causal relationship between representations, which is guaranteed by the following theorem.

Theorem 1. *Suppose Assumptions A1-A4 hold. Let* $\{a_t, z_t \backslash z_t^i\} = \{a_t, z_t^1, ..., z_t^{i-1}, z_t^{i+1}, ..., z_t^{d_z}\}$. *For any two latent variables* z_t^i *and* z_t^j, *if* $z_t^i \not\perp\!\!\!\perp z_{t+1}^j | \{a_t, z_t \backslash z_t^i\}$, *then* $z_t^i \to z_{t+1}^j$.

Proof. We need to show that if $z_t^i \nrightarrow z_{t+1}^j$, then the condition $z_t^i \not\perp\!\!\!\perp z_{t+1}^j | \{a_t, z_t \backslash z_t^i\}$ or the assumption is violated. Assume there is no direct edge from z_t^i to z_{t+1}^j but they are dependent, i.e., $z_t^i \nrightarrow z_{t+1}^j$ and $z_t^i \not\perp\!\!\!\perp z_{t+1}^j$, then there is a latent confounder variable Q'_t with the confounding path $z^i \leftarrow\!\!- Q'_t -\!\!\rightarrow z_{t+1}^j$ where $t' < t$ as we assume there is no simultaneous edge and backward edge. If Q'_t exists in this case, then it violates the condition because we condition on $\{a_t, z_t \backslash z_t^i\}$ which block all possible paths for $Q'_t -\!\!\rightarrow z_{t+1}^j$ and thus d-separate z_t^i and z_{t+1}^j.

Therefore we show that if $z_t^i \not\perp\!\!\!\perp z_{t+1}^j | \{a_t, z_t \backslash z_t^i\}$, then $z_t^i \to z_{t+1}^j$.

This theorem demonstrates that the causal relationship between two variables can be inferred by the Conditional Independence Test (CIT) method. Several CIT methods can be employed, including conditional mutual information and Kernel-based Conditional Independence Test (KCIT) [24].

In this paper, we use PCMCI (PC with Momentary Conditional Independence test) [18] to recover the causal graph. The consistency of the learned causal graph and the true causal graph is guaranteed by the following theorem, which has been proved in [18].

Theorem 2. (Consistency) *Let* **X** *be a Markov decision process with true transition causal graph* $\mathcal{G}$ *as defined in Definition 2 and* $\hat{\mathcal{G}}$ *be the estimated graph by PCMCI method with a consistent conditional independence test. Suppose Assumptions A1-A4 hold,* $\hat{\mathcal{G}} = \mathcal{G}$.

Previous research has shown that the learned causal graph among representations can be connected to the transition in reinforcement learning [11,13]. Inspired by [5], we define the transition causal graph as follows.

Definition 2. (Transition Causal Graph) *We define a Transition Causal Graph as a directed graph denoted by $\mathcal{G}$, where the vertices are divided into two disjoint sets: $\mathcal{U} = \{\mathcal{A}_t, \mathcal{Z}_t\}$ and $\mathcal{V} = \{\mathcal{Z}_{t+1}\}$. Let $\mathcal{A}_t$ represent action nodes at time step t, $\mathcal{Z}_t$ represent latent representation variables at step t, and $\mathcal{Z}_{t+1}$ denote the latent representation variable at step $t+1$. All edges start from set $\mathcal{U}$ and end in set $\mathcal{V}$.*

The Transition Causal Graph captures the causal relationships between latent representation variables at two consecutive time steps. It signifies that the values of a latent representation variable at time step $t+1$ depend on the values of the same latent representation variable at time step t. This temporal dependency in the graph reflects the dynamics of the system and how the latent representations evolve over time. By modeling these causal relationships, we can better understand and predict the changes in latent variables from one time step to the next, which is essential for planning, decision-making, and learning in reinforcement learning scenarios. Combined with the Definition 1, the latent representation transition, $p(z_{t+1}|z_t, a_t)$, can be expressed as follows:

$$p(z_{t+1}|z_t, a_t) = \prod_{i=1}^{N} p(z_{t+1}^i | PA(z_{t+1}^i), a_t)) \tag{2}$$

where $PA(z_{t+1}^i)$ is a set of parent of node z_{t+1}^i in the transition causal graph $\mathcal{G}$. From this equation, we can find that $p(z_{t+1}|z_t, a_t)$ essentially approximates a collection of functions f_i following the transition Causal Graph $\mathcal{G}$, which take as input the values of $PA(z_{t+1}^i)$ and output the value of z_i.

In practice, we use a collection of neural networks $\{f_{\theta_i}\}_{i=1}$ to model the latent transition corresponding to transition Causal Graph $\mathcal{G}$, which is formalized as:

$$z_{t+1}^i = f_i(PA(z_t^i), a_t, N_i), \tag{3}$$

where $PA(z_t^i)$ represents the values of all parents of node z_t^i in transition Causal Graph $\mathcal{G}$, and N_i is noise terms that are assumed to follow a Gaussian distribution, e.g. $N_i \sim \mathcal{N}(0, I)$.

Based on the above analysis, the loss function for learning the latent dynamics model is as follows:

$$\mathcal{L}(\theta, \phi) = \sum_{t=1}^{T} [||f_\theta(z_t, a_t, \mathcal{G}), g_\theta(s_{t+1})||_2 + ||r_\phi(z_t, a_t), r_t||_2] \tag{4}$$

where $||.||_2$ represents Mean-Square Error, f_θ denotes the transition model of latent representation z_t with parameters θ, r_ϕ denotes the predictive model of r_t with parameters ϕ. Notice that only the initial time, e.g. s_0 has a gradient during the encoder stage, whereas other time states have no gradient during the decoder stage.

Thus, the model can be trained by the following procedure. The procedure starts with a completely directed graph, and takes the state at the initial moment

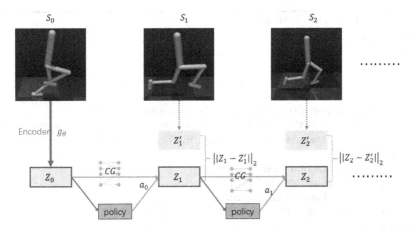

Fig. 2. Latent Causal Dynamics Models (LCDM) learning procedure. The first observation s_0 is encoded into a latent representation z_0 by encoder g_θ. Then the causal structure CG is used as a constraint in the subsequent prediction of latent representations $z_1, z_2, ..., z_T$.

to obtain the initial latent representation through the encoder g_θ. Then, it utilizes the causal graph as a constraint to learn the dynamics model between the latent representations. Additionally, it learns a reward model based on latent representations. The agent takes action to interact with the environment and generates data to update our causal graph. This subsequent process is repeated until convergence. This training procedure is characterized in Fig. 2.

3.2   Policy Learning

The learned causal dynamics model plays a crucial role in improving policy learning. Based on the learned model, we employ the model predictive control (MPC) [1] as the planning algorithm. MPC is an iterative and model-based control approach. After taking an action, MPC first generates a set of candidate action sequences. Then, it evaluates how good the result of each candidate sequence can be based on the current state. Finally, it picks the first action of the action sequence with the best result to execute. By iteratively repeating this process, the agent can make informed decisions at each time step, optimizing its actions based on the learned model's predictions of the future states and rewards.

Specifically, we use Model Predictive Path Integral (MPPI) [22] control algorithm. MPPI is a variation of the MPC algorithm that iteratively adjusts parameters for a range of distributions. It does so by employing an importance-weighted average of the expected returns of the top-k sampled trajectories. Because MPPI allows for changes in the drift and diffusion terms of stochastic diffusion processes, it can help mitigate the rollout error even when our learned latent causal dynamics model has a slight distribution shift with respect to the ground truth transition distribution.

Algorithm 1. LCDM Training

Require: Latent Transition model f_θ, Encoder g_θ
 1: Initialize latent causal graph $\mathcal{G}$ as a complete directed graph
 2: **while** θ not converged **do**
 3:      // Policy learning from planning
 4:      **for** step t=0...T **do**
 5:          Select action given by $a_t = Planner(f_\theta, g_\theta^t(s_t))$.
 6:          Execute a_t in the environment and observe reward r_t and new state s_{t+1}.
 7:          Store the transition(s_t, a_t, r_t, s_{t+1}) in Trajectory Buffer $\mathcal{B}_\tau$.
 8:      **end for**
 9:      // Latent transition model learning
10:      Update $f_\theta(\mathcal{G})$ and g_θ via Eq. (4) with $\mathcal{B}_\tau$
11:      // Estimate Latent Causal graph
12:      latent causal graph $\mathcal{G} \leftarrow PCMCI(\mathcal{B}_\tau)$.
13: **end while**

Therefore, combining the latent causal dynamics model learning and policy learning, our proposed method is summarized in Algorithm 1.

4   Experiments

In this section, we conducted experiments to evaluate our proposed method on some diverse and challenging continuous tasks from different environments that are described in Sect. 4.1. We aim to answer these questions:

$Q1$. How does the performance of our method compare to that of state-of-the-art model-based and model-free methods in different environments?
$Q2$. How useful are the learned representations for our dynamics model?
$Q3$. How helpful is learning the causal structure between representations for our method?

Evaluation Metric. We use the average cumulative reward across 5 seeds with random initialization to measure the performance of methods.

4.1   Environments and Baselines

Environment Details. We use environments (given in Fig. 3) from Deep-Mind (DM) Control Suite [21] that is a standard benchmark for continuous control. In these experiments, we set the episode length as 1000 steps and repeat an action twice for all tasks. Every experiment is run with different random seeds.

Baselines. We evaluate our method against the following methods as baselines:

– **SAC** [7]: Soft Actor-Critic is a model-free algorithm derived from maximum entropy RL. We choose SAC as our main point of comparison due to its popularity and strong performance on DMControl.

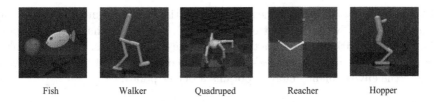

Fish   Walker   Quadruped   Reacher   Hopper

Fig. 3. An illustration of DM Control Suite environments that we used. Our criteria for selecting the environment is to choose according to the complexity of the state and action space, from low to high.

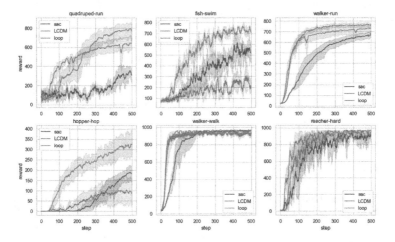

Fig. 4. DMControl tasks. Average returns of our method and baselines on 6 tasks from DMControl. Shaded areas are 95% confidence intervals.

- **LOOP**[19]: A hybrid algorithm that combines H-step lookahead policies with a learned model and a learned terminal value function using a model-free off-policy algorithm.
- **LCDM w/o representation**: This variant removes the representation learning procedure by setting the encoder g_θ as a identify function.
- **LCDM w/o causal structure**: This variant removes the causal graph between the learned latent representations.

4.2   Q1: Performance Comparison

Figure 4 shows the reward of our method and baseline methods in five tasks. From the results, we observe that: 1) In the expected *Quadruped-run* task, all methods converge quickly. Our method achieves higher cumulative rewards than all baseline methods in all the environments, with faster convergence. 2) Our method performs better in both low-dimensional and high-dimensional state space environments. Since SAC is a model-free algorithm, it needs to interact with the environment many times and explore the environment fully before the performance gets better and better. From Fig. 4, we can see that SAC can learn a better policy with fewer interactions for simple environments. But for complex environments such as *Quadruped-run*, SAC needs a lot of interactions to fully explore the environment before it obtains a better policy.

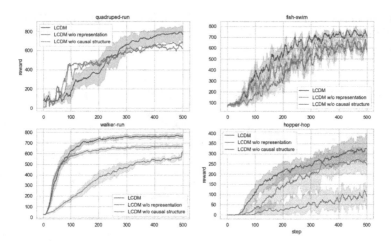

Fig. 5. Ablation Result. Average reward of our method and our ablation methods on 4 state-based continuous control tasks.

Our method can achieve better performance because it can filter out some irrelevant state information by the learned representations and obtain a powerful World Model by quickly learning the causal structure between representations, which helps to obtain the best reward.

4.3   Q2: Will the Learned Representations Be Useful for Our Dynamics Model?

We design an ablation version of our method, LCDM w/o representation, which uses the state space instead of the latent representations to build the world model and the reward predictor. In these experiments, the shaded areas are set as 95% confidence intervals. For each task, we run 5 trials and use the average reward of these trials as the result. The results are shown in Fig. 5. Compared with LCDM w/o representation, LCDM method earns a higher accumulative reward than its ablation version in all environments. In fact, although the LCDM w/o representation gets access to all observation state information, it may be confused or sensitive since part of the observable state dimensions may not be helpful for predicting rewards or performing tasks. Instead, the learned representations help to build a world model by ignoring some states that have no or little impact on the task and aggregating those that have a great impact on the task.

4.4   Q3: Is the Learned Causal Structure Between Representations Helpful for Our Method?

To analyze the impact of learning causal structure between representations, we construct an ablation version of our method, LCDM w/o causal structure, which used a fully connected structure to replace causal structure. According to the results shown in Fig. 5, our proposed method obtains a higher cumulative reward than LCDM w/o causal structure method. In fact, although we could not explain what representations we learned, the result reflects that there was a causal relationship between the representations. When we did not consider the causal relationship between them, the world model learned directly from the representations was incorrect, which led to a reduction in the generalization ability of the model. Thus, the model with causal structure among presentations helps to achieve better rewards while obtaining good generalizability.

5   Conclusion

In this paper, we present a novel model-based reinforcement learning method that learns a latent causal dynamics model from observations. Our method leverages both causal structure and representation learning to capture the essential dependencies information and mechanisms of the underlying system. We have shown that our method can learn accurate and robust world models that can generalize well to unseen states and be resilient to subtle changes. We have also demonstrated the effectiveness of our method on several challenging continuous control tasks, where it outperforms state-of-the-art model-based and model-free baselines. Our work opens up new possibilities for incorporating causal reasoning and representation learning in reinforcement learning.

References

1. Camacho, E.F., Alba, C.B.: Model Predictive Control. Springer, Heidelberg (2013). https://doi.org/10.1007/978-0-85729-398-5
2. Chen, W., Cai, R., Zhang, K., Hao, Z.: Causal discovery in linear non-gaussian acyclic model with multiple latent confounders. IEEE Trans. Neural Netw. Learn. Syst. **33**(7), 2816–2827 (2021)
3. Chua, K., Calandra, R., McAllister, R., Levine, S.: Deep reinforcement learning in a handful of trials using probabilistic dynamics models. In: Advances in Neural Information Processing Systems, vol. 31 (2018)
4. Deisenroth, M.P., Rasmussen, C.E., Fox, D.: Learning to control a low-cost manipulator using data-efficient reinforcement learning. Robot. Sci. Syst. **VII**(7), 57–64 (2011)
5. Ding, W., Lin, H., Li, B., Zhao, D.: Generalizing goal-conditioned reinforcement learning with variational causal reasoning. In: Advances in Neural Information Processing Systems (2022)
6. Ghugare, R., Bharadhwaj, H., Eysenbach, B., Levine, S., Salakhutdinov, R.: Simplifying model-based RL: learning representations, latent-space models, and policies with one objective. In: The Eleventh International Conference on Learning Representations (2022)

7. Haarnoja, T., Zhou, A., Abbeel, P., Levine, S.: Soft actor-critic: off-policy maximum entropy deep reinforcement learning with a stochastic actor. In: International Conference on Machine Learning, pp. 1861–1870. PMLR (2018)

8. Hafner, D., et al.: Learning latent dynamics for planning from pixels. In: International Conference on Machine Learning, pp. 2555–2565. PMLR (2019)

9. Hansen, N.A., Su, H., Wang, X.: Temporal difference learning for model predictive control. In: International Conference on Machine Learning, pp. 8387–8406. PMLR (2022)

10. Hu, A., et al.: Model-based imitation learning for urban driving. In: Advances in Neural Information Processing Systems, vol. 35, pp. 20703–20716 (2022)

11. Huang, B., et al.: Action-sufficient state representation learning for control with structural constraints. In: International Conference on Machine Learning, pp. 9260–9279. PMLR (2022)

12. Kurutach, T., Clavera, I., Duan, Y., Tamar, A., Abbeel, P.: Model-ensemble trust-region policy optimization. In: International Conference on Learning Representations (2018)

13. Lu, C.: Learning causal representations for generalization and adaptation in supervised, imitation, and reinforcement learning. Ph.D. thesis, University of Cambridge (2022)

14. Mnih, V., et al.: Human-level control through deep reinforcement learning. Nature **518**(7540), 529–533 (2015)

15. Nagabandi, A., Kahn, G., Fearing, R.S., Levine, S.: Neural network dynamics for model-based deep reinforcement learning with model-free fine-tuning. In: 2018 IEEE International Conference on Robotics and Automation (ICRA), pp. 7559–7566. IEEE (2018)

16. Nguyen, H., La, H.: Review of deep reinforcement learning for robot manipulation. In: 2019 Third IEEE International Conference on Robotic Computing (IRC), pp. 590–595. IEEE (2019)

17. Pearl, J.: Causality: Models, Reasoning, and Inference. Cambridge University Press, Cambridge (2009)

18. Runge, J., Nowack, P., Kretschmer, M., Flaxman, S., Sejdinovic, D.: Detecting and quantifying causal associations in large nonlinear time series datasets. Sci. Adv. **5**(11), eaau4996 (2019)

19. Sikchi, H., Zhou, W., Held, D.: Learning off-policy with online planning. In: Conference on Robot Learning, pp. 1622–1633. PMLR (2022)

20. Spirtes, P., Glymour, C.N., Scheines, R., Heckerman, D.: Causation, Prediction, and Search. MIT Press, Cambridge (2000)

21. Tassa, Y., et al.: Deepmind control suite. arXiv preprint arXiv:1801.00690 (2018)

22. Williams, G., Aldrich, A., Theodorou, E.: Model predictive path integral control using covariance variable importance sampling. arXiv preprint arXiv:1509.01149 (2015)

23. Ye, W., Liu, S., Kurutach, T., Abbeel, P., Gao, Y.: Mastering Atari games with limited data. In: Advances in Neural Information Processing Systems, vol. 34, pp. 25476–25488 (2021)

24. Zhang, K., Peters, J., Janzing, D., Schölkopf, B.: Kernel-based conditional independence test and application in causal discovery. In: 27th Conference on Uncertainty in Artificial Intelligence (UAI 2011), pp. 804–813. AUAI Press (2011)

Gradient Coupled Flow: Performance Boosting on Network Pruning by Utilizing Implicit Loss Decrease

Jiaying Wu[1,2], Xiatao Kang[1(✉)], Jingying Xiao[1], and Jiayi Yao[1]

[1] School of Computer and Communication Engineering,
Changsha University of Science and Technology, Changsha, China
`kangxiatao@gmail.com`
[2] Hunan Provincial Key Laboratory of Intelligent Processing of Big Data
on Transportation, Changsha, China

Abstract. Network pruning prior to training makes generalization more challenging than ever, while recent studies mainly focus on the trainability of the pruned networks in isolation. This paper explores a new perspective on loss implicit decrease of the data to be trained caused by one-batch training during each round, whose first-order approximation we term gradient coupled flow. We thus present a criterion sensitive to gradient coupled flow (GCS), which is hypothesized to capture those weights most sensitive to performance boosting at initialization. Interestingly, our explorations show there exists a linear correlation between generalization and implicit loss decrease based measurements on previous works as well as GCS, which ideally describes causes of accuracy fluctuation in a fine-grained manner. Our code is made public at: https://github.com/kangxiatao/pruning_before_training.

Keywords: Deep Learning · Network Pruning · Pruning Before Training · Single-shot Pruning

1 Introduction

Due to the characteristics of over-parameterization, neural network pruning has always been an essential subject in the field of deep learning. In recent years, research interest has increasingly focused on theoretical understanding of the essence of neural networks for promoting pruning progresses [25,26]. One of the most talked-about issues is the Lottery Ticket Hypothesis [7], which has identified that there exist sparse subnetworks in randomly-initialized neural networks that, when trained in isolation, can match the test accuracy of the original network. Around this hypothesis, relative work has been carried out [8,22] and intensely studied [19,20], and quite a good progress has been achieved. More importantly, the existence of ideal sparse structures raised by this hypothesis motivates researchers to explore inherent mechanisms of neural networks in the scene of the pruning subject [14,17].

B. Luo et al. (Eds.): ICONIP 2023, LNCS 14448, pp. 231–243, 2024.
https://doi.org/10.1007/978-981-99-8082-6_18

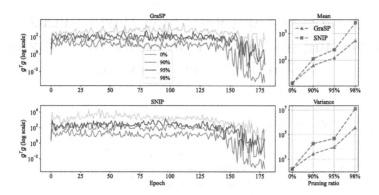

Fig. 1. Comparison of gradient norm squared at different pruning rates during training. The vertical axis uses a logarithmic scale. Experiments are performed under VGG19/CIFAR10, and samples are randomly selected before each training to calculate the loss gradient to obtain $g^{\top}g$.

We find that it is difficult to achieve structural deletion on channel/filter based level in a convolution layer before training, even at a very high compression rate for weight pruning. Thus the pruned weights organized by convolution structure are only in a zero-valued state for inferences and iterations, whose gradients keep changing during the whole training process. Furthermore, quite significant changes would occur on gradients of the retained weights with the increase of compression ratio (shown in Fig. 1). The drastic change of gradient norm is bound to bring non-negligible influence on norms of weights, loss reduction, and so on, which is the internal reason why relative kinds of literature [26,28] focus on the trainability issue of sparse subnetworks.

In addition, there is little literature on generalization improvement of network pruning before training. This paper explores a new perspective on implicit loss decrease of the data to be trained caused by one-batch training during each round, whose first-order approximation we term gradient coupled flow. We found a specific relationship between the coupled flow and the model generalization. Based on this, inspired by loss-sensitive single-shot pruning network methods [15], we apply the implicit loss to pruning before training. Specifically, we choose the weights that significantly influence the coupled flow to construct a sparse network to ensure the model generalization while maintaining the network trainability.

The contributions of this paper are addressed as follows:

- The concept of gradient coupled flow is proposed. It can effectively describe the process of generalization improvement in model training. Our explorations show that there exists a linear correlation between generalization and our relevance evaluation.
- A metric sensitive to gradient coupled flow (GCS) is proposed. GCS applies pruning before training, does not require training and iteration of weights, and only uses a small number of data samples for several inferences.

– GCS has excellent performance as well as scalability and trainability. It is shown that GCS has certain advantages in both single pruning and iterative pruning. Furthermore, a variant of GCS called GCS-Group achieves better generalization at low compression rates through the expansion of the subdivided coupled flow.

2   Related Work

The earliest pruning algorithms were based on pre-trained models considering the removal of the weights with the least impact on performance [13]. Later, better performance was achieved based on amplitude pruning with constant fine-tuning and retraining [10]. Approaches to removing weights incrementally with the training process [9] further demonstrated the advantages of iterative pruning and allowed the retraining process to be removed. Moreover dynamic pruning methods were proposed as the cut part could participate in the training again [1,23], improving pruning efficiency. Tradeoff between performance and efficiency has been seriously investigated by sparsity training approaches [5,6,21] that enforced a constant sparsity rate in advance, and penalty-based approaches [18,24] that penalized unimportant weights to zero.

In order to accelerate the training and avoid extra computations caused by iterative pruning or sparse constraints, the lottery hypothesis [7] experimentally demonstrates the possibility of obtaining excellent sparse structures before training. Subsequent work gives more theoretical explanations and proofs [20,32]. Research from initialization and stability of sparse sub-networks [4,8] provides new insights. Methods such as early stopping [30] and selecting specific samples [31] reduce the computational effort of deep networks to find sparse networks.

In the application of pruning before training, a method called SNIP to select the weights most sensitive to loss [15] has been widely accepted in the metric design of single pruning and improved with iterative [3,27] and dynamic [12] pruning to achieve better performance. GraSP achieves higher compression ratios in a single pruning by maximizing the gradient flow [29] of sparse networks. SynFlow iterations maintain synaptic flow [26] to avoid layer collapse in the network, and the iterative process considers the effective compression rate [28] to reach the limit of compression. A series of researches have explored relative issues such as transferability on pre-trained models [2], inter-layer structure [14,25], and sparse connections [11,17].

3   Gradient Coupled Flow

To facilitate the description of the problem, we assume that the dataset has m training samples $\mathcal{D} = \{(x_i, y_i)\}_{i \in (1,m)}$. With respect to a batch data randomly sampled from $\mathcal{D}$, we denote the loss function and corresponding gradient as $\mathcal{L}(\cdot)$ and g. And the loss can be written by a second-order Taylor series as:

$$\mathcal{L}(\boldsymbol{\theta} - \varepsilon\boldsymbol{g}) \approx \mathcal{L}(\boldsymbol{\theta}) - \varepsilon\boldsymbol{g}^\top\boldsymbol{g} + \frac{1}{2}\varepsilon^2\boldsymbol{g}^\top\boldsymbol{H}\boldsymbol{g} \tag{1}$$

where θ is the parameter, ε is the step size, and H is the Hessian Matrix. For sparse networks, we use the mask c to mark whether the weights are deleted, and κ to set the desired degree of sparsity $\|c\|_0 \leq \kappa$.

3.1   Implicit Loss Decrease on Data to Be Trained

Assume that a batch $\mathcal{D}^b$ is divided into any two data sets called $\mathcal{D}_i^b, \mathcal{D}_j^b$, with corresponding gradient denoted as g_i, g_j. At each iteration, the first-order approximation of $\Delta \mathcal{L}(\mathcal{D}^b)$ is:

$$\varepsilon g^\top g = \frac{1}{4} \varepsilon \left(g_i^\top g_i + g_j^\top g_j \right) + \frac{1}{2} g_i^\top g_j \tag{2}$$

Note that Eq. (2) can be represented as the two parts, the first term is the first-order approximation of the loss produced on $\mathcal{D}_i^b$ itself, which is the product of the step size $\Delta \theta = -\frac{1}{2}\varepsilon g_i$ and the gradient g_i. The second term can be seen as the coupled effect on $\mathcal{D}_j^b$. Similar result can also be obtained with respect to $\mathcal{D}_j^b$. Since $g_i^\top g_j = g_j^\top g_i$, and both appear in the losses of $\mathcal{D}_i^b$ and $\mathcal{D}_j^b$, we call it **gradient coupled flow**. It worth noticing that a half of the gradient coupled flow is the first-order approximation of $\Delta \mathcal{L}_c(\mathcal{D}_j^b)$, namely implicit loss decrease on $\mathcal{D}_j^b$, as the loss change is not produced by g_j.

Gradient Coupled Flow Tends to Zero in Advance. We randomly extract samples to examine the changes of $g_i^\top g_j$ during the training process under different circumstances. The curve changes shown in Fig. 2 reveals a phenomenon: for any g_i, g_j, the coupled flow $g_i^\top g_j$ always tends to zero earlier than $g^\top g$ during training.

The phenomenon that the coupled flow tends to zero in advance motivates us to study the coupled flow in detail. When the coupled flow is zero, it can be known from the properties of the coupled flow that the generation of $\Delta \mathcal{L}(\mathcal{D}_i^b)$ will not bring about a decrease in the loss of $\mathcal{D}_j^b$. This means that although the batch gradient is not zero at this time, it has little effect on the improvement of generalization. Because at this time, the iteration of weights only has a positive effect on some data and does not help the loss of other data. We believe that the synchronous zero-trending feature of coupled flow and gradient is also one of the key reasons for its ideal generalization performance. Therefore, it is more scientific to use gradient coupled flow as a measure of weight sensitivity.

Gradient Coupled Flow at Different Scales. If the training batch is divided into two, each sub-batch contains samples of all categories, and the sample size of each category is half of the original batch. Then the gradient coupled flow can approximately reflect the continuous training process based on sub-batches. If a sub-batch with half the size of the original batch is used for training, and it is assumed that for a certain sub-batch such as the gradient g_i generated by $\mathcal{D}_i^b$, the change after one weight training is negligible. Then the gradient coupled

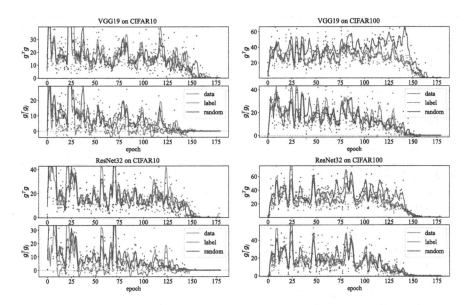

Fig. 2. The changing trend of $g^{\top}g$ and $g_i^{\top}g_j$ in the training process of VGG19 and ResNet32 on CIFAR10/100. There are three sampling methods: data (to ensure that the label of $\mathcal{D}_i^b, \mathcal{D}_j^b$ is the same when sampling), label (to ensure that the label of $\mathcal{D}_i^b, \mathcal{D}_j^b$ is different when sampling), and random (random sampling). Gradients are computed in eval mode before each training epoch. CIFRA10 draws 100 samples and CIFAR100 draws 200 samples.

flow reflects the interaction between the two sub-batches before and after, which essentially reflects the changing characteristics of the loss reduction in the sub-batch training.

3.2   A Criterion Sensitive to Gradient Coupled Flow

We set a pruning criterion based on the degree of weights sensitive to gradient coupled flow, that is, to filter the weight by calculating the influence of the removal weight on the gradient coupled flow, which we call gradient coupled sensitivity (GCS). The calculation method is similar to SNIP [15] and GraSP [29] and is summarized as synaptic saliency $\mathcal{S}(\theta) = \frac{\partial \mathcal{R}}{\partial \theta} \odot \theta$ in SynFLow [26], which is used as a score metric for weights.

Specifically, after network initialization, a batch of data $\mathcal{D}^b$ is randomly extracted from the training dataset, and divided into the two equal parts dented as $\mathcal{D}_i^b, \mathcal{D}_j^b$. As the loss is generally approximated as $\Delta\mathcal{L} \cong \varepsilon g^{\top}g$, the gradient coupled flow can be expressed as:

$$g_i^{\top}g_j = 2\Delta\mathcal{L}\left(\mathcal{D}^b\right) - \frac{\Delta\mathcal{L}\left(\mathcal{D}_i^b\right) + \Delta\mathcal{L}\left(\mathcal{D}_j^b\right)}{2} \tag{3}$$

Algorithm 1. Gradient Coupled Sensitive (GCS)

Input: Loss function $\mathcal{L}(\cdot)$ with weights $\boldsymbol{\theta}$, batch data $\mathcal{D}^b$, and sparsity level κ.
Output: Mask **c**
1: $L_i, L_j = \mathcal{L}\left(\mathcal{D}_i^b, \mathcal{D}_j^b\right)$
2: $\boldsymbol{g}_i, \boldsymbol{g}_j = \text{grad}\left(L_i, L_j\right)$
3: $\boldsymbol{H}_i\boldsymbol{g}_j + \boldsymbol{H}_j\boldsymbol{g}_i = \text{grad}\left(\boldsymbol{g}_i^\top \boldsymbol{g}_j\right)$
4: $S(\boldsymbol{\theta}) = \left|\boldsymbol{\theta} \odot \left(\boldsymbol{H}_i\boldsymbol{g}_j + \boldsymbol{H}_j\boldsymbol{g}_i\right)\right|$
5: $\mathbf{c} = S(\boldsymbol{\theta}) > \text{top}(S(\boldsymbol{\theta}), \kappa)$

Algorithm 2. GCS-Group

Input: A batch of data $\mathcal{D}^b$, number of groups p, and sparsity level κ.
Output: Score $S(\boldsymbol{\theta})$
1: **for** $i = 0$ **to** p **do**
2:    **for** $j = i+1$ **to** p **do**
3:       $HgList.append(getHgCross(\mathcal{D}_i^b, \mathcal{D}_j^b))$
4:    **end for**
5: **end for**
6: **for** $i = 0$ **to** $len(HgList)$ **do**
7:    $Hg+ = HgList[i] **2$
8: **end for**
9: $S(\boldsymbol{\theta}) = |\boldsymbol{\theta} \odot sqrt(Hg)|$

Then the sensitivity of parameter $\boldsymbol{\theta}$ to $\boldsymbol{g}_i^\top \boldsymbol{g}_j$ is:

$$S(\boldsymbol{\theta}) = \left|\boldsymbol{\theta} \odot \nabla_{\boldsymbol{\theta}}\boldsymbol{g}_i^\top \boldsymbol{g}_j\right| = \left|\boldsymbol{\theta} \odot \left(\boldsymbol{H}_i\boldsymbol{g}_j + \boldsymbol{H}_j\boldsymbol{g}_i\right)\right| \tag{4}$$

After obtaining the metric score, for the network pruning task, given the degree of sparsity κ, use the sensitivity $S(\boldsymbol{\theta})$ as the metric to calculate the mask $\mathbf{c} = S(\boldsymbol{\theta}) > S_{\text{top } \kappa}$. where $S_{\text{top } \kappa}$ represents that the mask satisfies the threshold of sparsity κ under $S(\boldsymbol{\theta})$.

As mentioned in Sect. 3.1, GCS is actually weights that are sensitive to changes in the loss of sample common characteristics, in contrast, those weights that are sensitive to sample individual characteristics are preferentially removed. From an implicit loss perspective, a weight metric that is sensitive to gradient coupled flow is to capture those weights that are most sensitive to performance improvements at initialization time. See Algorithm 1 for the specific gradient coupled sensitive pruning method.

3.3 A Flattened Variant of GCS to Get Subdivided Coupled Flows

The weights chosen are sensitive to the gradient coupled flow, meaning that the compressed model will preserve as much information as possible about the gradient coupled flow. To better exploit the coupled loss, we propose a variant of GCS called GCS-Group to obtain richer coupled flows from pairwise combinations.

The partition subset is extended to partition p subsets $\mathcal{D}_p^b = \{(x_p, y_p)\}$, and there are $\boldsymbol{g}_1, \boldsymbol{g}_2, \cdots, \boldsymbol{g}_p$ accordingly. As the sensitivity with regard to an element of $\{g_i^\top g_j \mid i \neq j, i, j \in p\}$ has been derived from Eq. (4), an integrated criterion is formed by summing multiple groups of gradient coupled flows. At present, we set $p = 5$, and use the Euclidean distance to define a unified measure (Algorithm 2): $\mathcal{S}_{\text{group}}(\theta) = \sqrt{\sum_{i=1}^{p} \sum_{j=i+1}^{p} \mathcal{S}\left(\theta, g_i^\top g_j\right)^2}$.

Theoretically, GCS-Group has more capacity to achieve richer coupled information than the original one, as inherent features shared by samples are more likely embodied by means of pairwise combinations of subdivided groups. However, the trade-off between distinguishability and sensitiveness is required to be further considered, especially at extremely high compression ratios.

4   Experiments

Our experiments verify the effectiveness of GCS from multiple perspectives such as image classification task performance, visualization, and analysis of implicit loss. In the image classification task, a variety of recent pruning methods before training are investigated for comparisons: (i) Single-shot pruning of SNIP [15] and GraSP [29] with corresponding layer sparse quota random pruning. (ii) Iterative pruning SynFlow [26] and dynamic iterative [12,16]. In addition, implicit loss analysis reveal some interesting phenomena.

4.1   Performance of GCS in Image Classification

We use VGG and ResNet architectures to analyze the performance of GCS on image classification tasks with different pruning ratios. Specifically, we choose the full networks with L2 regularization as the baseline. First, the compression efficiency of GCS single pruning on VGG19, ResNet18, ResNet32, and LeNet5 is investigated on datasets such as CIFAR10, CIFAR100, Tiny-Imagenet, MNIST, and Fashion-MNIST. Then we apply GCS to iterative pruning before training to examine the performance of GCS.

Single-Shot Pruning. The performance comparison on VGG19, ResNet18, and LeNet5 is shown in Fig. 3. Note that GCS outperforms SNIP and GraSP in most cases. GCS and SNIP are closer to the baseline at low compression ratios, while at high compression ratios above 99%, GCS is on par with GraSP, which also proves that GCS still maintains strong trainability with little generalization loss. Due to the relatively compact structure of ResNet, the residual structure strengthens the stability of the sparse network, and the advantage of GCS is not apparent, but it is still slightly higher than other single pruning. From the perspective of random pruning under the corresponding layer sparse quota, GCS-Random is much lower than GCS at a compression rate above 95%. The performance is severely affected after random shuffling, which indicates that the weights selected by GCS are more representative and cannot be replaced randomly. More experimental data are in Table 1 and Table 2.

GCS is evaluated on Tiny-ImageNet dataset with VGG19 and ResNet32, and compared with SNIP and GraSP the pruning rates of 90%, 95%, 98% respectively, which is shown in Table 3. As expected, the performance of GCS on Tiny-ImageNet is similar to that on CIFAR10/100, and it slightly outperforms other methods overall. With regards to more complex datasets, the generalization boost brought by GCS further demonstrates the advantages of gradient coupled flow.

Iterative Pruning. Compared with single pruning, progressive iterative pruning can obtain a sparse network with a higher compression rate and more stability. We compare SNIP, GraSP, and GCS with SynFLow using the iterative pruning method in FORCE [12], and the experimental results are shown in Fig. 4.

Table 1. Performance comparison of pruned VGG19 and ResNet32 on CIFAR10/100. The bolded number is the one with the highest accuracy among SNIP, GraSP, and GCS. The # indicates that the model failed to converge.

Dataset	CIFAR10			CIFAR100		
VGG19	Acc: 94.20%			Acc: 74.16%		
Pruning ratio	95%	98%	99%	95%	98%	99%
SNIP	**93.71**	92.22	#	71.95	56.23	#
GraSP	93.07	92.26	91.57	71.62	70.15	66.72
GCS	93.66	**92.95**	**91.84**	**72.50**	**70.23**	**67.84**
ResNet32	Acc: 94.80%			Acc: 74.83%		
Pruning ratio	95%	98%	99%	95%	98%	99%
SNIP	91.51	88.36	85.12	65.03	52.19	36.04
GraSP	91.77	89.53	85.52	66.95	58.55	**49.10**
GCS	**92.08**	**89.96**	**85.60**	**67.28**	**58.62**	48.59

Table 2. Performance comparison of pruned LeNet5 on MNIST and Fashion-MNIST.

Dataset	MNIST					Fashion-MNIST				
LeNet5	Acc: 99.40%					Acc: 91.98%				
Pruning ratio	90%	95%	98%	99%	99.5%	90%	95%	98%	99%	99.5%
SNIP	**99.35**	99.17	98.99	98.49	96.06	90.61	89.56	87.55	84.23	75.17
GraSP	99.10	99.17	99.06	98.62	**96.19**	89.69	89.60	86.60	**85.91**	**79.11**
GCS	99.33	**99.20**	**99.13**	**98.79**	95.92	**90.83**	**89.78**	**88.39**	85.59	77.86

Table 3. Performance comparison of pruned VGG19 and ResNet32 on Tiny-ImageNet.

Network	VGG19: 63.29%			ResNet32: 63.86%		
Pruning ratio	90%	95%	98%	90%	95%	98%
SNIP	61.07	59.27	48.95	51.56	40.41	24.81
GraSP	60.26	59.53	56.54	54.84	48.45	**37.25**
GCS	**61.12**	**59.69**	**57.35**	**55.37**	**49.66**	35.89

It is an obvious phenomenon that the difference between various algorithms is minimal at low compression ratios. As the compression ratio increases, the performance of SynFLow remains better when the compression ratio is exceptionally high (the remaining rate is less than 10^{-3}), which is an advantage of inter-layer equalization. Nevertheless, it is worth noting that on ResNet, the negative impact caused by invalid retention [28] in sparse networks is small, and profit from the sensitivity of weights to implicit losses, GCS-Iterative is better than SynFlow on more complex datasets.

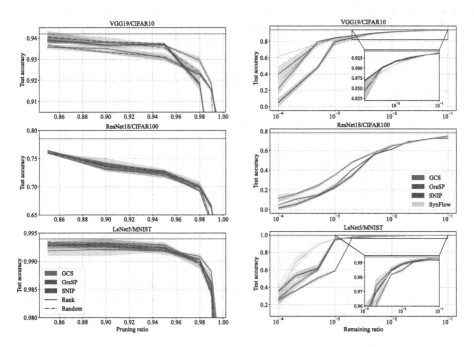

Fig. 3. Performance comparison of single pruning. Rank (solid line) sorts the selection weights according to the metrics, and Random (dotted line) prunes the selection weights under the corresponding layer sparse quota. The grey horizontal line represents the baseline, and the shaded area represents the range of deviations from replicate experiments.

Fig. 4. Performance comparison of iterative pruning. The grey horizontal line represents the baseline. The shaded area indicates the range of bias for replicate experiments.

4.2 Evaluate Model Generalization with Implicit Loss and Weight Reduction

The subdivision coupled flow mentioned in Sect. 3.1 can well explain the intrinsic reasons for the performance improvement achieved by methods such as small-batch training and L2 penalty. We argue that the implicit loss decline and weight change are the key factors affecting model generalization. We then further explore to quantitatively characterize generalization by utilizing the product of the cumulative implicit loss and the change of weight norms:

$$G_{CM} = (\|\boldsymbol{\theta}_T\| - \|\boldsymbol{\theta}_0\|) \int_0^T \Delta\mathcal{L}_c(\mathcal{D}_j^b)\, dt \qquad (5)$$

where T denotes the entire training time, and the cumulative implicit loss in integral form is constructed by relaxing the training rounds. Notes that calculations of the implicit loss between any two random batches are required to be performed before and after each round twice, which makes it impractical for application use in terms of computational and storage overhead. Alternatively, we take two randomly selected batches of data $\mathcal{D}_i^b, \mathcal{D}_j^b$ before each training round and use $\Delta\mathcal{L}_c(\mathcal{D}_j^b) = \mathcal{L}(\boldsymbol{\theta} - \epsilon\nabla_{\boldsymbol{\theta}}\mathcal{L}(\mathcal{D}_i^b); \mathcal{D}_j^b) - \mathcal{L}(\boldsymbol{\theta}; \mathcal{D}_j^b)$ as an approximate expression of the implicit loss at that time.

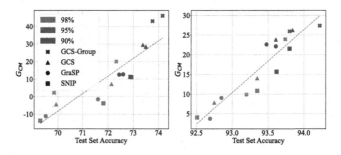

Fig. 5. Linear relationship between G_{CM} and test accuracy. Experiments are performed with VGG19 on CIFAR100/10 (corresponding to the left and right figures respectively), and the red dashed line shows a first-order linear fit. (Color figure online)

We calculate the corresponding G_{CM} for the mainstream algorithms such as SNIP, GraSP, and our GCS, GCS-Group. In order to avoid low stability due to one single experiment, we repeat it twice for all algorithms as shown in Fig. 5. Regardless of the existing work or our proposed GCS, there is a relatively obvious linear relationship between G_{CM} and test set accuracy.

Moreover, the fluctuation of performance is highly synchronized with the sum of the implicit loss as well as the compression ratio. Specifically, G_{CM} does provide a practical scheme to describe model stability and can be utilized to estimate in advance whether a model has excellent generalization ability or not.

5   Conclusion

In this work, we propose the concept of gradient coupled flow to describe loss implicit decrease of the data to be trained caused by one-batch training during each round. Based on the gradient coupled flow, a metric sensitive to gradient coupled flow (GCS) is proposed for single or iterative pruning before training and has successfully achieved excellent performance. In the follow-up research, we will try to use gradient coupled between network structures to achieve extreme compression ratio or obtain structured reduced networks. In addition, we also try to improve the generalization by NLP models based on the gradient coupled flow. The comprehensive correlation between subdivided coupled flow and model generalization is also an exciting direction that deserves further research.

References

1. Bellec, G., Kappel, D., Maass, W., Legenstein, R.: Deep rewiring: training very sparse deep networks. CoRR abs/1711.05136 (2017). https://openreview.net/forum?id=BJ_wN01C-

2. Chen, T., et al.: The lottery ticket hypothesis for pre-trained BERT networks. arXiv preprint arXiv:2007.12223 (2020). https://proceedings.neurips.cc/paper/2020/hash/b6af2c9703f203a2794be03d443af2e3-Abstract.html

3. Cho, M., Joshi, A., Hegde, C.: ESPN: extremely sparse pruned networks. CoRR abs/2006.15741 (2020). https://arxiv.org/abs/2006.15741

4. Desai, S., Zhan, H., Aly, A.: Evaluating lottery tickets under distributional shifts. EMNLP-IJCNLP 2019, p. 153 (2019). https://doi.org/10.18653/v1/D19-6117

5. Dettmers, T., Zettlemoyer, L.: Sparse networks from scratch: faster training without losing performance. CoRR abs/1907.04840 (2019). https://openreview.net/forum?id=K9bw7vqp_s

6. Evci, U., Gale, T., Menick, J., Castro, P.S., Elsen, E.: Rigging the lottery: making all tickets winners. In: ICML, Proceedings of Machine Learning Research, vol. 119, pp. 2943–2952. PMLR (2020). http://proceedings.mlr.press/v119/evci20a.html

7. Frankle, J., Carbin, M.: The lottery ticket hypothesis: finding sparse, trainable neural networks. In: ICLR (2019). https://openreview.net/forum?id=rJl-b3RcF7

8. I Frankle, J., Dziugaite, G.K., Roy, D.M., Carbin, M.: Stabilizing the lottery ticket hypothesis. arXiv preprint arXiv:1903.01611 (2019). https://doi.org/10.48550/arXiv.1903.01611

9. Gale, T., Elsen, E., Hooker, S.: The state of sparsity in deep neural networks. CoRR abs/1902.09574 (2019). http://arxiv.org/abs/1902.09574

10. Han, S., Mao, H., Dally, W.J.: Deep compression: compressing deep neural networks with pruning, trained quantization and Huffman coding. arXiv preprint arXiv:1510.00149 (2015)

11. Hayou, S., Ton, J., Doucet, A., Teh, Y.W.: Robust pruning at initialization. In: ICLR (2021). https://openreview.net/forum?id=vXj_ucZQ4hA

12. de Jorge, P., Sanyal, A., Behl, H.S., Torr, P.H.S., Rogez, G., Dokania, P.K.: Progressive skeletonization: trimming more fat from a network at initialization. In: ICLR (2021). https://openreview.net/forum?id=9GsFOUyUPi

13. LeCun, Y., Denker, J.S., Solla, S.A.: Optimal brain damage. In: Advances in Neural Information Processing Systems, pp. 598–605 (1990). http://papers.nips.cc/paper/250-optimal-brain-damage

14. Lee, N., Ajanthan, T., Gould, S., Torr, P.H.S.: A signal propagation perspective for pruning neural networks at initialization. In: ICLR (2020). https://openreview.net/forum?id=HJeTo2VFwH

15. Lee, N., Ajanthan, T., Torr, P.H.S.: Snip: single-shot network pruning based on connection sensitivity. In: ICLR (Poster) (2019). https://openreview.net/forum?id=B1VZqjAcYX

16. Liu, S., Yin, L., Mocanu, D.C., Pechenizkiy, M.: Do we actually need dense over-parameterization? In-time over-parameterization in sparse training (2021). https://doi.org/10.48550/ARXIV.2102.02887, https://arxiv.org/abs/2102.02887

17. Liu, T., Zenke, F.: Finding trainable sparse networks through neural tangent transfer. In: ICML, Proceedings of Machine Learning Research, vol. 119, pp. 6336–6347. PMLR (2020). http://proceedings.mlr.press/v119/liu20o.html

18. Liu, Z., Li, J., Shen, Z., Huang, G., Yan, S., Zhang, C.: Learning efficient convolutional networks through network slimming. In: ICCV, pp. 2755–2763. IEEE Computer Society (2017). https://doi.org/10.1109/ICCV.2017.298

19. Liu, Z., Sun, M., Zhou, T., Huang, G., Darrell, T.: Rethinking the value of network pruning. In: ICLR (Poster) (2019). https://openreview.net/forum?id=rJlnB3C5Ym

20. Malach, E., Yehudai, G., Shalev-Shwartz, S., Shamir, O.: Proving the lottery ticket hypothesis: pruning is all you need. In: ICML, Proceedings of Machine Learning Research, vol. 119, pp. 6682–6691. PMLR (2020). http://proceedings.mlr.press/v119/

21. Mocanu, D.C., Mocanu, E., Stone, P., Nguyen, P.H., Gibescu, M., Liotta, A.: Scalable training of artificial neural networks with adaptive sparse connectivity inspired by network science. Nat. Commun. 9(1), 1–12 (2018). https://doi.org/10.1038%2Fs41467-018-04316-3

22. Morcos, A.S., Yu, H., Paganini, M., Tian, Y.: One ticket to win them all: generalizing lottery ticket initializations across datasets and optimizers. In: NeurIPS, pp. 4933–4943 (2019). https://proceedings.neurips.cc/paper/2019/hash/a4613e8d72a61b3b69b32d040f89ad81-Abstract.html

23. Mostafa, H., Wang, X.: Parameter efficient training of deep convolutional neural networks by dynamic sparse reparameterization. In: ICML, Proceedings of Machine Learning Research, vol. 97, pp. 4646–4655. PMLR (2019). http://proceedings.mlr.press/v97/mostafa19a.html

24. Nowlan, S.J., Hinton, G.E.: Simplifying neural networks by soft weight-sharing. Neural Comput. 4(4), 473–493 (1992). https://doi.org/10.1162/neco.1992.4.4.473

25. Orseau, L., Hutter, M., Rivasplata, O.: Logarithmic pruning is all you need. In: NeurIPS (2020). https://proceedings.neurips.cc/paper/2020/hash/1e9491470749d5b0e361ce4f0b24d037-Abstract.html

26. Tanaka, H., Kunin, D., Yamins, D.L., Ganguli, S.: Pruning neural networks without any data by iteratively conserving synaptic flow. In: NeurIPS (2020). https://proceedings.neurips.cc/paper/2020/hash/46a4378f835dc8040c8057beb6a2da52-Abstract.html

27. Verdenius, S., Stol, M., Forré, P.: Pruning via iterative ranking of sensitivity statistics. CoRR abs/2006.00896 (2020). https://arxiv.org/abs/2006.00896

28. Vysogorets, A., Kempe, J.: Connectivity matters: neural network pruning through the lens of effective sparsity (2021). https://doi.org/10.48550/ARXIV.2107.02306, https://arxiv.org/abs/2107.02306

29. Wang, C., Zhang, G., Grosse, R.B.: Picking winning tickets before training by preserving gradient flow. In: ICLR (2020). https://openreview.net/forum?id=SkgsACVKPH

30. You, H., et al.: Drawing early-bird tickets: towards more efficient training of deep networks. CoRR abs/1909.11957 (2019). http://arxiv.org/abs/1909.11957

31. Zhang, Z., Chen, X., Chen, T., Wang, Z.: Efficient lottery ticket finding: less data is more. In: ICML, Proceedings of Machine Learning Research, vol. 139, pp. 12380–12390. PMLR (2021). http://proceedings.mlr.press/v139/zhang21c.html
32. Zhou, H., Lan, J., Liu, R., Yosinski, J.: Deconstructing lottery tickets: zeros, signs, and the supermask. In: NeurIPS, pp. 3592–3602 (2019), https://proceedings.neurips.cc/paper/2019/hash/1113d7a76ffceca1bb350bfe145467c6-Abstract.html

Motif-SocialRec: A Multi-channel Interactive Semantic Extraction Model for Social Recommendation

Hangyuan Du[1,2(✉)], Yuan Liu[1,2], Wenjian Wang[1,2], and Liang Bai[2,3]

[1] School of Computer and Information Technology, Shanxi University,
Taiyuan 030006, Shanxi, China
[2] Ministry of Education,
Key Laboratory of Computational Intelligence & Chinese Information Processing
(Shanxi University), Taiyuan 030006, Shanxi, China
duhangyuan@sxu.edu.cn
[3] Institute of Intelligent Information Processing, Shanxi University,
Taiyuan 030006, Shanxi, China

Abstract. To capture complex interaction semantics beyond pairwise relationships for social recommendation, a novel recommendation model, namely Motif-SocialRec, is proposed under the perspective of motif. It efficiently describes interaction pattern from multi-channel with different motifs. In the model, we depict a series of local structures by motif, which can describe the high-level interactive semantics in the fused network from three views. By employing hypergraph convolution network, representations that preserve potential semantic patterns can be learned. Additionally, we enhance the learned representations by establishing self-supervised learning tasks on different scales to further explore the inherent characteristics of the network. Finally, a joint optimization model is constructed by integrating the primary and auxiliary tasks to produce recommendation predictions. Results of extensive experiments on four real-world datasets show that Motif-SocialRec significantly outperforms baselines in terms of different evaluation metrics.

Keywords: Social recommendation · Motif · Multi-channel hypergraph convolution · Self-supervised learning

1 Introduction

In recent studies about recommendation systems, one idea is generally recognized that individual behaviors are driven by both intrinsic and social factors [1,2]. Consequently, social relationship between users are introduced into recommendation system to enhance the decision-making, that is called social recommendation (SocialRec). However, incorporating social relationships makes the entire

Supported by organization x.

heterogeneous interactive network more complex, in aspects of both structure and characteristics. Thus, how to efficiently extract valuable semantic patterns about user preference becomes an urgent problem in social recommendation.

Graph neural networks (GNNs) are widely used to learn node representation in social recommendation models. The information aggregation mechanism of GNNs can just fit the constrain of social relationships that correlated users always have common preference in the interaction network. GNNs always aggregate local information recursively relying on neighborhood structure in order to update node representations.

However, such a message propagation mechanism lacks the ability of modelling complex interaction patterns contained in high-level local structures of the network. In social recommendation problem, these local structures are always formed under the influence of social connections and behavior interactions, and imply abundant valuable information for capturing user preference. Consequently, how to model high-level local structures and learn representation conditioned on them is essential to produce reliable recommendation results.

Motif, as a particular local structure in network, provides a powerful modeling tool to address the aforementioned problem [15]. In this work, we use motif to depict the high-level local structure in the network result from the fusion of social relationships and user-item interaction network. Specifically, we extract three types of motifs to reflect different types of local connection mode in the network driven by diverse semantic patterns. By treating motif as carrier of information propagation, we utilize a hypergraph convolution network to extract different interactive semantics in representation learning in a multi-channel framework. Furthermore, we employ self-supervised learning (SSL) paradigm to enhance the learned representations by establishing contrastive tasks on multiple scales. In this design, user preference can be effectively recognized, and thus promoting a reliable recommendation prediction.

In brief, the main contributions of this paper can be summarized as follows:

- We extract a series of motifs to capture high-level interactive semantics in the information network.
- We propose a novel social recommendation model, namely Motif-SocialRec.
- Extensive experiments are conducted on four real-world datasets, demonstrating the superior performance of Motif-SocialRec.

2   Related Work

2.1   Social Recommendation

With the booming development of GNNs, various GNN-based SocialRec models emerge in recent years, whose performance largely depends on specific GNN encoders they adopt. Based on different GNN encoders, these models can be organized as follows: DiffNet [4] and GNNRec [6] employ Graph Convolutional Network (GCN) to learn user and item representations; GraphRec [7], DiffNet++ [8], and SGHAN [9] use Graph Attention Network (GAT) to recognize each user's preference on items or its influence on friends; DGRec [10] and GTN [11] utilize Graph Recurrent Neural Network (GRNN) to study both dynamic user interests and context-dependent social influences for online recommendation; CGL [12] and DcRec [13] learn user and item representations by linearly propagating them on the user-item interaction network based on LightGCN [5].

2.2 Motif

Motif, as a simple building block structure, is first introduced in [14]. It refers to the specific local connection mode occurring frequently in complex networks, and can be found in many complex systems modeled by network. Meanwhile, from a perspective of information processing, motifs that describe different physical meanings may play similar roles on defining universal characteristic of networks. Motif has been proved to be an effective basic building factor for analyzing formation rule in most networks. In this paper, high-level interactive semantics between users and items will be extracted and encoded from multi-channel under different motifs.

2.3 Self-supervised Learning

SSL is a popular learning paradigm for capturing implicit pattern or knowledge from abundant input data through well-designed pretext tasks without using manual labels. Driven by its successful application in computer vision and natural language processing, employing SSL to address the data sparsity problem and noise susceptibility in recommendation system holds bright prospect. In this work, we will perform cross-scale contrastive tasks to enhance representations by exploring inherent characteristics of the interaction patterns conditioned on different motifs.

3   Model

3.1   Preliminaries

For recommendation with user relations, let $U = \{u_1, u_2, \cdots, u_{|U|}\}$ be a set of users and $I = \{i_1, i_2, \cdots, i_{|I|}\}$ be a set of items. Let $R = \mathbb{R}^{|U| \times |I|}$ be a binary user-item interaction matrix, where $R_{u,i} = 1$ indicates that user u has an interaction with item i, otherwise $R_{u,i} = 0$. In addition, directed social relations are available, which are denoted by an asymmetric matrix $S \in \mathbb{R}^{|U| \times |U|}$, where $S_{j,k} = 1$ if user u_j is related to user u_k, otherwise $S_{j,k} = 0$.

Given the user-item interaction matrix R and the social relationship matrix S, the social recommendation task aims to predict whether user u has a potential interest in item i he has never interacted with. Finally, top-N items will be recommended to the user based on the predicted preference scores in descending order.

In a general graph, each edge can only connect two nodes, thus only pairwise relationships can be reflected. However, many relationships are not limited between two nodes in real-world applications, such as the stable triangular relationship in social networks, and the joint purchase relationship in e-commerce networks. In a hypergraph, this limitation can be broken. The hypergraph evolves an edge into a hyperedge, and a hyperedge can connect multiple nodes at the same time. A hypergraph can be denoted as $G = (V, E)$, where V is the set of nodes, and E is the set of hyperedges in the graph. The matrix $A \in \{0,1\}^{|V| \times |E|}$ records adjacency relationships between nodes, where each element represents whether hyperedge e contains node v:

$$a(v, e) = \begin{cases} 1, & \text{if } v \in e \\ 0, & \text{if } v \notin e \end{cases}. \tag{1}$$

The degree of a node in hypergraph can be represented by the number of hyperedges that contain the node, i.e., $d(v) = \sum_{\forall e \in E} a(v, e)$. Besides, since a hyperedge is no longer limited to connect only two nodes, its degree can be defined as the number of nodes it connects, i.e., $d(e) = \sum_{\forall v \in V} a(v, e)$. Then, the degrees of hyperedges and nodes in the hypergraph G can be recorded in the diagonal matrices $D_e \in R^{|E| \times |E|}$ and $D_v \in R^{|V| \times |V|}$ formed by $d(e)$ and $d(v)$, respectively.

3.2 Overall Framework

The overall framework of Motif-SocialRec is illustrated in Fig. 1, which consists of three components:

- **Motif information extraction module.** This module fuses social network information and user-item interactive information to construct a new heterogeneous network, then extracts the high-level local structures on the heterogeneous network under different motifs.
- **Representation generation module.** In this module, information of each motif is encoded by employing the hypergraph convolution network, in order to learn user representations. Then, item representations are acquired by leveraging user-item interactions.
- **Preference prediction module.** To enhance the learned representations, SSL task is performed in this module by establishing contrastive view across both micro-scale, meso-scale, and macro-scale. Then, model predictions are produced jointly driven by recommendation task and SSL task.

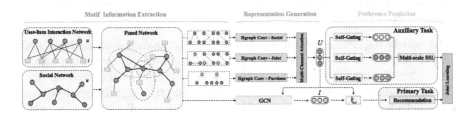

Fig. 1. Overall framework of Motif-SocialRec.

3.3   Motif Information Extraction

To preserve specific social connections between users and their corresponding interaction behaviors, three types of motifs, namely social motif, behavior motif and joint motif, are extracted from the fused heterogeneous network to model intricate relationships among users and their behaviors. Social motif organizes a group of interconnected users, in which explicit social relationships are reflected. Behavior motif models interaction patterns without constrain of social relationships in which users are correlated by the common item they both interact with. Joint motif describes reinforced local connection mode by incorporating social information and interaction information, where some potential relationships can be produced. These three types of motifs can reflect the aggregation of user interests under social relationship constraints and thereby uncover their behavior preferences.

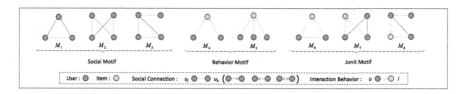

Fig. 2. Motifs extracted from the fused network.

Specifically, we totally construct eight fundamental motifs that can cover the above three types, as shown in Fig. 2. These motifs can be divided into two groups: 3-node motifs (M_1, M_4, and M_6) and 4-node motifs (M_2, M_3, M_5, M_7 and M_8). In Fig. 2, edges with different color represent social relationship and interaction behavior, respectively. In social motifs and joint motifs, the pink edge between two users u_j and u_k denotes three specific directed edges: $u_j \rightarrow u_k$, $u_j \leftarrow u_k$, and $u_j \leftrightarrow u_k$, corresponding to different social relationships between users. In behavior motifs and joint motifs, the yellow edge $u \rightarrow i$ represents the interaction behavior between user u and item i.

For 3-node motifs, three representative local structures can be extracted, which are illustrated as M_1, M_4, and M_6 in Fig. 2. These stable triangular relationships commonly exist in complex networks, and can be viewed as fundamental elements to constitute various structures. Compared with 3-node motif, the construction of 4-node motifs is more complicated:

- To model cyclic local structures composed of four nodes, we extend the 3-node motif M_1 by introducing another node and edge to capture more semantic relationships for the object user within its neighborhood, and design a *butterfly-shaped motif* (M_2) and a *star-shaped motif* (M_3), respectively.
- For behavior motifs, the 4-node motif (M_5) can be established by augmenting the 3-node motif (M_4) with another user who interacts with the same item.

- 4-node joint motifs can be built based on corresponding social motifs by incorporating interaction relationships. Specifically, we replace one arbitrary node in M_2 or M_3 whose degree equals or exceeds 2 with an item node, and substitute behavior edge $u \rightarrow i$ for the previous social edge linked with the removed user. Then, we obtain another *butterfly-shaped motif* M_7 and a *star-shaped motif* M_8.

By combining above 3-node motifs and 4-node motifs, we can model any complex connection patterns and local structures in the network. According to the principle proposed in [17], we induce several appropriate adjacency matrixes to reorganize the network based on these fundamental motifs. And, the element in these matrixes can be calculated as the co-occurrence of two nodes i and j in a certain motif:

$$[A_{M_m}]_{i,j} == \begin{cases} 1, & \text{if node } i \text{ and } j \text{ co-occur in the motif } M_m \\ 0, & \text{otherwise} \end{cases}, \qquad (2)$$

where A_{M_m} denotes the adjacency matrix defined under the motif M_m, and M_m $(m = 1, ..., 8)$ represents one of the eight motifs illustrated in Fig. 2. Further, these matrixes can be integrated from three views, i.e., adjacency matrix conditioned on social motifs: $A_S = \sum_{m=1}^{3} A_{M_m}$, behavior motifs: $A_B = \sum_{m=4}^{5} A_{M_m}$, and joint motifs: $A_J = \sum_{k=6}^{8} A_{M_m}$, respectively.

3.4 Representation Generation

In above design, the network is in fact transformed to a hypergraph conditioned on corresponding type of motifs. And, connection relationships preserved in the hypergraph focus on a certain aspect of inherent interaction patterns. To comprehensively capture these patterns, we construct a multi-channel model to learn user representations from the three views in parallel, in which motif is employed as the carrier of message propagation to gather high-level interactive semantics. In each channel, information of the corresponding motifs are encoded by the hypergraph convolutional neural network (HGNN) proposed in [18]:

$$H_k^{(l+1)} = \tilde{D}_k^{-1} A_k H_k^{(l)}, \qquad (3)$$

where $H_k \in \{H_S, H_B, H_J\}$ denotes the learned representations under the corresponding type of motifs, and $\tilde{D}_k$ represents the degree matrix of $A_k \in \{A_S, A_B, A_J\}$.

To overcome the over-smoothing problem occurred in information aggregation, outputs of all the layers in the HGNN are averaged as the representation in each channel:

$$\bar{H}_k = \frac{1}{L+1} \sum_{l=0}^{L} H_k^{(l)}, \qquad (4)$$

where L is the number of layers in HGNN, $\bar{H}_k$ denotes the output of HGNN in channel k. Then, attention mechanism is utilized to adaptively aggregate motif-based representations learned from different views, in order to obtain comprehensive user representations. For each user u, attention coefficients $(\alpha_S, \alpha_B, \alpha_J)$

are generated to describe weights of user representation under the three motifs, by following attention function:

$$\alpha_k = f_{att}\left(\bar{h}_k\right) = \frac{\exp\left(\gamma^{\mathrm{T}} \cdot W_{att}\bar{h}_k\right)}{\sum_{k' \in \{\mathrm{S, B, J}\}} \exp\left(\gamma^{\mathrm{T}} \cdot W_{att}\bar{h}_{k'}\right)}, \tag{5}$$

where $\gamma \in \mathbb{R}^d$ and $W_{att} \in \mathbb{R}^{d \times d}$ are learnable parameters of the attention function. Finally, the comprehensive user representation conditioned on motifs can be calculated as:

$$h^* = \sum_{k \in \{S,B,J\}} \alpha_k \bar{h}_k. \tag{6}$$

On this basis, the final item representation can be produced as follows:

$$E^{(l+1)} = D^{-1}R^{\mathrm{T}}H^{*(l)}, \tag{7}$$

where R is the interaction matrix between users and items, and D is the degree matrix of R.

3.5   Preference Prediction

In the preference prediction module, we devise a joint learning model, in which producing recommendation prediction based on the learned representations is set as the primary task, and a multi-scale contrastive mechanism is established as the auxiliary task to boost recommendation performance.

For the primary task, the BPR loss function is employed to optimize the model:

$$\mathcal{L}_P = \sum_{i \in I(u), o \notin I(u)} -\log \sigma\left(\hat{r}_{u,i} - \hat{r}_{u,o}\right) + \lambda \|\Theta\|_2^2, \tag{8}$$

where $I(u)$ represents the set of items that have interaction with user u, Θ denotes the set of trainable parameters in the model, and λ is the regularization coefficient. In the training progress, an optimization tuple is formed by three elements: a user u, a randomly sampled positive item $i \in I(u)$, and a negative item $o \notin I(u)$. $\hat{r}_{u,i}$ represents the recommendation result between u and i, which is produced by using their representations:

$$\hat{r}_{u,i} = h_u^{*\mathrm{T}} e_i. \tag{9}$$

To preserve the inherent characteristics of network distribution conditioned on different motifs, multi-scale SSL model is constructed to enhance the user representations generated under each motif. We abstract user representations into three scales to perform contrastive learning tasks, i.e., micro-scale, meso-scale and macro-scale. On micro-scale, representation of each user node learned by HGNN under each type of motifs is of interest. On meso-scale, representations are gathered within the first-order connected motifs for each user node. On macro-scale, the graph-level representation conditioned on the corresponding motifs is obtained by applying a global readout function.

To make the SSL task more effective, we firstly enhance the non-linear approximation ability and stability of the model by the self-gating activation function in Eq. (10).

$$\hat{H}_k = g_k(\bar{H}_k) = \bar{H}_k \odot \sigma(\bar{H}_k W_k^g + b_k), \tag{10}$$

where $W_k^g \in \mathbb{R}^{d \times d}$ and $b_k \in \mathbb{R}^d$ are learnable parameters, and $\sigma(\cdot)$ represents the sigmoid activation function. Then, we establish a two-level contrastive task for each type of motifs. The first level is the contrasting between micro-scale and meso-scale, and the second level is from meso-scale to macro-scale. Therefore, we define the SSL loss function as:

$$\mathcal{L}_A = -\sum_k \{\sum_U \log \sigma \left[f_d \left(\hat{h}_k, z_k \right) - f_d \left(\hat{h}_k, \tilde{z}_k \right) \right] + \sum_U \log \sigma \left[f_d \left(z_k, y_k \right) - f_d \left(\tilde{z}_k, y_k \right) \right] \}, \tag{11}$$

where $f_d(\cdot)$ is the discriminative function that measures the consistency between two representations, and user representation on the meso-scale is obtained using the readout function:

$$z_k = \text{READOUT} \left(\hat{h}_k, \mathbf{a}_k \right) = \frac{\hat{h}_k \mathbf{a}_k}{sum \left(\mathbf{a}_k \right)}. \tag{12}$$

In Eq. (12), $\mathbf{a}_k$ represents the row vector in matrix A_k corresponding to the target user, and $sum \left(\mathbf{a}_k \right)$ denotes the total number of connections in the meso-scale network. $\tilde{z}_k$ represents the negative sample created by performing shuffle disturbance on z_k. And, y_k is the user representation on the macro-scale obtained using the AveragePooling function in Eq. (13):

$$y_k = AveragePooling \left(\hat{H}_k \right). \tag{13}$$

3.6   Optimization Model

We integrate the primary task and auxiliary task to optimize the model:

$$\mathcal{L} = \mathcal{L}_P + \beta \mathcal{L}_A, \tag{14}$$

where β is the adjustment coefficient of the SSL loss.

4   Experiments and Results

To answer the following two research questions, we conduct a series of experiments on four real-world datasets:

- **RQ1:** How does the proposed Motif-SocialRec model perform compared with SOTA recommendation models?
- **RQ2:** How do the hyperparameters of Motif-SocialRec affect its recommendation performance?

4.1   Experimental Settings

Datasets. In the experiments, we select four public datasets, i.e., Douban, FilmTrust, LastFM, and Yelp [8,12], which consist of user ratings for items and directed social connections. The statistics of them are listed in Table 1.

Baselines. In the experiments, we compare the Motif-SocialRec with several SOTA recommendation models, which can be divided into three categories.

Table 1. Statistics of the datasets used in experiments.

Dataset	#Users	#Items	#Ratings	#Relation	Density
Douban	2848	39,586	894,887	35,770	0.79%
LastFM	1892	17,632	92,834	25,434	0.28%
FilmTrust	1508	2071	35,500	1854	1.14%
Yelp	19,539	21,266	450,884	864,157	0.11%

- The first category consists of two recommendation models based on matrix factorization (MF): **BPR** [19] and **SBPR** [20].
- The second category includes three GNN-based recommendation models: **NGCF** [3], **DiffNet** [4], and **LightGCN** [5].
- The third category consists of three recommendation models employing SSL auxiliary tasks: S^2-**MHCN** [15] and **SEPT** [16].

Metrics. To evaluate the Top-N recommendation results of different models, we select three commonly used metrics for recommendation system [4], including Precision, Recall, and Normalized Discounted Cumulative Gain (NDCG).

Model Settings. For each model in the experiments, the optimal iteration number of training process is selected by utilizing grid searching within the range of $[10, 100]$. And, the embedding dimension d is set to be 50, and the batch size is set to be 128. For the baselines, model parameters are set according to their authors' recommendation. In the proposed model, the number of network layer L is set to be 2, the regularization coefficient and the SSL loss coefficient are set as 0.005 and 0.2, respectively.

4.2   Recommendation Performance (RQ1)

In this section, we validate the superiority of Motif-SocialRec over baselines through performance comparison experiments. The experimental results are shown in Table 2, where the best performance is highlighted in bold, and the second-best performance is marked with underline. On the whole, Motif-SocialRec exhibits the best performance on different datasets. By further analysing and comparing, we have the following findings:

Table 2. Performance results obtained on the complete datasets.

Dataset	Metric	BPR	SBPR	NGCF	DiffNet	LightGCN	S^2-MHCN	SEPT	Motif-SocialRec	Improve
Douban	P@10	0.1628	0.1720	0.2056	0.1753	0.2203	0.2114	0.2326	**0.2367**	**1.76%**
	R@10	0.0380	0.0374	0.0428	0.0427	0.0510	0.0463	0.0570	**0.0571**	**0.18%**
	N@10	0.1763	0.1903	0.2044	0.1946	0.2439	0.2343	0.2602	**0.2624**	**0.85%**
FilmTrust	P@10	0.2663	0.2616	0.2615	0.2640	0.2666	0.2781	0.2848	**0.2883**	**1.23%**
	R@10	0.4936	0.4744	0.4840	0.4974	0.5094	0.5401	0.5499	**0.5581**	**1.49%**
	N@10	0.5101	0.5242	0.5267	0.5598	0.5713	0.6052	0.6045	**0.6113**	**1.01%**
LastFM	P@10	0.1095	0.1580	0.1654	0.1520	0.1644	0.1984	0.2012	**0.2060**	**2.39%**
	R@10	0.1121	0.1604	0.1679	0.1543	0.1669	0.2012	0.2042	**0.2090**	**2.35%**
	N@10	0.1372	0.1969	0.2018	0.1856	0.2039	0.2459	0.2471	**0.2533**	**2.51%**
Yelp	P@10	0.0140	0.0174	0.0213	0.0193	0.0183	0.0290	0.0285	**0.0299**	**3.10%**
	R@10	0.0357	0.0425	0.0546	0.0512	0.0497	0.0758	0.0725	**0.0767**	**1.19%**
	N@10	0.0252	0.0336	0.0416	0.0383	0.0358	0.0599	0.0574	**0.0607**	**1.34%**

- The GNN-based models generally outperform the MF-based models. And, models that can incorporate social relationships, such as SBPR, DiffNet, S^2-MHCN, SEPT, and Motif-SocialRec, achieve better performances compared to other models.
- Models that enhance the representation learning by SSL auxiliary task, like Motif-SocialRec, SEPT, and S^2-MHCN, outperform other models. It is worth noting that the Motif-SocialRec obtains better recommendation results than SEPT and S^2-MHCN, indicating that motif can effectively describe high-level semantic patterns.

4.3 Parameter Analysis (RQ2)

In this section, we test the impact of hyperparameters on performance of the top three models, i.e., Motif-SocialRec, SEPT, and S^2-MHCN. Figure 3 illustrate the recommendation results of the three models under different numbers of layer L. Figure 4 presents the performance of Motif-SocialRec with different values of β evaluated by three metrics.

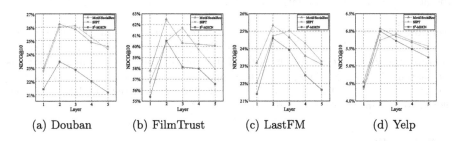

(a) Douban   (b) FilmTrust   (c) LastFM   (d) Yelp

Fig. 3. Results of three models under different numbers of layer.

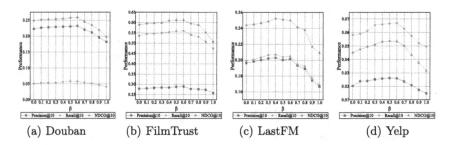

Fig. 4. Impact of parameter β on performances of three models.

- Results in Fig. 3 indicate that Motif-SocialRec achieves optimal performances with two layers on different datasets, and it outperforms other two models in most situations. These results testify that the proposed model can produce satisfying recommendation result by just employing a simple network structure.
- From the Fig. 4, it can be observed that recommendation results in terms of three metrics produced by the Motif-SocialRec are relevant to β on both four datasets. These results demonstrate that the inherent pattern seized by SSL paradigm can provide reinforcements to representation learned by the encoder in some extent.

5   Conclusion and Future Work

In this work, we propose a novel model Motif-SocialRec for the recommendation problem constrained with social relationships. In the model, multi-channel hypergraph convolution network is constructed to learn appropriate representations by utilizing motif as the unit of interaction modelling and message propagation, which can effectively capture high-level semantics in the interaction network. With the help of contrastive learning paradigm, we further enhance the recommendation producing by explore latent user preference from multiple scales. Extensive experiments demonstrate that the proposed model can improve the accuracy of recommendation results.

Acknowledgements. The authors are very grateful to the anonymous reviewers and editors. Their helpful comments and constructive suggestions helped us to significantly improve this work. We also wish to thank the authors of the compared algorithms for sharing their codes. This work was supported by the National Natural Science Foundation of China (U21A20513, 62076154, 62022052), and the Key R&D Program of Shanxi Province (202202020101003).

References

1. Tang, J., Hu, X., Gao, H., et al.: Exploiting local and global social context for recommendation. In: Proceedings of International Joint Conference on Artificial Intelligence, pp. 2712–2718. World Scientific, Chiyoda City, Tokyo (2013)

2. Gao, C., Zheng, Y., Li, N., et al.: A survey of graph neural networks for recommender systems: challenges, methods, and directions. ACM Trans. Recommender Syst. **1**(1), 1–51 (2023)
3. Wang, X., He, X., Wang, M., et al.: Neural graph collaborative filtering. In: Proceedings of the 42nd International ACM SIGIR Conference on Research and Development in Information Retrieval, pp. 165–174. Association for Computing Machinery, New York, NY, United States (2019)
4. Wu, L., Sun, P., Fu, Y., et al.: A neural influence diffusion model for social recommendation. In: Proceedings of the 42nd International ACM SIGIR Conference on Research and Development in Information Retrieval, pp. 235–244. Association for Computing Machinery, New York, NY, United States (2019)
5. He, X., Deng, K., Wang, X., et al.: LightGCN: simplifying and powering graph convolution network for recommendation. In: Proceedings of the 43rd International ACM SIGIR Conference on Research and Development in Information Retrieval, pp. 639–648. Association for Computing Machinery, New York, NY, United States (2020)
6. Liu, C., Li, Y., Lin, H., et al.: GNNRec: gated graph neural network for session-based social recommendation model. J. Intell. Inf. Syst. **60**(1), 137–156 (2023)
7. Bai, T., Zhang, Y., Wu, B., et al.: Temporal graph neural networks for social recommendation. In:2020 IEEE International Conference on Big Data, pp. 898–903. Institute of Electrical and Electronics Engineers, Piscataway, NJ (2020)
8. Wu, L., Li, J., Sun, P., et al.: Diffnet++: a neural influence and interest diffusion network for social recommendation. IEEE Trans. Knowl. Data Eng. **34**(10), 4753–4766 (2020)
9. Wei, C., Fan, Y., Zhang, J.: Time-aware service recommendation with social-powered graph hierarchical attention network. IEEE Trans. Serv. Comput. **16**(3), 2229–2240 (2022)
10. Song, W., Xiao, Z., Wang, Y., et al.: Session-based social recommendation via dynamic graph attention networks. In: Proceedings of the Twelfth ACM International Conference on Web Search and Data Mining, pp. 555–563. Association for Computing Machinery, New York, NY, United States (2019)
11. Hoang, T.L., Pham, T.D., Ta, V.C.: Improving graph convolutional networks with transformer layer in social-based items recommendation. In: Proceedings of the 13th International Conference on Knowledge and Systems Engineering, pp. 1–6. Institute of Electrical and Electronics Engineers, Piscataway, NJ (2021)
12. Zhang, Y., Huang, J., Li, M., et al.: Contrastive graph learning for social recommendation. Front. Phys. **10**, 35 (2022)
13. Wu, J., Fan, W., Chen, J., et al.: Disentangled contrastive learning for social recommendation. In: Proceedings of the 31st ACM International Conference on Information and Knowledge Management, pp. 4570–4574. Association for Computing Machinery New York, NY, United States (2022)
14. Milo, R., Shen-Orr, S., Itzkovitz, S., et al.: Network motifs: simple building blocks of complex networks. Science **298**(5594), 824–827 (2002)
15. Yu, J., Yin, H., Li, J., et al.: Self-supervised multi-channel hypergraph convolutional network for social recommendation. In: Proceedings of the Web Conference, pp. 413–424. Association for Computing Machinery, New York, NY, United States (2021)
16. Yu, J., Yin, H., Gao, M., et al.: Socially-aware self-supervised tri-training for recommendation. In: Proceedings of the 27th ACM SIGKDD Conference on Knowledge Discovery and Data Mining, pp. 2084–2092. Association for Computing Machinery, New York, NY, United States (2021)

17. Zhao, H., Xu, X., Song, Y., et al.: Ranking users in social networks with motif-based pagerank. IEEE Trans. Knowl. Data Eng. **33**(5), 2179–2192 (2019)
18. Feng, Y., You, H., Zhang, Z., et al.: Hypergraph neural networks. In: Proceedings of the 33rd Association for the Advancement of Artificial Intelligence, vol. 33, pp. 3558–3565. (2019)
19. Rendle, S., Freudenthaler, C., Gantner, Z., et al.: BPR: Bayesian personalized ranking from Implicit Feedback. UAI, 452–461 (2012)
20. Zhao, T., McAuley, J., King, I.: Leveraging social connections to improve personalized ranking for collaborative filtering. In: Proceedings of the 23rd ACM International Conference on Conference on Information and Knowledge Management, pp. 261–270. Association for Computing Machinery, New York, NY, United States (2014)

Dual Channel Graph Neural Network Enhanced by External Affective Knowledge for Aspect Level Sentiment Analysis

Hu Jin[1], Qifei Zhang[1(✉)], Xiubo Liang[1], Yulin Zhou[2], and Wenjuan Li[3]

[1] School of Software, Zhejiang University, Ningbo, China
{jinhu,cstzhangqf,liangxb}@zju.edu.cn
[2] Ningbo Innovation Center, Zhejiang University, Ningbo, China
zhou.yulin@zju.edu.cn
[3] School of Engineering, Hangzhou Normal University, Hangzhou, China
liwenjuan_jd@sjtu.edu.cn

Abstract. Aspect-level sentiment analysis is a prominent technology in natural language processing (NLP) that analyzes the sentiment polarity of target words in a text. Despite its long history of development, current methods still have some shortcomings. Mainly, they lack the integration of external affective knowledge, which is crucial for allocating attention to aspect-related words in syntactic and semantic information processing. Additionally, the synergy between syntactic and semantic information is often neglected, with most approaches focusing on only one dimension. To address these issues, we propose a knowledge-enhanced dual-channel graph neural network. Our model incorporates external affective knowledge into both the semantic and syntactic channels in different ways, then utilizes a dynamic attention mechanism to fuse information from these channels. We conducted experiments on Semeval2014, 2015, and 2016 datasets, and the results showed significant improvements compared to existing methods. Our approach bridges the gaps in current techniques and enhances performance in aspect-level sentiment analysis.

Keywords: Aspect Level Sentiment Analysis · Graph Neural Network · External Knowledge

1 Introduction

Aspect-level sentiment analysis is a fine-grained technique that analyzes the sentiment polarity of target words in a text. It provides more detailed and accurate sentiment information compared to traditional methods. This technology has applications in improving service quality, organizing feedback, and offering personalized recommendations.

Aspect level sentiment analysis is still a hot topic in the field of NLP. In recent years, researchers have proposed various novel models and methods, gradually improving its accuracy. However, after conducting a survey of existing methods, we found that most of them have limited usage of external affective knowledge and rarely utilize

© The Author(s), under exclusive license to Springer Nature Singapore Pte Ltd. 2024
B. Luo et al. (Eds.): ICONIP 2023, LNCS 14448, pp. 257–274, 2024.
https://doi.org/10.1007/978-981-99-8082-6_20

both semantic and syntactic information simultaneously, they often overlook either the semantic dimension [1] or the syntactic dimension [2]. Additionally, few studies consider the correlation between the semantic and syntactic dimensions.

Inspired by [3], the dual-channel model can effectively combines the information from both semantic and syntactic channels, resulting in a synergistic fusion of complementary insights. Furthermore, it leverages the interplay between these channels to exploit the underlying relationships.

Inspired by [4], external knowledge is an intuitive and effective source of information that can help reduce attention allocation errors. In the field of sentiment analysis, external affective knowledge can help model pay more attention to target-related words and sentiment-related words.

Therefore, we propose a new dual-channel graph neural network enhanced by external affective knowledge, called DCGK.

The contributions of this paper can be summarized as follows:

- We utilize a dual-channel graph neural network model that includes both semantic and syntactic information and investigates their correlation through dynamic attention mechanism. This approach allows for a more comprehensive analysis of information from different dimensions.
- To address the limitations of current semantic networks in considering target-related words, we integrate external affective knowledge into the semantic channel. This integration is achieved by embedding and attention mechanism. External knowledge provides sentiment information for each word and relation between words.
- To overcome the lack of commonsense sentiment information in current syntactic networks, we integrate external affective knowledge into the syntactic channel. This integration is achieved by the use of sentiment dictionary and preprocessing of adjacency matrix. This approach enables the model to focus on strongly emotional words.
- We use the attention mechanism for dynamic fusion of the above two channels. This method not only complements the information from the semantic and syntactic channels but also uncovers the inherent relationship between the two channels.

We conducted multiple experiments on datasets of Semeval 14, 15 and 16 [5–7], and the results showed that our model has achieved significant performance on most of these datasets.

2   Related Work

2.1   Early Machine Learning Methods

In the early stages, researchers primarily relied on hand-crafted features, such as word frequency [8], data analysis [9], and rule-based approaches [10]. The models employed included Naïve Bayes and support vector machines (SVM) [7, 11, 12]. However, the effectiveness of these methods heavily relied on feature engineering, which demanded a lot of labor costs.

2.2 Neural Network Methods

Neural networks have significantly reduced labor costs by automating feature learning. Researchers often encode sentences into low-dimensional word vectors, such as word2vec [13] and Glove [14], which are then fed into a neural network. The model then learns various implicit features automatically.

Tang et al. [15] introduced TD-LSTM, which utilizes a combination of word embeddings and target embeddings to capture the contextual meaning of words in relation to the target.

Tang et al. [16] developed MemNet, which incorporates a multi-hop attention mechanism to focus on words that are more relevant to the target word in the context. Wang et al. proposed ATAE-LSTM, based on an attention mechanism, allowing it to handle different parts of sentences when different aspects are involved.

Ma et al. [17] presented IAN, where they discovered the interactive learning of target and context attention. This led to the generation of corresponding attention representations based on this interaction. it.

2.3 Pre-trained Models

In recent years, the rise of pre-trained language models has revolutionized various NLP tasks. Particularly, models like BERT [18] have shown remarkable advancements in the field.

Song et al. [19] applied BERT to encode both the target word and its corresponding context. Sun et al. [20] proposed a method that utilizes auxiliary sentences to transform aspect-level sentiment analysis tasks into sentence-pair classification tasks. Li et al. [21] adopted the BERT model and global labels to accomplish aspect-level sentiment analysis in an "end-to-end" manner.

2.4 Graph Neural Network

Sun et al. [22] introduced a parsing tree-based convolutional network, incorporating Graph Convolutional Networks (GCN) to simulate the syntactic dependency structure in sentences. Zhang et al. [2] utilized syntactic information and word dependencies to construct a graph structure, leveraging GCN to simulate this structure. Liang et al. [23] proposed an interactive GCN model that extracts emotional features of specific target words by iteratively capturing relationships between aspect words and different aspects in the context.

Wang et al. [24] devised a Relational Graph Attention Network (R-GAT). By pruning and reshaping the parsing tree, they obtained a tree structure that prioritizes attention to the target word. A novel GAT model was proposed to encode dependencies and establish semantic relationships between target words and related words.

2.5 External Commonsense Knowledge

External Commonsense knowledge is an intuitive and effective auxiliary tool in many NLP works [4].

SenticNet [25] is a knowledge base that provides semantic, emotional, and polar information related to language concepts. Each word in SenticNet represents a sentiment polarity ranging from -1 to 1. SenticNet has shown promising results in the field of sentiment representation learning [26, 27].

Ma et al. [28] improved the gating unit of LSTM to integrate external knowledge. SenticNet is embedded into the neural network as word embeddings, effectively extracting features at both the target word level and sentence level. Li et al. [29] proposed a novel graph structure model that incorporates relationship information extracted from external knowledge graphs. Liang et al. [1] enhanced the parsing tree by incorporating SenticNet's affective knowledge, thus integrating emotional information in the process of generating the graph structure.

AffectiveSpace [30] is a mapping of SenticNet onto low-dimensional continuous vectors, creating a vector space model. By reducing the dimensionality of emotional commonsense knowledge, this model can summarize semantic features associated with concepts, enabling intuitive clustering of concepts based on their semantic and emotional relevance. This emotional intuition allows for the inference of emotions and polarity conveyed by multi-word expressions, thus achieving effective conceptual-level analysis.

3 Methodology

3.1 Problem Definition

Given a sentence $S = \{w_1, \ldots, w_i, \ldots, w_n\}$ and a target word $T = \{w_i, w_{i+1}, \ldots, w_{i+m}\}$ within this sentence, the goal is to determine the associated sentiment polarity $P \in \{Positive, Neutral, Negative\}$.

3.2 Model Overview

Our proposed model, as depicted in Fig. 1, consists of two channels and five modules: the semantic channel, the syntactic channel, the encoding input module, the knowledge-enhanced semantic module, the knowledge-enhanced syntactic module, the attention fusion module, and the classifier module.

The semantic channel and the syntactic channel serve as the two fundamental components of our model. The encoding input module processes the input text and prepares it for further analysis. The knowledge-enhanced semantic module and the knowledge-enhanced syntactic channel integrate external affective knowledge into the semantic channel and the syntactic channel, thereby enhancing the model's understanding of target-related words and sentiment-related words.

To leverage the combined benefits of both syntactic and semantic information, the attention fusion module employs a dynamic attention mechanism to merge the information from both channels. This fusion also enables the model to capture the synergistic effects of these two dimensions. Finally, the classifier module utilizes the fused information to classify the sentiment polarity of aspect words.

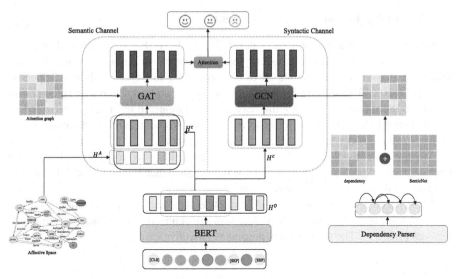

Fig. 1. Model Overview.

3.3 Input Module

The input module serves as the entry point of our model, we leverage BERT [18] as the upstream network, owing to their impressive encoding ability. We arrange the input in the following format:

$$[CLS] + \text{sentence} + [SEP] + \text{aspect} + [SEP] \tag{1}$$

where $[CLS]$ and $[SEP]$ are special tokens, $[CLS]$ represents the classification result of the "next sentence prediction" in the BERT pre-training task, while $[SEP]$ represents the token that separates the first and the second sentence.

For example, given the input sentence "The pancakes were delicious, but the service was terrible" and the target word "service", the input is formatted as follows: "$[CLS]$ The pancakes were delicious, but the service was terrible $[SEP]$ service $[CLS]$". This sentence is then fed into BERT, and a vectorized result can be obtained:

$$O = \text{BERT(W)} \tag{2}$$

After obtaining the initial encoding result O, we apply a masking algorithm on it to extract the part between the $[CLS]$ and the first $[SEP]$:

$$H^C = \text{MASK}(O) \tag{3}$$

H^C represents the encoding result of the original sentence, with special consideration given to the aspect target word placed after the first $[SEP]$. By focusing on the target aspect word, BERT can more effectively capture the information related to the specific aspect being analyzed. We then use H^C as the input of the downstream semantic and syntactic channel.

3.4 Knowledge-Enhanced Semantic Channel

Knowledge Embedding: Previous studies [31, 32] have shown that the self-attention mechanism of BERT often focuses only on local context information, while in actual sentiment analysis tasks, opinion words corresponding to target words may appear in distant contexts. To overcome this limitation, we propose adding knowledge graph information to the model by utilizing the conceptual relationship information provided by external knowledge. This approach enables the model to consider relevant opinion words in the global context when calculating the attention score.

We use AffectiveSpace [30] to obtain the embedding vectorized results of each word. AffectiveSpace is a 100-dimensional vector space that includes 200,000 concepts, forming a vast affective knowledge graph.

Principal component analysis (PCA) is applied to the matrix representation of Affect-Net [33], a semantic network that links commonsense concepts to semantic and emotional features (Table 1). The result of this process is AffectiveSpace.

Table 1. AffectNet Samples

AffectNet	Is A-pet	KindOf-food	Arises-joy	...
dog	0.981	0.000	0.789	...
cupcake	0.000	0.922	0.910	...
songbird	0.672	0.000	0.862	...

For the results from encoding input module, which is $H^C \in R^{n \times d_h}$, we use Affective Space to encode it again, then we get the AffectiveSpace encoding results:

$$H^A = \text{AffectiveSpace}(H^C) \# \tag{4}$$

here $H^A \in R^{n \times 100}$, then H^A and H^C are concatenated to obtain the vectorized representation of the semantic channel input:

$$H^{sem} = [H^A; H^C] \tag{5}$$

here ";" represents the concatenation operation, and $H^{sem} \in R^{l \times (d_h + 100)}$.

Specifically, for words that do not have corresponding embeddings in the AffectiveSpace, we assign them a zero vector $\vec{0} \in R^{100}$. Because we concatenate the AffectiveSpace encoding result with the original word embeddings to form a new vector, which is then used for attention calculation. Therefore, if the AffectiveSpace encoding result is zero, it implies that we are disregarding the attention contribution from the AffectiveSpace.

Graph Attention Network: The Graph Attention Network (GAT) is a deep learning model designed to update the information of each node in a graph by aggregating the information of its adjacent nodes. This is achieved through a layer-by-layer process, where each layer updates the node information based on the attention matrix. By stacking multiple GAT layers, information can be propagated to distant nodes in the graph, allowing for more comprehensive and accurate modeling of complex relationships within the graph.

In GAT, each node in the graph corresponds to the tokens in the sentence. The input of GAT is a set of vectors, denoted as $O^{GAT} = \{g_1, g_2, \ldots, g_n\}$, which is consistent with the semantic channel input $H^{sem} = \{h_1, h_2, \ldots h_n\}$. The set of neighbor nodes corresponding to a given node i is represented as $N(i)$.

The operation in a GAT layer can be expressed as follows:

$$g_i = \sum_{j \in N(i)} \|_{k=1}^{K} \alpha_{i,j}^{l,k} w_k^l g_j^{l-1} \tag{6}$$

here $\|$ represents the concatenation operation, K is the attention head number, $w_k^l \in R^{d_{head} \times d_h}$ represents the kth trainable matrix in the lth layer, $g_j^{l-1} \in R^{d_h}$ is the vectorized representation of the jth node in the $l-1$th layer, and $\alpha_{i,j}^{l,k} \in R^{n \times n}$ is the attention matrix obtained as follows:

$$\alpha_{i,j}^{l,k} = \frac{exp\left(s\left(g_i^{l-1}, g_j^{l-1}\right)\right)}{\sum_{q \in N(i)} s\left(g_i^{l-1}, g_q^{l-1}\right)} \tag{7}$$

here, $s(\cdot)$ is the dot product function, this is an effective function for calculation attention scores, its computing process is as follows:

$$s\left(g_i^{l-1}, g_j^{l-1}\right) = \frac{\left(W_Q^{l,k} g_i^{l-1}\right)^\top \left(W_K^{l,k} g_i^{l-1}\right)}{\sqrt{\frac{d}{head}}} \tag{8}$$

here, W_Q and $W_K \in R^{d_k \times d}$. They are both trainable parameter matrices, which are used to project the node vectors into a lower-dimensional space before computing the dot product.

3.5 Knowledge-Enhanced Semantic Channel

Referring to [35], the understanding of the syntactic structure in the last layer of BERT output is poor. So we designed an additional syntactic channel as a supplement.

Dependency Tree: Following [2, 22], we utilize the parsing syntax tree of the sentence $S = \{w_1, w_2, \ldots, w_n\}$ to construct a dependent graph structure $G = V, E$, where V represents the vertex set and E represents the edge set. Each token in sentence S is considered as a vertex, and the dependency relationship between nodes is represented as an edge, thus constructing a basic dependency graph structure. This graph structure serves as a basic reference for the computation of graph convolution network.

The dependency relationship between nodes is generated by Spacy[1], which is an NLP library based on Python.

---

[1] https://spacy.io/.

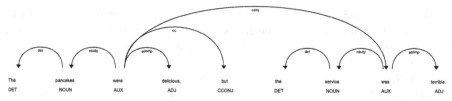

Fig. 2. Dependency Tree.

Figure 2 shows the parsing tree visualization results of the sentence "The pancakes were delicious, but the service was terrible", where the arrows represent dependencies.

The graph structure derived from the dependency tree is represented by the adjacency matrix. Each value in the adjacency matrix $D \in R^{n \times n}$ satisfies the following relationship:

$$D_{i,j} = \begin{cases} 1 \ if \ w_j \ depends \ on \ w_i \\ 1 \ if \ i == j \\ 0 \ others \end{cases} \tag{9}$$

here, we added self-link edges, corresponding to the case where $i == j$, to ensure that each token retains its own information in the forward computation process.

The dependency tree generated by Spacy is actually directed, and unlike the approach in [17], we retain its directed form.

External Affective Knowledge: To enhance the impact of emotional words related to the target, we integrated sentiment dictionary information from SenticNet into the adjacency matrix of GCN. SenticNet [25] assigns a sentiment score to each word, which ranges from -1 to 1. A score close to -1 indicates negative sentiment, a score close to 1 indicates positive sentiment, and a score of 0 indicates neutral sentiment. Table 2 presents some of the words in SenticNet along with their corresponding sentiment scores.

Table 2. SenticNet Samples

Word	*SenticNet* Score
good	0.849
bad	−0.800
smile	0.997
cry	−0.670
healthy	0.769

We assign the sentiment relationship between two dependent words w_i and w_j in the sentence based on SenticNet as follows:

$$T_{i,j} = \begin{cases} 1, \ if \ w_j \ depends \ on \ w_i \ and \ w_i \in aspect \\ 0, \ others \end{cases} \tag{10}$$

The final adjacency matrix can be expressed as $A = D + T$, where $+$ denotes the addition operation of the corresponding matrix elements.

Graph Convolution Network: We use the output H^C of the input module as the input for the first layer of the graph convolution network, and represent the graph structure using the adjacency matrix A described above. The input for each layer in GCN is derived from the output of the previous layer. In GCN, the convolution operation updates the information of each node based on the information of its neighboring nodes. By stacking L layers of GCN, each node can aggregate information up to nodes that are $L - 1$ hops away. The process of updating each node can be described as follows:

$$h_i^l = \sigma\left(\sum A_{i,j} W^l h_j^{l-1} + b^l\right) \tag{11}$$

here, $h_i^l \in R^{d_h}$ is the output of the i-th node in the l-th layer, $A_{i,j}$ is the value corresponding to i, j in the adjacency matrix A, and $W^l \in R^{d_h \times d_h}$ is the trainable parameter matrix shared in the l-th layer. Its function is similar to the convolution kernel in CNN. $b_l \in R^{d_h}$ is the bias, and $\sigma(\cdot)$ denotes the activation function.

3.6 Handling the Conflict of Different Word Segmentation Methods

The word segmentation results obtained from Spacy differ from the WordPiece strategy used in BERT word segmentation. For instance, for the same input "Dieters don't stick to salads or indulge in vegetarian platters", the word segmentation results obtained from Spacy and BERT WordPiece segmentation are different, as shown in Table 3.

Table 3. Different segmentation results of Spacy and WordPiece

strategy	Results
Spacy	['dieters', 'do', "n't", 'stick', 'to', 'salads', 'or', 'indulge', 'in', 'vegetarian', 'platters']
WordPiece	['dieter', '##s', 'don', '"', 't', 'stick', 'to', 'salad', '##s', 'or', 'ind', '##ul', '##ge', 'in', 'vegetarian', 'platt', '##ers']

As a result of different word segmentation strategies, the output H^C of the input module cannot be directly matched with the words in the parsing tree of two channels. Therefore, it is necessary to map these two segmentation results. We utilize pytokenizations[2] to perform the mapping.

After processing, we obtain the mapping results for the different word segmentation strategies. Figure 3 presents a visual diagram of the mapping.

The vector representation of Spacy's segmentation results is obtained simply by averaging the multiple vectors mapped to it in WordPiece.

---

[2] https://github.com/explosion/tokenizations.

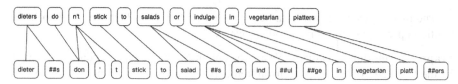

Fig. 3. Segment Results Mapping

3.7 Channel Fusion

Referring to [3], one of the key advantages of fusing the results from the semantic and syntactic channels is the ability to capture a more comprehensive representation of the input text.

After obtaining the outputs from the syntactic and semantic channels, the fusion process is carried out using an attention mechanism. We have the output of the syntactic channel $H^{syn} = \{h_1, h_2, \ldots h_n\}$, and the output of the semantic channel $H^{sem} = \{h_1, h_2, \ldots h_n\}$. We take the output of the syntactic channel H^{syn} as the query Q matrix and the output of the semantic channel H^{sem} as the key K matrix. By multiplying the Q matrix with the transpose of the K matrix, we obtain the attention score matrix. The attention score matrix can be obtained as follows:

$$\alpha_{i,j}^k = \frac{\exp(s(h_i, h_j))}{\sum_{q=1}^n \exp(h_i, h_q)} \tag{12}$$

here, $s(\cdot)$ denotes dot product attention function, which is computed as:

$$s(h_i, h_j) = \frac{\left(W_Q^k h_i\right)^\top (W_K^k h)}{\sqrt{\frac{d}{head}}} \tag{13}$$

where k is the attention head index, $head$ is the total amount of attention heads, W_Q and W_K are both trainable parameters in the k-th head.

In addition, H^{syn} needs to be masked, and the part of aspect word is used alone to query the K matrix. The masking process is as follows:

$$\alpha_{i,j} = \begin{cases} 0, & if \ i < \tau \\ \alpha_{i,j}, & if \ \tau \le i \le \tau + k \\ 0, & if \ i < \tau + k \end{cases} \tag{14}$$

$$h_i = \sum_j \|_{k=1}^K \alpha_{i,j}^k h_j \tag{15}$$

$$h^{merge} = \frac{\sum_{i=0}^n h_i}{k+1} \tag{16}$$

here, $\left[h_\tau, \ldots, h_{\{\tau+k\}}\right]$ is the aspect word, which may contain multiple tokens. Then we have the masked result $H^{merge} = \left[0, 0, \ldots, h_\tau, \ldots, h_{\tau+k}, \ldots, 0\right]$, where $h^{merge} \in R^{d_h}$.

Given the impressive performance of BERT in next sentence prediction tasks, we incorporate the output result pooler of $[CLS]$ as an additional component to complement the sentiment judgement. Specifically, we employ a gated network to fuse h^{pooler} and h^{merge}. The gated mechanism is implemented as follows

$$\alpha = W\left[h^{pooler}; h^{merge}\right] \tag{17}$$

$$h^{out} = \alpha \times h^{pooler} + (1 - \alpha) \times h^{merge} \tag{18}$$

here $W \in R^{2d_h \times 1}$ are trainable parameters.

3.8 Classification Model

This is a classification problem with three possible outcomes. To classify the final sentiment, we employ a fully connected network followed by a softmax activation function. The classification process can be expressed as:

$$y = softmax\left(Wh^{out} + b\right) \tag{19}$$

where $W \in R^{3 \times d_h}$ and $b \in R^{d_h}$.

3.9 Loss Function

As the objective function for optimization, we selected the cross-entropy loss function, which helps measure the dissimilarity between the predicted and correct sentiment distributions. Additionally, we incorporated L2 regularization to mitigate overfitting. The loss function is structured as follows:

$$L = -\sum_{i=1}^{T} \sum_{j=1}^{C} y_i^j log p_i^j + \lambda \|\Theta\|^2 \tag{20}$$

In Eq. (19), y_i represents the correct sentiment distribution, p_i represents the predicted sentiment distribution, T represents the training set, C represents the number of sentiment classification types, λ represents the L2 regularization coefficient, and Θ represents the training parameters.

4 Experiments

4.1 Data Sets

We conducted experiments and analysis on four publicly available datasets: the restaurant and laptop datasets from SemEval 2014 Task 4 [5], the restaurant datasets from SemEval 2015 Task 12 [6], and the restaurant datasets from SemEval 2016 Task 5 [7]. Each sample in these datasets consists of a comment, several aspect words, and their corresponding sentiment polarity. For instance, consider the sentence: "The tech guy then said the service center does not do 1-to-1 exchange, and I have to direct my concern to the 'sales' team, which is the retail shop where I bought my netbook from." In this example, there are three aspect words: "service center", "sales team", and "tech guy", corresponding to sentiment polarities of "negative", "negative", and "neutral", respectively.

The distribution of data sets is shown in Table 4.

Table 4. Distribution of Datasets

Datasets	Positive		Neutral		Negative	
	Train	Test	Train	Test	Train	Test
Rest14	2164	728	637	196	807	196
Lap14	994	341	464	169	870	128
Rest15	1178	439	150	35	382	328
Rest16	1620	597	88	38	709	190

4.2 Implementation Details

We utilized the basic BERT model, specifically Bert-Base-Uncased, as the initial word embedding input. The dimension of the word embeddings is set to 768, and the BERT model consists of 12 attention heads.

For the semantic channel, we employed AffectiveSpace encoding with a dimension of 100. Two layers of Graph Attention Network (GAT) were used for superposition. In the syntactic channel, we utilized two layers of Graph Convolutional Network (GCN) for superposition.

The learning rate was set to 0.00002, and the batch size was set to 32. We employed the Adam optimizer to optimize the model's parameters, with a decay rate of 0.001. To prevent overfitting, we applied L2 regularization with a weight of 0.001. Furthermore, the parameters W and b in all neural network layers were randomly initialized using a uniform distribution.

4.3 Baselines

We selected following models to be our comparisons:

- ATAE-LSTM [34]: ATAE-LSTM is based on LSTM and can focus on different contexts according to different aspect words.
- CDT [22]: GCN is used to process the syntactic information of dependency tree, and the structure of dependency tree is simulated by GCN.
- R-GAT [24]: R-GAT focuses on the complexity of sentence structures and multi-aspect words. The model constructs a dependency tree with aspect words as roots by reconstructing and pruning the dependency tree, and then uses this dependency tree as the basis of GAT operation.
- BERT [18]: Fine-tuning BERT.
- BERT-SPC [19]: The ABSA task is converted into a sentence pair classification task by the input '[CLS] + sentence + [SEP] + aspect + [SEP]' in the form of the BERT 'next sentence prediction' task, and uses the results of the pooler as output.
- AEN-BERT [19]: Considering the cost of obtaining labeled datasets, AEN-BERT proposes a form of adversarial network to generate data similar to reality reviews, thereby expanding the dataset and making the model more robust.
- R-GAT-BERT [24]: R-GAT-BERT replaces Bi-LSTM in R-GAT with BERT. is the model R-GAT with the Bi-LSTM replaced by BERT.

- DualGCN + BERT [3]: This model takes into account the complementarity of syntactic structure and semantic association, uses the syntactic network SynGCN and the semantic network SemGCN, and finally integrates through a BiAffine module.
- Sentic GCN-BERT [1]: This model takes into account the importance of external affective knowledge, uses SenticNet as the injection of affective knowledge, and uses sentiment dictionary to preprocess the adjacency matrix of GCN.

Table 5. Results Comparison

Models		Rest14		Lap14		Rest15		Rest16	
		Acc	F1	Acc	F1	Acc	F1	Acc	F1
RNN	ATAE-LSTM	77.20	67.02	68.70	63.93	78.48	60.53	83.77	61.71
Graph	CDT	82.30	74.02	–	–	77.19	72.99	85.58	69.93
	R-GAT	83.30	76.08	77.42	73.76	80.83	64.17	88.92	70.89
BERT	BERT	84.11	76.68	77.59	73.28	83.48	66.18	90.10	74.16
	BERT-SPC	84.32	77.15	78.54	75.26	–	–	–	–
	AEN-BERT	84.29	77.22	76.96	73.67	–	–	–	–
	R-GAT-BERT	86.60	81.35	78.21	74.07	83.22	69.73	89.71	76.62
	DualGCN + BERT	87.13	81.16	77.40	76.02	–	–	–	–
	Sentic GCN-BERT	86.92	81.03	**82.12**	**79.05**	85.32	71.28	91.97	79.56
Ours	DCGK-BERT	**87.70**	**82.10**	81.81	78.41	**85.72**	**71.70**	**92.21**	**80.21**

Table 6. Ablation Results

Model Branch	Rest14		Lap14		Rest15		Rest16	
	Acc	F1	Acc	F1	Acc	F1	Acc	F1
DCGK w/o AS	86.20	80.67	80.80	77.84	84.51	70.11	90.12	78.77
	↓1.50	↓1.43	↓1.01	↓0.57	↓1.21	↓1.59	↓2.09	↓1.44
DCGK w/o SN	85.79	79.95	80.92	77.96	84.83	69.98	90.79	78.89
	↓1.91	↓2.15	↓0.89	↓0.45	↓0.89	↓1.72	↓1.42	↓1.32
DCGK	87.70	82.10	81.81	78.41	85.72	71.70	92.21	80.21

4.4 Results Comparison

The results are presented in Table 5. Our model demonstrates superior performance in the Rest14, Rest15, and Rest16 datasets compared to the current models. Particularly, the Rest14 dataset exhibits a significant improvement, while the Rest15 and Rest16 datasets show a smaller improvement.

However, our model's performance on the Lap14 dataset is slightly below the best reported results. We assume that this discrepancy can be attributed to the inherent class imbalance in the dataset. Additionally, the syntactic information in the Lap14 dataset may carry more importance. As a result, the model solely focuses on enhancing the syntactic channel, achieves more notable improvements in this particular dataset.

4.5 Ablation Experiment

To evaluate the significance of each knowledge enhancement module in our proposed DCGK model, we conducted an ablation experiment, and the results are presented in Table 6. Specifically, we focused on the importance of the knowledge-enhanced semantic module (primarily AffectiveSpace) and the knowledge-enhanced syntactic module (primarily SenticNet). To examine the role of these modules, we designed two different model branches: DCGK w/o AS and DCGK w/o SN.

From the experimental results displayed in Table 6, we observe that both the knowledge-enhanced semantic module and the knowledge-enhanced syntactic module have an impact on accuracy (Acc) and F1 score. What's more, the impact of the semantic module is slightly more pronounced.

4.6 Case Study

In this section, we examined the influence of external knowledge on model attention allocation. We take the review "The pancakes were delicious, but the service was terrible." as the case. The goal is to predict the sentiment polarity of the aspect word "service."

Figure 4 illustrates the three attention distribution visualization results, respectively corresponding to the cases when the semantic enhancement module is removed, when the syntactic enhancement module is removed and when the complete network is used.

The darker color indicates a higher level of attention allocated in forward encoding of the aspect word.

Note that we may have several attention heads, so we display an average of the attention values from all attention heads.

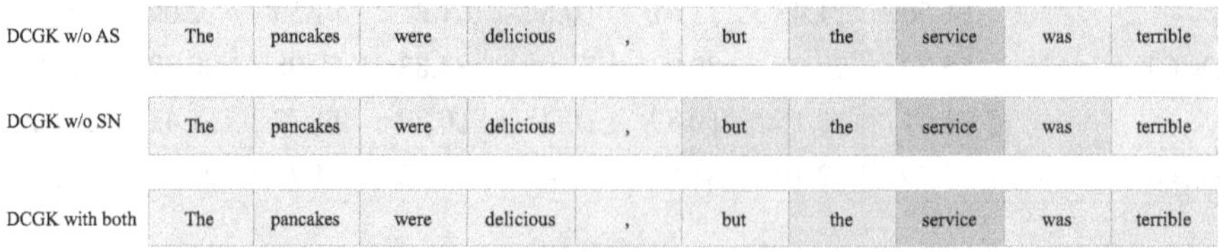

Fig. 4. Attention Visualization

When the semantic enhancement module is removed, it can be observed that some unrelated words receive attention. For example, in the figure, "delicious" is wrongly given too much attention, for it should be describing "pancakes". This misallocation of attention to irrelevant words may lead to errors in sentiment polarity prediction. However, with the semantic enhancement module and the integration of external knowledge, the relevance between "service" and "delicious" becomes less significant. Consequently, the model reduces its attention to "service", avoiding the misunderstanding of target-related words. Additionally, the semantic enhancement module gives more attention to relevant words. By using external knowledge to guide attention allocation, the model can focus on relevant words in distant contexts.

When the syntactic enhancement module is removed, the model fails to allocate sufficient attention to the crucial opinion words associated with "service". By integrating external affective knowledge and including the sentiment polarity of words in the GCN's adjacency matrix, we can assign higher weights to opinion words with strong emotions. This helps model to pay more attention to sentiment-related words.

By combining both modules, the model not only consider sentiment-related words but also consider the target-related words, leading to more accurate associations. As depicted in the Fig. 4, the model accurately focuses on the keyword "terrible" and is not influenced by the sentiment word "delicious" corresponding to another target word.

5 Conclusion

This paper proposes a novel aspect level sentiment analysis model named DCGK. We utilize a dual-channel graph neural network combination approach, integrating external affective knowledge into both channels, and finally fuse them with attention mechanism.

Our motivation stems from several shortcomings in current aspect level sentiment analysis methods. We found that most of them have limited usage of external affective knowledge and rarely utilize both semantic and syntactic information simultaneously, they often overlook either the semantic dimension or the syntactic dimension. Also, few studies consider the complementarity and inner correlation between the semantic and syntactic dimensions.

Our model has achieved significant performance on various datasets. We can also learn from ablation experiment that external knowledge plays an important role in improving the accuracy of attention allocation.

Additionally, we observed poor performance on the Lap14 dataset and assume that the lack of sufficient syntactic information could be a contributing factor. In future research, we plan to explore the incorporation of additional syntactic information, such as types of dependent edges and part-of-speech, to further improve the performance of the model.

References

1. Liang, B., Su, H., Gui, L., Cambria, E., Xu, R.: Aspect-based sentiment analysis via affective knowledge enhanced graph convolutional networks. Knowl.-Based Syst. **235**, 107643 (2022). https://doi.org/10.1016/j.knosys.2021.107643
2. Zhang, C., Li, Q., Song, D.: Aspect-based sentiment classification with aspect-specific graph convolutional networks. In: Proceedings of the 2019 Conference on Empirical Methods in Natural Language Processing and the 9th International Joint Conference on Natural Language Processing (EMNLP-IJCNLP) , pp. 4567–4577. Association for Computational Linguistics, Hong Kong, China (2019). https://doi.org/10.18653/v1/D19-1464

3. Li, R., Chen, H., Feng, F., Ma, Z., Wang, X., Hovy, E.: Dual graph convolutional networks for aspect-based sentiment analysis, p. 11 (2011)
4. Xie, Y., Pu, P.: How commonsense knowledge helps with natural language tasks: a survey of recent resources and methodologies. arXiv, 10 August 2021. https://doi.org/10.48550/arXiv.2108.04674
5. Pontiki, M., Galanis, D., Pavlopoulos, J., Papageorgiou, H., Androutsopoulos, I., Manandhar, S.: SemEval-2014 task 4: aspect based sentiment analysis (2016)
6. Pontiki, M., Galanis, D., Papageorgiou, H., Manandhar, S., Androutsopoulos, I.: SemEval-2015 task 12: aspect based sentiment analysis. In: Proceedings of the 9th International Workshop on Semantic Evaluation (SemEval 2015), June 2015, pp. 486–495. Association for Computational Linguistics, Denver, Colorado (2015). https://doi.org/10.18653/v1/S15-2082
7. Pontiki, M., et al.: SemEval-2016 task 5: aspect based sentiment analysis, p. 12 (2016)
8. Brun, C., Popa, D.N., Roux, C.: XRCE: hybrid classification for aspect-based sentiment analysis. In: Proceedings of the 8th International Workshop on Semantic Evaluation (SemEval 2014), pp. 838–842. Association for Computational Linguistics Dublin, Ireland (2014). https://doi.org/10.3115/v1/S14-2149
9. Vo, D.-T., Zhang, Y.: Target-dependent Twitter sentiment classification with rich automatic features, p. 7 (2015)
10. Ding, X., Liu, B., Yu, P.S.: A holistic lexicon-based approach to opinion mining. In: Proceedings of the International Conference on Web Search and Web Data Mining - WSDM '08, p. 231, Palo Alto, California, USA. ACM Press (2008). https://doi.org/10.1145/1341531.1341561
11. Liu, B., Blasch, E., Chen, Y., Shen, D., Chen, G.: Scalable sentiment classification for Big Data analysis using Naive Bayes Classifier. In: 2013 IEEE International Conference on Big Data, Silicon Valley, CA, USA. IEEE, October 2013, pp. 99–104 (2013). https://doi.org/10.1109/BigData.2013.6691740
12. Pang, B., Lee, L., Vaithyanathan, S.: Thumbs up?: Sentiment classification using machine learning techniques. In: Proceedings of the ACL-02 Conference on Empirical Methods in Natural Language Processing - EMNLP '02, pp. 79–86. Association for Computational Linguistics (2002). https://doi.org/10.3115/1118693.1118704
13. Mikolov, T., Chen, K., Corrado, G., Dean, J.: Efficient estimation of word representations in vector space. arXiv, 06 September 2013. http://arxiv.org/abs/1301.3781. Accessed 08 December 2022
14. Pennington, J., Socher, R., Manning, C.: Glove: global vectors for word representation. In: Proceedings of the 2014 Conference on Empirical Methods in Natural Language Processing (EMNLP), pp. 1532–1543. Association for Computational Linguistics, Doha, Qatar (2014). https://doi.org/10.3115/v1/D14-1162
15. Tang, D., Qin, B., Feng, X., Liu, T.: Effective LSTMs for target-dependent sentiment classification, p. 10 (2015)
16. Tang, D., Qin, B., Liu, T.: Aspect level sentiment classification with deep memory network. In: Proceedings of the 2016 Conference on Empirical Methods in Natural Language Processing, pp. 214–224. Association for Computational Linguistics, Austin, Texas (2016). https://doi.org/10.18653/v1/D16-1021
17. Ma, D., Li, S., Zhang, X., Wang, H.: Interactive attention networks for aspect-level sentiment classification. arXiv, 04 September 2017. http://arxiv.org/abs/1709.00893. Accessed 14 Dec 2022
18. Devlin, J., Chang, M.-W., Lee, K., Toutanova, K.: BERT: pre-training of deep bidirectional transformers for language understanding. arXiv, 24 May 2019. http://arxiv.org/abs/1810.04805. Accessed 14 Dec 2022
19. Song, Y., Wang, J., Jiang, T., Liu, Z., Rao, Y.: Attentional encoder network for targeted sentiment classification, pp. 93–103 (2019). https://doi.org/10.1007/978-3-030-30490-4_9

20. Sun, C., Huang, L., Qiu, X.: Utilizing BERT for aspect-based sentiment analysis via constructing auxiliary sentence (2019)
21. Li, X., Bing, L., Zhang, W., Lam, W.: Exploiting BERT for end-to-end aspect-based sentiment analysis. In: Proceedings of the 5th Workshop on Noisy User-Generated Text (W-NUT 2019), pp. 34–41. Association for Computational Linguistics, Hong Kong, China (2019). https://doi.org/10.18653/v1/D19-5505
22. Sun, K., Zhang, R., Mensah, S., Mao, Y., Liu, X.: Aspect-level sentiment analysis via convolution over dependency tree. In: Proceedings of the 2019 Conference on Empirical Methods in Natural Language Processing and the 9th International Joint Conference on Natural Language Processing (EMNLP-IJCNLP), pp. 5678–5687. Association for Computational Linguistics, Hong Kong, China (2019). https://doi.org/10.18653/v1/D19-1569
23. Liang, B., Yin, R., Gui, L., Du, J., Xu, R.: Jointly learning aspect-focused and inter-aspect relations with graph convolutional networks for aspect sentiment analysis. In: Proceedings of the 28th International Conference on Computational Linguistics, pp. 150–161. International Committee on Computational Linguistics, Barcelona, Spain (Online) (2020). https://doi.org/10.18653/v1/2020.coling-main.13
24. Wang, K., Shen, W., Yang, Y., Quan, X., Wang, R.: Relational graph attention network for aspect-based sentiment analysis. In: Proceedings of the 58th Annual Meeting of the Association for Computational Linguistics, pp. 3229–3238. Association for Computational Linguistics (2020). https://doi.org/10.18653/v1/2020.acl-main.295
25. Cambria, E., Poria, S., Hazarika, D., Kwok, K.: SenticNet 5: discovering conceptual primitives for sentiment analysis by means of context embeddings. In: AAAI, vol. 32, no. 1, April 2018. https://doi.org/10.1609/aaai.v32i1.11559
26. Yang, P., Li, L., Luo, F., Liu, T., Sun, X.: Enhancing topic-to-essay generation with external commonsense knowledge. In: Proceedings of the 57th Annual Meeting of the Association for Computational Linguistics, pp. 2002–2012. Association for Computational Linguistics, Florence, Italy (2019). doi: https://doi.org/10.18653/v1/P19-1193
27. Li, Y., Pan, Q., Yang, T., Wang, S., Tang, J., Cambria, E.: Learning word representations for sentiment analysis. Cogn. Comput. 9(6), 843–851 (2017). https://doi.org/10.1007/s12559-017-9492-2
28. Ma, Y., Peng, H., Cambria, E.: Targeted aspect-based sentiment analysis via embedding commonsense knowledge into an attentive LSTM. In: AAAI, vol. 32, no. 1, April 2018. https://doi.org/10.1609/aaai.v32i1.12048
29. Li, Y., Sun, X., Wang, M.: Embedding extra knowledge and a dependency tree based on a graph attention network for aspect-based sentiment analysis. In: 2021 International Joint Conference on Neural Networks (IJCNN), pp. 1–8. IEEE, Shenzhen, China July 2021. https://doi.org/10.1109/IJCNN52387.2021.9533695
30. Cambria, E., Fu, J., Bisio, F., Poria, S.: AffectiveSpace 2: enabling affective intuition for concept-level sentiment analysis. In: AAAI, vol. 29, no. 1, February 2015. https://doi.org/10.1609/aaai.v29i1.9230
31. Goldberg, Y.: Assessing BERT's syntactic abilities. arXiv, 16 January 2019. http://arxiv.org/abs/1901.05287. Accessed 29 June 2023
32. Htut, P.M., Phang, J., Bordia, S., Bowman, S.R.: Do attention heads in BERT track syntactic dependencies? arXiv, 27 November 2019. http://arxiv.org/abs/1911.12246. Accessed 29 June 2023
33. Cambria, E., Hussain, A.: Sentic Computing. SpringerBriefs in Cognitive Computation, vol. 2. Springer, Dordrecht (2012). https://doi.org/10.1007/978-94-007-5070-8

34. Wang, Y., Huang, M., Zhu, X., Zhao, L.: Attention-based LSTM for aspect-level sentiment classification. In: Proceedings of the 2016 Conference on Empirical Methods in Natural Language Processing, pp. 606–615. Association for Computational Linguistics, Austin, Texas (2016). https://doi.org/10.18653/v1/D16-1058
35. Jawahar, G., Sagot, B., Seddah, D.: What Does BERT Learn about the Structure of Language? In: Proceedings of the 57th Annual Meeting of the Association for Computational Linguistics, Florence, Italy, pp. 3651–3657 (2019). https://doi.org/10.18653/v1/P19-1356

New Predefined-Time Stability Theorem and Applications to the Fuzzy Stochastic Memristive Neural Networks with Impulsive Effects

Hui Zhao[1], Lei Zhou[1], Qingjie Wang[1,2(✉)], Sijie Niu[1], Xizhan Gao[1], and Xiju Zong[1]

[1] Shandong Provincial Key Laboratory of Network Based Intelligent Computing, School of Information Science and Engineering, University of Jinan, Jinan 250022, China
qj_paper@163.com
[2] College of Information Science and Engineering, Northeastern University, Shenyang 110819, China

Abstract. The paper mainly investigates the issue of achieving predefined-time synchronization for fuzzy memristive neural networks with both impulsive effects and stochastic disturbances. Firstly, due to the fact that the existed predefined-time stability theorems can hardly be applied to systems with impulsive effects, a new predefined-time stability theorem is proposed to solve the stability problem of the systems with impulsive effects. The theorem is flexible and can guide impulsive stochastic fuzzy memristive neural network models to achieve predefined-time synchronization. Secondly, due to the limitation problems for sign function that it can easily lead to cause the chattering phenomenon, resulting in undesirable results such as decreased synchronization performance. A novel and effective feedback controller without the sign function is designed to eliminate this chattering phenomenon in the paper. In addition, The paper overcomes the comprehensive influence of fuzzy logic, memristive state dependence and stochastic disturbance, and gives the effective conditions to ensure that two stochastic systems can achieve the predefined-time synchronization. Finally, the effectiveness of the proposed theoretical results is demonstrated in detail through a numerical simulation.

Keywords: Predefined-time Synchronization · Stochastic Disturbance · Fuzzy Neural Networks · Impulsive Effects

1 Introduction

In recent decades, neural networks have been extensively studied for their wide applications in information processing and pattern recognition [1–3]. The models of neural networks include inertial neural networks, reaction diffusion, memristor neural networks and so on. In particular, since the memristor neural network

B. Luo et al. (Eds.): ICONIP 2023, LNCS 14448, pp. 275–289, 2024.
https://doi.org/10.1007/978-981-99-8082-6_21

can better simulate the structure and function of the human brain, the synchronization of neural networks based on memristive has attracted the attention of many researchers [3–5].

In practice, systems are often required to achieve synchronization within a finite time. In order to achieve synchronization as soon as possible, kamenkov proposed the concept of finite-time stability [6]. The setting time of finite-time synchronization depends on the initial state of the system [7–9]. But in many systems with unknown initial states, finite time synchronization is no longer applicable. Fixed-time stability is proposed and applied to fixed-time synchronous applications without initial value dependence [10]. In recent years, many research achievements on fixed-time synchronization have been obtained [11–13]. Although fixed-time synchronization has high performance, it is difficult to control the synchronization time within the range we need. Predefined-time synchronization is a more flexible way than fixed-time synchronization. The synchronization time is a parameter in the controller and can be set freely. Recently, predefined-time synchronization has attracted extensive attention from researchers [14–16].

In practical applications, network nodes are inevitably subject to stochastic disturbances during the process of transmitting signals. This may result in the loss of information in the transmitted signal and significantly impact the neural network's behavior. Therefore, to study and simulate real neural networks, stochastic disturbances' influence must be considered when establishing neural network models. In addition, for many real networks, the state of nodes is often subject to instantaneous disturbances caused by factors such as frequency disturbances and burst noise, resulting in the state of nodes undergoing a sudden change at certain moments, i.e., forming the impulsive effect. There have been many papers on the predefined-time synchronization of neural networks under the influence of stochastic disturbances or impulsive effects [17–19]. However, few papers have investigated the problem of predefined-time synchronization of neural networks under the combined influence of both stochastic disturbances and impulsive effects.

In mathematical modeling of practical problems, unfavorable factors such as fuzziness and uncertainty are often encountered. Fuzzy logic theory, which can approximate any nonlinear function, has become a universal approximation method. This theory was first applied to cellular neural networks in references [20, 21], playing an important role in solving uncertainty and fuzziness. In recent years, synchronization of various fuzzy neural networks has received widespread attention [22, 23]. However, due to the complexity of predefined time synchronization of memristive fuzzy neural networks, few related papers have been published.

Inspired by the above discussions, this paper investigates the problem of achieving predefined-time synchronization of fuzzy memristive neural networks under both inpulsive effects and stochastic disturbances. The main contributions of this paper are summarized as follows:

(1) A new predefined-time stability theorem has been proposed, which is flexible and applicable to impulsive systems, filling the gap in existing research.
(2) A simple and effective controller is designed. The controller does not need to use sign function to avoid chattering. By adding a custom preset parameter T_c to the controller, the stability time of the system can be flexibly adjusted.
(3) A new criterion for achieving predefined-time synchronization of stochastic memristive neural networks with impulsive effects has been proposed.

The organization of this paper is as follows: In Sect. 2, the fuzzy stochastic memristive neural network model with impulsive effects and relevant preliminary knowledge are introduced. Section 3 reports the main results of this paper. Section 4 designs numerical simulation experiments and presents the experimental results. Finally, in Sect. 5, the conclusions are drawn.

2   Problem Statement and Preliminaries

Consider a class of fuzzy stochastic memristive neural networks with impulsive effects

$$
\begin{cases}
dx_i(t) = [-d_i x_i(t) + \displaystyle\sum_{j=1}^{n} a_{ij}(x_i(t))\hat{f}_j(x_j(t)) + \bigwedge_{j=1}^{n} b_{ij}\hat{f}_j(x_j(t)) \\
\qquad + \bigvee_{j=1}^{n} \hat{b}_{ij}\hat{f}_j(x_j(t)) + I_i]dt + \hat{g}_i(t, x_i(t))d\omega(t), \ t \geq 0, t \neq t_k, \\
\Delta x_i(t_k) = \mu_k x_i(t_k^-), \ t = t_k, k \in N,
\end{cases}
\tag{1}
$$

where $i = 1, ..., n$, $x_i(t)$ is the state of ith neuron at time t, $d_i > 0$ represents the self-inhibitions of ith neuron, b_{ij} and $\hat{b}_{ij}$ are elements of fuzzy feedback MIN template and fuzzy feedback MAX template, $\bigvee$ and $\bigwedge$ are fuzzy AND and fuzzy OR operations, $\hat{f}_j(\cdot)$ is the activation function, I_i stands for the external input, $\hat{g}_i(t, x_i(t)) \in R^n$ denote noise intensity function, ω is n-dimensional Brownian motion defined on $(\Omega, \mathcal{F}, \mathcal{P})$. μ_k is the strength of impulses at time t_k. The sequence $\{t_1, t_2, t_3...\}$ denotes the impulsive instants and $0 \leq t_1 < t_2 < ... < t_k < ...(t_k \to \infty$ as $k \to \infty)$, $\Delta x_i(t_k) = x_i(t_k^+) - x_i(t_k^-)$, $x_i(t_k) = x_i(t_k^+) = lim_{t \to t_k^+} x_i(t)$, $x_i(t_k^-) = lim_{t \to t_k^-} x_i(t)$, $a_{ij}(x_i(t))$ represents the memristor connection weight and is defined as:

$$
a_{ij}(x_i(t)) = \begin{cases} \grave{a}_{ij}, & |x_i(t)| \leq T_i, \\ \acute{a}_{ij}, & |x_i(t)| > T_i, \end{cases}
$$

in which the switching jump $T_i > 0$ and $\grave{a}_{ij}, \acute{a}_{ij}$ are known constants. Let $a_{ij}^m = \max\{|\grave{a}_{ij}|, |\acute{a}_{ij}|\}$.

System (1) is considered as a driving system, then the corresponding response system is described as:

$$\begin{cases} dy_i(t) = [-d_i y_i(t) + \sum_{j=1}^{n} a_{ij}(y_i(t)) \hat{f}_j(y_j(t)) + \bigwedge_{j=1}^{n} b_{ij} \hat{f}_j(y_j(t)) \\ \qquad + \bigvee_{j=1}^{n} \hat{b}_{ij} \hat{f}_j(y_j(t)) + u_i(t) + I_i] dt + \hat{g}_i(t, y_i(t)) d\omega(t), \ t \geq 0, t \neq t_k, \\ \Delta y_i(t_k) = \mu_k y_i(t_k^-), \ t = t_k, k \in N, \end{cases} \quad (2)$$

where $i = 1, ...n$, $y_i(t)$ denotes the state of ith neuron of response system at time t, $u_i(t)$ represents the appropriate control input to be designed, which will be given later.

Next, let the convex closure hull of set E be denoted as $\bar{co}\{E\}$. The set valued mapping is as follows

$$K[a_{ij}(x_i(t))] = \begin{cases} \grave{a}_{ij}, & |x_i(t)| < T_i, \\ \bar{co}\{\grave{a}_{ij}, \acute{a}_{ij}\}, & |x_i(t)| = T_i, \\ \acute{a}_{ij}, & |x_i(t)| > T_i, \end{cases}$$

Based on the theories of set-valued maps and differential inclusions, the system (1) can be described by the following differential inclusion:

$$\begin{cases} dx_i(t) \in [-d_i x_i(t) + \sum_{j=1}^{n} K[a_{ij}(x_i(t))] \hat{f}_j(x_j(t)) + \bigwedge_{j=1}^{n} b_{ij} \hat{f}_j(x_j(t)) \\ \qquad + \bigvee_{j=1}^{n} \hat{b}_{ij} \hat{f}_j(x_j(t)) + I_i] dt + \hat{g}_i(t, x_i(t)) d\omega(t), \ t \geq 0, t \neq t_k, \\ \Delta x_i(t_k) = \mu_k x_i(t_k^-), \ t = t_k, k \in N. \end{cases} \quad (3)$$

Equivalently, there exist $\breve{a}_{ij} \in K[a_{ij}(x_i(t))], \tilde{a}_{ij} \in K[a_{ij}(y_i(t))]$, such that

$$\begin{cases} dx_i(t) = [-d_i x_i(t) + \sum_{j=1}^{n} \breve{a}_{ij} \hat{f}_j(x_j(t)) + \bigwedge_{j=1}^{n} b_{ij} \hat{f}_j(x_j(t)) \\ \qquad + \bigvee_{j=1}^{n} \hat{b}_{ij} \hat{f}_j(x_j(t)) + I_i] dt + \hat{g}_i(t, x_i(t)) d\omega(t), \ t \geq 0, t \neq t_k, \\ \Delta x_i(t_k) = \mu_k x_i(t_k^-), \ t = t_k, k \in N. \end{cases} \quad (4)$$

Similarly, the system (2) can be rewritten as follows

$$\begin{cases} dy_i(t) = [-d_i y_i(t) + \sum_{j=1}^{n} \tilde{a}_{ij} \hat{f}_j(y_j(t)) + \bigwedge_{j=1}^{n} b_{ij} \hat{f}_j(y_j(t)) + \bigvee_{j=1}^{n} \hat{b}_{ij} \hat{f}_j(y_j(t)) \\ \qquad + u_i(t) + I_i] dt + \hat{g}_i(t, y_i(t)) d\omega(t), \ t \geq 0, t \neq t_k, \\ \Delta y_i(t_k) = \mu_k y_i(t_k^-), \ t = t_k, k \in N. \end{cases} \quad (5)$$

Let $e_i(t) = y_i(t) - x_i(t)$ is the synchronization error. Based on the systems (4) and (5), the error system can be derived as follows:

$$
\begin{cases}
de_i(t) = [-d_i e_i(t) + \displaystyle\sum_{j=1}^{n} (\tilde{a}_{ij} \hat{f}_j(y_j(t)) - \check{a}_{ij} \hat{f}_j(x_j(t))) + \bigwedge_{j=1}^{n} b_{ij} f_j(e_j(t)) \\
\quad + \displaystyle\bigvee_{j=1}^{n} \hat{b}_{ij} f_j(e_j(t)) + u_i(t)] dt + g_i(t, e_i(t)) d\omega(t), \ t \geq 0, t \neq t_k, \\
\Delta e_i(t_k) = \mu_k e_i(t_k^-), \ t = t_k, k \in N,
\end{cases}
\tag{6}
$$

where $f_j(e_j(t)) = \hat{f}_j(y_j(t)) - \hat{f}_j(x_j(t)), g_i(t, e_i(t)) = \hat{g}_i(t, y_i(t)) - \hat{g}_i(t, x_i(t))$.

Assumption 1. *An impulsive sequence* $\zeta = \{t_k, k \in N^+\}$ *belongs to* $\Omega \triangleq \{t_k, k \in N^+ | t_1 < t_2 < \dots < t_k < \dots, \lim_{k \to +\infty} t_k = +\infty\}$ *and have a average impulsive interval* τ_a, *then assume that there exists a positive constant* N_0 *such that*

$$
\frac{t-s}{\tau_a} - N_0 \leq N_\zeta(s, t) \leq \frac{t-s}{\tau_a} + N_0,
$$

where $N_\zeta(s, t)$ *is the number of impulsive times of impulsive sequences* ζ *in time interval* (s, t).

Assumption 2. *For each j, activation function $\hat{f}_j$ is bounded and satisfies Lipschitz continuous. There exists a positive constant l_j such that:*

$$
|\hat{f}_j(y_j) - \hat{f}_j(x_j)| \leq l_j |e_j(t)|.
$$

Assumption 3. *Suppose the noise intensity function $g_i(t, e_i(t))$ satisfies the uniformly Lipschitz continuous conditions, and there exist positive constant λ_i such that*

$$
trace[g_i^T(t, e_i(t)) g_i(t, e_i(t))] \leq \lambda_i e_i^T(t) e_i(t).
$$

Definition 1 ([24]). *If the system (6) is said to be globally stochastic fixed-time stability, and there exists a system parameter T_c such that the established time function $T(e_0, \omega) \leq T_c$. Then the system (6) is said to be globally stochastic predefined-time stability.*

Lemma 1 ([20]). *Suppose $b_{ij}, \hat{b}_{ij}, x_j, y_j \in R$, $f_j : R \to R$ be continuous functions, then we have*

$$
|\bigwedge_{j=1}^{n} b_{ij} f_j(y_j) - \bigwedge_{j=1}^{n} b_{ij} f_j(x_j)| \leq \sum_{j=1}^{n} |b_{ij}| |f_j(y_j) - f_j(x_j)|,
$$

$$
|\bigvee_{j=1}^{n} \hat{b}_{ij} f_j(y_j) - \bigvee_{j=1}^{n} \hat{b}_{ij} f_j(x_j)| \leq \sum_{j=1}^{n} |\hat{b}_{ij}| |f_j(y_j) - f_j(x_j)|.
$$

Lemma 2 ([25]). *Under Assumption 2, for any $i, j = 1, 2, \cdots, n$, if $f_j(\pm T_j) = 0$, then*

$$
|\tilde{a}_{ij} \hat{f}_j(y_j(t)) - \check{a}_{ij} \hat{f}_j(x_j(t))| \leq a_{ij}^m l_j |e_j(t)|.
$$

Lemma 3 ([26]). *If* $x_1, x_2, ..., x_n \geq 0$, $0 < p < 1$, $q > 1$ *then*

$$\sum_{i=1}^{n} x_i^p \geq (\sum_{i=1}^{n} x_i)^p, \quad \sum_{i=1}^{n} x_i^q \geq n^{1-q}(\sum_{i=1}^{n} x_i)^q.$$

3   Main Results

3.1   Predefined-Time Stability

Theorem 1. *Suppose that Assumption 1 holds. If there exists a positive definite and radial unbounded function* $V(\cdot) : R^n \rightarrow R_+ \bigcup \{0\}$, *and the following two conditions hold:*

1. *When* $t \neq t_k$, *there exist* ξ_1, ξ_2, ξ_3, $T_c > 0$, $0 < \alpha < 1$, $\beta > 1$, $W_1 = \frac{2}{\rho^{2N_0(1-\alpha)}\xi_1(1-\alpha)}$, $W_2 = \frac{2}{\rho^{N_0(\beta-1)}\xi_2(\beta-1)}$, *satisfies*

$$\mathcal{L}V(x,t) \leq -\frac{W_1}{T_c}\xi_1 V^\alpha(x,t) - \frac{W_2}{T_c}\xi_2 V^\beta(x,t) - \xi_3 V(x,t).$$

2. *When* $t = t_k$, *there exists* $0 < \rho \leq 1$,

$$V(x,t_k) \leq \rho V(x,t_k^-).$$

   Then the system (6) *is globally stochastic predefined-time stability within the time* T_c.

Proof. For simplicity, denoted by $V(t)$ as $V(x,t)$.
    According to *Itô* formula

$$dV(t) = \mathcal{L}V(t)dt + \sum_{i=1}^{n} V_{x_i}(t)g_i(t, e_i(t))d\omega(t),$$

where $V_{x_i}(t) = \frac{\partial V(t)}{\partial x_i}$, for $t \in [t_{k-1}, t_k)$. Take the mathematical expectation E on both sides respectively, and we can get

$$\frac{dEV(t)}{dt} = E\mathcal{L}V(t).$$

When $t \neq t_k$, it can be obtained that

$$\frac{dEV(t)}{dt} \leq \begin{cases} -\frac{W_1}{T_c}\xi_1 EV^\alpha(t) - \xi_3 EV(t), & 0 < V(t) < 1, t \neq t_k, \\ -\frac{W_2}{T_c}\xi_2 EV^\beta(t) - \xi_3 EV(t), & V(t) \geq 1, t \neq t_k, \\ 0, & V(t) = 0, t \neq t_k. \end{cases} \quad (7)$$

When $t = t_k, EV(t_k) \leq \rho EV(t_k^-)$.

Now consider the following auxiliary equation

$$\begin{cases} \dot{r}(t) = \begin{cases} -\dfrac{W_1}{T_c}\xi_1 r^\alpha(t) - \xi_3 r(t), & 0 < r(t) < 1, t \neq t_k, \\[2mm] -\dfrac{W_2}{T_c}\xi_2 r^\beta(t) - \xi_3 r(t), & r(t) \geq 1, t \neq t_k, \\[2mm] 0, & r(t) = 0, t \neq t_k, \end{cases} \\[6mm] r(t_k) = \rho r(t_k^-), t = t_k, \\[2mm] r(0) = r_0 = V(0). \end{cases} \tag{8}$$

Compared Eqs. (7) and (8), we can see that $0 \leq EV(t) \leq r(t), t \geq 0$. Consequently, if there exists $T > 0$ such that $r(t) \equiv 0$, for $t > T$, then $EV(t) \equiv 0$, for $t > T$. Hence, the stability of Eq. (8) implies stability of the zero solution of Eq. (7). Furthermore, the predefined-time stability of the system (6) can be obtained.

It is obtained from Eq. (8) that $r(t)$ is strictly decreasing on $[0, +\infty)$, if $0 < \rho \leq 1$. If $r_0 > 1$ and there exist constants T_1 and T_2 such that $r(t)$ reaches 1 at time T_1 and 0 from 1 at time T_2. Then $r(t)$ can goes to 0 in the time $T_1 + T_2$. In order to get T_1 and T_2, we divide the proof into two cases: $0 < \rho < 1$ and $\rho = 1$. Case 1: $0 < \rho < 1$.

When $r(t) \geq 1$, let $\eta(t) = r^{1-\beta}(t)$. It can be deduced that $\eta \to 1$ when $r \to 1$ and $\eta \to 0$ when $r \to +\infty$. Then, we can get

$$\begin{cases} \dot{\eta}(t) = \xi_3(\beta - 1)\eta(t) + \dfrac{W_2}{T_c}\xi_2(\beta - 1), 0 < \eta(t) \leq 1, \ t \neq t_k, \\[2mm] \eta(t_k) = \rho_1 \eta(t_k^-), & t = t_k, \\[2mm] \eta(0) = \eta_0 = r_0^{1-\beta}, \end{cases} \tag{9}$$

where $\rho_1 = \rho^{1-\beta} > 1$, we derive from Eq. (9) that

$$\eta(t) = e^{\xi_3(\beta-1)t}\rho_1^{N_\varsigma(0,t)}\eta(0) + \dfrac{W_2}{T_c}\xi_2(\beta - 1)\int_0^t e^{\xi_3(\beta-1)(t-s)}\rho_1^{N_\varsigma(s,t)}ds. \tag{10}$$

Since $\eta(0) = \rho_1^{N_\varsigma(0,0)}\eta(0) = \eta_0 < 1$, $\lim_{t \to +\infty}\eta(t) = +\infty$ and $\eta(t)$ is monotonously increasing when $t \geq 0$, there exists T_1 such that $\lim_{t \to T_1}\eta(t) = 1$ and $0 < \eta(t) < 1$ for $0 < t < T_1$. According to Eq. (10), we can get

$$e^{\xi_3(\beta-1)t}\rho_1^{N_\varsigma(0,t)}\eta(0) + \dfrac{W_2}{T_c}\xi_2(\beta - 1)\int_0^t e^{\xi_3(\beta-1)(t-s)}\rho_1^{N_\varsigma(s,t)}ds = 1. \tag{11}$$

Since $e^{\xi_3(\beta-1)t}\rho_1^{N_\varsigma(0,t)}\eta(0) \geq 0$ and $\rho_1^{\frac{t-s}{\tau_a} - N_0} \leq \rho^{N_\varsigma(s,t)}$, we have

$$\int_0^t e^{\xi_3(\beta-1)(t-s)}\rho_1^{\frac{t-s}{\tau_a}}ds \leq \dfrac{T_c\rho_1^{N_0}}{W_2\xi_2(\beta - 1)}. \tag{12}$$

Combined with inequality $ln(1+x) \leq x, x \geq -1$, solve the Eq. (12), we have

$$t \leq \frac{\tau_a}{\xi_3\tau_a(\beta-1)+ln\rho_1}ln(1+\frac{T_c\rho_1^{N_0}(ln\rho_1+\tau_a\xi_3(\beta-1))}{W_2\xi_2\tau_a(\beta-1)}) \leq \frac{T_c\rho_1^{N_0}}{W_2\xi_2(\beta-1)}.$$

Substituting $\rho_1 = \rho^{1-\beta}$ and $W_2 = \frac{2}{\rho^{N_0(\beta-1)}\xi_2(\beta-1)}$, we get $T_1 = t \leq \frac{T_c}{2}$. Thus, when $t \to T_1$, we can get $\eta(t) \to 1$ implies $r(t) \to 1$.

Next, we need to estimate the time T_2 such that $r(t)$ reaches 0 from 1. We suppose $\eta(t) = r^{1-\alpha}(t)$, when $0 < r(t) < 1$. It follows that $\eta(t) \to 1$ when $r(t) \to 1$ and $\eta \to 0$ when $r(t) \to 0$, therefore

$$\begin{cases} \dot{\eta}(t) = \xi_3(\alpha-1)\eta(t) + \frac{W_1}{T_c}\xi_1(\alpha-1), 0 < \eta(t) < 1, t \neq t_k, \\ \eta(t_k) = \rho_2\eta(t_k^-), t = t_k, \\ \eta(T_1) = r^{1-\alpha}(T_1) = 1, \end{cases} \tag{13}$$

where $\rho_2 = \rho^{1-\alpha}$ implies $\rho_2 \in (0,1)$. Similar to Eq. (10), one has, from Eq. (13), that

$$\eta(t) = e^{\xi_3(\alpha-1)(t-T_1)}\rho_2^{N_\varsigma(T_1,t)}\eta(T_1) - \frac{W_1}{T_c}\xi_1(1-\alpha)\int_{T_1}^t e^{\xi_3(\alpha-1)(t-s)}\rho_2^{N_\varsigma(s,t)}ds.$$

It can be concluded that $\eta(t)$ is decreasing on $[T_1, +\infty]$. Since $0 < \rho_2 < 1$, one has $\rho_2^{\frac{t-s}{\tau_a}+N_0} \leq \rho_2^{N_\varsigma(s,t)} \leq \rho_2^{\frac{t-s}{\tau_a}-N_0}$. Let $\eta(T_1) = 1, \eta(t) = 0$, then we can obtain

$$0 = \eta(t) = e^{\xi_3(\alpha-1)(t-T_1)}\rho_2^{N_\varsigma(T_1,t)} - \frac{W_1}{T_c}\xi_1(1-\alpha)\int_{T_1}^t e^{\xi_3(\alpha-1)(t-s)}\rho_2^{N_\varsigma(s,t)}ds,$$

$$\leq e^{\xi_3(\alpha-1)(t-T_1)}\rho_2^{\frac{t-T_1}{\tau_a}-N_0} - \frac{W_1}{T_c}\xi_1(1-\alpha)\int_{T_1}^t e^{\xi_3(\alpha-1)(t-s)}\rho_2^{\frac{t-s}{\tau_a}+N_0}ds,$$

$$= e^{\xi_3(\alpha-1)(t-T_1)}\rho_2^{\frac{t-T_1}{\tau_a}}(\rho_2^{-N_0} + \frac{W_1\xi_1\rho_2^{N_0}(1-\alpha)\tau_a}{T_c(\xi_3(1-\alpha)\tau_a - ln\rho_2)}) - \frac{W_1\xi_1\rho_2^{N_0}(1-\alpha)\tau_a}{T_c(\xi_3(1-\alpha)\tau_a - ln\rho_2)}. \tag{14}$$

Let $h(t) = e^{\xi_3(\alpha-1)(t-T_1)}\rho_2^{\frac{t-T_1}{\tau_a}}(\rho_2^{-N_0} + \frac{W_1\xi_1\rho_2^{N_0}(1-\alpha)\tau_a}{T_c(\xi_3(1-\alpha)\tau_a - ln\rho_2)}) - \frac{W_1\xi_1\rho_2^{N_0}(1-\alpha)\tau_a}{T_c(\xi_3(1-\alpha)\tau_a - ln\rho_2)}$. Since $h(T_1) > 0, h(+\infty) < 0$ and $\dot{h}(t) < 0$, there exists a unique time $t > 0$ such that $h(t) = 0$. Then we have

$$e^{\xi_3(\alpha-1)(t-T_1)}\rho_2^{\frac{t-T_1}{\tau_a}}(\rho_2^{-N_0} + \frac{W_1\xi_1\rho_2^{N_0}(1-\alpha)\tau_a}{T_c(\xi_3(1-\alpha)\tau_a - ln\rho_2)}) - \frac{W_1\xi_1\rho_2^{N_0}(1-\alpha)\tau_a}{T_c(\xi_3(1-\alpha)\tau_a - ln\rho_2)} = 0.$$

By solving the above inequality in combination with inequality $ln(1+x) \leq x, x \geq -1$, we can obtain

$$t - T_1 = \frac{\tau_a}{(1-\alpha)\xi_3\tau_a - ln\rho_2}ln(1+\frac{T_c(\xi_3\tau_a(1-\alpha)-ln\rho_2)}{W_1\rho_2^{2N_0}\xi_1(1-\alpha)\tau_a}) \leq \frac{T_c}{W_1\rho_2^{2N_0}\xi_1(1-\alpha)}.$$

Substitute $\rho_2 = \rho^{1-\alpha}$ and $W_1 = \frac{2}{\rho^{2N_0(1-\alpha)}\xi_1(1-\alpha)}$, we can get $T_2 = t - T_1 \leq \frac{T_c}{2}$. Hence, when $0 < \rho < 1$, we can conclude that $r(t) \equiv 0$ within the time $T_1 + T_2 \leq T_c$.

Case 2: $\rho = 1$. It is easy to observe $\rho_1 = \rho_2 = 1, W_1 = \frac{2}{\xi_1(1-\alpha)}, W_2 = \frac{2}{\xi_2(\beta-1)}$. When $r(t) \geq 1$, applying the same analytical method as case 1, we can obtain $T_1' \leq \frac{T_c}{2}$.

when $0 < r(t) < 1$, according to Eq. (14), we can get

$$0 = e^{\xi_3(\alpha-1)(t-T_1')} - \frac{W_1}{T_c}\xi_1(1-\alpha)\int_{T_1'}^{t} e^{\xi_3(\alpha-1)(t-s)}ds.$$

Solving above equation gives

$$t - T_1' = T_2' = \frac{1}{\xi_3(1-\alpha)}ln(1 + \frac{T_c\xi_3}{W_1\xi_1}) \leq \frac{T_c}{W_1\xi_1(1-\alpha)} = \frac{T_c}{2}.$$

Therefore, when $\rho = 1$, we can obtain $r(t) \equiv 0$ within the time $T_1' + T_2' \leq T_c$.

Based on the above discussion, it can be concluded that $r(t) \equiv 0$ when $t \geq T_c$. That is to say, the system (6) is globally stochastic predefined-time stability within the time T_c. The proof is completed.

3.2 Predefined-Time Synchronization

In order to achieve predefined-time synchronization between systems (1) and (2), we designed a non-chattering state feedback controller

$$u_i(t) = -k_{1i}e_i(t) - \frac{W_1}{T_c}k_{2i}e_i(t)^{\frac{s}{m}} - \frac{W_2}{T_c}k_{3i}e_i(t)^{\frac{p}{q}}, \tag{15}$$

where $i = 1, 2, ..., n$, T_c is an adjustable constant, $W_1 = \frac{2}{\rho^{2N_0(1-\alpha)}\xi_1(1-\alpha)}$, $W_2 = \frac{2}{\rho^{N_0(\beta-1)}\xi_2(\beta-1)}, \alpha = \frac{s+m}{2m}, \beta = \frac{p+q}{2q}, \xi_1 = \min_i\{2k_{2i}\}, \xi_2 = \min_i\{2k_{3i}\}n^{\frac{q-p}{2q}}$, s, m, p, q are odd numbers and satisfy $s < m$, $p > q$, k_{1i} is constant to be determined, k_{2i} and k_{3i} are positive constants.

Theorem 2. *If Assumptions 1-3 hold, and k_{1i} satisfies the following inequality*

$$k_{1i} > -d_i + \frac{1}{2}\sum_{j=1}^{n}\left((a_{ij}^m + |b_{ij}| + |\hat{b}_{ij}|)l_j + (a_{ji}^m + |b_{ji}| + |\hat{b}_{ji}|)l_i\right) + \frac{1}{2}\lambda_i.$$

Then, under the action of the controller (15), the systems (1) and (2) can achieve globally stochastic predefined-time synchronization within the time T_c.

Proof. The Lyapunov function is constructed as $V(e(t)) = \sum_{i=1}^{n} e_i^2(t)$. Denote $V(t)$ as $V(e(t))$. When $t \neq t_k$, by *Itô's* formula, we can get

$$\mathcal{L}V(t) = 2\sum_{i=1}^{n} e_i(t)\Bigg[-d_i e_i(t) + \sum_{j=1}^{n}(\tilde{a}_{ij}\hat{f}_j(y_j(t)) - \check{a}_{ij}\hat{f}_j(x_j(t)))$$

$$+ \bigwedge_{j=1}^{n} b_{ij}f_j(e_j(t)) + \bigvee_{j=1}^{n} \hat{b}_{ij}f_j(e_j(t)) - k_{1i}e_i(t) - \frac{W_1}{T_c}k_{2i}e_i(t)^{\frac{s}{m}} \quad (16)$$

$$- \frac{W_2}{T_c}k_{3i}e_i(t)^{\frac{p}{q}}\Bigg] + \sum_{i=1}^{n} trace[g_i^T(t,e_i(t))g_i(t,e_i(t))].$$

By Assumption 3, we get the following inequality

$$\sum_{i=1}^{n} trace[g_i^T(t,e_i(t))g_i(t,e_i(t))] \leq \sum_{i=1}^{n}\lambda_i e_i^T(t)e_i(t) = \sum_{i=1}^{n}\lambda_i e_i^2(t). \quad (17)$$

Based on Lemma 1, we can get

$$2\sum_{i=1}^{n} e_i(t)\bigwedge_{j=1}^{n} b_{ij}f_j(e_j(t)) \leq \sum_{i=1}^{n}\sum_{j=1}^{n}(|b_{ij}|l_j + |b_{ji}|l_i)|e_i(t)|^2,$$

$$2\sum_{i=1}^{n} e_i(t)\bigvee_{j=1}^{n} \hat{b}_{ij}f_j(e_j(t)) \leq \sum_{i=1}^{n}\sum_{j=1}^{n}(|\hat{b}_{ij}|l_j + |\hat{b}_{ji}|l_i)|e_i(t)|^2. \quad (18)$$

Similarly, according to Lemma 2, we have

$$2\sum_{i=1}^{n} e_i(t)\sum_{j=1}^{n}(\tilde{a}_{ij}\hat{f}_j(y_j(t)) - \check{a}_{ij}\hat{f}_j(x_j(t))) \leq \sum_{i=1}^{n}\sum_{j=1}^{n}a_{ij}^m l_j(|e_i(t)|^2 + |e_j(t)|^2),$$

$$= \sum_{i=1}^{n}\sum_{j=1}^{n}(a_{ij}^m l_j + a_{ji}^m l_i)|e_i(t)|^2. \quad (19)$$

Substituting Eqs. (17)-(19) into Eq. (16), we derive that

$$\mathcal{L}V(t) \leq -\sum_{i=1}^{n}\Bigg(2d_i - \sum_{j=1}^{n}((a_{ij}^m + |b_{ij}| + |\hat{b}_{ij}|)l_j + (a_{ji}^m + |b_{ji}| + |\hat{b}_{ji}|)l_i)$$

$$-\lambda_i + 2k_{1i}\Bigg)e_i^2(t) - \frac{W_1}{T_c}\sum_{i=1}^{n} 2k_{2i}(e_i^2(t))^{\frac{s+m}{2m}} - \frac{W_2}{T_c}\sum_{i=1}^{n} 2k_{3i}(e_i^2(t))^{\frac{p+q}{2q}}.$$

Let $\Lambda_i = 2d_i - \sum_{j=1}^{n}((a_{ij}^m + |b_{ij}| + |\hat{b}_{ij}|)l_j + (a_{ji}^m + |b_{ji}| + |\hat{b}_{ji}|)l_i) - \lambda_i + 2k_{1i}$. since $k_{1i} > -d_i + \frac{1}{2}\sum_{j=1}^{n}((a_{ij}^m + |b_{ij}| + |\hat{b}_{ij}|)l_j + (a_{ji}^m + |b_{ji}| + |\hat{b}_{ji}|)l_i) + \frac{1}{2}\lambda_i$, $\Lambda_i > 0$. Then we have

$$\mathcal{L}V(t) \le -\sum_{i=1}^{n} \Lambda_i e_i^2(t) - \frac{W_1}{T_c} \sum_{i=1}^{n} 2k_{2i}(e_i^2(t))^{\frac{s+m}{2m}} - \frac{W_2}{T_c} \sum_{i=1}^{n} 2k_{3i}(e_i^2(t))^{\frac{p+q}{2q}},$$

$$\le -\min_i\{\Lambda_i\} \sum_{i=1}^{n} e_i^2(t) - \frac{W_1}{T_c} \min_i\{2k_{2i}\} \sum_{i=1}^{n} (e_i^2(t))^{\frac{s+m}{2m}}$$

$$- \frac{W_2}{T_c} \min_i\{2k_{3i}\} \sum_{i=1}^{n} (e_i^2(t))^{\frac{p+q}{2q}}.$$

Since $s < m$, $p > q$, we get $0 < \frac{s+m}{2m} < 1$, $\frac{p+q}{2q} > 1$. From Lemma c3, we derive that

$$\mathcal{L}V(t) \le -\min_i\{\Lambda_i\} \sum_{i=1}^{n} e_i^2(t) - \frac{W_1}{T_c} \min_i\{k_{2i}\} (\sum_{i=1}^{n} e_i^2(t))^{\frac{s+m}{2m}}$$

$$- \frac{W_2}{T_c} \min_i\{k_{3i}\} n^{\frac{q-p}{2q}} (\sum_{i=1}^{n} e_i^2(t))^{\frac{p+q}{2q}}.$$

Let $\xi_1 = \min_i\{2k_{2i}\}, \xi_2 = \min_i\{2k_{3i}\} n^{\frac{q-p}{2q}}, \xi_3 = \min_i\{\Lambda_i\}$, we can get

$$\mathcal{L}V(t) \le -\frac{W_1}{T_c} \xi_1 (\sum_{i=1}^{n} e_i^2(t))^{\frac{s+m}{2m}} - \frac{W_2}{T_c} \xi_2 (\sum_{i=1}^{n} e_i^2(t))^{\frac{p+q}{2q}} - \xi_3 \sum_{i=1}^{n} e_i^2(t),$$

$$= -\frac{W_1}{T_c} \xi_1 V^{\frac{s+m}{2m}}(t) - \frac{W_2}{T_c} \xi_2 V^{\frac{p+q}{2q}}(t) - \xi_3 V(t).$$

When $t = t_k$, we have

$$V(t_k) = \sum_{i=1}^{n} e_i^2(t_k) = \sum_{i=1}^{n} e^2(t_k^+) = \sum_{i=1}^{n}(1+\mu_k)^2 e_i^2(t_k^-) = (1+\mu_k)^2 V(t_k^-).$$

Let $0 < (1 + \mu_k)^2 = \rho_k \le 1$. According to Theorem 1, the systems (1) and (2) can achieve globally stochastic predefined-time synchronization within the time T_c. The proof is completed.

4   Numerical Examples

Consider system (1) and system (2) as the driving system and response system respectively. The system parameters are shown as follows:

$$a_{11}(x_1(t)) = \begin{cases} 2, & |x_1(t)| \le 1, \\ 1.9, & |x_1(t)| > 1, \end{cases} \quad a_{12}(x_1) = \begin{cases} -3, & |x_1(t)| \le 1, \\ -4, & |x_1(t)| > 1, \end{cases}$$

$$a_{21}(x_2) = \begin{cases} -2, & |x_2(t)| \le 1, \\ -2.1, & |x_2(t)| > 1, \end{cases} \quad a_{22}(x_2) = \begin{cases} 1, & |x_2(t)| \le 1, \\ 1.8, & |x_2(t)| > 1, \end{cases}$$

$${b_{ij}} = \begin{pmatrix} 0.5 & 0.6 \\ 0.9 & 0.3 \end{pmatrix}, \ {\hat{b}_{ij}} = \begin{pmatrix} 4.2 & -3.5 \\ 6.1 & 1.2 \end{pmatrix}, \ i = 1, 2, j = 1, 2.$$

The initial values $x(0) = [1.2, 0.3]^T$ and $y(0) = [-0.4, 1.1]^T$, $d_1 = 0.8$, $d_2 = 0.9$, $\hat{f}_j(\cdot) = tanh(\cdot)$, $\mu_k = -0.4$, $\hat{g}(t, x(t)) = (\hat{g}_1(t, x_1(t)), \ \hat{g}_2(t, x_2(t)))^T = 0.4 diag(x_1(t), x_2(t))$, $\hat{g}(t, y(t)) = (\hat{g}_1(t, y_1(t)), \ \hat{g}_2(t, y_i(t)))^T = 0.4 diag(y_1(t), y_2(t))$, $t_{k+1} - t_k = 0.9$, $k \in N^+$, $I_1 = I_2 = 0$. Then we have $N_0 = 1$, $l_1 = l_2 = 1$, $\lambda_1 = \lambda_2 = 0.16$, $\rho = 0.36$. Figure 1 shows the state trajectories of the system (1). Figure 2 shows the synchronization error trajectories without controller. It can be seen from Fig. 2, system (1) and system (2) cannot achieve synchronization.

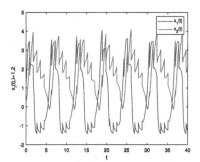

Fig. 1. State trajectories $x_i(t)(i = 1, 2)$ of drive system (1) with initial values $x(0) = [1.2, 0.3]^T$.

Fig. 2. Synchronization error trajectories of systems (1) and (2) without controller.

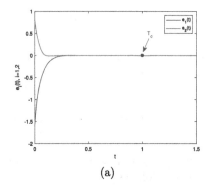
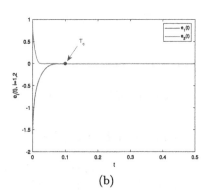

(a)   (b)

Fig. 3. (a) represents synchronization error trajectories of $T_c = 1$ with controller (15). (b) represents synchronization error trajectories of $T_c = 0.1$ with controller (15).

Now, we choose the controller parameters $s = 3$, $m = 5$, $p = 3$, $q = 1$, $k_{11} = k_{12} = 15$, $k_{21} = k_{22} = k_{31} = k_{32} = 1$. It's easy to calculate $\xi_1 = 2$, $\xi_2 = 1$, $W_1 = 5.9772$, $W_2 = 3.125$, k_{11} and k_{12} satisfy the conditions of Theorem 2. Figure 3 shows the state error trajectories at $T_c = 1$ and $T_c = 0.1$ under the action of controller (15) respectively. The error trajectories can converge to zero within the predefined times, which illustrate the validity of Theorem 2.

Then, consider five groups of two-dimensional fuzzy stochastic memristive neural networks with impulsive effects. The initial values of the drive-response system are initialized randomly. To facilitate observation, we constrain the range of random initialization of parameters: $x_i(0)$, $y_i(0) \in [-10, 10]$, $i = 1, 2$. Other parameter values remain unchanged. Synchronization error trajectories of drive-response system with randomly initial values are shown in Fig. 4.

It can be observed that the five groups of drive-response systems with randomly initial values can achieve synchronization within the time T_c. Therefore, the simulation verify the validity of the theoretical results.

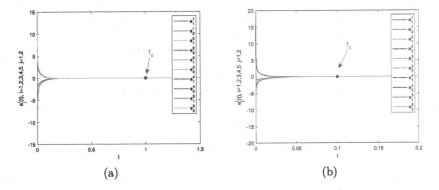

(a)   (b)

Fig. 4. (a) represents error trajectory of drive-response system with random initial values when $T_c = 1$. (b) represents error trajectory of drive-response system with random initial values when $T_c = 0.1$.

5 Conclusion

This paper investigates the problem of predefined-time synchronization in fuzzy stochastic memristive neural networks with impulsive effects. A new theorem for the predefined-time stability of the systems with impulsive effects is obtained using some simple inequality techniques. Based on this theorem and a non-chattering control method, a simple feedback control is designed to prevent the chattering phenomenon caused by the sign function, and a new criterion for achieving predefined-time synchronization in impulsive fuzzy stochastic memristive neural networks is presented. Finally, the effectiveness of the theoretical results is verified through simulation experiments. In the future, we will consider new solutions to reduce the degree of scaling in the derivation process.

Acknowledgements. This work is supported by the National Natural Science Foundation of China (Grant Nos.62103165, 62101213), and the Natural Science Foundation of Shandong Province (Grant No. ZR2022ZD01).

References

1. Liu, J., Shu, L., Chen, Q., Zhong, S.: Fixed-time synchronization criteria of fuzzy inertial neural networks via Lyapunov functions with indefinite derivatives and its application to image encryption. Fuzzy Sets Syst. **459**, 22–42 (2023)
2. Shanmugam, L., Mani, P., Rajan, R., Joo, Y.H.: Adaptive synchronization of reaction-diffusion neural networks and its application to secure communication. IEEE Trans. Cybern. **50**(3), 911–922 (2018)
3. Zhou, C., Wang, C., Yao, W., Lin, H.: Observer-based synchronization of memristive neural networks under DoS attacks and actuator saturation and its application to image encryption. Appl. Math. Comput. **425**, 127080 (2022)
4. Yao, W., Wang, C., Sun, Y., Gong, S., Lin, H.: Event-triggered control for robust exponential synchronization of inertial memristive neural networks under parameter disturbance. Neural Netw. **164**, 67–80 (2023)
5. Zhao, M., Li, H.L., Zhang, L., Hu, C., Jiang, H.: Quasi-synchronization of discrete-time fractional-order quaternion-valued memristive neural networks with time delays and uncertain parameters. Appl. Math. Comput. **453**, 128095 (2023)
6. Kamenkov, G.: On stability of motion over a finite interval of time. J. Appl. Math. Mech. **17**(2), 529–540 (1953)
7. Gao, L., Cai, Y.: Finite-time stability of time-delay switched systems with delayed impulse effects. Circuits Syst. Signal Process. **35**, 3135–3151 (2016)
8. Abdurahman, A., Jiang, H., Teng, Z.: Finite-time synchronization for memristor-based neural networks with time-varying delays. Neural Netw. **69**, 20–28 (2015)
9. Yang, X., Lu, J.: Finite-time synchronization of coupled networks with Markovian topology and impulsive effects. IEEE Trans. Autom. Control **61**(8), 2256–2261 (2015)
10. Polyakov, A.: Nonlinear feedback design for fixed-time stabilization of linear control systems. IEEE Trans. Autom. Control **57**(8), 2106–2110 (2011)
11. Fang, J., Zhang, Y., Liu, P., Sun, J.: Fixed-time output synchronization of coupled neural networks with output coupling and impulsive effects. Neural Comput. Appl. **33**(24), 17647–17658 (2021). https://doi.org/10.1007/s00521-021-06349-0
12. Chen, C., Li, L., Peng, H., Yang, Y., Mi, L., Zhao, H.: A new fixed-time stability theorem and its application to the fixed-time synchronization of neural networks. Neural Netw. **123**, 412–419 (2020)
13. Kong, F., Zhu, Q., Huang, T.: New fixed-time stability lemmas and applications to the discontinuous fuzzy inertial neural networks. IEEE Trans. Fuzzy Syst. **29**(12), 3711–3722 (2020)
14. Sánchez-Torres, J. D., Gómez-Gutiérrez, D., López, E., Loukianov, A. G.: A class of predefined-time stable dynamical systems. IMA Journal of Mathematical Control and Information 35(Supplement_1), i1–i29 (2018)
15. Assali, E.A.: Predefined-time synchronization of chaotic systems with different dimensions and applications. Chaos, Solitons Fractals **147**, 110988 (2021)
16. Chen, C., Mi, L., Liu, Z., Qiu, B., Zhao, H., Xu, L.: Predefined-time synchronization of competitive neural networks. Neural Netw. **142**, 492–499 (2021)
17. Wan, P., Sun, D., Zhao, M.: Finite-time and fixed-time anti-synchronization of Markovian neural networks with stochastic disturbances via switching control. Neural Netw. **123**, 1–11 (2020)
18. Li, L., Xu, R., Gan, Q., Lin, J.: A switching control for finite-time synchronization of memristor-based BAM neural networks with stochastic disturbances. Nonlinear Anal.: Modelling Control **25**(6), 958–979 (2020)

19. Fan, H., Xiao, Y., Shi, K., Wen, H., Zhao, Y.: μ-synchronization of coupled neural networks with hybrid delayed and non-delayed impulsive effects. Chaos, Solitons Fractals **173**, 113620 (2023)
20. Yang, T., Yang, L.B., Wu, C.W., Chua, L.O.: Fuzzy cellular neural networks: theory. In: 1996 Fourth IEEE International Workshop on Cellular Neural Networks and Their Applications Proceedings (CNNA-96), pp. 181–186 (1996)
21. Yang, T., Yang, L.B., Wu, C.W., Chua, L.O.: Fuzzy cellular neural networks: applications. In: 1996 Fourth IEEE International Workshop on Cellular Neural Networks and Their Applications Proceedings (CNNA-96), pp. 225–230 (1996)
22. Sadik, H., Abdurahman, A., Tohti, R.: Fixed-time synchronization of reaction-diffusion fuzzy neural networks with stochastic perturbations. Mathematics **11**(6), 1493 (2023)
23. Liu, Y., Zhang, G., Hu, J.: Fixed-time stabilization and synchronization for fuzzy inertial neural networks with bounded distributed delays and discontinuous activation functions. Neurocomputing **495**, 86–96 (2022)
24. Sánchez-Torres, J.D., Sanchez, E.N., Lou kianov, A.G.: A discontinuous recurrent neural network with predefined time convergence for solution of linear programming. In: 2014 IEEE Symposium on Swarm Intelligence, pp. 1–5 (2014)
25. Chen, J., Zeng, Z., Jiang, P.: Global Mittag-Leffler stability and synchronization of memristor-based fractional-order neural networks. Neural Netw. **51**, 1–8 (2014)
26. Hardy, G.H., Littlewood, J.E., Pólya, G.: Inequalities. Cambridge University Press, Cambridge Mathematical Library (1934)

FE-YOLOv5: Improved YOLOv5 Network for Multi-scale Drone-Captured Scene Detection

Chen Zhao[1,2,3], Zhe Yan[1,2,3], Zhiyan Dong[1,2,3]([✉]), Dingkang Yang[1,2,3], and Lihua Zhang[1,2,3,4]

[1] Academy for Engineering and Technology, Fudan University, Shanghai, China
{21210860101,22210860063}@m.fudan.edu.cn
[2] Engineering Research Center of AI and Robotics, Ministry of Education, Beijing, China
[3] Institute of AI and Robotics, Fudan University, Shanghai, China
{dongzhiyan,dkyang20,lihuazhang}@fudan.edu.cn
[4] Jilin Provincial Key Laboratory of Intelligence Science and Engineering, Changchun, China

Abstract. Due to the different angles and heights of UAV shooting, the shooting environment is complex, and the shooting targets are mostly small, so the target detection task in the drone-captured scene is still challenging. In this study, we present a highly precise technique for identifying objects in scenes captured by drones, which we refer to as FE-YOLOv5. First, to optimize cross-scale feature fusion and maximize the utilization of shallow feature information, we propose a novel feature pyramid model called MSF-BiFPN as our primary approach. Furthermore, to improve the fusion of features at different scales and boost their representational power, our innovative approach proposes an adaptive attention module. Moreover, we propose a novel feature enhancement module that effectively strengthens high-level features before feature fusion. This module effectively minimized feature loss during the fusion process, ultimately resulting in enhanced detection accuracy. Finally, the utilization of the normalized Wasserstein distance serves as a novel metric for enhancing the model's sensitivity and accuracy in detecting small targets. The experimental results of FE-YOLOv5 on the VisDrone data set show that mAP 0.5 has increased by 7.8%, and mAP 0.5:0.95 increased by 5.7%. At the same time, the training results of the model at 960×960 image resolution are better than the current YOLO series models, among which mAP 0.5 can reach 56.3%. Based on the experiments conducted, it has been demonstrated that the FE-YOLOv5 model effectively enhances the accuracy of object detection in UAV capture scenes.

Keywords: Adaptive attention · Feature enhancement · Multi-scale targets · YOLOv5

1 Introduction

Recently, as the Unmanned Aerial Vehicle (UAV) field continues to evolve, drones equipped with cameras have been widely used in agricultural inspection, urban

B. Luo et al. (Eds.): ICONIP 2023, LNCS 14448, pp. 290–304, 2024.
https://doi.org/10.1007/978-981-99-8082-6_23

inspection [1,2], aerial photography, ecological protection [3], etc. Hence, automatic and efficient object detection is increasingly important in drone-captured scene parsing.

Fig. 1. The intuitive scenario shows three important problems encountered in object detection in the UAV scene. The scenarios in (a), (b), and (c) respectively illustrate the problems of complex background, large target scale variation, and high density in drone-captured scenes.

With the advent of convolutional neural networks (CNNs), numerous approaches have emerged in the field of target detection, including the two-stage detectors RCNN [4], Fast RCNN [5], Faster RCNN [6], etc. the one-stage detectors SSD [7], Retina Net [8], and YOLO [9], etc. These methods have achieved good detection results on some benchmark datasets like MS COCO [10] and PASCAL VOC [11], and have promoted the development of object detection technology. Nevertheless, directly applying these methods to drone-captured scenes for object detection makes it difficult to achieve more precise results. Because most of the previous methods are designed to deal with images in natural scenes. Different from the target detection of natural scene images, object detection in the UAV scene has the following problems. Some cases in Fig. 1 intuitively illustrate the difficulties encountered. Figure 1(a) shows that there are many complex and disturbing backgrounds in the drone-captured image, including a large number of buildings occluded and a dark environment at night. Figure 1(b) shows that due to the different flying heights of the UAV, the scale of objects varies greatly, for example, the size of nearby vehicles is large, and the size of distant vehicles and pedestrians is small. Finally, Fig. 1(c) shows that there are high-density distributions of objects in the drone-captured scene and the problem of mutual occlusion. The above three problems all make it difficult for the target detection task in the UAV scene.

This paper follows the one-stage design and proposes an FE-YOLOv5 network to solve the problems mentioned above. Like the baseline network, we use CSPDarknet53 [12,13] as the backbone of the FE-YOLOv5 network. In the neck part, We propose a new Weighted Bidirectional Feature Pyramid Network named MSF-BiFPN to achieve more efficient cross-scale feature fusion and fuller utilization of shallow features. In the head part, we add a new detection head to detect smaller targets. To enhance the integration of multi-scale feature information and enhance feature representation capabilities, we propose a novel adaptive attention module (AAM). Additionally, a Feature Enhancement Module (FEM) has been proposed to minimize feature loss and enhance the quality of feature information. By replacing Intersection over Union (IoU) with Normalized Wasserstein Distance (NWD) [16] as a new evaluation indicator, the model becomes more sensitive to small targets, leading to improved performance of the FE-YOLOv5 model.

The main contributions of this article are summarized as follows:

- We propose MSF-BiFPN, which fuses more shallow features based on BiFPN while achieving more effective cross-scale feature fusion.
- To enhance the performance of the model, we incorporated a smaller detection head and introduced the Normalized Wasserstein Distance (NWD) as an alternative evaluation metric. This modification aims to improve the model's sensitivity towards smaller objects and ultimately enhance the accuracy of object detection.
- We propose a new Adaptive Attention Module (AAM), which extracts channel attention and spatial attention separately by fusing receptive fields of different sizes in the feature map, thereby enhancing the representation ability of features while achieving multi-scale feature fusion.
- We propose a new Feature Enhancement Module (FEM), which enhances feature information by fusing the contextual information of receptive fields of different sizes in the feature map and then replaces upsampling operations with sub-pixel convolutions to reduce feature loss.

2   Related Works

2.1   Object Detection in Drone-Captured Scenes

With the widespread application of drones and the advancement of target detection technology, more and more scholars have devoted themselves to the target detection task in the drone-captured scene to solve various problems in this task, thereby improving detection accuracy, and letting drones serve us better.

Among them, Que et al. [17] solved the noise problem caused by the complex environment in the UAV scene by adopting the strategy of mixed training of two kinds of data sets with noise and noise-free, which can make the model even noisy images. Can have a high detection accuracy. Literature [18] proposed a new model based on YOLOv5: ViT-YOLO. This article proposes an improved MHSA-Darknet backbone network based on Transformer, which improves model performance by retaining sufficient global context information and utilizing a multi-head attention mechanism to extract more features. Meanwhile, temporal test augmentation (TTA) and heavy box fusion (WBF) [19] are integrated into the model to enhance its performance and robustness of the model.

2.2 Muti-scale Feature Fusion

In the scene captured by the UAV, the scale of the target changes drastically, so effectively processing multi-scale features is also one of the difficulties in object detection in this scenario. A significant contribution to the field was made by Literature [21], which introduced the Feature Pyramid Structure (FPN). This framework effectively addresses the problem of feature information loss, resulting in improved accuracy in detection tasks. By merging high-level features with rich spatial information and shallow features with abundant semantic information, FPN successfully enhances the utilization of feature information. Following that, PANet [15] introduced a backbone network called bidirectional fusion based on FPN, incorporating both top-down and bottom-up approaches. This innovative approach enhanced the utilization of features to a greater extent. Efficient-Det [14] proposed BiFPN, which enhances the representation ability of features through effective bidirectional cross-scale connections and adds weight to the fusion of each scale feature through weighted feature fusion, thereby adjusting the contribution of each scale. Literature [22] proposes a new channel enhancement feature pyramid network (CE-FPN). This network proposes a novel fusion approach, building on the concept of sub-pixel convolution [23].

3   Proposed Method

YOLOv5 [11] is currently one of the most widely used and accurate one-stage detectors, so we choose YOLOv5 as the benchmark network. YOLOv5 includes five different models: YOLOv5n, YOLOv5s, YOLOv5m, YOLOv5l, and YOLOv5x. From the training results on the VirDrone2021 dataset [24], although the effect of YOLOv5x is the best among the above five models, the training computation cost is more than that of the other four models. At the same time, the results of YOLOv5l are slightly lower than that of YOLOv5x, but the parameter amount is only about half of that of YOLOv5x. Therefore, Considering the performance-overhead trade-off, we choose YOLOv5l as our final baseline.

3.1   The FE-YOLOv5 Network Framework

The fundamental architecture of YOLOv5 consists of Input, Backbone, Neck, and Head components. The Backbone section primarily comprises Convolution modules, CSP modules, and SPPF [25] modules, and Path Aggregation Network

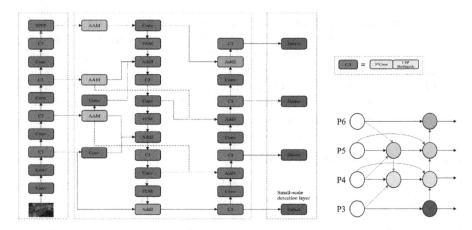

Fig. 2. The architecture of the FE-YOLOv5. The C3 module is composed of 3 convolution modules and a CSP Bottleneck. Add2 and Add3 refer to two and three-weighted feature maps to perform the Add operation. The lower right corner of the figure is the architecture of MSF-BiFPN. Among them, p3, p4, p5, and p6 respectively refer to four feature maps of different sizes. In addition, the red line indicates the shallow feature fusion supplemented based on BiFPN. (Color figure online)

(PANet) [15] is an integral part of the model's neck section. In the proposed FE-YOLOv5 model, we aim to enhance object detection in drone capture scenarios. The model employs a backbone network to extract features from input images, producing feature maps of varying sizes. These feature maps are then processed in the model's neck, where feature fusion and enhancement techniques are applied. As a result, three feature maps (P3, P4, and P5) are generated. Finally, the model's head performs target detection on these feature maps, completing the object detection process and generating the desired output. Figure 2 illustrates the framework of the FE-YOLOv5 model, which incorporates advancements specifically targeted at solving the difficulties posed by object detection in drone capture scenarios.

3.2 Detection Head for Tiny Objects

We analyzed the images taken by drones and found that many small targets are challenging to detect in the images, such as pedestrians, bicycles, etc. We added an extra detection layer to the original YOLOv5 model to better detect these small targets. As shown in Fig. 2, this added detection head is generated from shallow feature maps with a higher resolution because shallow features have smaller receptive fields. The feature becomes more sensitive to small objects and gains better detection capabilities as the receptive field size decreases. Compared to the original model, the implementation of four detection heads effectively reduces the influence of changes in object scale. This, in turn, enhances the model's capability to detect small targets.

3.3   MSF-BiFPN

In this section, we propose a Bidirectional Feature Pyramid Network (MSF-BiFPN) capable of integrating more shallow features, which incorporates more shallow features on top of BiFPN [14]. Compared to PANet [15], BiFPN uses a weighted feature fusion method to give different weights to the features in the fusion process by learning the contribution of different input features. Because feature maps of different sizes have different receptive field sizes, they contribute differently to the fusion process for objects of different sizes. For instance, shallow high-resolution feature maps possess smaller receptive fields, thereby enhancing their sensitivity towards small targets. Because in small object detection, the role of shallow feature maps is more significant. Therefore, we propose MSF-BiFPN based on BiFPN. The structure diagram of MSF-FPN is shown in Fig. 2. By using the shallow feature information of the P3 layer and P4 layer multiple times, it can more effectively extract the features of small targets while achieving more effective cross-scale feature fusion.

3.4   Adaptive Attention Module (AAM)

Due to the reduction of channels during feature extraction, contextual information is lost. To overcome this limitation, we introduce an innovative Adaptive Attention Module (AAM). This module leverages the fusion of feature information across various receptive fields in the feature map. Through the extraction of both channel attention and spatial attention, we achieve multi-scale feature fusion and improve the expressive power of the features.

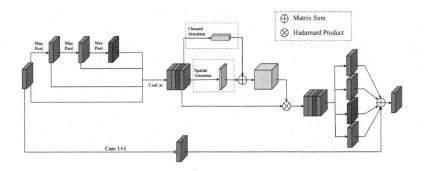

Fig. 3. The architecture of AAM.

The specific structure of AAM is shown in Fig. 3. Firstly, the input feature map is subjected to a series of three max pooling operations. Each pooling operation uses a convolutional kernel with a size of 4 × 4, a stride of 1, and applies a padding of 2 pixels. Afterward, the original feature map is combined with the feature maps obtained from the three pooling operations. By adopting this approach, the fusion of contextual features from various scales is achieved.

Subsequently, the fused feature map is fed into two separate branches, in one branch, channel attention maps of features are extracted, while in the other branch, spatial attention maps are extracted. The two branches' mapping results are combined by performing a matrix sum operation, resulting in a fused 3D attention feature map. This fused feature map is then multiplied element-wise with the previously obtained 3D attention map, and the result is separated by channels and added to the 1×1 convolution result of the original feature map thus enabling the aggregation of contextual information to acquire the ultimate feature map. The final result has feature information of different scales, which not only realizes the fusion of different scale features but also enhances the representation ability of features.

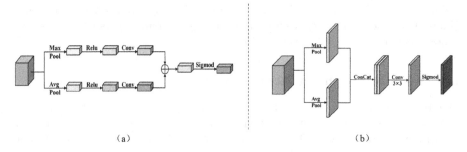

Fig. 4. (a) is the architecture of channel attention module and (b) is the architecture of spatial attention module.

The channel attention module [26] in AAM is shown in Fig. 4(a). The first half of the AAM module generates an enhanced feature map by fusing feature information under different receptive fields. Every channel within the feature map corresponds to a particular feature response, which is applied to emphasize the specific features that a channel concentrates on. As a result, the enhanced feature map encompasses feature maps of different channels associated with distinct receptive field sizes. Since the scale of the target in the drone-captured scene changes drastically, it is necessary to assign different weights to the channels corresponding to the receptive fields of different sizes through the attention channel module, to complete the detection task more accurately.

The spatial attention module [20] of AAM is shown in Fig. 4(b). The difference between the spatial attention module and the channel attention module is that the spatial attention module focuses on where rather than what the effective information is on the feature map. Due to various factors such as excessive noise, low light conditions at night, challenging geographical features, and interference from nearby buildings, drone-captured scenes often present difficulties for accurate detection. However, with the implementation of a spatial attention module, the model can effectively concentrate on the specific object location within the area and enhance the overall detection performance of the model.

3.5   Feature Enhancement Module (FEM)

During the process of fusing features, semantic information may be lost when transforming high-level low-resolution feature maps into low-level high-resolution feature maps. To address this issue, we draw inspiration from the CE-FPN [22] model and incorporate sub-pixel convolution [23]. Our proposed FEM module enhances the features that require upsampling by merging contextual information from various receptive fields in the feature map. This result in improved accuracy for multi-scale target detection.

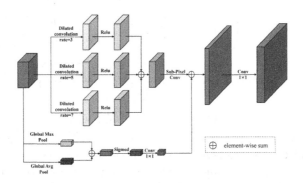

Fig. 5. The architecture of FEM.

As shown in Fig. 5, the FEM module has two branches. The first branch mainly obtains receptive fields of different sizes through dilated convolutions [27] of different rates while extracting local information from feature maps. First, the feature map ($2W \times 2H \times 8C$) undergoes hollow convolution with rates of 3, 5, and 7 respectively to obtain 3 feature maps of the same size ($2W \times 2H \times 4C$), then, the three feature maps are computed using the Relu function and then aggregated by element-wise summation. This fusion process combines local information from receptive fields of various sizes and generates a feature map with dimensions of $2W \times 2H \times 4C$. Next, the feature map is subjected to double-scale sampling through sub-pixel convolution, and finally, a feature map with a size of $4W \times 4H \times C$ is obtained. The primary purpose of the second branch is to acquire global information within the feature map. In this branch, the feature map ($2W \times 2H \times 8C$) respectively undergoes a global maximum pooling and a global average pooling. Two global maps of $1 \times 1 \times 8C$ size are obtained by simplification, and then these two global maps are added and then passed through the Sigmod function and the 1×1 convolutional layer respectively to obtain a $1 \times 1 \times C$ feature map. In the end, the feature map with dimensions $1 \times 1 \times C$ acquired from the second branch is expanded to a size of $4W \times 4H \times C$ through broadcasting, and then element-wise summed with the feature map obtained from the first branch, and the result is passed through 1×1 The convolutional layer finally obtains the upsampling result after the initial feature is enhanced.

3.6 Normalized Wasserstein Distance (NWD)

It is pointed out in NWD [16] that the offset IOU [28] of the boxes of small objects has a greater impact. The method proposed by NWD is, first, to remodel the box into a two-dimensional Gaussian distribution to make it more in line with the characteristics of small targets. Simultaneously transform the IoU of the ground-truth box and the prediction box into the similarity of the two distributions; secondly, Normalized Wasserstein Distance (NWD) is used as a new evaluation index, and NWD is employed to quantify the similarity between two distributions; the loss function of NWD is as follows:

$$W_2^2(N_a, N_b) = \left\| \left(\left[cx_a, cy_a, \frac{w_a}{2}, \frac{h_a}{2} \right]^{\mathrm{T}}, \left[cx_b, cy_b, \frac{w_b}{2} \cdot \frac{h_b}{2} \right]^{\mathrm{T}} \right) \right\|_2^2, \quad (1)$$

$$NWD(N_a, N_b) = \exp \left(-\frac{\sqrt{W_2^2(N_a, N_b)}}{C} \right), \quad (2)$$

$$L_{NWD} = 1 - NWD(N_a, N_b), \quad (3)$$

where A=$(x_a, y_a, w_a, h_a)$is the prediction box, B=(x_b, y_b, w_b, h_b) is the ground-truth box, and C is a constant closely related to the dataset. However, it was discovered during the experiment that when only NWD is utilized for loss calculation, the model fitting process becomes significantly sluggish, so we add the loss of L_{NWD} and IoU, the formula is as follows:

$$L = (1 - NWD(N_a, N_b)) * \alpha + (1 - IoU) * (1 - \alpha), \quad (4)$$

$$IoU = \frac{|A \cap B|}{|A \cup B|}, \quad (5)$$

where α is a custom coefficient, in this paper $\alpha = 0.6$.

4   Experiments and Analysis

We assess the model's performance using the VisDrone2021 dataset [24], which exclusively includes images captured by drones. This dataset comprises ten distinct categories of images with a grand total of 2.6 million labels. Among the 10,209 static images in the dataset, 6,471 are allocated for training, 3,190 for testing, and 548 for validation purposes.

4.1 Experimental Setting and Evaluation Criteria

This paper utilizes the Ubuntu 20.04 operating system, with the experimental environment consisting of cuda 11.3, python 3.8, and PyTorch 1.11.0. All models are trained and assessed using the same hyperparameters on NVIDIA RTX 3090 GPUs. Specifically, the epoch, batch size, and image size are set to 200, 8, and 960 × 960, respectively. In addition, in this article, we choose Precision, mAP 0.5(mean Average Precision), mAP 0.5:0.95, Params, and FLOPs as the evaluation indicators of the model.

4.2 Ablation Studies

In order to visually illustrate the improved accuracy of target detection using our proposed method for drone-captured scenes, we conducted ablation experiments using the VisDrone2021 dataset. The outcomes of these experiments are presented in Table 1, highlighting the improved detection accuracy achieved by our method.

Table 1. Ablation study

	Models	mAP 0.5(%)	mAP 0.5:0.95(%)	FLOPs(G)
A	YOLOv5l(Baseline)	48.5	29.5	107.8
B	+New detection head	50.6	31.0	132.6
C	+MSF-BiFPN	51.5	31.5	137.8
D	+AAM	53.4	32.3	155.4
E	+FEM	55.6	34.4	232.4
F	Ours	**56.3**	**35.2**	232.4

Table 1 shows the ablation results of adding module training based on the YOLOv5l model. It can be seen from the results that the mAP 0.5 and mAP 0.5:0.95 of the YOLOv5l model as the Baseline on the VisDrone dataset are 48.5% and 29.5%, respectively. The results of experiment B show that adding additional detection heads can increase mAP 0.5 and mAP 0.5:0.95 by 2.1% and 1.5%, respectively, which improves the detection accuracy of small targets. In Experiment C, by adding a small target detection head, the PANet of the neck part of the YOLOv5 model was replaced by MSF-BiFPN. Compared with B, the experimental results increased by 0.9% and 0.5% respectively for mAP 0.5 and mAP 0.5:0.95, suggesting that the integration of MSF-BiFPN leads to enhanced cross-scale feature fusion and superior model performance. The model of experiment D added the AAM module we proposed based on the model of experiment C, and the experimental results were improved by 1.9% and 0.8% respectively, suggesting that the AAM module is capable of integrating features of various scales, thereby enhancing the model's detection performance. Compared with Experiment D, mAP 0.5 and mAP 0.5:0.95 in Experiment E are improved by 2.2% and 2.1% respectively, indicating that the FEM module can enhance features, reduce feature loss and optimize the model. Experiment F is to combine all improvements, including NWD, and the final model improves by 7.8% and 5.7% on mAP 0.5 and mAP 0.5:0.95, respectively, which is the best performance of all models. While the integration of all these enhancements may result in a substantial increase in computational requirements for the model, in terms of results, both mAP 0.5 and mAP 0.5:0.95 have greatly improved, improving the effect of target detection in drone-captured scenes.

Table 2. Comparative experiments

Models	mAP 0.5(%)	mAP 0.5:0.95(%)	Params(M)	FLOPs(G)	FPS
YOLOv5l	48.5	29.5	46.2	107.8	76.9
DETR	20.5	8.4	60.0	253.0	10.7
Faster R-CNN	35.7	20.1	60.0	246.0	20.2
YOLOv5x	50.3	30.9	86.3	204.8	47.9
YOLOv6	47.6	29.1	58.5	144.0	64.9
YOLOv7	**56.3**	33.3	71.3	189.9	48.7
YOLOv8	54.0	33.9	68.1	257.4	27.8
Ours	**56.3**	**35.2**	86.2	232.4	28.2

4.3 Comparative Experiments

This comparative experiment aims to demonstrate the superiority of the FE-YOLOv5 model over other recent YOLO target detection models. It compares the performance of these models and quantitatively analyzes the target detection results of a scene captured by a drone, the FE-YOLOv5 model is compared with the Baseline YOLOv5l, YOLOv5x, YOLOv6 [29], YOLOv7 [30], and YOLOv8 models, the two-stage classic model Faster RCNN [6] and the DETR [31] model that has attracted much attention. The YOLOv6, YOLOv7, and YOLOv8 models all choose the largest network structure. At the same time, the model selected by Faster R-CNN is the Faster RCNN-R101-FPN network model, and the network model selected by DETR is the DETR-DC5-R101 network model, both of which are also the largest in their respective network structures. The epoch, batch size, and image size of all models in the comparative experiment were set to 200, 8, and 960 × 960 respectively, and the results obtained from the experiments can be observed in Table 2.

As can be seen from Table 2, the FLOPs of our model are 232.4G, which is similar to several other models except that it is much higher than baseline YOLOv5l, YOLOv6, Faster RCNN, and DETR. In terms of performance, our method reached 56.3 mAP on the VirDrone2021 dataset, which is the same as YOLOv7, but our method's mAP 0.5:0.95 reached 35.2, which is 1.9 higher than the mAP 0.5:0.95 of YOLOv7, which is the highest in the model, and 1.3 higher than the second highest YOLOv8 result. From the aforementioned findings, it is evident that our proposed method has exhibited enhancements over YOLOv5l. Although the amount of calculation has increased, its calculation amount is not much different from the calculation amount of the largest network structure of several current YOLO series models. While the enhanced model presented in this research has experienced a significant increase in computational requirements, its real-time performance may not be optimal. However, compared to the currently prominent YOLO series models, the improved model demonstrates superior target detection capabilities at an equivalent computational level in UAV shooting scenarios. Additionally, our proposed model has great practical significance for improving object detection accuracy in UAV scenarios.

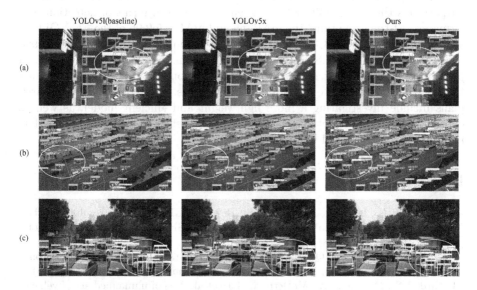

Fig. 6. Comparison chart of test results. (a) (b) (c) in the figure respectively represent the three problems encountered in target detection in the drone-captured scene, that is, the complex environment, dense distribution of targets, and large changes in target scale.

To demonstrate the detection capability of FE-YOLOv5 in real scenarios, we use Baseline (YOLOv5l), YOLOv5x, and FE-YOLOv5 to test the representative images in the VisDrone2021 dataset respectively, and the detection results are shown in Fig. 6.

In Fig. 6, the three groups of pictures (a), (b), and (c) respectively represent three problems encountered in target detection in the drone-captured scene. Figure 6(a) shows the complex background brought by the dark environment at night and the surrounding buildings, and Fig. 6(b) and Fig. 6(c) respectively show the high-density distribution of cars on the road and the target object changes drastically in scale. It can be seen from the highlighted results in the red circle in Fig. 6 that the FE-YOLOv5 method can not only detect small targets that are not detected by the two methods of YOLOv5l and YOLOv5x but also improve the detection accuracy.

5   Conclusion

In this paper, we present the FE-YOLOv5 algorithm as a promising approach for object detection in UAV shooting situations. Our main goal is to improve the accuracy of detecting complex backgrounds with high-density, multi-scale objects. Compared to the existing YOLO models, our proposed model achieves superior performance in terms of detection accuracy.

The experimental results demonstrate that our proposed enhanced model outperforms the current YOLO series' largest model in terms of detection performance. At the same time, the mAP 0.5 and mAP 0.5:0.95 of the FE-YOLOv5 algorithm can reach 56.3% and 35.2% under the input resolution of 960 × 960 respectively, which are 7.8% and 5.7% higher than the baseline. As the model performance improves, it increases the computational load, thereby decreasing the real-time performance of the detection process. Hence, our future research plans include investigating a more lightweight and efficient model, as well as exploring advanced detection models to achieve better target detection performance in scenes captured by drones.

References

1. Audebert, N., Le Saux, B., Lefèvre, S.: Beyond RGB: Very high resolution urban remote sensing with multimodal deep networks. ISPRS J. Photogramm. Remote. Sens. **140**, 20–32 (2018)
2. Gu, J., Su, T., Wang, Q., et al.: Multiple moving targets surveillance based on a cooperative network for multi-UAV. IEEE Commun. Mag. **56**(4), 82–89 (2018)
3. Hird, J.N., Montaghi, A., McDermid, G.J., et al.: Use of unmanned aerial vehicles for monitoring the recovery of forest vegetation on petroleum well sites. Remote Sensing **9**(5), 413 (2017)
4. Girshick R, Donahue J, Darrell T, et al. Rich feature hierarchies for accurate object detection and semantic segmentation. In: Proceedings of the IEEE Conference on Computer Vision and Pattern Recognition, pp. 580–587 (2014)
5. Girshick R. Fast r-cnn[C]//Proceedings of the IEEE international conference on computer vision. 2015: 1440–1448
6. Ren S, He K, Girshick R, et al. Faster r-cnn: towards real-time object detection with region proposal networks. Advances in neural information processing systems, 28 (2015)
7. Liu, W., Anguelov, D., Erhan, D., Szegedy, C., Reed, S., Fu, C.-Y., Berg, A.C.: SSD: single shot MultiBox detector. In: Leibe, B., Matas, J., Sebe, N., Welling, M. (eds.) ECCV 2016. LNCS, vol. 9905, pp. 21–37. Springer, Cham (2016). https://doi.org/10.1007/978-3-319-46448-0_2
8. Redmon, J., Divvala, S., Girshick, R., et al.: You only look once: unified, real-time object detection. In: Proceedings of the IEEE Conference on Computer Vision and Pattern Recognition, pp. 779–788 (2016)
9. Lin, T.-Y., et al.: Microsoft COCO: common objects in context. In: Fleet, D., Pajdla, T., Schiele, B., Tuytelaars, T. (eds.) ECCV 2014. LNCS, vol. 8693, pp. 740–755. Springer, Cham (2014). https://doi.org/10.1007/978-3-319-10602-1_48
10. Everingham, M., Van Gool, L., Williams, C.K.I., et al.: The pascal visual object classes (voc) challenge. Int. J. Comput. Vision **88**, 303–338 (2010)
11. Jocher, G., Stoken, A., Borovec, J., et al.: ultralytics/yolov5: v5. 0-YOLOv5-P6 1280 models, AWS, Supervise. ly and YouTube integrations. Zenodo (2021)
12. Bochkovskiy, A., Wang, C.Y., Liao, H.Y.M.: Yolov4: optimal speed and accuracy of object detection. arXiv preprint arXiv:2004.10934 (2020)

13. Wang, C.Y., Liao, H.Y.M., Wu, Y.H., et al.: CSPNet: a new backbone that can enhance learning capability of CNN. In: Proceedings of the IEEE/CVF Conference on Computer Vision and Pattern Recognition Workshops, pp. 390–391 (2020)
14. Tan, M., Pang, R., Le, Q.V.: Efficientdet: scalable and efficient object detection. In: Proceedings of the IEEE/CVF Conference on Computer Vision and Pattern Recognition, pp. 10781–10790 (2020)
15. Liu, S., Qi, L., Qin, H., et al.: Path aggregation network for instance segmentation. In: Proceedings of the IEEE Conference on Computer Vision and Pattern Recognition, pp. 8759–8768 (2018)
16. Wang, J., Xu, C., Yang, W., et al.: A normalized Gaussian Wasserstein distance for tiny object detection. arXiv preprint arXiv:2110.13389 (2021)
17. Que, J.F., Peng, H.F., Xiong, J.Y.: Low altitude, slow speed and small size object detection improvement in noise conditions based on mixed training. J. Phys. Conf. Ser. IOP Publishing **1169**(1), 012029 (2019)
18. Zhang, Z., Lu, X., Cao, G., et al.: ViT-YOLO: transformer-based YOLO for object detection. In: Proceedings of the IEEE/CVF International Conference on Computer Vision, pp. 2799–2808 (2021)
19. Solovyev, R., Wang, W., Gabruseva, T.: Weighted boxes fusion: ensembling boxes from different object detection models. Image Vis. Comput. **107**, 104117 (2021)
20. Woo, S., Park, J., Lee, J.Y., et al.: Cbam: convolutional block attention module. In: Proceedings of the European Conference on Computer Vision (ECCV), pp. 3–19 (2018)
21. Lin, T.Y., Dollár, P., Girshick, R., et al.: Feature pyramid networks for object detection. In: Proceedings of the IEEE Conference on Computer Vision and Pattern Recognition, pp. 2117–2125 (2017)
22. Luo, Y., Cao, X., Zhang, J., et al.: CE-FPN: enhancing channel information for object detection. Multimed. Tools Appl. **81**(21), 30685–30704 (2022)
23. Shi, W., Caballero, J., Huszár, F., et al.: Real-time single image and video super-resolution using an efficient sub-pixel convolutional neural network. In: Proceedings of the IEEE Conference on Computer Vision and Pattern Recognition, pp. 1874–1883 (2016)
24. Zhu, P., Wen, L., Bian, X., et al.: Vision meets drones: a challenge. arXiv preprint arXiv:1804.07437 (2018)
25. He, K., Zhang, X., Ren, S., et al.: Spatial pyramid pooling in deep convolutional networks for visual recognition. IEEE Trans. Pattern Anal. Mach. Intell. **37**(9), 1904–1916 (2015)
26. Hu, J., Shen, L., Sun, G.: Squeeze-and-excitation networks. In: Proceedings of the IEEE Conference on Computer Vision and Pattern Recognition, pp. 7132–7141 (2018)
27. Yu, F., Koltun, V.: Multi-scale context aggregation by dilated convolutions. arXiv preprint arXiv:1511.07122 (2015)
28. Rezatofighi, H., Tsoi, N., Gwak, J.Y., et al.: Generalized intersection over union: a metric and a loss for bounding box regression. In: Proceedings of the IEEE/CVF Conference on Computer Vision and Pattern Recognition, pp. 658–666 (2019)
29. Li, C., Li, L., Jiang, H., et al.: YOLOv6: a single-stage object detection framework for industrial applications. arXiv preprint arXiv:2209.02976 (2022)

30. Wang, C.Y., Bochkovskiy, A., Liao, H.Y.M.: YOLOv7: trainable bag-of-freebies sets new state-of-the-art for real-time object detectors. In: Proceedings of the IEEE/CVF Conference on Computer Vision and Pattern Recognition, pp. 7464–7475 (2023)

31. Zhu, X., Su, W., Lu, L., et al.: Deformable detr: deformable transformers for end-to-end object detection. arXiv preprint arXiv:2010.04159 (2020)

An Improved NSGA-II for UAV Path Planning

Wei Hang Tan[2], Weng Kin Lai[1(✉)], Pak Hen Chen[2], Lee Choo Tay[1],
and Sheng Siang Lee[3]

[1] Faculty of Engineering and Technology, Centre for Multimodal Signal Processing, Tunku Abdul Rahman University of Management and Technology, Kuala Lumpur, Malaysia
laiwk@tarc.edu.my

[2] Department of Electrical and Electronics Engineering, Faculty of Engineering and Technology, Tunku Abdul Rahman University of Management and Technology, Kuala Lumpur, Malaysia

[3] Aonic (HQ), Taman Perindustrian UEP, Subang Jaya, Selangor, Malaysia

Abstract. Palm oil is an edible vegetable oil that can be used in a wide range of products across different industries ranging from food and beverages, personal care and cosmetics, animal feed, industrial products, to biofuel. The palm oil industry contributes slightly less than 4% of Malaysia's overall GDP, making it the country's second-largest producer and exporter of palm oil worldwide. In Malaysia, it has been estimated that there are around 500,000 plantation workers in palm oil industries. In addition to getting a sufficient and steady supply of such usually low skilled workers, there are also issues related to the limits of the human body in performing tough physical work. As a result, UAVs may be utilized to support some of the processes in the palm oil businesses. However, the power of the batteries used in these UAVs is finite before they need to be recharged. Hence, the flight path for the UAV should be optimally computed for it to be able to cover the area it is assigned. In this paper, an improved *Non-Dominated Sorting Genetic Algorithm II* (NSGA-II) was developed to compute the optimal flight path of UAVs which also includes the turning angle and elevation. Enhancements to the algorithm is done by improving the selection, crossover, and mutation operations of the genetic algorithm which helps to improve the convergence and diversity of the algorithm beside avoiding getting trapped in local optimal solutions. In the majority of the tests, the improved *NSGA-II* was able to generate paths that are better than those identified by the human expert. Moreover, the proposed improved *NSGA-II* algorithm was able to compute good paths in less than the threshold of 10 min.

Keywords: Path planning · UAVs · metaheuristics · genetic algorithm · smart agriculture

1 Introduction

1.1 Research Background

Malaysia is the second largest producer and exporter of palm oil in the world [1]. In 2020, Malaysia's palm oil industry's export revenue contributed approximately 4% of Malaysia's total GDP [1]. To produce high-quality palm oil fruit many aspects of its

B. Luo et al. (Eds.): ICONIP 2023, LNCS 14448, pp. 305–316, 2024.
https://doi.org/10.1007/978-981-99-8082-6_24

value chain from planting to harvesting, and milling has to be carefully managed. It is a labourious and taxing process especially many aspects are being done manually, without the help of machines and automation. As a result, it is not uncommon for palm oil plantations to have a large labour force. In Malaysia, it has been estimated that there are around 500,000 plantation workers, with the majority of workers are low-skilled foreign workers [2, 3]. During the COVID-19 pandemic, travel restrictions prevented the entry of migrant labour which led to a severe shortage of plantation workers and consequently, huge financial losses to the palm oil sector [4]. Furthermore, humans cannot maintain consistent productivity due to the physical limitations of the human body as well as judgment calls. Humans are known to be inconsistent when it comes to quality assessment in some aspects of the palm oil production value chain [2]. Moreover, the new workers need to be adequately trained and this can be both cost and time inefficient. Lastly, the salary of the workers also needs to be adjusted upwards periodically to help motivate towards better productivity [2]. Therefore, UAVs may be used to assist the palm oil industry to manage some aspects of the growth of the palm oil tree in the plantations. This can involve aerial dispensing of insecticides, fertilizer, detecting unhealthy trees, etc. [5, 6]. Unfortunately, UAVs have a finite battery life as they cannot fly indefinitely to cover large tracts of land before stopping to recharge [7]. Therefore, proper path planning is required for better utilization of these UAVs when used to provide various services to the palm oil industry. When used to service the palm oil industry, the flight path planner will need to consider the overall distance flown, the turning angle, the maximum battery life, and its flight elevation. Hence a good flight path must include all these aspects to determine its quality. In this paper, the battery used for the UAVs can only sustain a flight time that cannot exceed 10 min.

2 Review of Prior Work

There has been some prior related work involving the use of swarm intelligence for UAV path planning. One of the earlier work by XiaoWei Jiang, Qiang Zhou, and Ying Ye investigated how using UAV systems may lower a logistics company's operating costs while increasing the transport efficiency [8]. In this study, the authors solved the vehicle routing problem with time windows (VRPTW) task model assignment with several constraints, including weight coefficients, time-window constraints, UAV constraints, and others by using an improved PSO method that can handle complicated combinatorial optimization problems. The performance of both the PSO and GA algorithms in handling the UAV task assignment was also compared.

On the other hand, Z. Wang, G. Liu, and A. Li [9, 10] used a hybrid algorithm which is a combination of simulated annealing and particle swarm optimization (SA-PSO) to solve three-dimensional path planning of UAVs. They deployed Simulated Annealing (SA) to cancel out the limitations of the PSO which frequently converges to local optimal solutions. The proposed new approach was able to offer shorter paths than the traditional PSO algorithm. Furthermore, Agnihotri and Gupta [11] used a hybrid algorithm by combining Particle Swarm Optimization with Genetic Algorithm to solve the routing of the wireless sensor network (WSN). Their PSO-GA method combines the benefits of each approach, resulting in a fast convergence rate and a high level of difficulty in falling in local optima. As a result, over generations, the PSO-GA algorithm can gradually increase the number of good individuals.

Huang et al. also studied this same hybrid algorithm to replace the traditional PSO algorithm with improvements in solving the UAV multi-target path planning [10]. Their approach has the ability to balance the global search with a local search of the particles to produce results that are superior in terms of accuracy and convergence speed.

Zhao et al. [12] investigated the use of *Ant Colony Optimization* (ACO) to enhance the effectiveness of the prevalent cost minimization model in the cold chain logistics distribution process. In determining the best path for the vehicles, Zhao created an improved version of the ACO with multiple objective functions. They were able to achieve lower delivery costs and carbon emissions as well as improved customer satisfaction. Their approach demonstrated that ACO can be used to address such difficult problems, but adjustments have to be made as the basic ACO can become stuck in the local optimal solutions.

NSGA-II is a modified version of NSGA [13] that was first introduced by Deb et al. [14] for solving multi-objective constrained. The difference between these two versions is that NSGA-II is faster, has better sorting algorithm, includes elitism which preserves an already founded Pareto optimal from deleting, uses explicit diversity preserving mechanism. The complexity of NSGA-II is at most $O(MN^2)$, while the complexity of NSGA is $O(MN^3)$ - where M is the number of objectives and N is the population size. NSGA-II incorporates standard GA (select, crossover, and mutation) with non-dominated sorting and new fitness value "*Crowding Distance*" which is assigned in order to measure the density of solutions surrounding a particular solution.

Shuai et al. [15] improved the NSGA-II algorithm by improving the local and global search capability of GA by designing the crossover and mutation operators to solve the multi-traveling salesman problem so that the salesman could visit every city while covering the fewest possible distance and maintaining the fewest possible range between salesmen. They used 6 different computational test conditions to analyze the effectiveness of the proposed algorithm and compared the results with 4 other algorithms. They showed that their proposed NSGA-II can increase the effectiveness in deriving the dominant front by providing the results with a good amount of variations.

Panda [16] investigated the performance of GA, PSO, and SA in solving the Traveling Salesman Problem (TSP) with different number of nodes. PSO showed the best performance in finding the shortest distance between nodes compared with either GA or SA but it takes a longer execution time. Another similar piece of research by Lathee et al. [17] compared the performance of GA and ACO in solving the TSP and their results showed that GA requires lesser execution time to find the shortest path compared to ACO but with a weaker performance.

UAVs when deployed in the palm oil estates to either spray pesticides, fungicides, fertilizers, or just to identify the diseased trees can help improve productivity but an optimal flight path is important for efficient use of their battery power. From the review of prior work, NSGA-II has been shown to be efficient and effective in finding the optimal paths. Hence it will be used to compute the optimal path of the palm oil estates in this study. Nevertheless, as there are now more constraints to be fulfilled, some improvements to the basic NGSA-II is necessary.

3 Methodology

3.1 NSGA-II Algorithm

The NSGA-II algorithm is a population-based meta-heuristic algorithm that grows throughout the solution space to determine a set of non-dominated solutions. It is a well-known acceptable way to achieve suitable solutions to difficult multi-objective problems. The NSGA-II algorithm's main objective is to locate a collection of solutions organized along fronts utilizing the Pareto dominance principle [18]. The Pareto dominance concept is to compare multiple solutions candidates and find the best solution that can dominate others and classify the solutions candidates by separating solutions candidates with different ranks (where 1 is the highest, 2 the next highest etc. until the lowest). For two solutions A and B, if solution A is to dominate solution B, then the following rules must be satisfied:

1. All objectives of solution A perform better compared to solution B
2. One of the objectives of solution A is to perform better than solution B and the other objectives of solution A have the same performance as solution B.

For a non-dominated set, when the objectives in solution A is weaker than solution B but the other is better, then this is a *non-dominated* set between A and B [18]. The best dominant solutions are used to create the next population once all the solutions have been categorized according to the above criteria and this ensures the convergence of the NSGA-II algorithm. The NSGA-II algorithm also has *a crowding distance sorting* process after the *non-dominated sorting* process which is used to estimate the density of the candidate solutions for a particular solution by taking into account the mean distances between the two neighbouring solutions. The solution that has a larger mean distance between the two neighbouring solutions will be selected [18].

General Framework

The flowchart of NSGA-II with an improved genetic algorithm is shown in Fig. 1(a).

In NSGA-II, the process of generating the new population, P^{t+1} is shown in Fig. 1(b). There are two parts of the process which are *non-dominated sorting and crowding distance sorting* in the selection operator. First, the population, P^t is randomly generated. The generated population P^t is used by the *Genetic Algorithm* to produce the improved population Q^t. The current population, P_c then becomes $P^t + Q^t$ [14]. To produce the boundary set R = r1, r2, r3,..., the current population, P_c, is put through the *non-dominated sorting* process (categories the current population, P_c into several ranks, the highest rank, i.e. rank 1 is the non-dominated population from the current population). The boundary sets are added to the new population, P^{t+1} based on the ranks of the boundary sets which starts from the highest rank (1) to the next highest (2) and so on until the lowest rank. If the size of the boundary set, r is larger than the difference $P^t - P^{t+1}$, then the boundary set, r is passed through the *crowding distance sorting* to calculate the mean distance between the two neighbouring solutions on the same non-dominated set. The higher value will be selected and added to P^{t+1} [15]. The solutions in P^{t+1} will be used for comparison in the following iterations to improve the quality of the solutions. The solutions generated will be updated when better solutions are found. After finishing all iterations, the best solution found will be the final solution.

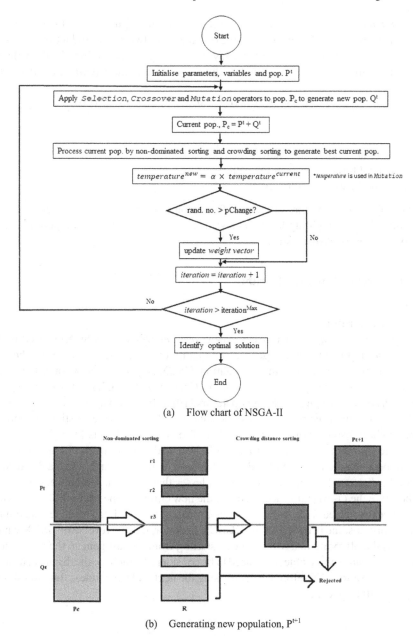

(a)   Flow chart of NSGA-II

(b)   Generating new population, P^{t+1}

Fig. 1. NSGA-II process of generating new populations

For NGSA-II, the temperature will be decreased for each iteration and its final value is then used in the mutation operator. Furthermore, a randomly generated number will be checked against *pChange* to decide if the weight vectors will be changed. At the end of the iterations, the *Pareto-dominated* set which is the final solution is determined. The

Pareto-dominated set will be checked to see if it has any segments that intersect in the path by using the method that identifies the orientation of two cross-product vectors. This leaves only the path that does not have any intersecting segments. Finally, the voltage, calculated from Eq. (1) is used to find the lowest voltage in the *Pareto-dominated* set.

$$voltage = 0.0000007f(D) + 0.00056f(\theta) + 0.021f(E) \tag{1}$$

where

$f(D)$ is the total length of the path,
$f(\theta)$ is the total turning angle of the path,
$f(E)$ is the total elevation of the path,

To avoid any one feature of the path from monopolising the solutions, each feature is multiplied by a weighting factor. The values used in each of these weighting factors was found after extensive testing.

3.2 Improvements to the Algorithm

Crossover Operator
A *Comprehensive Sequential Constructive Crossover Operator* (CSCX) is used in the algorithm to improve the coverage rate and quality of the algorithm in finding the best solution for multi-objective path optimization. CSCX is a crossover operator that combines *Reverse Greedy Sequential Constructive Crossover* (RGSCX) and *Greedy Sequential Constructive Crossover* (GSCX) operators [19] to form two new offsprings. For the GSCX operator, the least next node will be chosen and added to the existing incomplete offspring chromosome. For example, if P1 is {1, 6, 3,.....} and P2 is {1, 5, 7, 6, 8, ...}, the first gene and the next gene of P1 would be 1 and 6 respectively. Furthermore the distance between them is 9. As for P2, the first two genes are 1 and 5 with a distance of 13. As the distance between 1 to 6 of P1 is lesser than the distance between 1 to 5 of P2, therefore the (1–6) pair is accepted and the new chromosome becomes {1, 6}. After that, the next selected node in P1 becomes 3 while it is 8 for P2. This is repeated until the new chromosome is built. For the RGSCX operator, it is the reverse of GSCX which constructs the offspring in a reverse order, starting from the last gene to the first gene.

In this research, three objectives need to be considered which are the distance, turning angle, and elevation. Therefore when choosing the next node, the fitness function shown in Eq. (2) will be used.

$$F = w_d d + w_{\Delta z}\Delta z + w_\theta \theta \tag{2}$$

The value of w_d, $w_{\Delta z}$ and w_θ will change only if the randomly generated value is greater than $pChange$ at the end of each iteration to improve the quality of the solutions.

Equations (3), (4) and (5) describe the change of the weight of distance, turning angle, and elevation for each iteration. For w_d (weight of distance), the value is decreased while for turning angle and elevation, values are increased if the randomly generated value is larger than $pChange$ at the end of each iteration. With this, the algorithm is able to find the path without any segment intersections while reducing the turning angle and elevation of the path at the same time.

$$w_{d+1} = w_d \times \beta \qquad (3)$$

$$w_{\Delta z+1} = \frac{w_{\Delta z}}{mega} \qquad (4)$$

$$w_{\theta+1} = \frac{w_\theta}{mega} \qquad (5)$$

where β is the decrement factor of the weight of distance and $mega$ is the increment factor used in both the turning angle and elevation.

Mutation Operator

Three mutation methods namely SWAP, REVERSE and INSERTION were investigated. Each method has its own probability of being selected but together they must be equal to 1, i.e. *Swap + Insertion + Reverse = 1*. The selection of the mutation method is based on the *Roulette Wheel Selection* which randomizes the selection to improve the diversity. For the SWAP mutation, it will randomly select two nodes and swap these the locations of these two nodes to form a new chromosome. In the REVERSE mutation, two nodes are randomly selected, and then their arrangement reversed in the new chromosome. The last method, which is INSERTION, two nodes are randomly selected. The first node is then inserted next to the second node. Here the mutation is similar to simulated annealing which has adaptive acceptance that depends on the temperature. There are two ways for the new chromosome to replace the current chromosome. It can happen if the fitness of the new chromosome is better than the current chromosome, or if the P value calculated from Eq. (7) is larger or equal to the randomly generated value. This can improve the convergence rate while improving the diversity of the technique in finding good solutions.

$$\delta = \frac{fitness_{new} - fitness_{current}}{fitness_{current}} \qquad (6)$$

$$P = e^{(\delta/temperature)}. \qquad (7)$$

$$temperature^{new} = \alpha \times temperature^{current} \qquad (8)$$

In order to find the value for P, Eq. (6) is used to find δ that is used in Eq. (7). The temperature is decreased after each iteration as per Eq. (8) where α is the temperature decrement factor.

Selection Operator

The tournament selection method is used to select the better individuals to produce the better next generation (in this research the equation to calculate the fitness is the same as Eq. (2) as three objectives are considered). First, the tournament size, N is selected. $N = 2$ is used here which means that two individuals will be randomly chosen from the population to compare every time and the individual with the better fitness will be chosen. In the event both have the same fitness, then the individual selected is chosen randomly instead [20].

4 Results

The new NGSA-II was used to test 202 maps that came from three large areas as shown in Fig. 2. Notice that M2 has the smallest number of sub sections of 50 while M1 has the largest with 90. M3 has a total of 62 sub sections making a total of 202 individual maps from all these three estates. The paths from each of these sub sections have been manually identified by the trained operators and the results will be compared with those computed by NGSA-II.

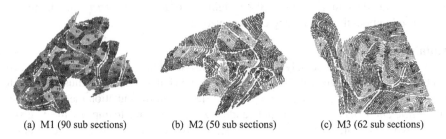

(a) M1 (90 sub sections)   (b) M2 (50 sub sections)   (c) M3 (62 sub sections)

Fig. 2. Three estates of M1, M2 and M3

An important criterion for the paths is that they need to be *"smooth"* which basically means that the paths should not have any intersections. Figure 3 illustrate paths that is either **not** *smoot*h (Fig. 3 (a)) or does not have any intersections – *smooth* (Fig. 3. (b)).

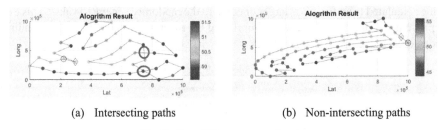

(a)   Intersecting paths   (b)   Non-intersecting paths

Fig. 3. Two different types of paths

After executing the 202 maps by using the proposed algorithm, 201 out of the 202 or 99.5% smooth paths were obtained with the proposed algorithm of which 99.0% or 199 of these 201 smooth paths have a smaller power usage compared to the manually generated ones. Furthermore, the execution time of the proposed algorithm is lower than 10 min to find the optimal solutions for individual maps. This was done on an Intel® Core™ i5-10500 CPU @3.10 GHz. Table 1 shows the execution times for a sample of the maps (F1–F10) from M2.

Table 2 shows that for most of the factors that cause the voltage to decrease are those paths that have smaller length, angle, and elevation compared to those generated manually. Furthermore, it also showed that the improvements for multiple objectives are larger compared to improvements for only single objectives. Lastly, this also illustrates

Table 1. Execution time of the proposed algorithm for M2

	execution time			execution time
F1	8 mins 59.2 secs		F6	8 mins 53.3 secs
F2	9 mins 12.0 secs		F7	8 mins 54 secs
F3	8 mins 6.6 secs		F8	9 mins 7.0 secs
F4	8 mins 53.2 secs		F9	8 mins 45.1 secs
F5	8 mins 31.9 secs		F10	9 mins 3.1 secs

that the proposed algorithm can find smooth paths with smaller voltages compared to those generated by the human experts.

Table 2. The factors that reduce the voltage of the path

Factors	No. of Maps	Estimate Range of Voltage Improvement
Smaller length, angle, and elevation	120	0.7040
Smaller length and angle	26	0.3626
Smaller length and elevation	8	0.2677
Smaller elevation and angle	34	0.3325
Smaller length	3	0.1065
Smaller angle	8	0.1944
Smaller elevation	0	-

The results from testing three randomly chosen samples are summarized in Tables 3 and 4. They show that the path's voltage generated by the proposed algorithm is smaller than those paths that were generated manually. This shows that the proposed algorithm was able to generate paths with smaller voltages.

Table 3. Results from three randomly chosen samples

	NSGA-II				manually generated			
	voltage	length	angle	elevation	voltage	length	angle	elevation
M1 F16	6.3614	6057499	2224	41.71	6.4535	5944496	2667	38.04
M1 F17	6.8115	5827678	2716	57.66	7.2437	5522551	3536	66.56
M2 F46	7.1851	6798781	2734	42.62	7.8188	6651666	3584	55.04

Furthermore, the limitation of the proposed algorithm was investigated too. The additional tests with a range of different number of nodes were chosen for each of the 9 ranges shown in Table 5. As the number of nodes from each data set is lesser than 90, the maps that have larger number of nodes would need to be created by combining

Table 4. Results from three randomly chosen samplesPercentage changes from three examples

	voltage change (%)	length change (%)	angle change (%)	elevation change (%)
M1 F16	-1.43	1.90	-16.62	9.65
M1 F17	-5.98	5.53	-23.18	-13.36
M2 F46	-8.10	2.21	-23.72	-22.57

an additional appropriate number of nodes from neighbouring areas in M1 (Fig. 2 (a)). For each of the remaining 8 ranges with 91 to 160 nodes, the new maps were generated by combining data from F1 with an appropriate number of additional nodes from its neighbours in M1. Each data set was tested 10 times to determine the consistency of the proposed algorithm in solving these problems.

Table 5. Results for different number of nodes

Number of nodes	Number of non-intersecting paths from 10 runs (minimum number)
40 – 80	10
81 – 90	8
91 – 100	4
101 – 110	2
111 – 120	2
121 – 130	2
131 – 140	1
141 – 150	2
151 – 160	0

Table 5 shows that as the node number increases to the upper range of (151–160), the proposed algorithm is unable to find the path without intersecting segments for 10 runs. Moreover, it can clearly be seen that the proposed algorithm would not have more than 20% chance of getting a smooth path when the number of nodes increases beyond 103 to 143. However, for 93 nodes our proposed NGSA-II algorithm has a better chance at 40% to generate a smooth path. The results also showed that this becomes significantly better for areas with smaller number of nodes to traverse. In conclusion, the performance of the proposed algorithm tapers off for the same set of hyperparameters when the number of nodes increases. Improvements by tuning these hyperparameters can help ensure the algorithm is able to produce paths that satisfy all its multiple objectives for large number of nodes. Nevertheless, the maximum number of nodes for the 202 cases tested here is only 85. This reinforced the fact that the proposed NGSA-II is able to generate paths with better fitness which is a combination of the total path distance, turning angle and voltage, as its performance will only start to deteriorate significantly when the number of nodes is more than 90.

5   Conclusion

This paper proposed a new NGSA-II to generate the flight paths of UAVs which were used to undertake a range of services in large palm oil plantations that must fulfill multiple objectives. In addition to the standard total distance travelled, the paths must have a minimum number of sharp angular turns as well as changes in the elevation which is measured in terms of the power consumption – voltage. Results from extensive testing of over 200 maps showed that it is able to find the good flight paths with a lower voltage compared to the manually generated paths at 98.5% of the test set. Furthermore, the time required for the proposed algorithm in generating these paths is within the 10 min threshold set by the end users. However, the proposed NGSA-II has some limitations as it did not perform very well in finding the best solution when the number of nodes in the problem exceeds 90.

With such metaheuristics, the quality of the solutions is sensitive to the initial solution generated. Hence intelligent initialization of the solution can both lead to better and faster solutions. Secondly, optimum values for the population size, mutation rate, and crossover rate can affect the local and global search of the algorithm in solving the problem as it is well known that these can have a significant effect on the genetic algorithm in which the NSGA-II is based. Furthermore, the value of the weight vectors and probability of the weight vectors changes, initial temperature, and decrement and increment factors can also affect the quality of the solutions. Prior work has also shown that including local search to help it escape local optima, can also make a big difference in the quality of the results too.

Acknowledgements. WH Tan, WK Lai, PH Chen and LC Tay are grateful to TAR UMT for both financial and material support in the work reported here.

References

1. Palm Oil - What is it and why is it important? - Green Campus Initiative - University of Maine. https://umaine.edu/gci/2021/10/25/palm-oil-what-is-it-and-why-is-it-important/. Accessed 28 Apr 2023
2. Migrant workers in oil palm plantations deserve better treatment—Tenaganita | Malay Mail. https://www.malaymail.com/news/what-you-think/2018/09/25/migrant-workers-in-oil-palm-plantations-deserve-better-treatment-tenaganita/1676208. Accessed 28 Apr 2023
3. Protecting Local Labour Rights in the Palm Oil Sector - Roundtable on Sustainable Palm Oil (RSPO). https://rspo.org/protecting-local-labour-rights-in-the-palm-oil-sector/. Accessed 28 Apr 2023
4. Palm oil industry in Malaysia - statistics & facts | Statista. https://www.statista.com/topics/5814/palm-oil-industry-in-malaysia/#topicHeader__wrapper. Accessed 28 Apr 2023
5. How Are Drones Used in Oil Palm Plantations. https://www.droneacademy-asia.com/post/how-are-drones-used-in-oil-palm-plantations. Accessed 28 Apr 2023
6. The Pros And Cons Of Drones In Agriculture - DroneGuru. https://www.droneguru.net/the-pros-and-cons-of-drones-in-agriculture/. Accessed 28 Apr 2023
7. 10 Limitations Of Drones | Grind Drone. https://grinddrone.com/features/10-limitations-of-drones. Accessed 28 Apr 2023

8. Jiang, X., Zhou, Q., Ye, Y.: Method of task assignment for UAV based on particle swarm optimization in logistics (2017). https://doi.org/10.1145/3059336.3059337

9. Wang, Z., Liu, G., Li, A.: Three-dimensional path planning of UVAs based on simulated annealing and particle swarm optimization hybrid algorithm. In: Proceedings of 2021 IEEE 3rd International Conference on Civil Aviation Safety and Information Technology, ICCASIT 2021, pp. 522–525 (2021). https://doi.org/10.1109/ICCASIT53235.2021.9633423

10. Huang, Q., Sheng, Z., Fang, Y., Li, J.: A simulated annealing-particle swarm optimization algorithm for UAV multi-target path planning. In: 2022 2nd International Conference on Consumer Electronics and Computer Engineering, ICCECE 2022, pp. 906–910 (2022). https://doi.org/10.1109/ICCECE54139.2022.9712678

11. Agnihotri, A., Gupta, I.K.: A hybrid PSO-GA algorithm for routing in wireless sensor network. In: 2018 4th International Conference on Recent Advances in Information Technology (RAIT), Dhanbad, India, pp. 1–6 (2018). https://doi.org/10.1109/RAIT.2018.8389082

12. Zhao, B., Gui, H., Li, H., Xue, J.: Cold chain logistics path optimization via improved multi-objective ant colony algorithm. IEEE Access **8**, 142977–142995 (2020). https://doi.org/10.1109/ACCESS.2020.3013951

13. Deb, K.: An efficient constraint handling method for genetic algorithm. Comput. Methods Appl. Mech. Eng. **186**, 311–338 (2000). https://doi.org/10.1016/S0045-7825(99)00389-8

14. Deb, K., Pratap, A., Agarwal, S., Meyarivan, T.: A fast and elitist multiobjective genetic algorithm: NSGA-II. IEEE Trans. Evol. Comput. **6**(2), 182–197 (2002). https://doi.org/10.1109/4235.996017

15. Shuai, Y., Yunfeng, S., Kai, Z.: An effective method for solving multiple travelling salesman problem based on NSGA-II. Syst. Sci. Control Eng. **7**(2), 121–129 (2019). https://doi.org/10.1080/21642583.2019.1674220

16. Panda, M.: Performance comparison of genetic algorithm, particle swarm optimization and simulated annealing applied to TSP. Int. J. Appl. Eng. Res. **13**(9), 6808–6816 (2018). http://www.ripublication.com. Accessed 07 Sept 2022

17. Lathee, M.G., Zhan, G.H., Salam, Z.A.B.A.: Thinakaranthinakaran, R.: Comparative analysis of ant colony optimization and genetic algorithm on solving symmetrical travelling salesman problem. J. Adv. Res. Dyn. Control Syst. **12**(7 Special Issue), 2629–2635 (2020). https://doi.org/10.5373/JARDCS/V12SP7/20202399

18. Bolaños, R.I., Echeverry, M.G., Escobar, J.W.: A multiobjective non-dominated sorting genetic algorithm (NSGA-II) for the multiple traveling salesman problem. Decis. Sci. Lett. **4**(4), 559–568 (2015). https://doi.org/10.5267/J.DSL.2015.5.003

19. Ata Al-Furhud, M., Hussain Ahmed, Z.: Genetic algorithms for the multiple travelling salesman problem. IJACSA: Int. J. Adv. Comput. Sci. Appl. **11**(7), 2020 (2022). www.ijacsa.thesai.org. Accessed 07 Sept 2022

20. Shi, L., Li, Z.: An improved pareto genetic algorithm for multi-objective TSP. In: 5th International Conference on Natural Computation, ICNC 2009, vol. 4, pp. 585–588 (2009). https://doi.org/10.1109/ICNC.2009.510

Reimagining China-US Relations Prediction: A Multi-modal, Knowledge-Driven Approach with KDSCINet

Rui Zhou[✉], Jialin Hao, Ying Zou, Yushi Zhu, Chi Zhang, and Fusheng Jin

Beijing Institute of Technology, Beijing, China
{zhourui,hjlin,1120203433,1120202344,1120200837,jfs21cn}@bit.edu.cn

Abstract. Statistical models and data driven models have achieved remarkable results in international relation forecasting. However, most of these models have several common drawbacks, including (i) rely on large amounts of expert knowledge, limiting the objectivity, applicability, usability, interpretability and sustainability of models, (ii) can only use structured unimodal data or cannot make full use of multimodal data. To address these two problems, we proposed a Knowledge-Driven neural network architecture that conducts Sample Convolution and Interaction, named KDSCINet, for China-US relation forecasting. Firstly, we filter events pertaining to China-US relations from the GDELT database. Then, we extract text descriptions and images from news articles and utilize the fine-tuned pre-trained model MKGformer to obtain embeddings. Finally we connect textual and image embeddings of the event with the structured event value in GDELT database through multihead attention mechanism to generate time series data, which is then feed into KDSCINet for China-US relation forecasting. Our approach enhances prediction accuracy by establishing a knowledge-driven temporal forecasting model that combines structured data, textual data and image data. Experiments demonstrate that KDSCINet can (i) outperform state-of-the-art methods on time series forecasting problem in the area of international relation forecasting, (ii) improving forecasting performance through the use of multimodal knowledge.

Keywords: Knowledge-driven · China-US relation · Multimodal data · Time-series forecasting

1 Introduction

International relations have a significant impact on the economy, politics, and security [13,16,26]. Factors such as power, interdependence, ideology, and conflicts play a crucial role in shaping international relations [8,9]. Forecasting

Supported by Beijing Institute of Technology.

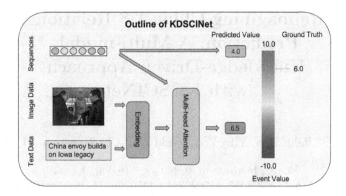

Fig. 1. The illustration of the multimodal knowledge supporting time series forecasting.

changes in international relations is therefore highly valuable. Existing forecasting methods, including statistical modeling and natural language processing with
text mining techniques [14,21,22], have shown promise in this area. However,
these methods often rely heavily on expert knowledge, which can be influenced
by data quality and structural characteristics. Moreover, they may suffer from
limited domain knowledge and model generalization capabilities, necessitating a
large amount of labeled data.

Motivated by the above, this paper introduces CU-GDELT, a comprehensive
multimodal knowledge graph that incorporates diverse and abundant images into
the textual entities of GDELT. Furthermore, we propose a novel neural network
architecture called KDSCINet (Knowledge-Driven neural network architecture
that conducts Sample Convolution and Interaction) to tackle the challenge of
international relation forecasting. We select the Sample Convolution and Interaction Network (SCINet) [17] as our framework of choice, as it outperforms other
deep neural network models for time series forecasting, including Transformer-
based models [24] and temporal convolutional networks [1].

To tackle the prediction challenge, we leverage structured data from GDELT[1]
and gather corresponding image and textual data. After preprocessing, we construct a multimodal Knowledge Graph named CU-GDELT. We individually vec-
torize the structured data, textual data and image data, and then concatenate
them. These embeddings are inputted into our KDSCINet, where an attention
mechanism aggregated the knowledge and a SCINet-based forecasting model is
employed.

The KDSCINet approach, depicted in Fig. 1, demonstrates how multimodal
knowledge can assist in forecasting the development of event values. When we
observe an image showing a meeting between leaders of China and the United
States, it implies positive progress in China-US relations. Similarly, when we
read news about "China envoy building on Iowa legacy" it reinforces the same

---

[1] https://www.gdeltproject.org/.

idea. Our model can capture the correlation between multimodal knowledge and structured data, leading to more accurate China-US relation forecasting.

To the best of our knowledge, the main contributions of this paper are as follows:

- KDSCINet is the first one to utilize the Sample Convolution and Interaction Network to forecast international relation, integrating structured data, textual data and image data. It is a hierarchical time series forecasting framework with a multi-head attention-based input layer, which effectively combines knowledge within the same time period, resulting in improved performance in forecasting international relations.
- We present CU-GDELT, a comprehensive multimodal knowledge graph that incorporates various data modalities. The integration of structured data, textual data and image data enhances the richness and depth of the knowledge graph, further contributing to the accuracy of our forecasting.

Experimental results show that KDSCINet outperforms other state-of-the-art models in forecasting the development of China-US relations. We attribute this improvement to three factors: (i) Multimodal Knowledge enrichs the contextual information, enhancing the model's understanding of temporal data and improving prediction accuracy. (ii) The multi-head attention mechanism effectively integrates knowledge from the multimodal graph by learning different weight distributions.

2 Related Work

In this section, we review the existing literature on forecasting international relations and discuss the limitations of both traditional statistical models and more recent data-driven models. We then introduce the concept of knowledge-driven models and highlight their potential to overcome the challenges posed by previous approaches.

2.1 Statistical Models

Extensive research has focused on forecasting international relations, initially relying on traditional statistical models and rule-based approaches using historical data and expert knowledge. However, these early efforts [19,20] failed to materialize due to the multitude of factors involved and the limitations of statistical modeling and computational statistics. The Logit model, commonly used for binary classification [14], has limitations in international relations forecasting. It helps identify significant factors and quantify their impacts on specific events, but it struggles with incomplete and inaccurate data, which can distort its forecasts. Additionally, the Logit model fails to capture the complexity and nonlinear nature of international relations. The Probit model, also used for international relations analysis and prediction [22], falls short in accounting for interdependencies implied by the theory, making interpretation of parameter estimation challenging.

2.2   Data-Driven Models

To address the limitations of statistical models, data-driven models utilizing natural language processing (NLP) and sentiment analysis have been introduced. In the study by [2], sentiment analysis of social media data was employed to establish potential connections between country relationships and the level of cooperation among them. In another study by [22], NLP and sentiment analysis were used to classify sentiment and extract keywords from news texts, enabling the forecasting of geopolitical risk changes. However, both have limitations due to data quality, coverage, and domain-specific context. [15] utilizes the AvgTone variable from the GDELT dataset as structured data and applies sliding window time series method to forecast international material cooperation and diplomatic cooperation. The results demonstrate the predictive capability of data-driven forecasting models in specific contexts. However, the method is limited by the small size of the dataset and reliance on a single variable for prediction. Therefore, further research is needed to expand the dataset size and incorporate multimodal data to enhance the accuracy and reliability of the model.

2.3   Knowledge-Driven Models

Efforts in the field of international relations forecasting have been limited by the subjectivity of approaches relying solely on news sources or the objectivity of focusing solely on evaluation indicators. Incorporating knowledge has been demonstrated to be crucial for capturing the inconsistent evolution of streaming data and improving accuracy in forecasting [3]. Integrating knowledge graphs (KG) into the learning process of event embeddings enables the encoding of valuable background knowledge [6]. Utilizing Multimodal Knowledge Graphs (MKG) and Multimodal Graph Convolution Networks (MGCN) has proven effective in establishing reliable image-text connections and achieving state-of-the-art performance in image-text retrieval [7]. Enhancements through knowledge integration, such as the combination of RNN-based networks with Key-Value Memory Networks (KV-MN), have been proposed for sequential recommendation systems [12]. Deep knowledge-aware networks (DKN) that incorporate KG have shown promise in news recommendation [25]. In the domain of stock trend forecasting and explanation, the integration of Knowledge-Driven Temporal Convolutional Networks with structured tuples from financial news, leveraging background knowledge from KG, has been developed [5].

Traditional statistical models and current deep models have failed to achieve promising results in international relations forecasting due to limited expressiveness and lack of high-quality datasets. To overcome these challenges, we introduce KDSCINet, a novel framework, along with the CU-GDELT multimodal knowledgegraph.

3   The KDSCINet Framework

In this section, we first present the CU-GDELT multimodal knowledge graph in Sect. 3.2. Then we present the overall framework of KDSCINet, which is a

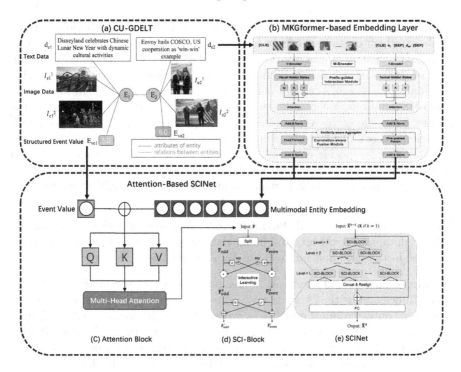

Fig. 2. The overall architecture of KDSCINet framework.

general framework that can be applied to widespread knowledge-driven temporal modeling and forecasting tasks and we specifically apply it to forecast China-US relations. To facilitate understanding, we introduce its detailed implementation, including the MKGformer-based embedding layer in Sect. 3.3 and the attention-based SCINet in Sect. 3.4.

3.1 Overview

The overview of KDSCINet architecture is shown in Fig. 2. After constructing a knowledge graph based on the GDELT database, we utilize MKGformer to extract multimodal embeddings from the text descriptions and images of entities, then concatenate them with structured event values, creating an entity representation that is rich in multimodal information. Then, we feed it into attention block to combine entities with the same time features. Finally, the combined vectors are input into a SCINet-based model for China-US relation forecasting.

3.2 CU-GDELT

To explicitly represent the umplicit relations between the image and the text, we build the CU-GDELT, shown in Fig. 2(a)

Data Collection. Our study utilizes data from the GDELT 2.0 global events database, which contains a comprehensive collection of worldwide news reports. Specifically, we utilize data from the year 2022 at 15-minute intervals. We obtain total of 34,953 time periods of event data, comprising a total of 36,720,483 news data, the size of the data compression package is 2.21GB. In forecasting China-US relations, we mainly leverage the "GoldsteinScale" field within the dataset. This field serves as an indicator of the potential impact of events on both countries. To complement structured data, we extract news texts and images corresponding to specific entities based on the url fields stored for each piece of data in GDELT. The collected data is then preprocessed to create the multimodal knowledge graph, CU-GDELT, which is utilized by our KDSCINet model.

Data Preprocessing and Formatting. We select and clean the original GDELT data, and crawl text and image data using different strategies to reduce the noise in the data and improve the data quality:

1. **Time-series Event Data** $\mathcal{X}$ When processing the raw GDELT data, we carefully examine the "Actor1Code" and "Actor2Code" fields in the data information, only retaining event data where these fields correspond to "CHN", "USA", "USA", and "CHN". After applying the filtering to the raw GDELT data, we obtain an event value dataset that consists of value records collected at 15-minute intervals from the GDELT dataset. The dataset covers a timespan from January 1, 2022, to December 31, 2022.
2. **Textual News Data** $\mathcal{T}$ The textual news datasets consist of historical news headlines extracted from GDELT. After completing the screening process, we acquire 150,968 pieces of textual news data.
3. **News Image Data** $\mathcal{I}$ When utilizing a crawler to fetch images, we apply specific criteria to filter out images with dimensions below 200×200 pixels and screen the image attributes to identify and exclude irrelevant images such as logos and news thumbnails. By implementing these filtering criteria, we obtain a total of 5,107 images for our news image data.

Construction of CU-GDELT. We link image data and textual data with other entities as entities in a triple structure, stored in the form of *<entity, relation, entity>* and *<entity, attribute, attribute value>*.

Schema. We use a bottom-up approach to build schema, first defining specific concepts and their attributes and relationships related to the analysis of China-US geopolitical events, then clustering concepts with highly overlapping attributes or relations. In CU-GDELT, five concepts of Event, Actor, EventAction, EventGeography, and MultimodalData are respectively established. After that, we define the entities, attributes and relationships in each concept of the ontology, and the relations include 15 different relations.

Data. The data layer of CU-GDELT is mainly used to store and manage a large amount of data obtained in the data collection and preprocessing stages. Except for the keyword attribute in the concept of MultimodalData, the attribute data of other concepts are mainly obtained through Data Collection combined with Data Preprocessing and Formatting. The corresponding fields in the structured data are obtained by entity extraction. For the keyword in the concept of MultimodalData, the obtained EventTitle attribute data is extracted through the TextRank algorithm, and the extraction result is used as the keyword attribute data.

Compared to previous studies that primarily relied on the basic GDELT dataset for forecasting, such as using GDELT data for macroeconomic forecasting [23] and investigating violence-related issues in Iraq [10], our research takes a more comprehensive approach. We delve deeper into the extensive information contained within the GDELT dataset by incorporating textual data and images, thereby expanding the range of available data modalities.

3.3   MKGformer-Based Embedding Layer

Hybrid Transformer Architecture. MKGformer is a unified multimodal knowledge graph completion framework based on the transformer network architecture [4]. The overall framework is depicted in the Fig. 2(b). The hybrid transformer architecture of MKGformer mainly includes three stacked modules: (i) visual encoder, namely (V-Encoder), (ii) textual encoder, namely (T-Encoder), and (iii) the upper multimodal encoder, namely (M-Encoder). We denote the number of V-Encoder layers as L_V, the number of T-Encode layers as L_T and the number of M-Encoder layers as L_M.

Multimodal Entity Modeling. Unlike previous approaches that merely concatenate or fuse specific visual and textual features of entities, we fully leverage the "masked language modeling" (MLM) ability of pre-trained transformers to model multimodal representations of the entities in the knowledge graph that incorporate image-text integration. We input the entity description $d_{e_i} = (w_1, \ldots, w_n)$ and its corresponding multiple images $I_{e_i} = \{I^1, I^2, \ldots, I^o\}$ from CU-GDELT to the visual side of hybrid transformer architecture and convert the textual side input sequence of hybrid transformer architecture as follows:

$$T_{e_i} = [CLS] \; d_{e_i} \; [SEP] \; is \; the \; description \; of \; [MASK][SEP] \qquad (1)$$

We extend the word embedding layer of BERT to treat each token embedding as the corresponding multimodal representation E_{e_i} of i-th entity e_i. Then we train the MKGformer to predict the [MASK] in the multimodal entity embedding E_{e_i} using cross entropy loss for classification:

$$\mathcal{L}_{link} = -log(p(e_i|(T_{e_i}))) \qquad (2)$$

Additionally, in MKGformer, the entire model, except for the newly added multimodal entity embedding parameters, is frozen. This is done to guide the MKGformer model in selectively fusing textual and visual information into the multimodal entity embeddings.

Multimodal Information Extraction. For the task of multimodal information extraction, we fine-tune the input sequence of the T-Encoder to better
capture the textual description information of the entities. The input sequence
format is transformed as follows:

$$T_{e_i} = [CLS]e_i[SEP]d_{e_i}[SEP] \tag{3}$$

Given the presentation format of entities' textual data in our dataset, we
apply mean pooling after obtaining token embeddings to obtain the multimodal
embedding representations E_{e_i} for entities. Subsequently, we combine them with
the entities' event values E_{ve_i} that we have extracted from the GDELT dataset,
resulting in the multimodal information representation of the entity, $I_{e_i} = \{E_{e_i}, E_{ve_i}\}$.

3.4   Attention-Based SCINet

We refer to the SCINet architecture proposed by [17]. To handle multi-entity
data obtained from multi channel concatenation, we enhance the existing SCINet
model architecture by adding an attention block, as shown in Fig. 2(c). We utilize
a multi-head attention mechanism to combine the data and obtain time series
data as input. The resulting time series data, obtained through the attention
mechanism, is subsequently fed into the SCINet model for further processing.

Attention Block. The attention block Fig. 2(c) is the input layer of SCINet,
which combines the input multi time series $F[M \times N]$ into time series $F'[M]$
(M is the length of time series, N is the dimension of the embedding after
multi-channel splicing at the same time) through multi-head attention mechanism [24], allowing the model to learn different behaviors based on the same
attention mechanism. These different behaviors are then combined as knowledge to capture dependencies among the vectors. Formally, the attention function maps a sequence of query $Q = \{Q_1, \ldots, Q_N\}$ and a set of key-value pairs
$\{K, V\} = \{(K_1, V_1), \ldots, (K_M, V_M)\}$ to outputs, where $Q \in \mathbb{R}^{J \times d}, K, V \in \mathbb{R}^{M \times d}$.
More specifically, the multi-head attention model first transforms Q, K and V
into H subspaces with different, learnable linear projections, namely:

$$Q_h, K_h, V_h = QW_h^Q, KW_h^K, VW_h^V \tag{4}$$

where $\{Q_h, K_h, V_h\}$ are respectively the query, key, and value representations of
the h-th head. $\{W_h^Q, W_h^K, W_h^V\} \in \mathbb{R}^{d \times \frac{d}{H}}$ denote parameter matrices associated
with the h-th head, where d represents the dimension of the model hidden states.
Furthermore, H attention functions are applied in parallel to produce the output
states $\{O_1, \ldots, O_H\}$, among which:

$$O_h = A_h V_h \, with \, A_h = softmax(\frac{Q_h K_h^T}{\sqrt{d_k}}) \tag{5}$$

Here, A_h is the attention distribution produced by the h-th attention head. Finally, the output states are concatenated to produce the final state:

$$Concat : \hat{O} = [O_1, \ldots, O_H], Linear : O = \hat{O}W^O \tag{6}$$

where $O \in \mathbb{R}^{\mathbb{J}\times}$ denotes the final output states, and $W^O \in \mathbb{R} \times$ is a trainable matrix. The conventional multi-head attention uses a straightforward concatenation and linear mapping to aggregate the output representations of multiple attention heads.

By performing the aforementioned operations, we have successfully merged the input features into temporal features, transforming the multi time series F into F', which is then fed into SCINet for forecasting.

SCI-Block. The SCI-Block in Fig. 2(d) is the basic module of the SCINet, which decomposes the input feature F into two sub-features F'_{odd} and F'_{even} through the operations of *Spliting* and *Interactive learning*.

SCINet. The entire structure of SCINet is a binary tree, shown in Fig. 2(e). In each SCI-Block, the time series is divided into two parts. As the depth of the binary tree increases, the "temporal resolution" of the time series decreases. In this way, both short-term and long-term dependencies in the time series are captured. After passing through L layers of SCI-Blocks, the time series is reassembled to generate a new sequence, which is then added to the original sequence through *residual connections* [11] for forecasting. Finally, a fully connected layer is applied to obtain the forecasting results or features.

3.5 Complexity Analysis

The time complexity analysis of each component in the KDSCINet model can be determined based on several factors, including the size of the look-back window denoted as T, the number of entities per day represented by M, the entity embedding dimension denoted as d, the length of the entity sequence represented by n, as well as the number of entity image blocks indicated as p and the number of attention denoted as H. The breakdown of time complexity for each component is as follows:

Within the *T-Encoder*, considering the number of entities as $T \times M$, the time complexity for processing each entity is $\mathcal{O}(n^2 \times d)$. Thus, the overall time complexity is $\mathcal{O}(T \times M \times n^2 \times d)$. Similarly, in the *V-Encoder*, the time complexity remains $\mathcal{O}(T \times M \times n^2 \times d)$. In *M-Encoder*, it comprises the PGI module and CAF module. In the PGI module, employing a multi-head mechanism necessitates calculating both the text head and the visual head, resulting in a time complexity of $\mathcal{O}(T \times M \times (m + p)^2 \times d)$. In the CFM module, the time complexity for this module is $\mathcal{O}(T \times M \times (mpd + np + nd))$. In our fine-tuned model, the entity embedding dimension is set to 768, the length of the entity sequence is extended to 32, and each image is divided into 50 patches. Summing up, the total time complexity of the Embedding layer is $\mathcal{O}(T \times M)$.

In the attention-based SCINet, the number of entities per day is denoted as D, and the input length is $T \times D$. After combining the event value, the dimension becomes N, resulting in a time complexity of $\mathcal{O}((T \times D)^2 \times N)$. In our model, we set D to be the maximum number of entities in a day and fill all values with 0 for time periods with less than D. Additionally, we set N to be 770. Therefore, the time complexity of the Attention layer is $\mathcal{O}(T^2)$. The worst-case time complexity for SCINet is $\mathcal{O}(T \log T)$ [17].

To sum up, the time complexity for KDSCINet is $\mathcal{O}(T^2)$, which is similar to that of vanilla Transformer-based solutions: $\mathcal{O}(T^2)$.

4   Experiments

In this section, we will first illustrate the baselines and settings of our experiment. Next we show the quantitative comparisons with the state-of-the-art models for time series forecasting performed on CU-GDELT. Additionally, we will present comprehensive experiments to evaluate the effectiveness of multimodal knowledge and different components in KDSCINet.

4.1   Baselines and Settings

We consider baseline model variations as shown in Table 1. In the first column, the prefix TB means textual embeddings, IB means image embeddings. The difference between TBIB-SCINet and TBIB-KDSCINet is that in TBIB-SCINet, we replaced the multi-head attention layer with an average pooling layer used in [5]. To ensure a fair comparison, we perform a grid search across all essential hyperparameters using the held-out validation set of the dataset and use Mean Absolute Errors (MAE) and Mean Squared Errors (MSE) for model comparison.

Table 1. Baseline models with different input. In the first column, prefix TB means text embeddings, IB means image embeddings. TBIB-SCINet utilizes an average layer as input layer while TBIB-KDSCINet and TB-KDSCINet utilizes attention block as input layer.

Model	Processed training data
TCN [1]	Structured data($\mathcal{X}$)
Informer [28]	Structured data($\mathcal{X}$)
Prayformer [18]	Structured data($\mathcal{X}$)
Autoformer [27]	Structured data($\mathcal{X}$)
SCINet [17]	Structured data($\mathcal{X}$)
TB-KDSCINet(ours)	Structured data + Textual data($\mathcal{X} + \mathcal{T}$)
TBIB-KDSCINet(ours)	Structured data + Textual data + Image data($\mathcal{X} + \mathcal{T} + \mathcal{I}$)
TBIB-SCINet(ours)	Structured data + Textual data + Image data($\mathcal{X} + \mathcal{T} + \mathcal{I}$)

4.2   Result Analysis

The performance of KDSCINet is examined across three progressive aspects: (i) evaluating of the optimized pre-trained model MKGformer, (ii) evaluating of the basic SCINet architecture, (iii) assessing of the influence of different model inputs with SCINet, and (iv) examining the impact of different input layers added to SCINet.

Multimodal and Unimodal Inputs for MKGformer. To verify that multimodal information enhances entity embedding representation for our dataset, we compared the effects of unimodal (textual data only) and multimodal (textual + image data) inputs in the task of predicting masked entities during the pre-training of MKGformer, as shown in Table 2. The experimental results show that the multimodal model surpasses the unimodal model in all evaluation indicators, indicating the multimodal model's superior ability in entity embedding representation and prediction accuracy.

Table 2. Comparison of Multimodal and Unimodal Inputs in the Task of Predicting Masked Entities During the Pre-training of MKGformer.

Modal	Hits@1	Hits@3	Hits@10	MR
Unimodal (Textual data)	0.3472	0.7062	0.9103	88.7
Multimodal (Textual data + Image data)	0.3581	0.7175	0.9167	82.91

Basic Evaluation for SCINet. To verify the claim that the generic SCINet architecture can outperform other traditional forecasting models, we compare its performance with Transformer-based methods [18,27,28] and TCN [1], since they are more popular in recent long-term TSF research. As can be seen from Table 3, SCINet achieves state-of-the-art performance in most forecasting length settings. Overall, SCINet yeilds an average improvement of 13.78% on MSE and 14.54% on MAE under 24, 48, 168 horizons. This is attributed to the fact that the proposed SCINet can better capture both short (local temporal dynamics) and long (trend, seasonality) term temporal dependencies. Consequently, we select SCINet as our fundamental forecasting model in this paper. It is important to note that all experiments reported in this section only utilize structured event values as input.

Different Model Inputs with KDSCINet. To validate the effectiveness of integrating structured data, textual data and image data in predicting international relations, we compare the performance of models with different inputs, as presented in Table 4 and 5. Overall, TB-KDSCINet, which incorporates only structured data and textual data, demonstrates an average improvement of

Table 3. International relation prediction performance over the CU-GDELT multimodal Knowledge graph with different basic prediction models. IMP shows the improvement of SCINet over the best model.

Model	Metrics	Horizon		
		24	48	168
TCN [1]	MSE	1.3270	1.5798	1.5901
	MAE	0.9599	1.1540	1.0752
Informer [28]	MSE	0.9408	0.9852	1.1749
	MAE	0.7306	0.7752	0.7820
Prayformer [18]	MSE	0.9465	0.9643	1.0354
	MAE	0.7216	0.7503	0.7737
Autoformer [27]	MSE	0.9585	0.9634	0.9775
	MAE	0.7371	0.7371	0.7412
SCINet [17]	MSE	0.9294	0.9302	0.9351
	MAE	0.7189	0.7232	0.7267
IMP	MSE	11.81%	10.62%	18.91%
	MAE	9.51%	18.11%	16.06%

43.89% on MSE compared to the original SCINet, that relies solely on structured data. Moreover, TBIB-KDSCINet, which incorporates structured data, textual data and image data, achieves an average improvement of 48.61% on MSE compared to the original SCINet, surpass TB-KDSCINet by an average of 8.45%. These results indicate that using multimodal data contributes to enhancing forecasting performance. In summary, these findings substantiate the validity of integrating multimodal knowledge and structured event value as inputs for the model.

Table 4. International relation prediction performance over the CU-GDELT Knowledge graph with different inputs. SCINet takes structured data as input, TB-KDSCINet takes structured data and textual data as input and TBIB-KDSCINet takes structured data, textual data and image data as input.

Model	Metrics	Horizon		
		24	48	168
SCINet [17]	MSE	0.9294	0.9302	0.9351
	MAE	0.7189	0.7232	0.7267
TB-KDSCINet(ours)	MSE	0.4627	0.5328	0.5727
	MAE	0.4319	0.4918	0.5331
TBIB-KDSCINet(ours)	MSE	0.4213	0.4846	0.5304
	MAE	0.3819	0.4339	0.4956

Table 5. Average improvement of KDSCINet in MSE under different inputs. TB-KDSCINet takes structured data and textual data as input and TBIB-KDSCINet takes structured data, textual data and image data as input.

Comparison Other	SCINet	TB-KDSCINet(ours)	TBIB-KDSCINet(ours)
SCINet			
TB-KDSCINet(ours)	43.89%		
TBIB-KDSCINet(ours)	48.61%	8.45%	

Different Input Layer with SCINet. To assess the impact of replacing the average layer with an attention block, we compare the prediction performance of models utilizing different input layers, as shown in Table 6. As observed, TBIB-KDSCINet, which incorporates the multi-head attention mechanism in the input layer, outperforms TBIB-SCINet, which employs an average layer [5], by an average of 36.56% on MSE and 27.82% on MAE. We attribute the significant improvement of KDSCINet to the following factors: (i) Multi-head attention mechanism can capture diverse relationships. (ii) Multi-head attention mechanism can enhance representation learning. (iii) Multi-head attention mechanism can handle varing importance.

Table 6. International relation prediction performance over the CU-GDELT dataset with different input layer. TBIB-SCINet utilizes an average layer as input and TBIB-KDSCINet utilizes attention block as input.

Model	Metrics	Horizon		
		24	48	168
TBIB-SCINet(ours)	MSE	0.7223	0.7667	0.7713
	MAE	0.5351	0.5638	0.5653
TBIB-KDSCINet(ours)	MSE	0.4213	0.4846	0.5304
	MAE	0.3819	0.4339	0.4956
IMP	MSE	41.67%	36.79%	31.23%
	MAE	40.10%	29.93%	14.06%

5   Conclusion

This article introduces KDSCINet, a novel Knowledge-Driven neural network architecture for China-US relation forecasting. We create the CU-GDELT multimodal knowledge graph to capture implicit relations between images and text. By integrating the MKGformer-based embedding layer, we generate embedded representations of entities and relations. These representations, combined with structured event values, yield informative temporal features. Through a multi-head attention mechanism, our model fuses features and extracts valuable information from entity embeddings, focusing on diverse subspaces and relations. SCINet is then utilized to forecast future event value trends. Extensive experiments validate our approach, showcasing its effectiveness and superior performance.

Future prospects include extending our method to tasks like stock trend prediction and other time series forecasting. Exploiting multimodal knowledge relations in CU-GDELT for enhanced performance and interpretability is important. Moreover, the integration of entity linking technology holds potential to further enrich the available multimodal information.

Acknowledgements. This study was funded by National Natural Science Foundation of China(No. 62272045).

References

1. Bai, S., Kolter, J.Z., Koltun, V.: An empirical evaluation of generic convolutional and recurrent networks for sequence modeling. arXiv preprint arXiv:1803.01271 (2018)
2. Chambers, N., et al.: Identifying political sentiment between nation states with social media. In: Proceedings of the 2015 Conference on Empirical Methods in Natural Language Processing, pp. 65–75 (2015)
3. Chen, J., Lécué, F., Pan, J., Chen, H.: Learning from ontology streams with semantic concept drift. In: Twenty-Sixth International Joint Conference on Artificial Intelligence. International Joint Conferences on Artificial Intelligence Organization (2017)
4. Chen, X., et al.: Hybrid transformer with multi-level fusion for multimodal knowledge graph completion. In: Proceedings of the 45th International ACM SIGIR Conference on Research and Development in Information Retrieval, pp. 904–915 (2022)
5. Deng, S., Zhang, N., Zhang, W., Chen, J., Pan, J.Z., Chen, H.: Knowledge-driven stock trend prediction and explanation via temporal convolutional network. In: Companion Proceedings of The 2019 World Wide Web Conference, pp. 678–685 (2019)
6. Ding, X., Zhang, Y., Liu, T., Duan, J.: Knowledge-driven event embedding for stock prediction. In: Proceedings of Coling 2016, The 26th International Conference on Computational Linguistics: Technical Papers, pp. 2133–2142 (2016)
7. Feng, D., He, X., Peng, Y.: MKVSE: multimodal knowledge enhanced visual-semantic embedding for image-text retrieval. ACM Trans. Multimedia Comput. Commun. Appl. **19**, 1–21 (2023)
8. Gilpin, R.: War and Change in World Politics. Cambridge University Press (1981)
9. Goldstein, J., Keohane, R.O.: 1. ideas and foreign policy: an analytical framework. In: Ideas and Foreign Policy, pp. 3–30. Cornell University Press (2019)
10. González, M., Alférez, G.H.: Application of data science to discover violence-related issues in Iraq. arXiv preprint arXiv:2006.07980 (2020)
11. He, K., Zhang, X., Ren, S., Sun, J.: Deep residual learning for image recognition. In: Proceedings of the IEEE Conference on Computer Vision and Pattern Recognition, pp. 770–778 (2016)

12. Huang, J., Zhao, W.X., Dou, H., Wen, J.R., Chang, E.Y.: Improving sequential recommendation with knowledge-enhanced memory networks. In: The 41st International ACM SIGIR Conference on Research & Development in Information Retrieval, pp. 505–514 (2018)

13. Keohane, R.O., Nye, Jr., J.S.: Power and interdependence. Survival **15**(4), 158–165 (1973)

14. King, G., Zeng, L.: Explaining rare events in international relations. Int. Organ. **55**(3), 693–715 (2001). https://doi.org/10.1162/00208180152507597

15. Kocyigit, T.: Forecasting the nature of country relations using sentiment analysis, Ph. D. thesis, Tilburg University (2020)

16. Lake, D.A.: Hierarchy in International Relations. Cornell University Press (2011)

17. Liu, M., et al.: SciNet: time series modeling and forecasting with sample convolution and interaction. Adv. Neural. Inf. Process. Syst. **35**, 5816–5828 (2022)

18. Liu, S., et al.: Pyraformer: Low-complexity pyramidal attention for long-range time series modeling and forecasting. In: International Conference on Learning Representations (2021)

19. Metternich, N., Gleditsch, K., Dworschak, C.: Forecasting in international relations (2016). https://doi.org/10.1093/obo/9780199743292-0179

20. Organski, A.F.K.: Population dynamics and international violence: Propositions, insights and evidence by nazli choucri. (lexington, mass.: D.c. heath and company, 1974. pp. 281.). Am. Polit. Sci. Rev. **71**(2), 814–816 (1977)

21. Shukla, D., Unger, S.: Sentiment analysis of international relations with artificial intelligence. Athens J. Sci. **9**, 1–16 (2022)

22. Signorino, C.S.: Strategic interaction and the statistical analysis of international conflict. Am. Polit. Sci. Rev. **93**(2), 279–297 (1999). https://doi.org/10.2307/2585396

23. Tilly, S., Ebner, M., Livan, G.: Macroeconomic forecasting through news, emotions and narrative. Expert Syst. Appl. **175**, 114760 (2021)

24. Vaswani, A., et al.: Attention is all you need. In: Advances in neural information processing systems, vol. 30 (2017)

25. Wang, H., Zhang, F., Xie, X., Guo, M.: DKN: deep knowledge-aware network for news recommendation. In: Proceedings of the 2018 world wide web conference, pp. 1835–1844 (2018)

26. Wendt, A.: Social Theory of International Politics, vol. 67. Cambridge University Press (1999)

27. Wu, H., Xu, J., Wang, J., Long, M.: Autoformer: decomposition transformers with auto-correlation for long-term series forecasting. Adv. Neural. Inf. Process. Syst. **34**, 22419–22430 (2021)

28. Zhou, H., et al.: Informer: beyond efficient transformer for long sequence time-series forecasting. In: Proceedings of the AAAI Conference on Artificial Intelligence, vol. 35, pp. 11106–11115 (2021)

A Graph Convolution Neural Network for User-Group Aided Personalized Session-Based Recommendation

Hui Wang[ID], Hexiang Bai[✉][ID], Jun Huo[ID], and Minhu Yang[ID]

Key Laboratory of Computational Intelligence and Chinese Information Processing
Ministry of Education, School of Computer and Information Technology,
Shanxi University, Taiyuan 030006, China
bai_research@163.com

Abstract. Session-based recommendation systems aim to predict the next user interaction based on the items with which the user interacts in the current session. Currently, graph neural network-based models have been widely used and proven more effective than others. However, these session-based models mainly focus on the user-item and item-item relations in historical sessions while ignoring information shared by similar users. To address the above issues, a new graph-based representation, User-item Group Graph, which considers not only user-item and item-item but also user-user relations, is developed to take advantage of natural sequential relations shared by similar users. A new personalized session-based recommendation model is developed based on this representation. It first generates groups according to user-related historical item sequences and then uses a user group preference recognition module to capture and balance between group-item preferences and user-item preferences. Comparison experiments show that the proposed model outperforms other state-of-art models when similar users are effectively grouped. This indicates that grouping similar users can help find deep preferences shared by users from the same group and is instructive in finding the most appropriate next item for the current user.

Keywords: Recommendation system · Session-based recommendation · Graph convolution neural network

1 Introduction

As an important branch of recommendation systems, session-based recommendation systems (SBR) mainly focus on generating the most possible next item for the active session. It has been widely used in online service platforms [18]. Since no profiles and long-term behavior records of users are available, the interactions of restricted length within the active session are the only information available [3]. Currently, the effective inference of the correct next item only using these interactions has become a challenging task and attracts a great deal of attention.

B. Luo et al. (Eds.): ICONIP 2023, LNCS 14448, pp. 332–345, 2024.
https://doi.org/10.1007/978-981-99-8082-6_26

Generally, the relations among multiple sequential items are complex in SBR. Most related works attempt to model these relations from different perspectives. Traditionally, the transition probability was used to model these relations under the framework of Markov Chains [14,15,24]. However, the relations hidden in user sessions consist not only of direct relations which can be modeled using transition probability but also indirect high-order relations and relations concerning hidden variables. To address this issue, many researchers have attempted to model these complex relations with the help of Recurrent Neural Networks (RNNs) [2,4,7,13] or CNN [16]. These models have been proven more effective than transition probability-based models. Nevertheless, user preferences always change over time and can hardly be effectively captured by RNN or CNN.

In recent years, more and more SBRs are designed based on Graph Neural Networks (GNN) [10,11,19], which topology structure is well suited for the representation and inference of complex relations among items. Generally, there are three major learning strategies for updating GNN-based networks: Gated Graph Neural Networks (GGNN) [5], GAT [17], and GCN [20]. GGNN focuses on the input and output edges of the current node, and GAT mainly uses the information from the instant neighboring nodes. Different from the above two updating strategy, GCN can learn high-order neighboring node relations in a single update and have been widely adopted by most recent GNN-based SBRs.

Currently, besides taking account of both within-session information, some GCN-based [10] approaches attempted to introduce the user's historical sessions into the network to provide personalized recommendations. These methods need all the sessions available related to a user to infer the next item for current sessions. In the early stages, some models, such as H-RNN [12] and A-PGNN [23], collected all historical sessions of the current user to help predict the next item. However, these models ignored the general preference for the natural sequence of items that are shared by different users. For example, almost every user is going to order a mouse after buying a laptop. This is a general preference shared almost by all users even the current user did not buy any laptops in historical sessions. To fully capture these natural sequential relations among items across all user historical sessions, HG-GNN [10] first proposed using a heterogeneous global graph neural networks(HGNN) to encode all user and item nodes. In this way, the global item transition patterns were embedded in generating a personalized SBR.

Although the HGNN representation is much more informative than previous models, it is more vulnerable to the well-known over-smoothing effects in GCN due to its complex graph structure. In the HG-GNN, each user node and event node not only have connections to other nodes in the current session or user-specific historical sessions but also are connected to items from historical sessions from other users. Therefore, the local graph of each node is much more complex than before. As a result, after several updates, the node-specific personalized information will be smoothed by global item-transition pattern information. This will in turn lead to poor performance.

To address this issue, a new graph-based representation of the interactions among users and items, the User-item Group Graph (UGG), is developed in this paper since most natural sequential relations only take effect within a group of users rather than all users in practical applications. For example, boys prefer to buy some beer after a football ticket, while girls would like to get lipstick after a handbag. The grouping of users with similar item sequences is sufficient for modeling natural sequences of items and is effective in reducing the graph redundancy at the same time. The UGG creates several sub-graphs for different groups of users with similar item-sequences and each sub-graph is constructed follows the same way as the HG-GNN. Comparison and abolition experiments have been conducted on three real-world datasets to validate the effectiveness of grouping similar users during recommendations as well as its limitations. In summary, the main contributions of this work are as follows:

1) A new user-item group graph representation is proposed for reducing redundancy in modeling natural item relations. This new representation is more conformed to the reality of SBR than HG-GNN and alleviates the over-smoothing issue as the graph convolution is only performed within the respective subgraph.
2) A hierarchical gating mechanism is proposed to effectively capture user preferences through modeling the group-items relations besides the user-item and item-item relations.

2   Related Works

Deep Learning-Based Methods. As deep learning methods have become increasingly popular, researchers have turned to RNNs to effectively model sequential data patterns and achieve remarkable outcomes in recommender systems. GRU4REC [2] broke new ground in session recommendation by introducing the application of RNNs. In contrast to conventional session-based recommendation models that overlook distant historical behavioral information, GRU4REC adeptly models users' behavioral sequences and effectively captures long-term dependencies. NARM [4] captures the user's main purpose in an active session by incorporating an attention mechanism with RNN.

GNN-Based Methods. Graph neural networks(GNNs) have gained considerable attention in recent years as they offer a powerful way to model complex relations among items. SR-GNN [21], leverages the potential of GNNs by considering items in a session sequence as nodes. It constructs a session graph based on the sequence of clicked items and employs a gated graph neural network to propagate information across the session graph. Based on SR-GNN, GC-SAN [22] uses a self-attention mechanism to capture long-range dependencies between items in a session. SGNN-HN [9] adds extra nodes to the session graph to connect all non-adjacent items. For the same graph structure, there may be different session sequences, and LESSR [1] preserves information about the order of the edges in the session graph. GCE-GNN [19] proposes a representation of global graph-enhanced nodes that takes into account all historical sessions from similar item pairwise transitions. However, these methods assume that the user is anonymous and treat multiple sessions of the current user as multiple unrelated sessions and therefore do not take into account the user's historical sessions.

Personalized Session-Based Recommendation. Most methods of personalized session-based recommendation take into account historical sessions for the current user. H-RNN [12] designs a hierarchical RNN mechanism including session-level and user-level to extract users' interest preferences over time. Similar to the way SR-GNN constructs a session graph, A-PGNN [23] involves all sessions of the current user in the construction of the graph, further strengthening the connection between different sessions of current users. Consider that valuable transitions between items exist not only for the current user but also for other user interactions that may contain valuable item transitions. HG-GNN [10] proposes to construct a global heterogeneous graph to exploit the historical sessions of all users, further extending the scope of item transformation relationships. However, these methods ignore the item sequence information shared by similar users. The grouping of similar users is expected to improve the performance of personalized session-based recommendations further.

3 Method

3.1 Preliminaries

Assuming U and V are the set of all users and all items. For any $u \in U$ and $v \in V$, denote $\boldsymbol{e_u} \in \mathbb{R}^d$ and $\boldsymbol{e_v} \in \mathbb{R}^d$ as user and item initial embedding vectors, respectively, where d is the dimension of the embedding. Each u generally has multiple historical sessions, and the interaction in the jth session is denoted as $S_{u,j} = \{v_{u,j,1}, \cdots, v_{u,j,n}\}$, where $v_{u,j,l} \in V$ denotes the lth interaction item for user u, and n is the length of $S_{u,j}$. All historical session interactions of u constitute a set $S_u = \{S_{u,1}, S_{u,2}, S_{u,3}, \cdots\}$.

The difference between different GNN-based SBRs is the way that users and items are connected to a graph. Take HG-GNN as an example. User u and item v have an edge between them if and only if $v \in S_{u,j}$ and $S_{u,j} \in S_u$. Given any user u, all items in each session $S_{u,j}$ are connected with each other by an edge. The weight between $v_{u,j,i}$ and $v_{u,j,i+1}$ is count of the observations of these two items occur in this strict sequence in all sessions, and the weights between other edges are calculated using their co-occurrence frequency [10].

For the current session C_u of user u with t items, the Personalized SBR model is designed to predict the next possible items based on S_u and its current session $C_u = \{v_{u,c,1}, \cdots, v_{u,c,t}\}$. The model scores each candidate item $v \in V$ as $\hat{y} = \{\hat{y}_1, \hat{y}_2, \cdots, \hat{y}_{|V|}\}$, where $|V|$ is the number of items, Finally, the top-k most possible items are recommended to u. For example, HG-GNN uses three modules to help score all items. It first converts the graph representation into embedding representations of users and items through the HGNN module, then aggregates item relation information with user preferences with the help of the gating mechanism in the personalized session encoder module, finally all the items are scored using a softmax function.

3.2 User-Item Group Graph Based GCN

To make the best of the information shared by similar users, a new SBR model, User-item Group Graph based GCN (UGG-GCN), is proposed in this section. Different from other models, a User-item Group Graph Generation module (UGG-gen) is used to group similar users using the connections between users and items in the preprocessing stage. The general structure of UGG-GCN, which is similar to HG-GNN, is shown in Fig. 1. First of all, the model converts all the subgraphs into their embedding representations through several HGNN modules. Next, the item information from different subgraphs is aggregated through the Cross-group item information merging (CIM) to merge shared item information from different user-item groups. Subsequently, a User Group Preference Recognition (UGPR) module extracts user and group preferences on item relations through a gating mechanism. Finally, the score for each item is assigned using a softmax function in terms of the user and group preferences.

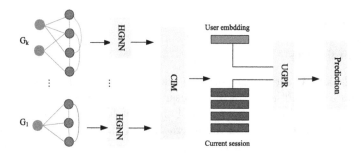

Fig. 1. Overall framework overview of the model.

User-Item Group Graph Generation. In personalized SBR, grouping similar users is instructive for effectively extracting user preferences. The grouping of similar users can avoid most of the conflicts of item sequences to a large extent which in turn alleviates the over-smoothing effect of the HGNN module. Inspired by the user grouping method for the bipartite graph [6], a subgraph generation module is used to group users and generate subgraphs for each group based on the graph representation of users and item sequences in HG-GNN. As is shown in Fig. 2, this module first constructs features for user from its ID and corresponding item sequences using $F_u = \sigma(W_1(e_u^{(0)} + e_u^{(1)}) + b_1)$, where σ is the activation function, $W_1 \in \mathbb{R}^{d \times d}$ and $b_1 \in \mathbb{R}^d$ are the weighting matrix and bias respectively, $e_u^{(0)}$ is the embedding of the user id, and $e_u^{(1)} = \sum_{v \in N_u} e_v^{(0)} / |N_v|$. Here N_u is the union of item sets in all historical sessions related to u, N_v is the number of item v neighbor nodes. $e_v^{(0)}$ is the embedding of the item v. Next, similar to [6], users are grouped using two layers of neural networks and. In this way, each user is grouped into one and only one group. After dividing users into k groups, the users as well as their item sequences are reorganized into k HGNN graphs for each group, namely the User-item Group Graph. Each node may eventually have multiple embeddings. In the subsequent chapters, the approach of taking the average of embeddings is applied to nodes with multiple embeddings to form their final representation. Furthermore, a higher number of constructed subgraphs leads to longer processing time.

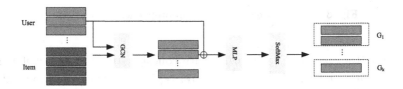

Fig. 2. User-item group graph generation module.

Cross-Group Item Information Merging. Although dividing users into k groups makes it possible to learn the embedding of nodes in consideration of group preferences, it brings two new issues to be addressed in item recommendations. One is the possible conflicts among item sequences from different subgraphs and the other is the missing of global item sequence information in the embedding of item nodes in each subgraph. Therefore, besides learning the embedding of items from the item sequences within each subgraph, it is important to merge cross-group item information for finding rational recommendations. Generally, denote $e_{vs}^{(0)}$ as the initial embedding vector of item v in the sth subgraph learned from UGG-gen. $e_{vs}^{(m)}$ is the result of m GCN convolutions of $e_{vs}^{(0)}$. Following the idea from [6], the information of item v from different subgraphs is merged as one embedding representation of v using $e_v^{(m)} = \sum_{s=1}^{k} e_{vs}^{(m)}$.

User Group Preference Recognition. Before scoring each item in UGG-GCN, it is important to take full advantage of user and group preferences. To this end, a gating system with three sub-modules, i.e., the UGPR module, is developed. These sub-modules are designated to capture group-item preferences, extract user-item preferences, and balance between the contributions of these two preferences. Because the embeddings for both users and items learned in the previous modules carry the group information, the group preferences are passed to both submodules by nature. The whole module is shown in Fig. 3. Similar to HG-GNN, the group-item preference capturing submodule

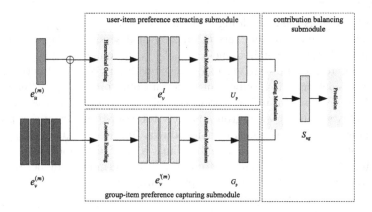

Fig. 3. The overview of user group preference recognition.

(see the bottom part of Fig. 3) begins with concatenating the item embedding with its location information by $e'^{(m)}_v = W_c \left[e^{(m)}_v \parallel l_v \right]$, where position embedding $l_v \in \mathbb{R}^d$ [10] and $W_c \in \mathbb{R}^{d \times 2d}$ are trainable parameters. As $e'^{(m)}_v$ is constructed from subgraphs for different groups, it carries not only item sequence information but also the group information compared with HG-GNN. Therefore, $e'^{(m)}_v$ partly captures the group-item preference. Next, a linear combination of the $e'^{(m)}_v$ of every item in the current session forms the session embedding S_g with group-item information using $S_g = \frac{1}{|C_u|} \sum_{v \in C_u} e'^{(m)}_v$. Subsequently, an attention mechanism is used to select important items $v \in C_u$ with group information for making the best prediction. The importance of item v is calculated as $\alpha_v = \text{softmax}_v (\omega_0^T \sigma (W_2 e'^{(m)}_v + W_3 S_g + b_2))$, where $\sigma (\cdot)$ is the activation function, $W_2, W_3 \in \mathbb{R}^{d \times d}$, $\omega_0 \in \mathbb{R}^d$, and $b_2 \in \mathbb{R}^d$ are trainable parameters. Finally, the current session group-item preference is $G_p = \sum_{v \in C_u} \alpha_v e'^{(m)}_v$.

The user-item preference extracting submodule (see the top part of Fig. 3) aims at extracting user-related item features for the current session. Following the ideal of [8], this submodule attempts to fuse the item embedding $e^{(m)}_v$ and user embedding $e^{(m)}_u$ to find the linear combination of users and items which best fits the correct recommendation, or the user-item preference $e^{F(m)}_v = e^{(m)}_v \cdot \sigma (W_{g_1} e^{(m)}_v + W_{g_2} e^{(m)}_u + b_g)$, where $W_{g_1} \in \mathbb{R}^{d \times d}$, $W_{g_2} \in \mathbb{R}^{d \times d}$, $b_g \in \mathbb{R}^d$ are trainable parameters, and $\cdot$ is the element-wise product. Similar to the group-item capturing submodule, the learned user-item preference $e^{F(m)}_v$ and $e^{(m)}_u$ undergoes an attention module to help extract salient user-item features for finding most possible recommendations using $e^I_v = e^{F(m)}_v \cdot \sigma (W_{g_3} e^{F(m)}_v + e^{(m)T}_u W_{g_4})$, where $W_{g_3} \in \mathbb{R}^{d \times d}$, $W_{g_4} \in \mathbb{R}^d$ are learnable parameters. Finally, a linear combination of e^I_v for each item in the current session is used as the session user-item preference $U_p = \sum_{v \in C_u} \beta_v e^I_v$, where $\beta_v = \text{softmax}_v (\omega_1^T \sigma (W_4 e^I_v + W_5 e^{(m)}_u + b_3))$ and $W_4, W_5 \in \mathbb{R}^{d \times d}$, $\omega_1 \in \mathbb{R}^d$, and $b_3 \in \mathbb{R}^d$ are trainable parameters.

In the end, the contribution balancing submodule (see the right part of Fig. 3) uses a gating mechanism to merge the user and group preference on the next items. It first concatenates the current session group-item preference $\boldsymbol{G_P}$ and the current session user-item preference $\boldsymbol{U_p}$ using $\boldsymbol{D_{ug}} = [\boldsymbol{U_p}||\boldsymbol{G_p}]$, then $\sigma(\boldsymbol{W_s D_{ug}})$ and $1\text{-}\sigma(\boldsymbol{W_s D_{ug}})$ are used as weights for $\boldsymbol{U_p}$ and $\boldsymbol{G_p}$, respectively, where $\boldsymbol{W_s} \in \mathbb{R}^{d \times 2d}$ is a trainable parameter, i.e., $\boldsymbol{S_{ug}} = \sigma(\boldsymbol{W_s D_{ug}}) \cdot \boldsymbol{U_p} + (1-\sigma(\boldsymbol{W_s D_{ug}})) \cdot \boldsymbol{G_p}$.

3.3 Prediction and Loss Function

After obtaining the final session embedding $\boldsymbol{S_{ug}}$, it is possible to calculate the probability distribution of the next item. The recommendation probability, denoted as $\hat{y}$, can be computed for candidate items for the current session:

$$\hat{y}_i = softmax\left(\boldsymbol{S}_{ug}^T \boldsymbol{e}_{v_i}^{(0)}\right) \tag{1}$$

where $\hat{y}_i \in \hat{\boldsymbol{y}}$ represents the likelihood that the user will click on item $v_i \in V$ in the next interaction. The optimization objective function can be formulated as the cross-entropy loss:

$$\mathcal{L}(\hat{y}) = -\sum_{i=1}^{|V|} y_i log(\hat{y}_i) + (1 - y_i) log(1 - \hat{y}_i) \tag{2}$$

where $\boldsymbol{y} \in \mathbb{R}^{|V|}$ is a one-hot vector of ground truth.

4   Experiments

4.1   Dataset

UGG-GCN is validated compared with other models on three publicly available datasets. The first dataset Last.fm contains historical song list records of users, in which all the records outsides the 40,000 most popular artists and groups are removed, and the records of the same user within 8 h constitute a session. The second dataset Xing is constituted by the interaction of 770,000 users with job postings over 80 d. And the third dataset Reddit is composed of tuples of user names, a subreddit where the user comments on a thread, and a timestamp for the interaction. Similar to [10,12], items with less than 3 occurrences are removed to preserve session information with session length greater than three. In all experiments, 80% of sessions are used as training data, and the remaining session is used as test data.

4.2   Evaluation Metrics

Following [19,21], Hit Ratio (HR@k) and Mean Reciprocal Rank (MRR@k) are used to evaluate the recommendation performance in the experiments, where k represents the number of most possible recommended items adopted in accuracy assessment. No matter which metric is used, a correct recommendation refers to the ground truth next item exists in the top k recommended items. HR@k refers to the ratio of correctly recommended items in the test dataset, while MRR@k indicates the average rank score of the correct items in the top k predicted items. The larger these two indices are, the better the recommendation results. In all experiments, both types of accuracy measures are used when k is 5 and 10.

4.3   Implementation Details

UGG-GCN is implemented using Pytorch-1.7.1 and DGL. The Adam optimizer is adopted in the training phase. The size of the embedding dimension of the item and user is set to 256. The range of the number of sub-graphs and graph convolutional layers are 2, 3, 4, 5, 6, and 7, respectively. Each experiment is trained for 20 epochs with a batch size of 648. The initial learning rate and the dropout are set to 0.01 and 0.5, respectively. All the experiments are carried out on a computer with an NVIDIA RTX3090 GPU.

4.4   Ablation Studies

The Last.fm dataset only collects 992 users, which is less than 1/10 users in the other two datasets. This may result in inadequate information available for recommendation in each subgraph after grouping. Therefore, this dataset is not suitable for evaluating the effectiveness of different modules of UGG-GCN. As a result, Xing and Reddit datasets are used to demonstrate the effectiveness of the UGG-gen and UGPR module in finding the best next item. HR@10 and MRR@10 are used to compare the identification accuracy. The experimental results for different components are shown in Table 1.

Table 1. Ablation studies on two datasets.

Methods	Xing		Reddit	
	H@10	MRR@10	HR@10	MRR@10
HG-GNN	20.30	12.79	60.51	36.89
+UGG-gen	22.13	13.79	61.48	37.93
+UGPR	21.66	13.13	61.19	37.83
+ALL	22.32	14.33	62.06	38.41

Effectiveness of UGG-Gen. The using of the UGG-gen module increases the accuracy measures by at least 1% for the Xing dataset and 0.97% for the Reddit dataset, and increases at most 1.83% for the Xing dataset and 1.04% for the Reddit dataset. All these results suggest the effectiveness of the UGG-gen module, which groups similar users and shares item sequence information within each subgraph. This is instrucLNCStive in embedding group preference on item sequences, which in turn improves the recommendation accuracy.

Effectiveness of UGPR. The HR@10 and MRR@10 increased by 1.36% and 0.34 respectively for the Xing dataset. And in the Reddit dataset, the HR@10 and MRR@10 increased by 0.68% and 0.94% respectively. This indicates that the UGPR module is superior to the personalized session encoder in HG-GNN. This is due to that the UGPR module encodes the user preference on items and item sequences with two individual submodules. This is more effective in merging user preference and item sequence information than directly processing both information simultaneously in one module. In addition, cooperating with the UGG-gen module makes it possible to add group preference information on items and item sequences into the UGPR module as the user and item embeddings from the UGG-gen module carry the corresponding group information. Therefore, as is shown in the last row of Table 1, the recommendation accuracy is the highest when both modules are used together.

4.5   Comparison with Other Models

UGG-GCN is compared with six popular SBR models on all three datasets. These models can be roughly divided into two groups. The first group includes three models with no personal preference, i.e. SR-GNN [21], LESSR [1], and GCE-GNN [19]. The remaining three models, H-RNN [12], A-PGNN [23], and HG-GNN [10], all take user preference into account. In the comparison, HR@5, HR@10, MRR@5, and MRR@10 are used to evaluate all seven models' performance.

Tables 2 and 3 show the comparison results on these datasets. UGG-GCN achieves the best performance both on the Xing and Reddit datasets. Compared to the non-personalized models, results from UGG-GCN increase by a minimum of 10.45% and a maximum of 19.68% on the Reddit datasets. On the Xing dataset, the minimum and maximum increments of accuracy are 1.46% and 5.99%, respectively. Compared with personalized recommendation models, UGG-GCN improves by a maximum of 8.62% and 8.65% on the Reddit and Xing datasets, respectively. In comparison with the most recent model, HG-GNN, the recommendation accuracy still increases by at least 1.4% in both datasets.

Table 2. Comparison with other SBR methods (HR@k).

Method	Reddit		Xing		Last.fm	
	HR@5	HR@10	HR@5	HR@10	HR@5	HR@10
SR-GNN	34.96	42.38	13.38	16.71	11.89	16.90
LESSR	36.03	43.27	14.84	16.77	12.96	17.88
GCE-GNN	36.30	45.16	16.98	20.86	12.83	18.28
H-RNN	44.76	53.44	10.72	14.36	10.92	15.83
A-PGNN	49.10	58.23	14.23	17.01	12.10	17.13
HG-GNN	51.08	60.51	17.25	20.30	13.09	19.39
UGG-GCN	52.48	62.06	19.37	22.32	12.95	18.70

Table 3. Comparison with other SBR methods (MRR@k).

Method	Reddit		Xing		Last.fm	
	MRR@5	MRR@10	MRR@5	MRR@10	MRR@5	MRR@10
SR-GNN	25.90	26.88	8.95	9.39	7.23	7.85
LESSR	26.45	27.41	11.98	12.13	8.24	8.82
GCE-GNN	26.65	27.70	11.14	11.65	7.60	8.32
H-RNN	32.13	33.29	7.22	7.74	6.71	7.39
A-PGNN	33.54	34.62	10.26	10.58	7.37	8.01
HG-GNN	35.46	36.89	12.23	12.79	7.35	8.18
UGG-GCN	37.10	38.41	13.84	14.33	7.51	8.27

These results show that grouping users by item sequences is conducive to extracting shared item sequence preferences from similar users. This group preference sheds light on the deep user preferences which are neglected by other methods. Therefore, the grouping of users contributes to finding better next items for the current user and can increase the recommendation accuracy. However, UGG-GCN relies on the effective grouping of users. When the number of users is not large enough to generate rational groups for the recommendation, the recommendation accuracy may decrease rather than increase.

An example is shown in the comparison experiments on the Last.fm dataset, in which the number of users is less than 1/10 of that from the Xing and Reddit dataset. According to the results shown in Tables 2 and 3, the introducing of user preference does not ensure more accurate results. The MRR@5 and MRR@10 both suggest that the non-personalized models LESSR and GCE-GNN both outperform all other models. This is because the limited number of users cannot support effective modeling of user preference. Similarly, it also puts a limit on correct modeling group information hidden in data. As a result, UGG-GNN is also inferior to non-personalized models on this dataset. In addition, compared with HG-GNN, UGG-GCN gains better performance on the MRR@k but worse performance on the HR@k metrics. This also indicates that ineffective grouping cannot ensure better performance.

4.6 The Best Number of Groups

The number of subgraphs is an important parameter for generating an effective grouping of users. A sensitive analysis is performed on all three datasets to find the best number of groups. During the experiment, the number of groups is set to two to seven and all the other parameters are the same as in previous experiments. The HR@10 is used as the accuracy measure. The trend of the accuracy over the number of subgraphs in each dataset is demonstrated in Fig. 4. In the Last.fm dataset, as stated before, it is hard to get effective user groups with a limited number of users. The prediction accuracy declines when the number of groups increases in general. However, when there are a sufficient number of users, for example, the Xing and Reddit dataset, the prediction accuracy increases and approaches maximum when the number of groups is three. When the number of groups keeps increasing, the prediction accuracy decreases steadily. The reason for the same best number of groups is that these two datasets contain a comparable number of users, 11,479 for the Xing dataset and 18,271 for the Reddit dataset. The last two datasets have a larger number of users, resulting in a larger optimal number of groups. This suggests that as the number of users increases, the optimal number of groups should also increase. In our future work, we will further explore and analyze this issue using more datasets.

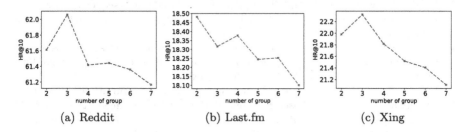

(a) Reddit   (b) Last.fm   (c) Xing

Fig. 4. Effectiveness of number of groups.

5   Conclusion

Based on the fact that grouping of users is instructive in revealing deep user-related item sequence preferences, a new SBR model, UGG-GCN, is proposed to help predict the most possible next item for the current session. In UGG-GCN, users with similar item sequences are grouped and constitute sever subgraphs. Next, the UGPR module attempts to capture group-item preferences, extract user-item preferences, and balance between the contributions of these two preferences based on the user and item embeddings learned from subgraphs. Comparison experiments show that UGG-GCN outperforms other personalized or non-personalized models as long as the group information is effectively captured.

Acknowledgements. The work is supported by the National Natural Science Foundation of China (No. 41871286) and the 1331 Engineering Project of Shanxi Province, China.

References

1. Chen, T., Wong, R.C.W.: Handling information loss of graph neural networks for session-based recommendation. In: Proceedings of the 26th ACM SIGKDD International Conference on Knowledge Discovery & Data Mining, pp. 1172–1180 (2020). https://doi.org/10.1145/3394486.3403170

2. Hidasi, B., Karatzoglou, A., Baltrunas, L., Tikk, D.: Session-based recommendations with recurrent neural networks. arXiv preprint arXiv:1511.06939 (2015). https://doi.org/10.48550/arXiv.1511.06939

3. Lai, S., Meng, E., Zhang, F., Li, C., Wang, B., Sun, A.: An attribute-driven mirror graph network for session-based recommendation. In: Proceedings of the 45th International ACM SIGIR Conference on Research and Development in Information Retrieval, pp. 1674–1683 (2022). https://doi.org/10.1145/3477495.3531935

4. Li, J., Ren, P., Chen, Z., Ren, Z., Lian, T., Ma, J.: Neural attentive session-based recommendation. In: Proceedings of the 2017 ACM on Conference on Information and Knowledge Management, pp. 1419–1428 (2017). https://doi.org/10.1145/3132847.3132926

5. Li, Y., Tarlow, D., Brockschmidt, M., Zemel, R.: Gated graph sequence neural networks. arXiv preprint arXiv:1511.05493 (2015). https://doi.org/10.48550/arXiv.1511.05493

6. Liu, F., Cheng, Z., Zhu, L., Gao, Z., Nie, L.: Interest-aware message-passing gcn for recommendation. In: Proceedings of the Web Conference 2021, pp. 1296–1305 (2021). https://doi.org/10.1145/3442381.3449986

7. Liu, Q., Zeng, Y., Mokhosi, R., Zhang, H.: Stamp: short-term attention/memory priority model for session-based recommendation. In: Proceedings of the 24th ACM SIGKDD International Conference on Knowledge Discovery & Data Mining, pp. 1831–1839 (2018). https://doi.org/10.1145/3219819.3219950

8. Ma, C., Kang, P., Liu, X.: Hierarchical gating networks for sequential recommendation. In: Proceedings of the 25th ACM SIGKDD International Conference on Knowledge Discovery & Data Mining, pp. 825–833 (2019). https://doi.org/10.1145/3292500.3330984

9. Pan, Z., Cai, F., Chen, W., Chen, H., De Rijke, M.: Star graph neural networks for session-based recommendation. In: Proceedings of the 29th ACM International Conference on Information & Knowledge Management, pp. 1195–1204 (2020). https://doi.org/10.1145/3340531.3412014

10. Pang, Y., et al.: Heterogeneous global graph neural networks for personalized session-based recommendation. In: Proceedings of the Fifteenth ACM International Conference on Web Search and Data Mining, pp. 775–783 (2022). https://doi.org/10.1145/3488560.3498505

11. Qiu, R., Li, J., Huang, Z., Yin, H.: Rethinking the item order in session-based recommendation with graph neural networks. In: Proceedings of the 28th ACM International Conference on Information and Knowledge Management, pp. 579–588 (2019). https://doi.org/10.1145/3357384.3358010

12. Quadrana, M., Karatzoglou, A., Hidasi, B., Cremonesi, P.: Personalizing session-based recommendations with hierarchical recurrent neural networks. In: Proceedings of the Eleventh ACM Conference on Recommender Systems, pp. 130–137 (2017). https://doi.org/10.1145/3109859.3109896

13. Ren, P., Chen, Z., Li, J., Ren, Z., Ma, J., De Rijke, M.: RepeatNet: a repeat aware neural recommendation machine for session-based recommendation. In: Proceedings of the AAAI Conference on Artificial Intelligence, vol. 33, pp. 4806–4813 (2019). https://doi.org/10.1609/aaai.v33i01.33014806

14. Rendle, S., Freudenthaler, C., Schmidt-Thieme, L.: Factorizing personalized markov chains for next-basket recommendation. In: Proceedings of the 19th International Conference on World Wide Web, pp. 811–820 (2010). https://doi.org/10.1145/1772690.1772773

15. Shani, G., Brafman, R.I., Heckerman, D.: An MDP-based recommender system. arXiv e-prints arXiv:1301.0600 (2012)

16. Tang, J., Wang, K.: Personalized top-n sequential recommendation via convolutional sequence embedding. In: Proceedings of the Eleventh ACM International Conference on Web Search and Data Mining, pp. 565–573 (2018). https://doi.org/10.1145/3159652.3159656

17. Veličković, P., Cucurull, G., Casanova, A., Romero, A., Lio, P., Bengio, Y.: Graph attention networks. arXiv preprint arXiv:1710.10903 (2017)

18. Wang, S., Cao, L., Wang, Y., Sheng, Q.Z., Orgun, M.A., Lian, D.: A survey on session-based recommender systems. ACM Comput. Surv. (CSUR) **54**(7), 1–38 (2021). https://doi.org/10.1145/3465401

19. Wang, Z., Wei, W., Cong, G., Li, X.L., Mao, X.L., Qiu, M.: Global context enhanced graph neural networks for session-based recommendation. In: Proceedings of the 43rd International ACM SIGIR Conference on Research and Development in Information Retrieval, pp. 169–178 (2020). https://doi.org/10.1145/3397271.3401142

20. Wu, F., Zhang, T., Holanda de Souza A., Jr., Fifty, C., Yu, T., Weinberger, K.Q.: Simplifying graph convolutional networks. arXiv e-prints arXiv:1902.07153 (2019)

21. Wu, S., Tang, Y., Zhu, Y., Wang, L., Xie, X., Tan, T.: Session-based recommendation with graph neural networks. In: Proceedings of the AAAI Conference on Artificial Intelligence, vol. 33, pp. 346–353 (2019). https://doi.org/10.1609/aaai.v33i01.3301346

22. Xu, C., et al.: Graph contextualized self-attention network for session-based recommendation. In: IJCAI International Joint Conference on Artificial Intelligence, p. 3940 (2019). https://doi.org/10.24963/ijcai.2019/547

23. Zhang, M., Wu, S., Gao, M., Jiang, X., Xu, K., Wang, L.: Personalized graph neural networks with attention mechanism for session-aware recommendation. IEEE Trans. Knowl. Data Eng. **34**(8), 3946–3957 (2020). https://doi.org/10.1109/TKDE.2020.3031329

24. Zimdars, A., Chickering, D.M., Meek, C.: Using temporal data for making recommendations. arXiv preprint arXiv:1301.2320 (2013)

Disentangling Node Metric Factors
for Temporal Link Prediction

Tianli Zhang[1,2], Tongya Zheng[1,3,4], Yuanyu Wan[1,3,4]([✉]), Ying Li[5],
and Wenqi Huang[6]

[1] Zhejiang University, Hangzhou 310027, China
`{zhangtianli,tyzheng,wanyy}@zju.edu.cn`
[2] Zhejiang University - China Southern Power Grid Joint Research Centre on AI,
Hangzhou 310058, China
[3] ZJU-Bangsun Joint Research Center, Hangzhou, China
[4] Shanghai Institute for Advanced Study of Zhejiang University, Hangzhou, China
[5] Zhejiang Bangsun Technology Co. Ltd., Hangzhou 310012, China
`li_ying@bsfit.com.cn`
[6] Digital Grid Research Institute, China Southern Power Grid, Guangzhou 510663,
China
`huangwq@csg.cn`

Abstract. Temporal Link Prediction (TLP), as one of the highly concerned tasks in graph mining, requires predicting the future link probability based on historical interactions. On the one hand, traditional methods based on node metrics, such as Common Neighbor, achieve satisfactory performance in the TLP task. On the other hand, node metrics overly focus on the global impact of nodes while neglecting the personalization of different node pairs, which can sometimes mislead link prediction results. However, mainstream TLP methods follow the standard paradigm of learning node embedding, entangling favorable and harmful node metric factors in the representation, reducing the model's robustness. In this paper, we propose a plug-and-play plugin called **N**ode **M**etric **D**isentanglement, which can apply to most TLP methods and boost their performance. It explicitly accounts for node metrics and disentangles them from the embedding representations generated by TLP methods. We adopt the attention mechanism to reasonably select information conducive to the TLP task and integrate it into the node embedding. Experiments on various *state-of-the-art* methods and dynamic graphs verify the effectiveness and universality of our NMD plugin.

Keywords: Dynamic Graphs · Temporal Link Prediction ·
Disentangled Representation Learning · Node Metric

1 Introduction

Graph data, particularly dynamic graphs, have grown exponentially in recent years, arousing widespread attention from researchers and practitioners [25, 27, 28]. Dynamic graphs can represent the changeable interactions of nodes with

© The Author(s), under exclusive license to Springer Nature Singapore Pte Ltd. 2024
B. Luo et al. (Eds.): ICONIP 2023, LNCS 14448, pp. 346–357, 2024.
https://doi.org/10.1007/978-981-99-8082-6_27

time elapsing in real-graph scenarios, such as social networks [23], academic networks [31], transaction networks [16,17], etc. One of the most challenging issues related to dynamic graphs is temporal link prediction (TLP), which aims to estimate the likelihood of paired nodes being connected in the future [6]. It helps to track the dynamic characteristics and reveal the evolutionary patterns of the system. Some applications of TLP tasks include analyzing community clusters in social networks [5], revealing author collaboration trends in academic networks [15], and predicting commercial intercourse in transaction networks [18].

Traditional link prediction methods utilize the property metrics of nodes to predict links, such as similarity-based algorithms using degree or common neighbors [19]. Preferential Attachment [3] explicitly considers the degree centrality of nodes. Several heuristic algorithms like the Jaccard Coefficient [2], Adamic Adar Index [1], and Resource Allocation index [37] consider the common neighbors to predict a connection. On the one hand, node metrics reflect the global impact and promote link prediction in specific scenarios. For example, a new manuscript may prefer papers with more citations in their references. On the other hand, node metrics neglect the personalization of different node pairs, such as music application users preferring niche songs over popular songs on the list. It is crucial to handle node metrics information in link prediction tasks properly.

The interaction of many complex factors, including node metrics, usually drives temporal link changes in dynamic graph systems. Existing *state-of-the-art* methods, including GCN-GRU [29], EvolveGCN [25], DySAT [28], etc., predict temporal link existence by learning node embedding from the ego network based on various specific Graph Neural Networks. However, these TLP methods mix different factors into a shared embedding representation space, which may introduce harmful node metrics to temporal link prediction tasks. Numerous works in multiple fields [21,34] show that disentangling various information factors in embedding is conducive to better representation learning and model generalization. It inspires us to disentangle node metrics from embedding representations and integrate information conducive to link prediction into the embedding by a fusion module. In this paper, we aim to design a disentanglement plugin that can be applied to most TLP methods, disentangling node metric factors and improving their performance on different dynamic graphs.

To this end, we propose a dual-branch framework consisting of a temporal link prediction (TLP) branch and a node metric disentanglement (NMD) branch. The former describes a general structural-temporal framework for existing TLP methods, which can generate node representations based on historical graph snapshots. The latter uses multiple optional types of node metrics as factors, such as degree centrality, closeness centrality [8], PageRank [24], etc. We introduce a similarity decoupling loss to disentangle the node representations of different time slots under multiple metrics. It helps two branch focus on edge-level personalization and node-level global functionality, respectively. The TLP branch is optimized for an edge-level classification objective, and the NMD branch is optimized for a node-level regression objective. Finally, an attention-based fusion module is adopted for the two sets of historical embedding obtained from the two branches, integrating information beneficial to the TLP task and exploring fine-grained representations. Our contributions are summarized as follows:

- To the best of our knowledge, we are the first to consider the positive and negative effects of node metric factors in representation learning on the temporal link prediction task.
- We propose a Node Metric Disentanglement (NMD) plugin that can disentangle node metrics in most TLP methods and generate fine-grained and more beneficial representations by the attention mechanism.
- Experiments on five dynamic graphs validate the universal improvement of our NMD plugin for existing SOTA methods. Elaborate ablation studies further verify the effectiveness of different components.

2   Related Work

Temporal Link Prediction. Some heuristics strategies based on graph topology are proposed to solve link prediction tasks, such as Katz [10], singular value decomposition (SVD) [7], and non-negative matrix factorization (NMF) [36]. Learning low-dimensional embedding on dynamic graphs is an emerging topic under investigation [4]. Dyngraph2Vec [9] uses recurrent neural networks (RNN) to learn each vertex's complex transformation. Inspired by the progress of Graph Neural Networks (GNNs) [12], EvolveGCN [25] trains RNN parameters at each time step to dynamically update GCN parameters. STGSN [22] introduces the attention mechanism to model social networks' spatial and temporal dynamicity.

Disentangled Representation Learning. [20] proposes a disentangled multichannel convolutional layer and devises a neighborhood routing mechanism to disentangle the underlying factors behind a graph. [33] establishes a set of intent-aware graphs and chunked representations, disentangling representations of users and items at the granularity of user intents. [21] proposes disentangled variational autoencoder and a beam-search strategy to explicitly models the separation of macro and micro factors. [34] utilizes a mechanism that incorporates the micro-and macro-disentanglement in knowledge graphs.

Node Metrics. Centrality is a classical measure of nodes in complex networks, including k-shell decomposition [13], closeness centrality [8], or betweenness centrality [8]. [32] uses the concept of *diversity entropy* to describe the relative frequency of access received by nodes and study the internal and external accessibility of nodes. [14] defines *dynamic influence* as the leading left eigenvector of a characteristic matrix that encodes the interaction between graph topology and dynamics. Adamic Adar index [1] is a similarity measure that assigns higher importance to common neighbors with lower degrees in a graph representation.

3   Method

3.1   Problem Statement

Most dynamic networks can be described as a weighted temporal graph $G(V, E)$ with a node set $V = \{v_i\}_{i=1}^{N}$ and an edge set $E \subseteq |V| \times |V|$, where $N = |V|$

is the number of unique vertices. An edge between nodes v_i and v_j with weight $w_{ij} \in \mathbf{R}$ at time step $t \in \mathbf{R}^+$ is written as $e_{ij} = (v_i, v_j, t, w_{ij}) \in E$. A graph G can be split into a series of snapshots $G = \{G_1, ..., G_T\}$. We transform E_t into an adjacency matrix $A_t \in \mathbf{R}^{N \times N}$ according to the rule: $(A_t)_{ij} = w_{ij}$ otherwise 0. The d-dimensional node feature matrix in G_t is defined as $X_t \in \mathbf{R}^{N \times d}$.

Temporal Link Prediction (TLP) task predicts future link status based on historical snapshots. Formally, given δ snapshots $\{A_\tau, X_\tau\}_{\tau=t-\delta}^{t-1}$ at time t, our goal is to learn a function $f(\cdot)$ that predicts the adjacency matrix A_t:

$$\{A_\tau, X_\tau\}_{\tau=t-\delta}^{t-1} \xrightarrow{f(\cdot)} A_t, t \in \{1 + \delta, ..., T\} \tag{1}$$

3.2   Temporal Link Prediction Branch

The general paradigm of TLP methods is a sequential combination of a multi-layer structural encoder, a temporal encoder, and a link predictor [27]. For each historical snapshot G_τ at time t, an l-th layer of the structural encoder takes the adjacency matrix A_τ and the $(l-1)$-th output embedding Z_τ^{l-1} as input:

$$Z_\tau^l = \text{StructuralEncoder}^l(A_\tau, Z_\tau^{l-1}), l = 1, ..., L, \tag{2}$$

where L is the number of layers, and the initial embedding matrix comes from the raw node features, *i.e.* $Z_\tau^0 = X_\tau$. Then a temporal encoder combines the L-th layer embedding of δ historical snapshots $\{Z_\tau^L\}_{\tau=t-\delta}^{t-1}$ and generates the current snapshot's node embedding Z_t for downstream tasks:

$$Z_t = \text{TemporalEncoder}(Z_{t-1}^L, ..., Z_{t-\delta}^L). \tag{3}$$

The TLP task can be regarded as a binary classification of positive and negative edges. As for an edge e_{ij} with its label $y = (A_t)_{ij} \in \{0, 1\}$, we use the following classification layer to obtain the prediction probability $p_{ij} \in [0, 1]$:

$$p_{ij}^t = \sigma\left(\text{MLP}\left(\mathbf{z}_i^t || \mathbf{z}_j^t\right)\right), \tag{4}$$

where σ is the softmax function, MLP is a multi-layer perceptron (MLP), and $||$ represents the concatenation operation. $\mathbf{z}_i^t = (Z_t)_i \in \mathbf{R}^{d_1}$ is the d_1-dimensional embedding of node v_i, *i.e.*, the i-th row of the embedding matrix Z_t. The TLP branch calculates the link prediction loss at t time by the Cross-Entropy (CE):

$$\mathcal{L}_{TLP}^t = K \sum_{e_{ij} \in E_t^+} -\log p_{ij}^t - \sum_{e_{ij} \in E_t^-} 1 - \log p_{ij}^t, \tag{5}$$

where E_t^+ and E_t^- are the set of positive and negative edges, respectively. K balances the loss of E_t^+ and E_t^-, usually set to the ratio of $|E_t^+|$ and $|E_t^-|$.

3.3   Node Metric Disentanglement Branch

As shown in Fig. 1, the proposed node metric disentanglement (NMD) branch includes a structural encoder with the same architecture as the TLP branch,

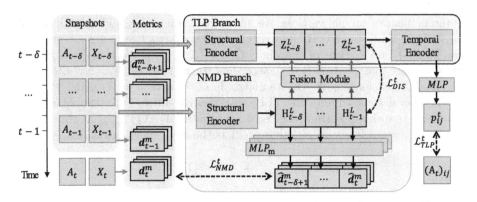

Fig. 1. Framework of our NMD plugin based on the structural-temporal paradigm. TLP Branch, applicable to any TLP method, predicts the link status at time t based on δ snapshots $\{A_\tau, X_\tau\}_{\tau=t-\delta}^{t-1}$. NMD Branch generates node-level embedding $\{H_\tau^L\}_{\tau=t-\delta}^{t-1}$ by predicting temporal node metrics $\{\mathbf{d}_\tau^m\}_{\tau=t-\delta}^t$. It disentangles the node representations of different time slots by the decoupling loss $\mathcal{L}_{DIS}^t$ and integrates beneficial factors into $\{Z_\tau^L\}_{\tau=t-\delta}^{t-1}$ by an attention-based fusion module.

multiple MLP predictors, and an attention-based fusion module. It uses multiple self-selectable node metrics as decoupling factors, such as degree centrality, closeness centrality [8], PageRank [24], etc.

Formally, given a node metric $m \in \mathcal{M}$, $\mathcal{M}$ is a set of various node metrics selected according to different contexts. We calculate the node metric vector $\mathbf{d}_\tau^m \in \mathbf{R}^{|V|}$ for each historical snapshot at time t and obtain the set of node metrics $\{\mathbf{d}_\tau^m\}_{\tau=t-\delta}^t$.

Then, we perform regression optimization on multiple temporal node metrics to explicitly model metric-aware representations. Precisely similar to the process described in Eq. 2, an L-layer structural encoder generates the representation of nodes in each historical snapshot G_τ a time t:

$$H_\tau^l = \text{StructEncoder}^l(A_\tau, H_\tau^{l-1}), l = 1, ..., L, \tag{6}$$

where H_τ^l is the l-th layer embedding matrix, and H_τ^0 is initialized with the raw feature matrix X_t. Our NMD branch tries to predict the node metric $\mathbf{d}_{\tau+1}^m$ of the next snapshot $G_{\tau+1}$ by each historical embedding H_τ^L:

$$\widehat{\mathbf{d}}_{\tau+1}^m = \text{MLP}_m(H_\tau^L), \tau = t - \delta, ..., t, \tag{7}$$

where MLP_m is a multi-layer perceptron predictor, which can project H_τ^L into representation spaces of different node metrics. Naturally, we calculate the average MSE (Mean Squared Error) between the ground-truth metric $\mathbf{d}_{\tau+1}^m$ and the predicted metric $\widehat{\mathbf{d}}_{\tau+1}^m$ over the period $[t - \delta, t]$:

$$\mathcal{L}_{NMD}^t = \frac{1}{|\mathcal{M}|} \sum_{m \in \mathcal{M}} \frac{1}{\delta} \sum_{\tau=t-\delta}^{t-1} \text{MSE}\left(\mathbf{d}_{\tau+1}^m, \widehat{\mathbf{d}}_{\tau+1}^m\right), \tag{8}$$

where $\mathcal{L}_{NMD}^{t}$ is our NMD branch loss at time t, $\mathcal{M}$ is the set of multiple optional types of node metrics, and δ is the number of historical snapshots. The above prediction strategy fully utilizes the multiple historical snapshots' information, favoring each historical embedding to establish a non-linear relationship with the node metric of the following snapshot. Our NMD branch builds a fine-grained metric-aware embedding $\{H_{\tau}^{L}\}_{\tau=t-\delta}^{t-1}$ that can represent multiple metric factors of different periods by the regression optimization.

The TLP branch's embedding is mixed with many complex factors, including node metrics. Considering that node metrics may mislead the link prediction results, we devise a decoupling loss to disentangle them from $\{Z_{\tau}^{L}\}_{\tau=t-\delta}^{t-1}$. It can use statistical measures such as distance correlation or mutual information as a regularizer to encourage specialization in representation space. Distance correlation can characterize the independence of any two paired vectors in their linear and nonlinear relationships. Here we calculate the decoupling loss $\mathcal{L}_{DIS}^{t}$ by the commonly-used cosine similarity:

$$\mathcal{L}_{DIS}^{t} = \sum_{\tau=t-\delta}^{t-1} \sum_{i=1}^{N} \frac{\mathbf{z}_{i}^{\tau} \cdot \mathbf{h}_{i}^{\tau}}{||\mathbf{z}_{i}^{\tau}|| \, ||\mathbf{h}_{i}^{\tau}||} \tag{9}$$

where $\mathbf{z}_{i}^{\tau} = (Z_{\tau})_{i} \in \mathbf{R}^{d_1}$, $\mathbf{h}_{i}^{\tau} = (H_{\tau})_{i} \in \mathbf{R}^{d_1}$, and $|| \cdot ||$ is the vector norm.

Under the guidance of $\mathcal{L}_{DIS}^{t}$, the representation function of $\{Z_{\tau}^{L}\}_{\tau=t-\delta}^{t-1}$ and $\{H_{\tau}^{L}\}_{\tau=t-\delta}^{t-1}$ are deconstructed and refined. The former focuses more on personalized aggregation information of the node's multi-hop neighbors, while the latter pays more attention to the global impact of the node itself. If node-level and edge-level factors are highly intertwined without processing, the inconsistency between the TLP and metric regression tasks will lead to a seesaw phenomenon [30]. The negative correlation information can degrade the representation ability, thereby damaging performance. In short, the TLP branch and the NMD branch are devoted to generating representations that do not contain mutual information as much as possible, achieving the disentanglement of the two levels of information.

3.4 Attention-Based Fusion Module

In this section, we need to integrate beneficial information of the metric-aware representation $\{H_{\tau}^{L}\}_{\tau=t-\delta}^{t-1}$ into the TLP embedding $\{Z_{\tau}^{L}\}_{\tau=t-\delta}^{t-1}$, as it only leaves non-metric information after disentangling. We employ the attention mechanism to learn the fusion mode between $\{Z_{\tau}^{L}\}_{\tau=t-\delta}^{t-1}$ and $\{H_{\tau}^{L}\}_{\tau=t-\delta}^{t-1}$, which can discard the damaging information and capture meaningful information.

For each historical snapshot G_{τ} at time t, we first learn N nodes' attention vector $\mathbf{a}^{Z}, \mathbf{a}^{H} \in \mathbf{R}^{N \times 1}$:

$$\mathbf{a}^{Z} = tanh(Z_{\tau}^{L}W)\mathbf{q}, \quad \mathbf{a}^{H} = tanh(H_{\tau}^{L}W)\mathbf{q}, \tag{10}$$

where $tanh$ is a activation function, $W \in \mathbf{R}^{d1 \times d2}$ and $\mathbf{q} \in \mathbf{R}^{d2 \times 1}$ are learnable weights. As for node $v_i \in V$, we normalize its two attention coefficients a_i^Z, a_i^H into 2-dimensional probability $\mathbf{p}_i \in \mathbf{R}^2$ by the softmax function.

$$\mathbf{p}_i = softmax([a_i^Z, a_i^H]), i = 1, ..., N, \tag{11}$$

Then we linearly combine v_i's two embeddings, $\mathbf{z}_i^\tau = (Z_\tau)_i \in \mathbf{R}^{d_1}$ and $\mathbf{h}_i^\tau = (H_\tau)_i \in \mathbf{R}^{d_1}$, with the probability $\mathbf{p}_i = [p_{i0}, p_{i1}]$:

$$\widetilde{\mathbf{z}}_i^\tau = p_{i0}\mathbf{z}_i^\tau + p_{i1}\mathbf{h}_i^\tau, i = 1, ..., N, \tag{12}$$

where $\widetilde{\mathbf{z}}_i^\tau$ is v_i's embedding after fusing, which Eq. 4 will use to predict the probability of temporal links.

Finally, our method collaboratively infers temporal links and node metrics and jointly optimizes the loss functions of two branches to simulate network evolution in an interdependent manner. The total loss is summarized as follows:

$$\mathcal{L} = \sum_{t=1+\delta}^{t=T} \mathcal{L}_{TLP}^t + \alpha\mathcal{L}_{NMD}^t + \beta\mathcal{L}_{DIS}^t, \tag{13}$$

where δ is the number of snapshots, $\mathcal{L}_{TLP}^t$ and $\mathcal{L}_{NMD}^t$ are the TLP and NMD branch's losses, $\mathcal{L}_{DIS}^t$ is the decoupling loss, and α and β are trade-off weights.

4   Experiment

4.1   Experimental Setup

Datasets and Baselines. We conducted experiments on five dynamic graphs, including UCI [23], BC-Alpha [16,17], BC-OTC [16,17], DBLP [31], and APS[1]. Aiming to evaluate our NMD plugin's broad applicability, the compared baselines include three different types of TLP methods: one RNN-Based method (Dyngraph2Vec [9]), two GCN-Based methods (GCN-GRU [29], EvolveGCN [25]), and three Attention-Based methods (STGSN [22], DySAT [28], HTGN [35]).

Training Details. The common-used deep learning framework PyTorch [26] is adopted to implement all our experiments. We split the datasets into the training, validation, and testing set with ratios 7:1:2 chronologically. Referring to the general setting of the link prediction tasks, we randomly select 100 negative samples for each edge in each snapshot during data loading. The parameters of model architectures are fixed: all methods' structural encoder layers are 2, and the hidden layer dimension is 128. For a fair comparison, the downstream classifier for all methods is a trainable two-layer perceptron. The Adam optimizer [11] and an early-stopping strategy are adopted for model training.

4.2   Overall Performance

Table 1 shows the performance comparison and improvement of the original TLP methods and our method on five dynamic graphs. The default node metric of the experiment is degree centrality. Our NMD plugin achieves significant and

---

[1] https://journals.aps.org/datasets.

Table 1. AUC (%) performance of the original temporal link prediction methods (**Base**) and our method (**Ours**) on five dynamic graphs. The **Gain (%)** row reports the mean gain percentage of the six baselines. The reported results are the average scores on all timestamps of the testing set.

Case	UCI		BC-Alpha		BC-OTC		DBLP		APS	
	Base	Ours	Base	Ours	Base	Ours	Base	Ours	Base	Ours
Dyn2Vec	$83.62_{\pm1.03}$	$\mathbf{89.08_{\pm1.12}}$	$83.18_{\pm1.00}$	$\mathbf{87.56_{\pm1.01}}$	$75.93_{\pm1.01}$	$\mathbf{84.24_{\pm1.06}}$	$72.31_{\pm1.22}$	$\mathbf{74.75_{\pm1.26}}$	$71.71_{\pm0.18}$	$\mathbf{76.24_{\pm0.21}}$
GCN-GRU	$86.73_{\pm1.10}$	$\mathbf{87.55_{\pm1.11}}$	$85.43_{\pm1.26}$	$\mathbf{91.57_{\pm1.43}}$	$76.06_{\pm1.10}$	$\mathbf{85.06_{\pm1.41}}$	$73.33_{\pm1.02}$	$\mathbf{74.73_{\pm1.02}}$	$73.17_{\pm0.14}$	$\mathbf{78.61_{\pm0.17}}$
EvolveGCN	$87.09_{\pm1.18}$	$\mathbf{89.95_{\pm1.12}}$	$82.37_{\pm1.29}$	$\mathbf{84.82_{\pm1.46}}$	$81.05_{\pm1.34}$	$\mathbf{82.65_{\pm1.29}}$	$75.06_{\pm0.10}$	$\mathbf{77.02_{\pm0.10}}$	$72.35_{\pm0.12}$	$\mathbf{77.25_{\pm0.13}}$
STGSN	$88.04_{\pm1.06}$	$\mathbf{90.76_{\pm1.02}}$	$80.48_{\pm1.03}$	$\mathbf{83.79_{\pm1.03}}$	$78.85_{\pm1.09}$	$\mathbf{81.92_{\pm1.13}}$	$78.39_{\pm0.09}$	$\mathbf{79.62_{\pm0.08}}$	$71.14_{\pm0.09}$	$\mathbf{77.91_{\pm0.11}}$
DySAT	$87.51_{\pm0.04}$	$\mathbf{89.61_{\pm1.14}}$	$82.19_{\pm1.16}$	$\mathbf{85.96_{\pm1.33}}$	$78.24_{\pm1.15}$	$\mathbf{83.19_{\pm1.25}}$	$73.89_{\pm1.08}$	$\mathbf{75.60_{\pm0.21}}$	$72.14_{\pm0.14}$	$\mathbf{78.77_{\pm0.24}}$
HTGN	$88.60_{\pm0.15}$	$\mathbf{91.36_{\pm1.16}}$	$90.84_{\pm1.18}$	$\mathbf{94.31_{\pm1.32}}$	$80.97_{\pm1.17}$	$\mathbf{84.94_{\pm1.24}}$	$73.44_{\pm0.12}$	$\mathbf{78.05_{\pm0.26}}$	$73.62_{\pm0.19}$	$\mathbf{79.97_{\pm0.29}}$
Gain(%)	+2.79		+3.92		+5.15		+2.22		+5.77	

consistent improvements in six different baselines, with the mean gain of 2.79%, 3.92%, 5.15%, 2.22%, and 5.77% on five datasets, respectively. Part of the reason for our success is a more fine-grained embedding representation than baselines, which fully exploits the advantages of node metric factors by the decoupling loss and the attention-based fusion module. It reveals that our work is conducive to promoting the development of link predictions and recommender systems.

Table 2. AUC(%) scores and performance improvement of three loss terms and three node metrics on UCI and BC-Alpha, where DG, CN, and PG are abbreviations for degree centrality, closeness centrality, and PageRank, respectively.

Loss Term			Node Metric			UCI			BC-Alpha		
$\mathcal{L}^t_{TLP}$	$\mathcal{L}^t_{NMD}$	$\mathcal{L}^t_{DIS}$	DG	CN	PG	Dyn2Vec	GCN-GRU	DySAT	Dyn2Vec	GCN-GRU	DySAT
✓						83.62	86.73	87.51	83.18	85.43	82.19
✓	✓		✓			86.46 (+2.84)	83.34 (−3.39)	88.38 (+0.87)	86.66 (+3.48)	89.57 (+4.14)	83.43 (+1.24)
✓	✓	✓	✓			89.08 (+5.46)	87.55 (+0.82)	89.61 (+2.10)	87.56 (+4.38)	91.57 (+6.14)	85.96 (+3.77)
✓	✓	✓		✓		88.65 (+5.03)	87.36 (+0.63)	89.44 (+1.93)	89.03 (+5.85)	92.05 (+6.62)	83.79 (+1.60)
✓	✓	✓			✓	89.48 (+5.86)	88.32 (+1.59)	90.76 (+3.25)	84.68 (+1.50)	90.47 (+5.04)	85.56 (+3.37)
✓	✓	✓	✓	✓		90.65 (+7.03)	88.23 (+1.50)	91.41 (+3.90)	89.13 (+5.95)	93.79 (+8.36)	87.98 (+5.79)
✓	✓	✓		✓	✓	90.08 (+6.46)	89.86 (+3.13)	92.14 (+4.63)	89.71 (+6.53)	93.73 (+8.30)	89.06 (+6.87)
✓	✓	✓	✓	✓	✓	90.16 (+6.45)	90.06 (+3.33)	93.33 (+5.82)	90.29 (+7.11)	94.18 (+8.75)	89.57 (+7.38)

4.3 Ablation Study

To intuitively understand the effect of each part of our NMD plugin, we implement the ablation experiment consisting of three loss terms and three kinds of node metrics. The former includes $\mathcal{L}^t_{TLP}$, $\mathcal{L}^t_{NMD}$, and $\mathcal{L}^t_{DIS}$. The latter includes degree centrality (DG), closeness centrality (CN), and PageRank (PR). Table 2 shows the AUC scores and performance improvement under different experimental settings on UCI and BC-Alpha datasets.

Using only $\mathcal{L}^t_{NMD}$ without $\mathcal{L}^t_{DIS}$ may lead to performance degradation, such as a decrease of 3.39% of the GCN-GRU baseline on the UCI dataset. It indicates a seesaw phenomenon [30] in the TLP and NMD tasks, i.e., a negative correlation between the edge-level and node-level tasks, affecting the quality of the fused embedding. Hence we decouple the global properties of the nodes from the edge-level embedding of the TLP branch by the decoupling loss $\mathcal{L}_{DIS}^{t}$, which brings 0.82% improvements of the GCN-GRU baseline on the UCI dataset. It emphasizes the importance of disentangling node metric factors. In addition, the performance of different node metrics is very similar due to the general consistency of node attribute information in graph data. From the results, we find that our plugin achieves greater relative improvement as the number of node metric factors increases. We conjecture that disentangling more node metrics can obtain more fine-grained embedding with better representation ability.

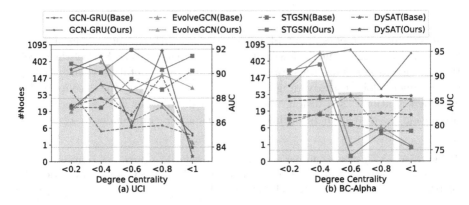

Fig. 2. AUC comparison over the degree centrality of node groups on UCI and BC-Alpha datasets, where the background histograms indicate the number of nodes (#Nodes) involved in each group. The dotted line and the solid line represent the TLP baselines and our improved methods, respectively.

4.4 Performance on Different Node Groups

To carefully explore the effectiveness of our plugin on nodes with different metrics values, we divide the node set into five groups according to the degree centrality of nodes. Figure 2 displays the AUC performance on five node groups with different levels on UCI and BC-Alpha datasets. As for the two GCN-Based methods (GCN-GRU and EvolveGCN), the performance improvement of medium to high degree nodes is more significant than those with low degree. While as for the two Attention-Based methods (STGSN and DySAT), the performance of the five groups of nodes has significantly improved. One possible reason is that the attention mechanism is better at capturing the correlation between node embedding and metrics in the historical link state to promote the generation of high-quality representations. In short, our plugin consistently promotes the representation learning of node groups with different metrics distribution ranges.

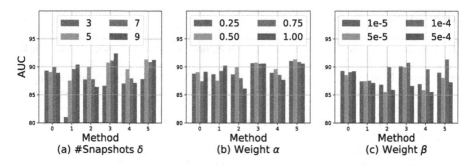

Fig. 3. Parameter sensitivity of six baselines based on our plugin. Methods are indexed in the following order: Dyn2vec, GCN-GRU, EvolveGCN, STGSN, DySAT, and HTGN.

4.5 Parameter Sensitivity

In this section, our parameter sensitivity experiment includes the number of historical snapshots (#Snapshots) and the trade-off weights α and β. The three hyper-parameters' default settings are 5, 0.50, and $1e^{-4}$. Figure 3 shows the performance comparison of different values over six baselines with our plugin on the UCI dataset. Firstly, most methods benefit from more historical snapshots because of more temporal information. Secondly, the trade-off weight α, defined as Eq. 13, controls the relative importance of the edge-level TLP loss and the node-level NMD loss in the dual-branch framework. The higher α performance is better, which proved that the NMD plugin achieved our expected effect, namely, learning fine-grained node embedding. Finally, the performance of different values of β is similar. It reflects that the decoupling loss can steadily improve the diversity and specialization of embedding representations.

5 Conclusion

In this paper, we point out the benefits and harmfulness of node metric factors in temporal link prediction representation. The proposed NMD plugin disentangles node metrics from the node embedding generated by most TLP methods. We further devise an attention-based module to explore fine-grain and high-quality embedding representation. In the future, we will extend our auxiliary components to predict continuous-valued timestamped links and explore the problem of mutual promotion and restriction between node-level and edge-level tasks.

Acknowledgements. This work is funded by the National Key Research and Development Project (Grant No: 2022YFB2703100), the Starry Night Science Fund of Zhejiang University Shanghai Institute for Advanced Study (Grant No. SN-ZJU-SIAS-001), and the Fundamental Research Funds for the Central Universities (2021FZZX001-23, 226-2023-00048).

References

1. Adamic, L.A., Adar, E.: Friends and neighbors on the web. Soc. Netw. **25**(3), 211–230 (2003)
2. Ahmad, I., Akhtar, M.U., Noor, S., Shahnaz, A.: Missing link prediction using common neighbor and centrality based parameterized algorithm. Sci. Rep. **10**(1), 1–9 (2020)
3. Barabási, A.L., Albert, R.: Emergence of scaling in random networks. Science **286**(5439), 509–512 (1999)
4. Barros, C.D., Mendonça, M.R., Vieira, A.B., Ziviani, A.: A survey on embedding dynamic graphs. ACM Comput. Surv. (CSUR) **55**(1), 1–37 (2021)
5. Bedi, P., Sharma, C.: Community detection in social networks. Wiley Interdiscip. Rev. Data Mining Knowl. Discov. **6**(3), 115–135 (2016)
6. Divakaran, A., Mohan, A.: Temporal link prediction: a survey. N. Gener. Comput. **38**, 213–258 (2020)
7. Dunlavy, D.M., Kolda, T.G., Acar, E.: Temporal link prediction using matrix and tensor factorizations. ACM Trans. Knowl. Discov. Data (TKDD) **5**(2), 1–27 (2011)
8. Freeman, L.C.: Centrality in social networks conceptual clarification. Soc. Netw. **1**(3), 215–239 (1978)
9. Goyal, P., Chhetri, S.R., Canedo, A.M.: Capturing network dynamics using dynamic graph representation learning. US Patent App. 16/550,771 (2020)
10. Katz, L.: A new status index derived from sociometric analysis. Psychometrika **18**(1), 39–43 (1953)
11. Kingma, D.P., Ba, J.: Adam: a method for stochastic optimization. arXiv preprint arXiv:1412.6980 (2014)
12. Kipf, T.N., Welling, M.: Semi-supervised classification with graph convolutional networks. arXiv preprint arXiv:1609.02907 (2016)
13. Kitsak, M., et al.: Identification of influential spreaders in complex networks. Nat. Phys. **6**(11), 888–893 (2010)
14. Klemm, K., Serrano, M., Eguíluz, V.M., Miguel, M.S.: A measure of individual role in collective dynamics. Sci. Rep. **2**(1), 1–8 (2012)
15. Kong, X., Shi, Y., Yu, S., Liu, J., Xia, F.: Academic social networks: modeling, analysis, mining and applications. J. Netw. Comput. Appl. **132**, 86–103 (2019)
16. Kumar, S., Hooi, B., Makhija, D., Kumar, M., Faloutsos, C., Subrahmanian, V.: Rev2: fraudulent user prediction in rating platforms. In: Proceedings of the Eleventh ACM International Conference on Web Search and Data Mining, pp. 333–341. ACM (2018)
17. Kumar, S., Spezzano, F., Subrahmanian, V., Faloutsos, C.: Edge weight prediction in weighted signed networks. In: 2016 IEEE 16th International Conference on Data Mining (ICDM), pp. 221–230. IEEE (2016)
18. Lin, D., Wu, J., Xuan, Q., Chi, K.T.: Ethereum transaction tracking: inferring evolution of transaction networks via link prediction. Phys. A **600**, 127504 (2022)
19. Lorrain, F., White, H.C.: Structural equivalence of individuals in social networks. J. Math. Sociol. **1**(1), 49–80 (1971)
20. Ma, J., Cui, P., Kuang, K., Wang, X., Zhu, W.: Disentangled graph convolutional networks. In: International Conference on Machine Learning, pp. 4212–4221. PMLR (2019)
21. Ma, J., Zhou, C., Cui, P., Yang, H., Zhu, W.: Learning disentangled representations for recommendation. In: Advances in Neural Information Processing Systems, vol. 32 (2019)

22. Min, S., Gao, Z., Peng, J., Wang, L., Qin, K., Fang, B.: Stgsn-a spatial-temporal graph neural network framework for time-evolving social networks. Knowl.-Based Syst. **214**, 106746 (2021)
23. Opsahl, T., Panzarasa, P.: Clustering in weighted networks. Soc. Netw. **31**(2), 155–163 (2009)
24. Page, L., Brin, S., Motwani, R., Winograd, T.: The pagerank citation ranking: bring order to the web. Technical report, Stanford University (1998)
25. Pareja, A., et al.: Evolvegcn: evolving graph convolutional networks for dynamic graphs. In: Proceedings of the AAAI Conference on Artificial Intelligence, vol. 34, pp. 5363–5370 (2020)
26. Paszke, A., et al.: Pytorch: an imperative style, high-performance deep learning library. In: Advances in Neural Information Processing Systems, vol. 32 (2019)
27. Qin, M., Yeung, D.Y.: Temporal link prediction: a unified framework, taxonomy, and review. arXiv preprint arXiv:2210.08765 (2022)
28. Sankar, A., Wu, Y., Gou, L., Zhang, W., Yang, H.: Dysat: deep neural representation learning on dynamic graphs via self-attention networks. In: Proceedings of the 13th International Conference on Web Search and Data Mining, pp. 519–527 (2020)
29. Seo, Y., Defferrard, M., Vandergheynst, P., Bresson, X.: Structured sequence modeling with graph convolutional recurrent networks. In: Cheng, L., Leung, A.C.S., Ozawa, S. (eds.) ICONIP 2018. LNCS, vol. 11301, pp. 362–373. Springer, Cham (2018). https://doi.org/10.1007/978-3-030-04167-0_33
30. Tang, H., Liu, J., Zhao, M., Gong, X.: Progressive layered extraction (PLE): a novel multi-task learning (MTL) model for personalized recommendations. In: Proceedings of the 14th ACM Conference on Recommender Systems, pp. 269–278 (2020)
31. Tang, J., Zhang, J., Yao, L., Li, J., Zhang, L., Su, Z.: Arnetminer: extraction and mining of academic social networks. In: Proceedings of the 14th ACM SIGKDD International Conference on Knowledge Discovery and Data Mining, pp. 990–998 (2008)
32. Travençolo, B.A.N., Costa, L.D.F.: Accessibility in complex networks. Phys. Lett. A **373**(1), 89–95 (2008)
33. Wang, X., Jin, H., Zhang, A., He, X., Xu, T., Chua, T.S.: Disentangled graph collaborative filtering. In: Proceedings of the 43rd International ACM SIGIR Conference on Research and Development in Information Retrieval, pp. 1001–1010 (2020)
34. Wu, J., et al.: Disenkgat: knowledge graph embedding with disentangled graph attention network. In: Proceedings of the 30th ACM International Conference on Information & Knowledge Management, pp. 2140–2149 (2021)
35. Yang, M., Zhou, M., Kalander, M., Huang, Z., King, I.: Discrete-time temporal network embedding via implicit hierarchical learning in hyperbolic space. In: Proceedings of the 27th ACM SIGKDD Conference on Knowledge Discovery & Data Mining, pp. 1975–1985 (2021)
36. Yu, W., Aggarwal, C.C., Wang, W.: Temporally factorized network modeling for evolutionary network analysis. In: Proceedings of the Tenth ACM International Conference on Web Search and Data Mining, pp. 455–464 (2017)
37. Zhou, T., Lü, L., Zhang, Y.C.: Predicting missing links via local information. Eur. Phys. J. B **71**, 623–630 (2009)

Action Prediction for Cooperative Exploration in Multi-agent Reinforcement Learning

Yanqiang Zhang, Dawei Feng$^{(\boxtimes)}$, and Bo Ding

National Laboratory for Parallel and Distributed Processing,
National University of Defense Technology, Changsha 410073, China
{zhangyq1119,dingbo}@nudt.edu.cn, davyfeng.c@qq.com

Abstract. Multi-agent reinforcement learning methods have shown significant progress, however, they continue to exhibit exploration problems in complex and challenging environments. To address the above issue, current research has introduced several exploration-enhanced methods for multi-agent reinforcement learning, they are still faced with the issues of inefficient exploration and low performance in challenging tasks that necessitate complex cooperation among agents. This paper proposes the prediction-action Qmix (PQmix) method, an action prediction-based multi-agent intrinsic reward construction approach. The PQmix method employs the joint local observation of agents and the next joint local observation after executing actions to predict the real joint action of agents. The method calculates the action prediction error as the intrinsic reward to measure the novel of the joint state and encourages agents to actively explore the action and state spaces in the environment. We compare PQmix with strong baselines on the MARL benchmark to validate it. The result of experiments demonstrates that PQmix outperforms the state-of-the-art algorithms on the StarCraft Multi-Agent Challenge (SMAC). In the end, the stability of the method is verified by experiments.

Keywords: Multi-agent Systems · Reinforcement Learning · Intrinsic Reward

1 Introduction

Cooperative multi-agent reinforcement learning (MARL) is required to solve many real-world control problems [2,15,20]. MARL faces two main challenges: scalability, which poses multiple challenges to the multi-agent algorithms as the number of agents increases; and local observation, where the communication constraints make it necessary for agents to execute policies exclusively based on their individual observations. Existing research algorithms tackle the above challenges through methods such as parameter sharing and centralized training distributed

B. Luo et al. (Eds.): ICONIP 2023, LNCS 14448, pp. 358–372, 2024.
https://doi.org/10.1007/978-981-99-8082-6_28

execution (CTDE). Based on these above methods, researchers have developed various classical MARL algorithms, including policy-based MADDPG [11], value function-based VDN [23], Qmix [18], QTRAN [21], QPLEX [25]. These approaches have shown progress and effective results in multi-agent tasks [10,11].

Although MARL has been successful in numerous tasks, these above methods often employ ϵ-greedy exploration of the environment and cannot efficiently solve complex and difficult cooperative tasks [21,25]. Current academics have conducted several studies in the field of multi-agent exploration and have proposed some classical exploration methods, such as EDTI [27] uses an information gain approach to quantify the mutual influence among agents to encourage agents to explore the action and state spaces. Even though the method shows good results, it requires calculating the interactions among agents, making it less scalable as the number of agents increases. Another approach called MAVEN [12], which introduces a hierarchically controlled potential space, mixes value function-based and policy-based approaches, and lets the hierarchical policy control the potential variables, which determine the actions of the value function-based agents, and the approach is not efficient in complex tasks with large joint state spaces. In addition, LIIR [4] parameterizes the intrinsic reward function for each agent to guide the diverse behavior of the agents to fully explore the environment, which is experimental in simple tasks and does not work well in more complex task environments. A relatively novel and effective multi-agent reinforcement learning exploration method is RODE [26], in which each agent makes decisions by first choosing a role, then selecting the policy based on the role to improve the exploration ability of multi-agents. Despite the effectiveness of current multi-agent exploration methods, there is still the problem of poor performance in complex and difficult cooperative tasks.

To address the above problem, based on the classical Qmix method, our paper proposes an intrinsic reward method based on action prediction, the PQmix method, to improve the processing ability of agents for complex cooperative tasks. The method includes a prediction model that takes the joint local observation and the next joint local observation of agents as input, and the model outputs the predicted joint action. The method uses the prediction error between the predicted joint action and the actual joint action as the intrinsic reward to encourage agents to explore the environment. The PQmix algorithm offers several advantages over previous approaches. It is scalable, with the number of agents having little impact on the computational resource consumption and its performance. In addition, the algorithm actively encourages agents to explore action and state spaces in environmental tasks that do not have extrinsic rewards that may lead to optimal performance. The algorithm's curiosity module, based on action prediction, evaluates the novelty of the joint state and adds the intrinsic reward to the novel action space and state space, thus enabling agents to discover better cooperative strategies and maximize their cumulative rewards.

In this paper, we compare the PQmix algorithm with other classical and effective methods in several complex cooperative tasks in the SMAC environment [19]. Further, we visualize the convergence and final performance of the

PQmix method and verify the significant effectiveness and advantages of the method through experimental results. Additionally, this paper experimentally verifies the algorithm's robustness under various circumstances by testing different parameter combinations.

2   Related Work

In the field of single-agent reinforcement learning, exploration methods have been well investigated in current research. A summary of such methods has been provided in the paper [7]. For example, as opposed to using count-based methods to assess the novelty of states [1], there are some studies that have attempted to use pseudo-count methods [24] to calculate intrinsic rewards to assess the novelty of states, and these methods have better relative to the former assessment ability; in addition, there are methods that use the prediction error of the prediction network as the intrinsic reward such as the ICM algorithm [16], which learns a forward model and a backward model, the former model serves to take the current state and the action of the agent to predict the next state and use that prediction error as the intrinsic reward, and the latter model serves to reduce the environmental random factors on the prediction. In addition to these, there are other methods for the exploration problem, such as feature control [3], and decision states [5], Bayesian networks [6], which uses the reduction of the uncertainty of the Bayesian network weights as the intrinsic reward to explore the environment.

Although exploration approaches have been very successful in the single-agent domain, multi-agent reinforcement learning exploration approaches are still under development. The LIIR approach [4] updates and guides the agents' algorithms by learning intra-personal rewards to improve the agents' exploration capabilities. Jaques [9] uses the interactions among agents as intrinsic rewards to encourage agents to perform behaviors that can influence other agents. The paper [27] proposed the EDTI and EITI algorithms, which use mutual information (MI) to measure the influence among agents on the transfer function and reward constructs respectively, to encourage the agents to fully explore the environment. An effective method RODE [26] improves agent exploration by giving roles to agents in order to assign specific strategies. These methods have achieved better results in complex tasks, but the exploration efficiency and performance of the above methods are reduced when faced with complex cooperative tasks, which is still one of the serious challenges of multi-agent exploration approaches. The PQmix method is based on the ICM method for the training of the inverse model and proposes a novel exploration in the MARL domain method as a way to improve the exploration ability of agents.

3   Notation

First, we model the multi-agent cooperation task as Dec-POMDP [13], which is defined as the tuple $G = <\mathcal{N}, \mathcal{S}, \mathcal{A}, P, R, \Omega, O, n, \gamma >$, where $\mathcal{N}$ is the set of n agents, $\mathcal{S}$ is the joint state space of the environment, and $\mathcal{A}$ is the joint action space. $\mathcal{A}^i$ is the action space of the agent i, and the relationship between them is $\mathcal{A} := \mathcal{A}^1 \times \cdots \times \mathcal{A}^n$. P denotes the state transfer probability matrix from $s \in \mathcal{S}$ state to any state s' according to the joint action a, denoted as $P : \mathcal{S} \times \mathcal{A} \times \mathcal{S}' \rightarrow [0,1]$. R defines the reward which the team receives from s, a to s', expressed by the formula: $\mathcal{S} \times \mathcal{A} \times \mathcal{S} \rightarrow \mathbb{R}$. The $\gamma \in [0,1)$ is the discount factor. De-POMDP is set to be partially observable, i.e., at each time step, the agent $i \in \mathcal{I}$ can only access the observation function $o_i \in \Omega$, which is derived from the $O(s,i)$ of the agent i, the joint local observation is defined as $o \equiv [o_i]_{i=1}^n$. In addition, each agent has an action observation history $\tau_i \in \mathcal{T} \equiv (\Omega \times \mathcal{A})^* \times \Omega$ and constructs the individual agent's strategy to jointly maximize the team reward. Each agent chooses an action $a_i \in \mathcal{A}$ to form a joint action $a \equiv [a_i]_{i=1}^n \in \mathcal{A}$ interacts with the environment to generate the team's $r = R(s,a)$ and the next state s' according to the state transfer probability matrix $P(s'|s,s)$. The objective function is to find a joint strategy π that maximizes a joint value function $V^\pi(s) = \mathbb{E}[\sum_{t=0}^\infty \gamma^t r_t | s = s_0, \pi]$, or a joint action-value function $Q^\pi(s,a) = r(s,a) + \gamma\mathbb{E}_{s'}[V^\pi(s')]$ maximizes.

The CTDE centralized training distributed approach is widely used in multi-agents deep reinforcement learning [14], where the agents' policies are trained by a centralized approach and use the global information during training, while the agents select actions based on local observation history and their own policies during sampling. During training, the global cooperative learning of the optimal joint action value function $Q_{\text{tot}}(s,a) = r(s,a) + \gamma\mathbb{E}_{s'}[\max_{a'} Q_{\text{tot}}(s',a')]$. Due to partial observation, this paper uses $Q_{\text{tot}}(\tau,a;\theta)$ instead of $Q_{\text{tot}}(s,a;\theta)$, where $\tau \in \mathcal{T}$. Then, the Q-value neural network is trained to minimize the following expected TD error.

$$L(\theta) = \mathbb{E}_{\tau,a,r,\tau' \in D} \left[r + \gamma V(\tau')\theta^-) - Q_{\text{tot}}(\tau,a;\theta) \right]^2, \qquad (1)$$

where D is the reply buffer and θ^- denotes the parameters of the target network, θ are the parameters of the training network that is interactively sampled with the environment, and the parameters of the target network are periodically updated by the training network θ. $V(\tau';\theta^-)$ is the expected future return of TD. The individual agent has access only to the local action and observation histories. The joint Q-value function requires inference based on the individual Q-value function $Q_i(\tau_i, a_i)$, so the relationship between the individual Q-value function $Q_i(\tau_i, a_i)$ and the joint Q-value function Q_{tot} is the factorization structure [18,23].

4 Algorithm

This section of the paper describes the framework of the algorithm, intrinsic reward, and implementation of PQmix. The algorithm uses prediction errors of joint actions as the intrinsic reward to assess the novelty of the joint state, thereby guiding agents to explore the environment adequately. After fully exploring the environment and collecting rich trajectory information, the agent algorithm facilitates the learning of states and actions with high team expectation rewards.

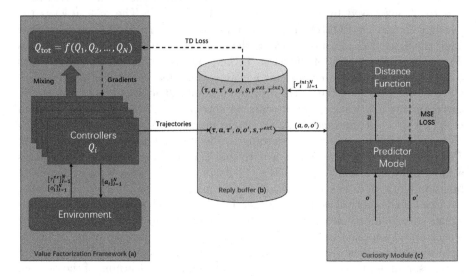

Fig. 1. The algorithm architecture of PQmix.

4.1   Framework of PQmix and Intrinsic Reward

Figure 1 (a) shows the value function module, an example of CDTE centralized training distributed execution. In this approach, local agents execute policies based on their own Q-value functions that take locally observed trajectories as input and are updated by a centralized module with global training information. The value function decomposition approach is chosen as the basis of this algorithm due to its higher robustness and effectiveness during experimentation. This section explores how to measure the novelty of the joint state of agents to encourage full exploration of the environment. The current joint local observation $(o_i, \cdots, o_n)$ and the next joint local observation $(o'_i, \cdots, o'_n)$ of the agents are connected to the joint action $(a_i, \cdots, a_n)$ taken by the agent. The ICM method models the environment dynamics by predicting actions between two states. Based on these concepts, this paper uses two consecutive joint local observations to predict joint actions that lead to changes in agents' local observations. The environmental dynamics model of the multi-agent task is learned based on the error of this prediction, and the novelty of the joint state is evaluated.

Figure 1 depicts the PQmix curiosity module combined with a value function-based decomposition approach. The curiosity module comprises a predictor model that evaluates the joint state's novelty and a distance function that utilizes samples from the replay buffer for training. The prediction model takes the current joint local observation o, the next joint local observation o', and the real joint action a as inputs, and updates the predictor model by the error between the predicted joint action $\bar{a} = [\bar{a}_i]_{i=1}^n$ and the actual joint action. The distance function calculates the distance between $\bar{a}$ and a, such as L_2 or L_1. In this paper, the model is trained using the mean squared error (MSE) of the minimized distance. The formula is:

$$L\phi = \frac{1}{N} \sum_{j=1}^{N} \frac{1}{n} \sum_{i=1}^{n} (a_{j,i} - \bar{a}_{j,i})^2, \tag{2}$$

where N is the number of samples sampled from the reply buffer, and ϕ are the parameters of the predictor model. The intrinsic reward is defined as

$$r_t^{int} = \frac{1}{n} \sum_{i=1}^{n} (a_{t,i} - \bar{a}_{t,i})^2. \tag{3}$$

The intrinsic reward r^{int} is combined with the extrinsic reward r^{ext} to form the combined reward value $r = r_t^{ext} + cr_t^{int}$, where c is the linear annealing parameter that gradually decreases over the training period. It encourages extensive exploration of the agent algorithm in the early training period while reducing the encouragement for environmental exploration as the training progresses to maximize the agent's cumulative rewards. The independence of curiosity provides an added advantage: the architecture of the PQmix method can be applied to many multi-agent algorithms using the CDTE method. Specifically, the generic function f in Fig. 1 can represent specific value decomposition structures such as VDN [23], Qmix [18], and QPLEX [25]. This paper uses the Qmix approach for the value function decomposition method. The curiosity module, based on action prediction, allows for diverse and effective exploration of the environment by introducing the curiosity-driven prediction error into the general multi-agent reinforcement learning (MARL) algorithm.

4.2 Implementation

The PQmix method is improved on the basis of the classical Qmix algorithm, the connection between individual Q values and global Q values is established by mixing network, and the loss function of Q_{tot} is:

$$L(\boldsymbol{\theta}) = \sum_{i=1}^{n} \left[\left(y_i^{tot} - Q_{tot}(\boldsymbol{\tau}, \boldsymbol{a}, \boldsymbol{s}; \boldsymbol{\theta}) \right)^2 \right], \tag{4}$$

where $\boldsymbol{\theta}$ are parameters of Q_{tot}, and where $y^{tot} = r + \gamma \max_{\boldsymbol{a}'} Q_{tot}(\boldsymbol{\tau}', \boldsymbol{a}', \boldsymbol{s}') \boldsymbol{\theta}^-)$, $\boldsymbol{\theta}^-$ are the parameters of the target Q network, γ is the discount factor, where r is the sum of extrinsic and intrinsic rewards and is controlled using linear annealing.

$$L = L(\boldsymbol{\theta}) + b * L(\boldsymbol{\phi}), \tag{5}$$

where b is the hyper-parameter that controls the update magnitude of the predictor model, setting different combinations of parameters b and c depending on the environmental task. The PQmix method uses Adam as the optimizer in the training process:

$$\boldsymbol{\theta}, \boldsymbol{\phi} = Adam(\boldsymbol{\theta}, \boldsymbol{\phi}, \nabla L, \eta). \tag{6}$$

This algorithm uses TD(λ) to replace the representation of the objective value function y^{tot}, $y_t^{tot} = \lambda \gamma y_{t+1}^{tot} + (r_t + (1 - \lambda)\gamma Q_{tot}(\boldsymbol{\tau}', \boldsymbol{a}', \boldsymbol{s}')\theta^-))$, where λ is set to 0.6, while the learning rate η of both the value function framework and the curiosity module is 0.001, and the number of samples n randomly selected from the replay buffer at each update of the algorithm is 128.

5   Experiments

In this section, experiments are conducted to evaluate the PQmix algorithm in several difficult tasks of SMAC. We compare the PQmix algorithm approach with the classical and effective multi-agent reinforcement learning approaches. Additionally, the robustness of PQmix method is evaluated.

5.1   Environment

The SMAC environment provides a diverse and differentiated set of agents, each with a different action, and the environment allows for complex cooperation of agents. In this paper, we evaluate the PQmix approach in the SMAC benchmark, where a multi-agent algorithm controls a set of units with mixed types that need to cooperate to defeat another set of enemy units with mixed types controlled by fixed heuristic rules. This experiment chooses SMAC version 4.6 and the adversary difficulty as 7 (the default difficulty of the SMAC environment). As in most methods, this paper uses the default environment settings of SMAC. Each agent is allowed to observe its own state within its observation range, the statistics of other units, such as blood, position, and type of other units. Agents can only attack enemies within firing range. During the battle, agents will be awarded both when the battle is won and the enemy unit is killed or wounded. Each task will have a step limitation, and the task may end without a victory or failure. The 2c_vs_64zg task and corridor task respectively are in Fig. 2. The 2c_vs_64zg (Fig. 2(a)) is an asymmetric confrontation task between different unit types, where we use 2 Colossi against 64 Zerglings with fast speed, lower the life, and lower the damage. The corridor (Fig. 2(b)) is a difficult confrontation task between different unit types, in which we use 6 Zealots against 24 Zerglings.

5.2   Baselines

We compare the PQmix method with several classical and state-of-the-art methods in a range of difficult multi-agent tasks. Qmix [18], a classical multi-agent reinforcement learning method, which uses networks to establish the relationship between individual Q and Q_{tot}, the method is effective. In addition, the

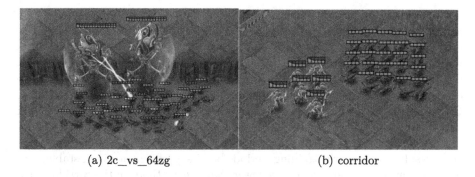

(a) 2c_vs_64zg   (b) corridor

Fig. 2. Example tasks of the SMAC environment.

article [8] improves Qmix by adding some tricks related to the performance improvement of the algorithm, so that it achieves more SOTA results in most environments. We take the Fintuned-Qmix (Qmix) as one of the comparison baselines. QTRAN [21] and WQmix [17] improve the expressiveness of Qmix using true Q-value networks and theoretical constraints, respectively. Moreover, we use the more effective centralized weight function for the WQmix method, i.e., the CWQmix method. The article [28] proposes the DOP algorithm, which combines a policy-based method (MADDPG) and Qmix. LICA method [29] is a policy-based multi-agent algorithm that uses a centralized Critic network and a hyper network to make full use of state information to make judgments on individual agents. This method also uses an improved adaptive entropy regularization method that makes the size of the entropy term gradient dynamically rescaled to improve the exploration ability of the agents. The Vmix method [22] addresses the problem that Qmix does not work well in parallel training, and extends the idea of value function-based decomposition to a policy-based approach, specifically, by combining the A2C and Qmix methods and extending the monotonicity constraint to the value network Critic. In addition, we compare our method with the advanced role-based exploration method RODE [26].

5.3   Comparison Experiment in SMAC Environment

In this section, we evaluate the performance of the PQmix algorithm and other algorithms in challenging SMAC tasks. Each algorithm is trained with six random seeds in each task. For the PQmix algorithm, we tested two sets of hyperparameter combinations: c = [0.005, 0.002, 0.001, 0.0008, 0.0005, 0.0002, 0.0001] and b = [1, 0.1, 0.01]. The superior combination of parameters is shown in Table 1. The reason we use the above set of range parameters is that excessively high parameter c will make the agent focus on exploration rather than solving the task, while too small the parameter c will fail to reflect the role of intrinsic rewards; similarly, excessively high parameter b will prevent the curiosity module from converging, while too small the parameter b will make it fall into the local optimum.

In Fig. 3, the horizontal coordinate is the number of training steps by the algorithm, the vertical coordinate is the average of win rates. The solid line is the average of the win rates, and the range of the region near the solid line is its range of variance. As results in Fig. 3 show, on one hand, PQmix outperforms QTRAN, DOP, Vmix, and LICA, which failed to learn valid information in more challenging tasks such as the corridor task. On the other hand, PQmix, Qmix, CWQmix, and RODE performed well in difficult environments. However, PQmix outperforms all other methods in all eight difficult tasks that need complex cooperation, particularly in the 3s_vs_5z task. In the 2c_vs_64zg task, RODE achieves good results in the early training period, but its performance is unstable, and its final convergence performance is lower than our method. In the corridor task, PQmix exhibits better convergence speed and effectiveness compared with other methods. In the 5m_vs_6m, 6h_vs_8z, 8m_vs_9m, 10m_vs_11m, and MMM2 tasks, PQmix significantly outperforms all other methods, especially in the 5m_vs_6m and 10m_vs_11m tasks where its final convergence performance is much better than others. In all tasks, PQmix exhibits lower variance during training, indicating its stable performance across different seed settings and challenging tasks. The experimental results mean that the curiosity module proposed in this paper improves the exploration capabilities of Qmix, making Qmix's performance close to or even exceed other advanced methods. In summary, the PQmix algorithm shows good exploration ability, better robustness, faster convergence speed, and excellent average convergence performance compared with other strong methods.

Table 1. Table of important parameter settings for PQmix in SMAC tasks.

Task	c	b
2c_vs_64zg	0.0001	0.1
3s_vs_5z	0.002	0.1
5m_vs_6m	0.002	0.1
6h_vs_8z	0.002	1
8m_vs_9m	0.005	0.01
10m_vs_11m	0.002	1
corridor	0.0001	1
MMM2	0.0008	0.1
3s5z_vs_3s6z	0.002	1

In summary, possible reasons for the effectiveness of the PQmix method in the above difficult tasks are as follows: Firstly, the curiosity module proposed in this paper is accurate and effective for the novelty of joint state assessment; Secondly, states without external rewards may be some beneficial states and worth being explored, for example, in the corridor task, agents are required to lure enemies and finally defeat them by cooperation among agents. The PQmix method encourages agents to explore the action and state space that have no

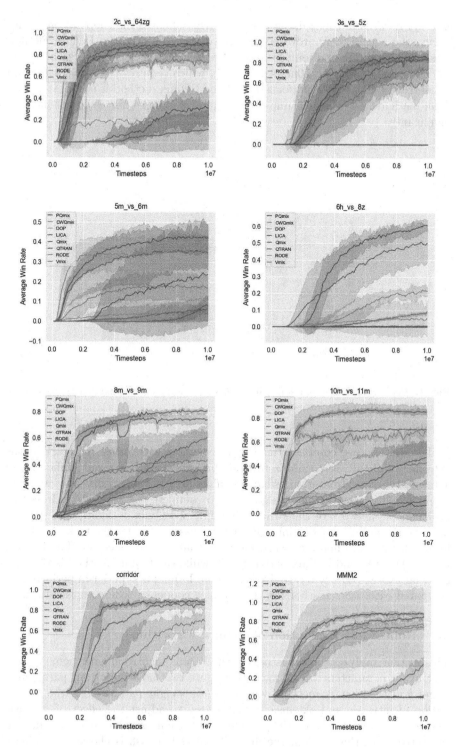

Fig. 3. Performance evaluation of PQmix. Compared with Qmix (Fintuned-Qmix), QTRAN, WQmix, DOP, LICA, Vmix, and RODE, PQmix makes great progress in sample efficiency and performance.

external rewards in the training period, which enhances the cooperation ability among the agents to improve their performance. Thirdly, the PQmix method uses the joint action and joint local observation, and when the number of agents increases, such as in the 10m_vs_11m task, which requires the agent algorithm to control 10 agents, the performance of the PQmix method is still better than other methods, and the consumption of computational resources do not increase significantly.

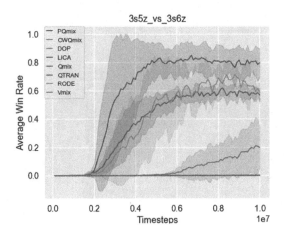

Fig. 4. Performance comparison with other methods in the extremely hard task 3s5z_vs_3s6z, PQmix performs pretty well.

In addition, we compare the PQmix method with other methods in the extremely difficult 3s5z_vs_3s6z task, despite the task being very difficult and a mixed-types confrontation, which requires strong exploratory and cooperative capabilities for the agent algorithm to complete the task. To address this problem, the network of the curiosity module of PQmix is slightly extended to improve the ability of the joint state novelty assessment for the task. As shown in Fig. 4, there are many classical methods that are less effective. The RODE, Qmix, QTRAN, and PQmix perform well, while the PQmix method is the only one that exceeds 80% win rate. In conclusion, PQmix is superior in performance compared with other methods in the complex cooperative 3s5z_vs_3s6z task.

5.4   Robustness Evaluation of PQmix Algorithm

This section analyzes the robustness of the PQmix algorithm when some important hyper-parameters are changed. There are two most important parameters c and b in the PQmix method, which are the weight parameter of the combination of intrinsic and extrinsic rewards used to adjust the effect of intrinsic rewards on exploration, and the parameter of the curiosity module loss function, respectively. In this section, the difficult 8m_vs_9m task is chosen to evaluate

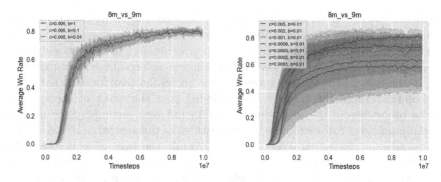

Fig. 5. The left and right figures fix the parameters c and b respectively, PQmix robustness evaluation analysis in the 8m_vs_9m task.

the robustness of the PQmix method for testing. 6 seeds are used for each curve in the two figures shown in Fig. 5. Specifically, as shown in Fig. 5(a) the PQmix method is adjusted for the parameter b of the gradient based on the three values [1, 0.1, 0.01] under the setting of fixed parameter $c = 0.005$, from which it can be found that the method is less sensitive to the parameter b under the fixed parameter c. In Fig. 5(b), we fix the gradient of parameter $b = 0.01$ and make the intrinsic reward parameter c adjusted according to [0.005, 0.002, 0.001, 0.0008, 0.0005, 0.0002, 0.0001], from which the analysis shows that the PQmix method is more sensitive to intrinsic reward parameter. In summary, from the combination of different parameters in the figure, we can know that the PQmix method is more sensitive to the change of the intrinsic reward parameter c, while the change of parameter b has less influence on the performance of the method, so the method has some robustness.

Table 2. The average performance results of PQmix with different parameter combinations after training 10M timesteps in the 8m_vs_9m task.

c	b		
	1	0.1	0.01
0.0001	0.73	0.76	0.67
0.0002	0.77	0.48	0.63
0.0005	0.68	0.72	0.59
0.0008	0.76	0.63	0.72
0.001	0.58	0.75	0.71
0.002	0.76	0.78	0.77
0.005	0.79	0.78	**0.81**

To further analyze the robustness of the algorithm when the parameters are changed, we subject the PQmix method to several experiments with other settings of parameter combination and draw a table comparing the average performance of the algorithm after 10M timesteps of training in the 8m_vs_9m task. As shown in Table 2, we combine the b and c parameters in the sets mentioned above, and record the average performance under each combined parameter setting in the table, with 6 seeds trained in the 8m_vs_9m task for each parameter combination. From the table, it is concluded that the effect of changes in b on the robustness of the algorithm is well for different settings of the intrinsic reward parameter c, for example, the robustness of the algorithm is stable when the parameter c is set as 0.005 and 0.002, while the average win rate of the algorithm after training is more variable under other parameter settings. From the analysis in the table, it can be obtained that the appropriate parameter c has some improvement for the robustness of the algorithm. From the table, it can be concluded that the average win rate of the PQmix method is more than 80% in the 8m_vs_9m task using the $b = 0.01, c = 0.005$ parameter combination, and the method is superior with the above parameter combinations. In this section, we make the PQmix method conduct extensive experiments with different parameter combination settings in various complex tasks and obtain excellent hyperparameter combinations as shown in Table 1.

6   Conclusion

This paper introduces the PQmix method, which utilizes action prediction to improve the efficiency of multi-agent algorithm exploration using a curiosity-driven module. PQmix makes changes to the classical Qmix method by adding the curiosity module to measure the novelty of the joint state and uses the error of action prediction as the intrinsic reward to encourage agents to fully explore the joint state space. More we evaluate the exploratory and cooperative capabilities of the PQmix method in the SMAC environment. The PQmix method is compared with other classical methods in nine complex cooperative tasks, and the results show that it has a significant advantage over others in terms of final performance and convergence speed. In addition, this paper measures the performance robustness of the PQmix method in harder tasks and verifies that the method has some robustness. In the future, we will experiment with more novel ways of measuring the novelty of joint states and will add diversity rewards among agents to encourage different agents to perform differentiated actions without increasing resource consumption.

Acknowledgements. This work is partially supported by the major Science and Technology Innovation 2030 "New Generation Artificial Intelligence" project 2020AAA0104803.

References

1. Bellemare, M., Srinivasan, S., Ostrovski, G., Schaul, T., Saxton, D., Munos, R.: Unifying count-based exploration and intrinsic motivation. In: Advances in Neural Information Processing Systems, vol. 29 (2016)

2. Buşoniu, L., Babuška, R., De Schutter, B.: Multi-agent reinforcement learning: an overview. In: Srinivasan, D., Jain, L.C. (eds.) Innovations in Multi-agent Systems and Applications-1, pp. 183–221. Springer, Heidelberg (2010). https://doi.org/10.1007/978-3-642-14435-6_7

3. Dilokthanakul, N., Kaplanis, C., Pawlowski, N., Shanahan, M.: Feature control as intrinsic motivation for hierarchical reinforcement learning. IEEE Trans. Neural Netw. Learn. Syst. **30**(11), 3409–3418 (2019)

4. Du, Y., Han, L., Fang, M., Liu, J., Dai, T., Tao, D.: LIIR: learning individual intrinsic reward in multi-agent reinforcement learning. In: Advances in Neural Information Processing Systems, vol. 32 (2019)

5. Goyal, A., et al.: Infobot: transfer and exploration via the information bottleneck. arXiv preprint arXiv:1901.10902 (2019)

6. Graves, A.: Practical variational inference for neural networks. In: Advances in Neural Information Processing Systems, vol. 24 (2011)

7. Hao, J., et al.: Exploration in deep reinforcement learning: from single-agent to multiagent domain. IEEE Trans. Neural Netw. Learn. Syst. (2023)

8. Hu, J., Jiang, S., Harding, S.A., Wu, H., Liao, S.W.: Rethinking the implementation tricks and monotonicity constraint in cooperative multi-agent reinforcement learning (2021)

9. Jaques, N., et al.: Social influence as intrinsic motivation for multi-agent deep reinforcement learning. In: International Conference on Machine Learning, pp. 3040–3049. PMLR (2019)

10. Liang, L., Ye, H., Li, G.Y.: Spectrum sharing in vehicular networks based on multi-agent reinforcement learning. IEEE J. Sel. Areas Commun. **37**(10), 2282–2292 (2019)

11. Lowe, R., Wu, Y.I., Tamar, A., Harb, J., Pieter Abbeel, O., Mordatch, I.: Multi-agent actor-critic for mixed cooperative-competitive environments. In: Advances in Neural Information Processing Systems, vol. 30 (2017)

12. Mahajan, A., Rashid, T., Samvelyan, M., Whiteson, S.: Maven: multi-agent variational exploration. In: Advances in Neural Information Processing Systems, vol. 32 (2019)

13. Oliehoek, F.A., Amato, C.: A Concise Introduction to Decentralized POMDPs. Springer, Cham (2016). https://doi.org/10.1007/978-3-319-28929-8

14. Oliehoek, F.A., Spaan, M.T., Vlassis, N.: Optimal and approximate Q-value functions for decentralized POMDPs. J. Artif. Intell. Res. **32**, 289–353 (2008)

15. Oroojlooy, A., Hajinezhad, D.: A review of cooperative multi-agent deep reinforcement learning. Appl. Intell. 1–46 (2022)

16. Pathak, D., Agrawal, P., Efros, A.A., Darrell, T.: Curiosity-driven exploration by self-supervised prediction. In: International Conference on Machine Learning, pp. 2778–2787. PMLR (2017)

17. Rashid, T., Farquhar, G., Peng, B., Whiteson, S.: Weighted QMIX: expanding monotonic value function factorisation for deep multi-agent reinforcement learning. Adv. Neural. Inf. Process. Syst. **33**, 10199–10210 (2020)

18. Rashid, T., Samvelyan, M., Schroeder, C., Farquhar, G., Foerster, J., Whiteson, S.: QMIX: monotonic value function factorisation for deep multi-agent reinforcement learning. In: International Conference on Machine Learning, pp. 4295–4304. PMLR (2018)

19. Samvelyan, M., et al.: The starcraft multi-agent challenge. arXiv preprint arXiv:1902.04043 (2019)

20. Shalev-Shwartz, S., Shammah, S., Shashua, A.: Safe, multi-agent, reinforcement learning for autonomous driving. arXiv preprint arXiv:1610.03295 (2016)

21. Son, K., Kim, D., Kang, W.J., Hostallero, D.E., Yi, Y.: Qtran: learning to factorize with transformation for cooperative multi-agent reinforcement learning. In: International Conference on Machine Learning, pp. 5887–5896. PMLR (2019)

22. Su, J., Adams, S., Beling, P.: Value-decomposition multi-agent actor-critics. In: Proceedings of the AAAI Conference on Artificial Intelligence, vol. 35, pp. 11352–11360 (2021)

23. Sunehag, P., et al.: Value-decomposition networks for cooperative multi-agent learning. arXiv preprint arXiv:1706.05296 (2017)

24. Tang, H., et al.: # exploration: a study of count-based exploration for deep reinforcement learning. In: Advances in Neural Information Processing Systems, vol. 30 (2017)

25. Wang, J., Ren, Z., Liu, T., Yu, Y., Zhang, C.: Qplex: duplex dueling multi-agent Q-learning. arXiv preprint arXiv:2008.01062 (2020)

26. Wang, T., Gupta, T., Peng, B., Mahajan, A., Whiteson, S., Zhang, C.: Rode: learning roles to decompose multi-agent tasks. In: Proceedings of the International Conference on Learning Representations. OpenReview (2021)

27. Wang, T., Wang, J., Wu, Y., Zhang, C.: Influence-based multi-agent exploration. arXiv preprint arXiv:1910.05512 (2019)

28. Wang, Y., Han, B., Wang, T., Dong, H., Zhang, C.: Off-policy multi-agent decomposed policy gradients. arXiv preprint arXiv:2007.12322 (2020)

29. Zhou, M., Liu, Z., Sui, P., Li, Y., Chung, Y.Y.: Learning implicit credit assignment for cooperative multi-agent reinforcement learning. Adv. Neural. Inf. Process. Syst. **33**, 11853–11864 (2020)

SLAM: A Lightweight Spatial Location Attention Module for Object Detection

Changda Liu[ID], Yunfeng Xu[✉][ID], and Jiakui Zhong[ID]

Hebei University of Science and Technology, Shijiazhuang 050000, Hebei, China
hbkd_xyf@hebust.edu.cn

Abstract. Aiming to address the shortcomings of current object detection models, including a large number of parameters, the lack of accurate localization of target bounding boxes, and ineffective detection, this paper proposes a lightweight spatial location attention module (SLAM) that achieves adaptive adjustment of the attention weights of the location information in the feature map while greatly improving the feature representation capability of the network by learning the spatial location information in the input feature map. First, the SLAM module obtains the spatial distribution of the input feature map in the horizontal, vertical, and channel directions through the average pooling and maximum pooling operations, then generates the corresponding location attention weights by computing convolution and activation functions, and finally achieves the weighted feature map by aggregating the features along the three spatial directions respectively. Extensive experiments show that the SLAM module improves the detection performance of the model on the MS COCO dataset and the PASCAL VOC 2012 dataset with almost no additional computational overhead.

Keywords: Attention mechanism · Spatial position · Object detection

1 Introduction

Object Detection is one of the core topics in the computer vision field and is the basis of many other vision tasks [19,32]. It aims to find out and localize specific targets in image information with the integration of cutting-edge technologies in many fields such as image processing, pattern recognition, feature extraction, and deep learning, making itself a very challenging task.

Deep learning techniques are favored by researchers for their powerful feature extraction capabilities. Convolutional neural networks (CNNs), as a commonly used method in deep learning, have been widely applied in the field of object detection. Significant progress has been made in the research of object detection algorithms based on CNNs, replacing traditional object detection methods [4]. These algorithms have been industrially applied in various fields such as security, industry, and automotive-assisted driving [5,13]. In the practical application of object detection, very few algorithms are able to achieve both speed and accuracy

B. Luo et al. (Eds.): ICONIP 2023, LNCS 14448, pp. 373–387, 2024.
https://doi.org/10.1007/978-981-99-8082-6_29

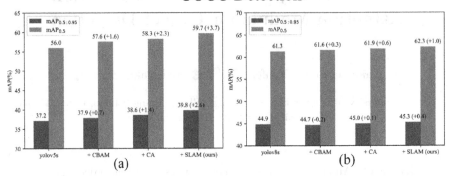

Fig. 1. Performance of different attention methods on COCO object detection dataset. The results of (a) are based on YOLOv5s and (b) are based on YOLOv8s. The Y-axis is mAP. Clearly, our approach performs better.

for that they are often contradictory. In general, the larger the number of model parameters a model has, the better it performs, but the longer the consumed inference time, the higher the required computational power of the device. And deep learning-based object detection methods typically employ deeper network structures and more complex feature fusion techniques to extract more expressive deep features and enhance the model's representation capability. This leads to most of the existing convolutional neural network-based target detection algorithms suffering from bloated network structures, a large number of parameters, slow inference speed, difficulty in real-time detection issues such as bloated network structures, excessively large parameter sizes, slow inference speeds, and difficulties in achieving real-time detection. Therefore, finding a model with a suitable balance between quality and computational requirements, particularly for devices with limited computing power, remains a challenging task.

In this paper, we propose the Spatial Location Attention Module (SLAM), a novel network module that enhances the accuracy of the model while incurring minimal computational cost. SLAM effectively incorporates spatial position information into the feature map and dynamically adjusts the significance of each position, thereby improving the localization and recognition of objects of interest. Our approach is flexible and lightweight, making it easily transferable to other CNNs. To achieve this, we employ average pooling and max pooling to aggregate the input feature map across the horizontal, vertical, and channel dimensions, resulting in three attention maps with directional sensitivity. These attention maps are then multiplied with the input feature map, generating a feature map enriched with spatial position information.

Extensive experiments have been conducted on the MS COCO dataset [16] and the PASCAL VOC 2012 dataset [3] to demonstrate the advantages of our proposed attention module over others. It turns out that the networks incorporating our attention module have an improvement in performance with negligible increase in parameter and computational complexity. As shown in Fig. 1, the network added with the SLAM achieved higher accuracy than the baseline network and outperformed other attention modules. Moreover, a visualization of the feature maps generated through different attention methods produced by Grad-CAM [25], shows that the network with the SLAM has a better focus on the target objects compared to the baseline network and other attention module networks.

The main contributions of this article are as follows:

(1) Propose a lightweight Spatial Location Attention Module (SLAM) that can be flexibly integrated into CNNs to enhance network performance.
(2) Validating the effectiveness of each component of the SLAM through ablation experiments.
(3) Experimental results have demonstrated that the insertion of SLAM at different network positions leads to improved detection performance on both the MS COCO and PASCAL VOC 2012 datasets.

2   Related Work

2.1   Fully Convolutional Object Detector

The R-CNN [8–10,24] series of algorithms are object detection algorithms that are built on region proposal boxes. They generate a series of candidate bounding boxes (i.e., proposal boxes) utilizing the RPN network. Subsequently, the candidate boxes are classified and regressed for determining the presence and location of the target objects within the image. The RPN network is a deep learning-based proposal box generation network that employs convolutional operations to generate the candidate boxes on feature maps. These candidate boxes are then mapped to the corresponding feature maps using an ROI (Region of Interest) Pooling layer. The R-CNN series of algorithms have demonstrated excellent detection performance, but they suffer from slow processing speed and struggle to meet the demands of real-time detection. In contrast, the YOLO (You Only Look Once) [1,7,14,21–23,28,31] series of algorithms are end-to-end object detection algorithms. These algorithms directly classify and regress the entire image using a neural network, resulting in speedy detection. YOLOv5 adopts an anchor-based approach, which involves generating anchor boxes on the input image and then using a classifier and regressor to predict whether the anchor box contains an object, as well as adjusting the position and size of the anchor box to better match the actual location of the object. YOLOv5 also employs a coupled head structure that integrates features from varying scales, leading to enhanced detection accuracy. On the other hand, YOLOv8 employs an anchor-free approach, eliminating the need to generate anchor boxes in advance. Instead, it densely predicts on the feature map, avoiding the process of anchor box generation and filtering. This not only reduces computational complexity but also minimizes memory usage. YOLOv8 also incorporates the proposed integral representation from the Distribution Focal Loss [15] into its regression branch. This integration allows for better handling of changes in target scale and aspect ratio, ultimately improving detection robustness. Furthermore, YOLOv8 adopts a decoupled head structure, which removes the objectness prediction branch and simplifies the model's structure while enhancing detection speed.

The YOLO series models offer a diverse range of applications, feature simple structures, and effectively fulfill real-time detection requirements. Consequently, we have selected YOLOv5 and YOLOv8 as our base models.

2.2 Attention Mechanisms

It is widely acknowledged that attention plays an important role in the human visual system [26]. By focusing solely on important target areas within a scene and channeling more attention resources to these regions, we can extract crucial information while disregarding irrelevant data. Through this mechanism, valuable information can be quickly filtered out from a large amount of data.

In the domain of computer vision, attention mechanisms also play a crucial role in processing various computer vision tasks [6,18,27]. Attention mechanisms measure the weight differences between information to quantify importance, allowing the model to focus on essential information and disregard unimportant details, thereby enhancing its efficiency. Hu et al. [12] proposed the Squeeze-and-Excitation Network (SENet), SE module performs a squeeze operation on the feature map to obtain channel-wise global features and weights for each channel. Then, it applies an excitation operation to the global features, allowing the model to learn the importance of different channel information and enhance the representational power of the basic modules in the network. Wang et al. [29] highlighted the significance of learning effective channel attention, avoiding dimensionality reduction, and fostering appropriate cross-channel interactions. To address this, they proposed ECA-Net, which exclusively employs a one-dimensional convolution operation to calculate the channel attention. Chen et al. [2] introduced SCA, a mechanism that integrates both spatial and channel attention to learn weights for each spatial position and channel. These learned weights are subsequently used to obtain an adaptive feature representation from the input feature map. Woo et al. [30] proposed CBAM, which combines channel and spatial attention mechanisms. The CBAM module sequentially infers attention weights along the channel and spatial dimensions and applies them to the input feature map, resulting in refined and adaptive features that improve the model's accuracy.

More close to our work, Hou et al. proposed the CA module [11]. This module decomposes channel attention by employing global average pooling and aggregating features along two spatial directions through one-dimensional feature encoding processes. However, experiments have revealed that incorporating max-pooling is crucial for acquiring more intricate channel attention, and the authors have overlooked the importance of attention in spatial directions. Thus, to overcome these limitations, in SLAM, we simultaneously aggregate features along all three spatial directions using both average pooling and max-pooling. Remarkably, experimental results have demonstrated the outstanding performance of SLAM in terms of object detection tasks on both the MS COCO dataset and the PASCAL VOC dataset.

3   Spatial Location Attention Module

SLAM can be considered a computational unit that enhances the network's focus on important feature information. Given an input feature map $F \in \mathbb{R}^{C \times H \times W}$, the SLAM module sequentially calculates the attention feature maps in the horizontal direction, denoted as $M_W \in \mathbb{R}^{C \times H \times 1}$, in the vertical direction as $M_H \in \mathbb{R}^{C \times 1 \times W}$, and in the channel direction as $M_S \in \mathbb{R}^{1 \times H \times w}$. The complete process can be summarized as follows:

$$F' = F \otimes M_W(F) \otimes M_H(F) \otimes M_C(F) \tag{1}$$

where, $\otimes$ denotes element-wise multiplication between the attention feature maps and the input feature map. During the multiplication process, the attention feature maps need to be broadcasted (copied) appropriately. F' is the final output.

3.1   Attention Module Review and Comparison

CBAM is an attention module that combines spatial and channel attention, emphasizing learning important features along both the channel and spatial dimensions. It utilizes both channel attention and spatial attention modules in a sequential manner. As shown in Fig. 2, in the CBAM attention module, the channel attention applies max pooling and average pooling operations to the input feature map $F \in \mathbb{R}^{C \times H \times W}$, followed by fully connected layers and corresponding activation functions to generate a 1D channel attention map $M_C \in \mathbb{R}^{C \times 1 \times 1}$. M_C multiplied element-wise with the feature map F to obtain the feature map F'. The spatial attention module applies max pooling and average pooling operations to the feature map F', followed by specific convolutions and activation functions to generate a spatial attention map $M_S \in \mathbb{R}^{1 \times H \times W}$. M_S multiplied element-wise with the feature map F' to obtain the feature map F''. The entire attention calculation process can be summarized as follows:

$$\begin{aligned} F' &= M_C(F) \otimes F \\ F'' &= M_S(F') \otimes F' \end{aligned} \tag{2}$$

where, $\otimes$ denotes element-wise multiplication, F'' is the final output.

CBAM considers attention in both spatial and channel dimensions but overlooks positional information, resulting in poorer capture of long-range dependency relationships.

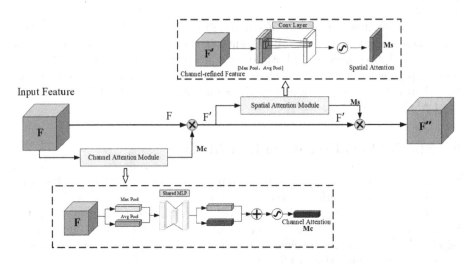

Fig. 2. CBAM's process and organization.

CA addresses the loss of positional information in two-dimensional global pooling by embedding positional information into channel attention. As shown in Fig. 3, the CA attention module divides the input feature map $F \in \mathbb{R}^{C \times H \times W}$ into width and height directions and performs global average pooling separately. This yields width-wise attention map $M_W \in \mathbb{R}^{C \times H \times 1}$ and height-wise attention map $M_H \in \mathbb{R}^{C \times 1 \times W}$. These maps then undergo specific convolutions and activation functions to obtain the width-wise attention map $M_W' \in \mathbb{R}^{C \times H \times 1}$ and height-wise attention map $M_H' \in \mathbb{R}^{C \times 1 \times W}$. The feature map F is multiplied element-wise with the attention maps M_W' and M_H' to obtain the final output feature map F'. The entire process can be summarized as follows:

$$F' = F \otimes M_W(F) \otimes M_H(F) \tag{3}$$

where, $\otimes$ denotes element-wise multiplication, F' is the final output. CA incorporates positional information into channel attention to capture long-range dependency relationships, but it overlooks spatial attention. It only uses average pooling for channel attention, lacking the maximum pooling that could infer more precise channel attention. Now let's introduce a new attention module that considers both channel and spatial positional relationships and embeds spatial positional information into attention.

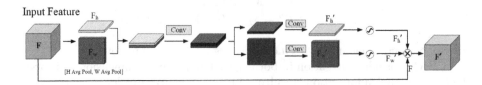

Fig. 3. CA's process and organization.

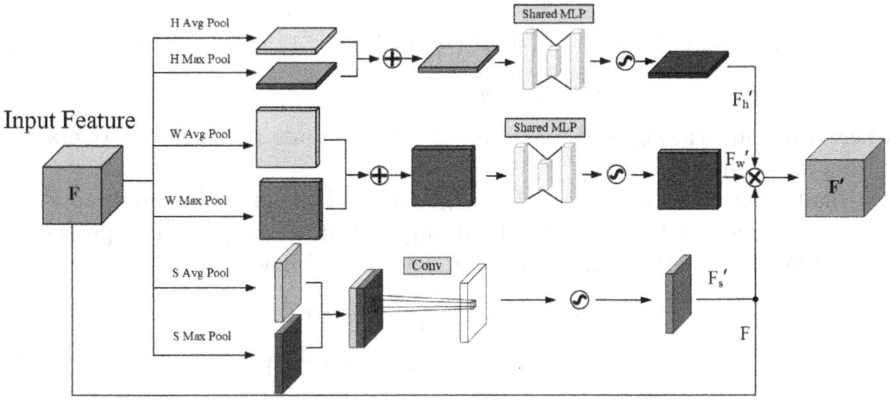

Fig. 4. SLAM's process and organization.

Our spatial positional attention generation involves three primary steps, as illustrated in Fig. 4. For the input feature map $F \in \mathbb{R}^{C \times H \times W}$, the first step entails applying average pooling and max pooling operations vertically, resulting in two attention maps: $M_{avg}^H \in \mathbb{R}^{C \times 1 \times W}$ and $M_{max}^H \in \mathbb{R}^{C \times 1 \times W}$. Subsequently, the two attention maps are summed together and passed through a fully connected layer followed by an appropriate activation function to generate the vertical channel attention map $M_H \in \mathbb{R}^{C \times 1 \times W}$. The process can be summarized as follows:

$$M_H(F) = \sigma(MLP(AvgPool_H(F) + MaxPool_H(F)))$$
$$= \sigma(W_1(W_0(F_{avg}^H + F_{max}^H))) \tag{4}$$

where σ denotes the sigmoid function, $W_0 \in \mathbb{R}^{C/r \times C}$ and $W_1 \in \mathbb{R}^{C \times C/r}$ are the weights of the MLP.

In the second step, average pooling and max pooling operations are performed along the horizontal direction of the feature map. This generates two attention maps: $M_{avg}^W \in \mathbb{R}^{C \times H \times 1}$ and $M_{max}^W \in \mathbb{R}^{C \times H \times 1}$. Then, these two attention maps are added together and passed through a fully connected layer and an appropriate activation function to generate the horizontal channel attention map $M_W \in \mathbb{R}^{C \times H \times 1}$. The specific process can be summarized as follows:

$$M_W(F) = \sigma(MLP(AvgPool_W(F) + MaxPool_W(F)))$$
$$= \sigma(W_1(W_0(F_{avg}^W + F_{max}^W))) \tag{5}$$

3.2   The Generation of Spatial Location Attention

Global pooling is used to encode spatial information globally. We adopt both average pooling and max pooling operations to aggregate the spatial information of the feature map. Average pooling is utilized to preserve the overall characteristics of the input data and the global information in the image. On the other hand, max pooling is employed to infer more fine-grained attention.

where σ denotes the sigmoid function, $W_0 \in \mathbb{R}^{C/r \times C}$ and $W_1 \in \mathbb{R}^{C \times C/r}$ are the weights of the MLP.

In the third step, we aggregate channel feature information using average pooling and max pooling operations to obtain two attention maps: $M_{avg}^S \in \mathbb{R}^{1 \times H \times W}$ and $M_{max}^S \in \mathbb{R}^{1 \times H \times W}$. Then, these two attention maps are concatenated along the channel dimension and passed through corresponding Conv and activation function operations to generate the spatial attention feature map $M_S \in \mathbb{R}^{1 \times H \times W}$. The specific process can be summarized as follows:

$$
\begin{aligned}
M_S(F) &= \sigma(f^{7 \times 7}(AvgPool_S(F) + MaxPool_S(F))) \\
&= \sigma(f^{7 \times 7}(F_{avg}^S + F_{max}^S))
\end{aligned}
\tag{6}
$$

where σ denotes the sigmoid function, $f^{(7 \times 7)}$ represents a convolution operation with a kernel size of 7×7.

Finally, the attention maps M_H, M_W, M_S and the input feature map F are multiplied together to obtain the final output feature map F'. The process of generating spatial attention can be summarized as follows:

$$
\begin{aligned}
F' = F &\otimes \sigma(W_1(W_0(F_{avg}^H + F_{max}^H))) \\
&\otimes \sigma(W_1(W_0(F_{avg}^W + F_{max}^W))) \\
&\otimes \sigma(f^{7 \times 7}(F_{avg}^S + F_{max}^S))
\end{aligned}
\tag{7}
$$

where σ denotes the sigmoid function, $W_0 \in \mathbb{R}^{C/r \times C}$ and $W_1 \in \mathbb{R}^{C \times C/r}$ are the weights of the MLP, $f^{(7 \times 7)}$ represents a convolution operation with a kernel size of 7×7, F' is the final output.

What sets our attention module apart from others is that it incorporates spatial position information into the feature map, allowing for adaptive adjustment of the importance of each position. This approach enables more precise localization of the exact positions of the regions of interest, facilitating better recognition by the network. We will provide detailed demonstrations in the experimental section.

3.3   Exemplars: SLAM_C3 and SLAM_C2f

SLAM, as a plug-and-play attention mechanism module, can be easily applied to most convolutional neural networks. In our experiments, we seamlessly integrated the SLAM module into the C3 module of YOLOv5 and the C2f module of YOLOv8, resulting in the SLAM_C3 and SLAM_C2f modules, as shown in Fig. 5.

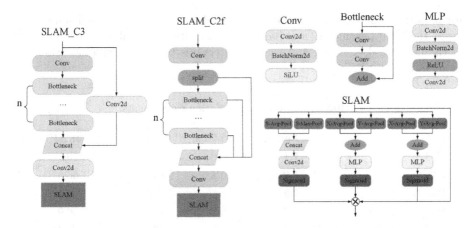

Fig. 5. We have depicted the processes of the SLAM_C3 and SLAM_C2f modules in the diagram, along with the detailed components of each module.

4   Experiments

In this section, we first describe the parameter settings of the experiments. Then, we conduct a series of ablation experiments to demonstrate the contributions of each component in the proposed SLAM to network performance. Next, we compare the attention method proposed in this paper with other existing attention methods. Finally, we compare the YOLOv5s_SLAM network with other state-of-the-art object detection networks in terms of performance.

4.1   Experiment Setup

We use the PyTorch framework [20] to implement all experiments and use the SGD optimizer to train the models. The initial learning rate is set to 0.01, momentum to 0.937, and weight decay to 0.0005. Training is performed using 8 NVIDIA GeForce RTX 3090 GPUs. The input image size for the model is 640×640, and the training batch size is set to 256. For YOLOv8, a series of experiments with the MS COCO dataset are conducted for 500 epochs, while other experiments are trained for 300 epochs.

4.2   Ablation Studies

We conducted a series of ablation experiments, and the results are shown in Table 1. Through the experiments, we have demonstrated that using both average pooling and max pooling can achieve more precise attention inference. Additionally, combining the attention in horizontal, vertical, and channel directions yields the best results. Firstly, we compared three variations of horizontal and vertical (H&W) attention: average pooling, max pooling, and the simultaneous use of both pooling methods. When both pooling operations are used simultaneously, the MLP parameters are shared. Subsequently, we combined the attention in the horizontal, vertical, and channel directions to create our final attention module, named SLAM.

Table 1. The results of ablation experiments.

Model	Param.(M)	FLOPs.(B)	$mAP_{0.5:0.95}(\%)$	$mAP_{0.5}(\%)$
YOLOv5s	7.1	15.9	46.4	66.1
+ HW_mean	7.2	16.1	46.4	66.5
+ HW_max	7.2	16.1	46.6	66.4
+ HW_mean&max	7.2	16.1	46.8	66.9
+ HWS_mean&max(SLAM)	7.2	16.1	**47.2**	**67.0**

From the data in Table 1, we can observe that both average pooling and max pooling in the horizontal and vertical directions improve the model's performance. However, when both pooling methods are used simultaneously, the model achieves even better results. Furthermore, it is evident that under comparable numbers of learnable parameters and computational costs, SLAM allows the model to learn information from different positions in the feature map. It captures spatial context, enhances spatial position awareness, and improves model performance.

4.3   Compare with Other Attention Methods

We conducted extensive experiments on the MS COCO dataset and the PASCAL VOC 2012 dataset to compare the performance of SLAM, CBAM, and CA in object detection tasks. The results demonstrate that our proposed attention method has distinct advantages over the others.

MS COCO Object Detection. We conducted experiments on the COCO 2017 dataset, which consists of 118,287 training images and 5,000 validation images. The evaluation metrics for the model include precision (P), recall (R), and mean average precision (mAP). The experimental results are shown in Table 2.

In Table 2, we present the experimental results of YOLOv5s and YOLOv8s networks with different attention methods added on the COCO 2017 validation dataset. Clearly, the performance of the networks improves when SLAM is added compared to the baseline. The number of model parameters remains almost unchanged, indicating the effectiveness of our proposed method.

Compared to other attention methods, our SLAM performs better with the highest mAP values. Especially when SLAM is added to YOLOv5s, the $mAP_{0.5:0.95}$ increases by 2.6% to 39.8%, and the $mAP_{0.5}$ increases by 3.7%,

Table 2. Object detection on the MS COCO validation set.

Model	Param.(M)	FLOPs.(B)	P	R	$mAP_{0.5:0.95}(\%)$	$mAP_{0.5}(\%)$
YOLOv5s	7.2	16.5	64.0	51.3	37.2	56.0
+ CBAM	8.0	17.5	67.4	52.2	37.9	57.6
+ CA	7.3	16.6	**68.2**	52.8	38.6	58.3
+ SLAM	7.3	16.7	68.1	**54.2**	**39.8**	**59.7**
YOLOv8s	11.2	28.6	68.2	**56.3**	44.9	61.3
+ CBAM	11.9	29.2	68.4	56.0	44.7	61.6
+ CA	11.2	28.8	**69.4**	55.3	45.0	61.9
+ SLAM	11.2	28.7	69.2	56.2	**45.3**	**62.3**

indicating a significant improvement in detection accuracy. Adding SLAM to YOLOv8s also results in a slight improvement in detection accuracy, with a 1.0% increase in $mAP_{0.5}$.

PASCAL VOC2012 Object Detection. We also conducted experiments on the PASCAL VOC 2012 dataset. In this experiment, we redefined the dataset, with a ratio of 8:2 for training to validation sets, including 13,700 training images and 3,425 validation images. The experimental results are shown in Table 3. We can clearly see that the parameter overhead introduced by adding SLAM to the network is negligible, while the network's performance is enhanced. This indicates that the improved network performance is not caused by the increase in model parameters, but rather the effectiveness of our SLAM method in refining features.

Table 3. Object detection on the PASCAL VOC2012 validation set.

Model	Param.(M)	FLOPs.(B)	P	R	$mAP_{0.5:0.95}(\%)$	$mAP_{0.5}(\%)$
YOLOv5s	7.1	15.9	**73.2**	61.1	46.4	66.1
+ CBAM	7.8	16.8	68.8	**64.2**	46.9	66.1
+ CA	7.2	16.1	71.7	63.1	47.0	66.9
+ SLAM	7.2	16.1	72.7	61.8	**47.2**	**67.0**
YOLOv8s	11.1	28.5	69.7	60.9	49.3	65.3
+ CBAM	11.9	29.1	69.1	**61.5**	49.7	65.5
+ CA	11.2	28.6	69.3	**61.5**	49.9	65.9
+ SLAM	11.2	28.6	**71.7**	61.3	**50.7**	**66.4**

The experiments on both the COCO and Pascal VOC datasets demonstrate that the attention method proposed in this paper performs better compared to other attention methods.

4.4　Network Visualization with Grad-CAM

To make the results more intuitive, we applied Grad-CAM to models with different attention methods using images from the MS COCO validation set. Grad-CAM is a method for visualizing convolutional neural networks predictions, which effectively highlights the regions of interest in the input image that the network focuses on. We compared the visualization results of YOLOv5s_SLAM, YOLOv5s_CBAM, YOLOv5s_CA, and YOLOv5s networks, as shown in Fig. 6.

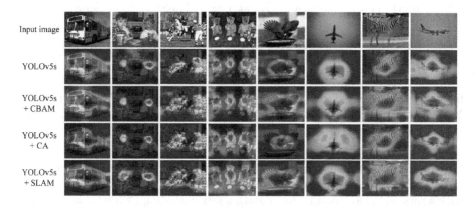

Fig. 6. Grad-CAM visualization results.

In Fig. 6, we can observe that compared to CBAM and CA, our SLAM can better locate the objects of interest. In other words, the SLAM-integrated network can better identify which regions in the input image are most important for the classification decisions in object detection.

4.5　Compare with Other Models

To further demonstrate the superiority of SLAM proposed in this paper, we compared it with YOLOv5s and other state-of-the-art object detection methods. We replaced the C3 module in the YOLOv5s network with the SLAM_C3 module, as shown in Fig. 5. The rest of the experimental settings followed the original files. The experimental results of each network on the COCO 2017 validation set are listed in Table 4.

As shown in Table 4, YOLOv5s_SLAM significantly improves the object detection results compared to the baseline network YOLOv5s. It also performs competitively when compared to other networks with similar or slightly higher parameters and computational complexities. It demonstrates good performance in terms of $mAP_{0.5}$ and $mAP_{0.5:0.95}$ metrics, especially excelling in $mAP_{0.5}$. In the comparison of the $mAP_{0.5:0.95}$ metric, we can see that YOLOv6-Tiny achieves slightly better results, but it has more than twice the parameters and computational complexity compared to our network. Overall, the YOLOv5s_SLAM model achieves excellent detection results while ensuring speed and accuracy.

Table 4. Table captions should be placed above the tables.

Model	Param.(M)	FLOPs.(B)	$mAP_{0.5:0.95}(\%)$	$mAP_{0.5}(\%)$
SSD [17]	44.1	387.0	26.8	46.5
YOLOv3	61.9	156.6	33.0	57.9
YOLOv4-S	8.3	21.2	38.9	57.7
YOLOv5-S	7.1	16.1	37.2	56.0
YOLOvX-S	9.6	26.8	39.6	57.7
YOLOv6-Tiny	15.0	36.7	**40.3**	56.6
YOLOv7-Tiny	6.2	13.9	37.5	52.5
ours	7.3	16.7	39.8	**59.7**

5 Conclusion

In this paper, we propose a lightweight spatial location attention module that can significantly improve network performance with almost no additional overhead. The method adaptively adjusts the importance of each location in the feature map by learning the spatial location information of the input feature map to better locate which regions in the input image are most important for the classification decision of target detection. We have conducted extensive experiments using different datasets comparing other commonly used attention methods and confirmed the effectiveness of our SLAM.

Our method is also more lightweight than other attention methods in terms of computational resource consumption, which makes our method suitable for use in computationally resource-constrained environments, such as mobile devices or embedded systems.

In summary, the proposed spatial location attention module is an effective lightweight attention mechanism that improves network performance in object detection tasks. We believe that this approach can find broader applications in future computer vision research and deliver better results in practical applications.

Acknowledgements. This paper is founded by Supported projects of key R & D programs in Hebei Province (No. 21373802D) and Artificial Intelligence Collaborative Education Project of the Ministry of Education (201801003011).

The GPU server in this paper is jointly funded by Shijiazhuang Wusuo Network Technology Co., LTD and Hebei Rouzun Technology Co., LTD.

References

1. Bochkovskiy, A., Wang, C.Y., Liao, H.Y.M.: Yolov4: optimal speed and accuracy of object detection. arXiv preprint arXiv:2004.10934 (2020)
2. Chen, L., et al.: SCA-CNN: spatial and channel-wise attention in convolutional networks for image captioning. In: Proceedings of the IEEE Conference on Computer Vision and Pattern Recognition, pp. 5659–5667 (2017)

3. Everingham, M., Eslami, S.A., Van Gool, L., Williams, C.K., Winn, J., Zisserman, A.: The pascal visual object classes challenge: a retrospective. Int. J. Comput. Vision **111**, 98–136 (2015)
4. Zhou, F.Y., Jin, L.P., Dong, J.: Review of convolutional neural networks. J. Comput. Sci. **40**(06), 1229–1251 (2017)
5. Feng, D., et al.: Deep multi-modal object detection and semantic segmentation for autonomous driving: datasets, methods, and challenges. IEEE Trans. Intell. Transp. Syst. **22**(3), 1341–1360 (2020)
6. Fu, J., et al.: Dual attention network for scene segmentation. In: Proceedings of the IEEE/CVF Conference on Computer Vision and Pattern Recognition, pp. 3146–3154 (2019)
7. Ge, Z., Liu, S., Wang, F., Li, Z., Sun, J.: Yolox: exceeding yolo series in 2021. arXiv preprint arXiv:2107.08430 (2021)
8. Girshick, R.: Fast R-CNN. In: Proceedings of the IEEE International Conference on Computer Vision, pp. 1440–1448 (2015)
9. Girshick, R., Donahue, J., Darrell, T., Malik, J.: Rich feature hierarchies for accurate object detection and semantic segmentation. In: Proceedings of the IEEE Conference on Computer Vision and Pattern Recognition, pp. 580–587 (2014)
10. He, K., Gkioxari, G., Dollár, P., Girshick, R.: Mask R-CNN. In: Proceedings of the IEEE International Conference on Computer Vision, pp. 2961–2969 (2017)
11. Hou, Q., Zhou, D., Feng, J.: Coordinate attention for efficient mobile network design. In: Proceedings of the IEEE/CVF Conference on Computer Vision and Pattern Recognition, pp. 13713–13722 (2021)
12. Hu, J., Shen, L., Sun, G.: Squeeze-and-excitation networks. In: Proceedings of the IEEE Conference on Computer Vision and Pattern Recognition, pp. 7132–7141 (2018)
13. Li, B., Ouyang, W., Sheng, L., Zeng, X., Wang, X.: GS3D: an efficient 3D object detection framework for autonomous driving. In: Proceedings of the IEEE/CVF Conference on Computer Vision and Pattern Recognition, pp. 1019–1028 (2019)
14. Li, C., et al.: Yolov6: a single-stage object detection framework for industrial applications. arXiv preprint arXiv:2209.02976 (2022)
15. Li, X., et al.: Generalized focal loss: learning qualified and distributed bounding boxes for dense object detection. Adv. Neural. Inf. Process. Syst. **33**, 21002–21012 (2020)
16. Lin, T.Y., et al.: Microsoft coco: common objects in context (2014). arXiv preprint arXiv:1405.0312 (2019)
17. Liu, W., et al.: SSD: single shot MultiBox detector. In: Leibe, B., Matas, J., Sebe, N., Welling, M. (eds.) ECCV 2016. LNCS, vol. 9905, pp. 21–37. Springer, Cham (2016). https://doi.org/10.1007/978-3-319-46448-0_2
18. Liu, Y., Shao, Z., Hoffmann, N.: Global attention mechanism: retain information to enhance channel-spatial interactions. arXiv preprint arXiv:2112.05561 (2021)
19. Lu, H., Zhang, Q.: Application of deep convolutional neural networks in computer vision. Data Acquisition Process. **31**(01), 1–17 (2016)
20. Paszke, A., et al.: Pytorch: an imperative style, high-performance deep learning library. In: Advances in Neural Information Processing Systems, vol. 32 (2019)
21. Redmon, J., Divvala, S., Girshick, R., Farhadi, A.: You only look once: unified, real-time object detection. In: Proceedings of the IEEE Conference on Computer Vision and Pattern Recognition, pp. 779–788 (2016)
22. Redmon, J., Farhadi, A.: Yolo9000: better, faster, stronger. In: Proceedings of the IEEE Conference on Computer Vision and Pattern Recognition, pp. 7263–7271 (2017)

23. Redmon, J., Farhadi, A.: Yolov3: an incremental improvement. arXiv preprint arXiv:1804.02767 (2018)
24. Ren, S., He, K., Girshick, R., Sun, J.: Faster R-CNN: towards real-time object detection with region proposal networks. In: Advances in Neural Information Processing Systems, vol. 28 (2015)
25. Selvaraju, R.R., Cogswell, M., Das, A., Vedantam, R., Parikh, D., Batra, D.: Gradcam: visual explanations from deep networks via gradient-based localization. In: Proceedings of the IEEE International Conference on Computer Vision, pp. 618–626 (2017)
26. Tsotsos, J.K.: A Computational Perspective on Visual Attention. MIT Press, Cambridge (2021)
27. Wang, W., Shen, J.: A review of visual attention detection. J. Softw. **30**(02), 416–439 (2019)
28. Wang, C.Y., Bochkovskiy, A., Liao, H.Y.M.: Yolov7: trainable bag-of-freebies sets new state-of-the-art for real-time object detectors. In: Proceedings of the IEEE/CVF Conference on Computer Vision and Pattern Recognition, pp. 7464–7475 (2023)
29. Wang, Q., Wu, B., Zhu, P., Li, P., Zuo, W., Hu, Q.: ECA-net: efficient channel attention for deep convolutional neural networks. In: Proceedings of the IEEE/CVF Conference on Computer Vision and Pattern Recognition, pp. 11534–11542 (2020)
30. Woo, S., Park, J., Lee, J.-Y., Kweon, I.S.: CBAM: convolutional block attention module. In: Ferrari, V., Hebert, M., Sminchisescu, C., Weiss, Y. (eds.) ECCV 2018. LNCS, vol. 11211, pp. 3–19. Springer, Cham (2018). https://doi.org/10.1007/978-3-030-01234-2_1
31. Xu, S., et al.: PP-YOLOE: an evolved version of yolo (2022)
32. Yin, H., Chen, B., Chai, Y., Liu, Z.: A review of vision-based target detection and tracking. Acta Automatica Sinica **42**(10), 1466–1489 (2016)

A Novel Interaction Convolutional Network Based on Dependency Trees for Aspect-Level Sentiment Analysis

Lei Mao[1]($\boxtimes$), Jianxia Chen[1], Shi Dong[2], Liang Xiao[1], Haoying Si[1], Shu Li[1], and Xinyun Wu[1]

[1] School of Computer Science, Hubei University of Technology, Wuhan, China
2219236005@qq.com

[2] School of Computer Science and Technology, Zhoukou Normal University, Zhoukou, China

Abstract. Aspect-based sentiment analysis aims to identity the sentiment polarity of a given aspect-based word in a sentence. Due to the complexity of sentences in the texts, the models based on the graph neural network still have issues in the accurately capturing the relationship between aspect words and viewpoint words in sentences, failing to improve the accuracy of classification. To solve this problem, the paper proposes a novel Aspect-level Sentiment Analysis model based on Interactive convolutional network with the dependency trees, named ASAI-DT in short. In particular, the ASAI-DT model first extracts the aspect words representation from the sentence representation trained by the Bi-GRU model. Meanwhile, the self-attention score of both the sentence and aspect representation are calculated separately by the self-attention mechanism, in order to reduce the attention to the irrelevant information. Afterward, the proposed model constructs the sub-tree of the dependency trees for the word, while the attention weight scores of the aspect representations will be integrated into the sub-tree. Therefore, the acquired comprehensive information about aspect words is processed by the graph convolutional network to maximize the retention of valid information and minimize the interference of noise. Finally, the effective information can be preserved more completely in the integrated information through the interactive network. Through a large number of experiments on various data sets, the proposed ASAI-DT model shows both the effectiveness and the accuracy of aspect sentiment analysis, which outperforms many aspect-based sentiment analysis models.

Keywords: Aspect-based Sentiment Analysis · Dependency Trees · Attention Mechanism · Graph Convolutional Network · Interactive Network

1 Introduction

Aspect-based sentiment analysis (ABSA) is one of a critical sentiment analysis problem that aims to capture the sentiment polarity of a given aspect word in a sentence, in which the sentiment polarity includes three categories such as the positive, the negative and the neutral sentiment. For example, there is a sentence, "The service is pretty good, but

B. Luo et al. (Eds.): ICONIP 2023, LNCS 14448, pp. 388–400, 2024.
https://doi.org/10.1007/978-981-99-8082-6_30

the food tastes bad.", the aspect word "service" has a positive sentiment polarity, for the aspect word "food", however, its sentiment polarity is negative. In other words, different sentimental aspects in a sentence need to be analyzed to distinguish various sentimental polarities between different aspects of various words.

One of the key points of ABSA is to establish a relationship between the aspect and the opinion words. Recently, with the rapid development of graph convolutional networks (GCNs), ABSA approaches usually combine GCNs with the syntactic structure of sentences to solve the relation problems between the aspects and the opinions [1–3]. For example, The AS-GCN [1] model integrated syntactic information into the GCNs; CDT [2] model incorporated the syntactic information into the word embeddings of the GCN [3] to improve the learning representation of aspects. However, these models still have a issue that since aspect words adjacent to each other were given the same weight score, the models were unable to distinguish between adjacent words, resulting in noisy data affecting the models accuracy.

From the above analysis, this paper proposes a novel Aspect-level Sentiment Analysis Interaction model based on the Dependency Tree, named ASAI-DT in short, which can effectively eliminate the noise of irrelevant sentiment words. In particular, the ASAI-DT model first extracts the aspect words representation from the sentence representation trained by the Bi-GRU model. Meanwhile, the self-attention score of both the sentence and aspect representation are calculated separately by the self-attention mechanism [4], which reduces the attention to the irrelevant information. Afterward, according to the grammatical distance from the aspect word, the proposed model constructs the sub-tree of the dependency trees (DTs) for the word, while the attention weight scores of the aspect representations will be integrated into the sub-tree. Therefore, the acquired comprehensive information about aspect words is processed by the GCN to maximize the retention of valid information and minimize the interference of noise. Finally, the effective information can be preserved more completely in the integrated information through the interactive network [5] in order to improve the proposed model performance. Therefore, the paper describes the main contributions as follows:

- Combine the DTs with the attention weight scores of the aspect representation to eliminate the noise of irrelevant sentiment words.
- Utilize an interactive network to minimize the influence of losing information via convolutions and the scattered sentence information.
- The results of experiments demonstrate that the ASAI-DT model outperforms baselines on the public datasets.

2 Related Work

2.1 Dependency Trees

Dependency parsing can be utilized to expose the syntactic structure of a sentence via the dependencies analysis. Dependency syntax can parse sentences as a dependency syntax trees, named DTs in short, to indicate syntactic relationships between related words semantically.

It is a popular trend of recent researches to utilize the DTs into the ABSA tasks. For example, Dong [6] proposed an adaptive recurrent neural network (AdaRNN) that can transfer the word sentiment to the target words according to the syntactic information. Li [7] obtained the lexical properties of words through the DTs, but ignored the semantic information of the sentences themselves. The CDT [2] directly integrated the DTs information into the word embedding to enrich the information of the word embedding. Zhang [8] obtained more contextual and grammatical information about the aspect words by reconstructing the original DTs generated from the sentences. However, these models are affected by the noisy data from the irrelevant words to decrease their performance.

Inspired by the CDT model [2], the syntactic information of the DTs has been utilized in the proposed model. Different from the CDT model, however, this paper splits the syntactic information according to the syntactic distance, to improve the efficiency of the syntactic information.

2.2   Graph Convolutional Networks

Since GCNs are effective feature extractor that continuously learns the feature information from the local topology of the previous layer to the next layer, the GCNs-based attention mechanisms have achieved excellent results in the area of ASBA. For example, a new graph attention network (GAT) [9] had been proposed to not only enrich the semantic information based on the syntactic information but also capture the semantic association of words. Huo [10] proposed the weighted relational graph network model (ERGAT) to obtain different types of dependencies of sentences. However, these models cannot make full use of semantic information.

Therefore, GCNs-based DTs have been developed to solve the above problem. For example, Li [11] proposed the DualGCN model to complement the syntactic and semantic knowledge relevance. However, the DualGCN model cannot utilize the syntactic information fully. Jiang [12] proposed a dependency weighting model based on a GAT by weighting the obtained DT structure, however, the model ignores the contextual information. Pang [13] improved the dependency relations between the sentence representations by dividing different dependency sub-trees. Wang [14] utilized two GCNs to encode the dependent edges and dependent labels separately. However, the interaction process cannot eliminate the irrelevant word data well.

3   Methodology

As shown in Fig. 1, the ASAI-DT is made up of five layers, including input and encoding, attention mechanism, dependency tree, interactive network and the GCN layer.

3.1   Input and Encoding Layer

The input & encoding layer aims to map sentences to the vector representations, which is input into the Bi-GRU model to preserve the order of each word vector.

Take a sentence s presented as $s = \{w_1, w_2, \ldots, a_1, \ldots, a_m, \ldots, w_n\}$, where $\{w_1, w_2, \ldots, \ldots, w_n\}$ is context word, $\{a_1, \ldots, a_m\}$ represents an aspect word. Input data s obtained by the input layer to the encoding layer. The word embedding x of

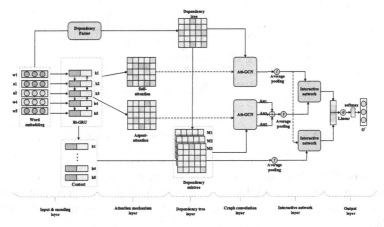

Fig. 1. ASAI-DT model architecture.

a sentence is defined as $x = \{x_1, x_2, \ldots, x_{a+1}, \ldots, x_{a+m}, \ldots, x_n\}$, where x_t represents the word embedding representation. The paper obtains the hidden state vector H as $H = \{h_1, h_2, \ldots, h_{a+1}, \ldots, h_{a+m}, \ldots, h_n\}$ through the Bi-GRU, where $h_t \in \mathbb{R}^{2d}$ defines the hidden state at the time step t. The d presents the the hidden state vector dimension in a one-way GRU. The processing of the word embedding representation x by the GRU unit is simply expressed by the following formula (1–3):

$$\overrightarrow{h_t} = \overrightarrow{GRU}(x_t) \tag{1}$$

$$\overleftarrow{h_t} = \overleftarrow{GRU}(x_t) \tag{2}$$

$$h_t = \left[\overrightarrow{h_t}; \overleftarrow{h_t}\right] \tag{3}$$

afterward, the hidden state vector c of the aspect word context is extracted. The c can be obtained by the following formula (4):

$$c = unmask(H) \tag{4}$$

where the *unmask* function is to set the aspect word vector representation in the hidden state to 0. The hidden state vector of the context is defined as $C = \{h_1, h_2, \ldots, 0_{a+1}, \ldots, 0_{a+m}, \ldots, h_n\}, h_i \in \mathbb{R}^{2d}$.

3.2 Attention Mechanism Layer

After the attention mechanism layer, the proposed model can obtain both the self-attention score (S_{att}) represented by the sentence, and the aspect-attention score(A_{att}) represented by the aspect word.

Therefore, the attention mechanism layer adopts two kinds of attention scores, one is the calculation aspect word score, another is the self-attention score of the sentence, as shown in Fig. 2.

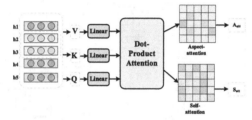

Fig. 2. Flowchart of the dot product attention mechanism.

It is given that a flowchart of a dot-product attention mechanism in Fig. 2, which has a key K, a query Q, and a value V, the dimension of the word representation in the sentence representation represented by d_w. The calculation process utilizes the following formula (5):

$$Attention(Q, K, V) = softmax\left(\frac{QK^T}{\sqrt{d_w}}\right)V \qquad (5)$$

Aspect Word Attention Score. Firstly, this paper obtains the hidden state H_a of the aspect word from the hidden state H of the sentence by the following formula (6):

$$H_a = mask(H) \qquad (6)$$

The *mask* function obtains the aspect word representation by setting the word representation that is not an aspect word to 0. $H_a = \{0, 0, \ldots, h_{a+1}, \ldots, h_{a+m}, \ldots, 0\}$, where $h_{a+i} \in \mathbb{R}^{2d}$. Afterward, assign H_a to Q, and H to K and V. Finally, through the $Attention(Q, K, V)$ resulted form the formula (5), the paper obtains the aspect-attention score of the aspect word representation. The calculation process is defined as the following formula (7):

$$A_{att} = Attention(H_a, H, H) \qquad (7)$$

Self-attention Score Calculation. Based on the calculation of $Attention(Q, K, V)$, the proposed model can obtain the self-attention score about the sentence representation. The calculation process is defined as the following formula (8):

$$S_{att} = Attention(H, H, H) \qquad (8)$$

3.3 Dependency Trees Layer

The dependency tree layer aims to capture the sentence syntactic knowledge according to the sentence's dependency tree structure. To capture the latent syntactic information, the proposed model needs a grammar parser to extract the dependencies of each word. A sentence can be formed a DT as shown in Fig. 3.

To get the most direct word-dependent information with respect to the aspect words, the paper defines the distance between two connected nodes as the score of 1 unit, and all syntactic distances between the aspect words and the remaining nodes are obtained

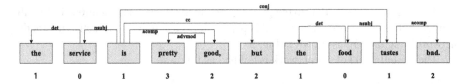

Fig. 3. The formation of the DT.

by traversing the DT. Therefore, Fig. 3 shows the grammatical distance between the aspect words "service" or "food" and other words in the sentence. The distance between "service" and itself is zero, the distance between "the" and "is" is 1, and so on. By dividing the words in the sentence according to the dependency distance, the influence of irrelevant words on the aspect words is effectively reduced.

In this way, the local DT of the aspect word "service" can be obtained, which generates a syntactic information graph M about the sentence based on the syntactic information of the sentence. Afterward, the aspect words are partitioned according to their grammatical distance from other words in the sentence to form a local DT $\{M_1, M_2, M_3\}$. Where M_1 and M_2 are generated by partitioning according to syntactic distances of 1 and 2, respectively. M_3 is generated according to the grammatical distance greater than or equal to 3 for the uniform segmentation, this is because if a distance is too long it will lose the dependency of the sentence itself.

3.4 Graph Convolution Network Layer

As shown in Fig. 4, the GCN layer develops a GCN-based attention mechanism model, named Att-GCN in short, to extract the effective information from the hidden state representation and syntactic graph representation of sentences.

The extracted knowledge is added to the attention score calculated by the attention layer to enhance the acquisition of the word representations associated with aspects. The integration process is defined as the following formula (9–10):

$$H^d = GCN(H_0, D) \tag{9}$$

$$A_d = H^d \otimes A \tag{10}$$

where:

- H_0 is the hidden state sentence representation.
- D presents the acquired syntax information representation.
- A is denoted as the corresponding attention score.

During the calculation process, the hidden state representation H and the syntactic graph representation M are input into the Att-GCN model. Integrating with the self-attention score S_{att} of the sentence s, an output $A_M \in \mathbb{R}^{2d}$ is obtained.

Moreover, this paper transfers the H and the $\{M_1, M_2, M_3\}$ into the Att-GCN model, which is integrated with the A_{att} to obtain the outputs $\{A_{M1}, A_{M2}, A_{M3}\} \in \mathbb{R}^{2d}$.

The interactive network is to integrate the obtained effective information. To fuse the extracted effective information well, this paper adopts a simple and effective interactive network structure [5], as shown in Fig. 5.

The formula (11) of the network is defined as follows:

$$X_{l+1} = X_0 X_l^T W_l + b_l + X_l \tag{11}$$

where:

- $X_l \in \mathbb{R}^d$ is the input of the l th layer of the interaction network.
- $X_{l+1} \in \mathbb{R}^d$ is the output of layer $l+1$.
- $W_l \in \mathbb{R}^d$ is the weight of l layer.
- $b_l \in \mathbb{R}^d$ is the bias of l layer.

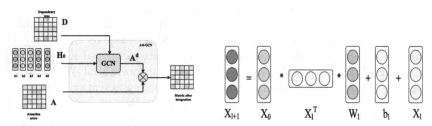

Fig. 4. Att-GCN structure diagram.   **Fig. 5.** Interactive network structure.

In the GCN layer, this paper sets up a multi-layer Att-GCN for the convolution operation. An interactive operation is performed in the middle of the convolutional layer. The A_M and $\{A_{M1}, A_{M2}, A_{M3}\} \in \mathbb{R}^{2d}$ obtained by the Att-GCN convolution are interacted with the H. The interaction process is defined as the following formula (12–13):

$$X_{l+1}^d = HA_M^T W_l + b_l + A_M \tag{12}$$

$$X_{l+1}^{\{M1,M2,M3\}} = HA_{\{M1,M2,M3\}}^T W_l + b_l + A_{\{M1,M2,M3\}}) \tag{13}$$

where:

- X_{l+1}^M is the input of the next layer of convolution.
- $\{X_{l+1}^{M1}, X_{l+1}^{M2}, X_{l+1}^{M3}\}$ is defined as the inputs to the next convolution layer.

After the multi-layer convolution, the paper obtains outputs A_{l+1}^M and $\{A_{l+1}^{M1}, A_{l+1}^{M2}, A_{l+1}^{M3}\}$ that are spliced together to be A_{asp}. The stitching process is defined as the following formula (14–15):

$$A_{self} = A_{l+1}^M \tag{14}$$

$$A_{asp} = \left[A_{l+1}^{M1} \oplus A_{l+1}^{M2} \oplus A_{l+1}^{M3} \right] \tag{15}$$

After splicing, the paper obtains an output $A_{asp} \in \mathbb{R}^{6d}$ about the aspect word representation. $A_{asp} \in \mathbb{R}^{6d}$ not only reduces the influence of noisy data, but also obtains the word representation of the sentence most related to the aspect representation.

3.5   Interactive Network Layer

Afterward, $A_{self} \in \mathbb{R}^{2d}$ and $A_{asp} \in \mathbb{R}^{6d}$ are averaged based on an average pooling layer. Meanwhile, the hidden state vector $c \in \mathbb{R}^{2d}$ is subjected to the same average pooling process. The processing procedure is defined as the following formula (16):

$$y = Averagepooling(x) \tag{16}$$

where:

- x is the vector representation of the input.
- y is the vector output after the average pooling operation.

In this paper, A_{self}, A_{asp} and C are input to the average pooling layer, afterward, A_{self}^o, A_{asp}^o and C^o are output to the interactive processing.

After to reduce the loss of effective information, the paper sends four parameters, including the average pooled data C^o resulted from the context hidden state vector C, the average pooled representation A_{self}^o of the sentence, and the average pooled representation of the aspect word A_{asp}^o, to the interaction network together. The process of interaction is defined as the following formula (17–18):

$$X_C = C^o (A_{self}^o)^T + A_{self}^o \tag{17}$$

$$X_{asp} = A_{asp}^o (A_{self}^o)^T + A_{self}^o \tag{18}$$

where:

- $X_C \in \mathbb{R}^a$ is the output of sentence representation and context representation respectively.
- $X_{asp} \in \mathbb{R}^a$ is the output of sentence representation and aspect word representation.

3.6   Output Layer

In this layer, two obtained outputs X_C and X_{asp} are firstly spliced and sent to the classifier. The output processes are shown in the formula (19–20):

$$o = [X_C, X_{asp}] \tag{19}$$

$$o\prime = softmax(Linear(o)) \tag{20}$$

where:

- $o \in \mathbb{R}^{2a}$ is the output after splicing.
- $o\prime \in \mathbb{R}^{out}$ is the final output.
- out is the final output classification number.

The proposed model utilizes the cross-entropy loss function (CrossEntropy Loss) as the loss function in the formula (21):

$$H(p, q) = -\sum_x p(x) \log q(x) + \lambda ||\theta|^2 \tag{21}$$

where:

- $p(x)$ is the true sample distribution.
- $q(x)$ is the sample prediction distribution.

4 Experiments

4.1 Datasets

To evaluate the experiments, this paper utilizes three public datasets, including the Twitter datasets [6], lap14 and rest14 of SemEval 2014 Task [15]. The details of the datasets are shown in Table 1.

4.2 Parameters Settings

In this paper, the word embedding of datasets is initialized as the 300 dimensions word vector by the GloVe [16]. The paper utilizes the Adma optimizer [17]. The Parameters settings are shown in Table 2. In addition, the paper utilizes the accuracy and Macro-F1 to evaluate the models' performance.

Table 1. Datasets statistics

Datasets	Positive		Neutral		Negative	
	Train	Test	Train	Test	Train	Test
Rest14	2164	728	637	196	807	196
Lap14	994	341	464	169	870	128
Twitter	1561	173	3127	346	1560	173

Table 2. Hyper-parameter settings

Hyper-parameter	Description	Value
dropout rate	Word embedding layer	0.5
batch_size	Mini-batch Size	32
r	Initial learning rate	0.001
d_e	Embedding layer Size	300
d_h	Hidden layer Size	300
l_2	L_2-Regularization weight	0.0001

4.3 Model Comparison

The ASAI-DT model has been compared with the following baseline models.

- AS-GCN [1] utilizes GCNs to learn the word dependencies.
- CDT [2] utilizes a DT to get representations of sentence features.
- SK-GCN [18] Two strategies are developed using GCNs to exploit syntactic dependency trees and commonsense knowledge.
- BiGCN [19] hierarchically models syntactic and lexical graphs for the ABSA.
- kumaGCN [20] utilizes the dependency graph to get syntactic features (Table 3).

Table 3. Comparison result of models

Models	Twitter Datasets		Lap14 Datasets		Rest14 Datasets	
	Accuracy(%)	Macro-F1(%)	Accuracy(%)	Macro-F1(%)	Accuracy(%)	Macro-F1(%)
SK-GCN	71.97	70.22	73.20	69.18	80.36	70.43
AS-GCN	72.15	70.40	75.55	71.05	80.96	72.21
CDT	74.66	73.66	77.19	72.99	82.30	74.02
BiGCN	74.16	73.35	74.59	71.84	81.97	73.48
kumaGCN	72.45	70.77	76.12	72.42	81.41	73.64
ASAI-DT	**76.16**	**74.74**	**77.43**	**73.46**	**83.04**	**76.33**

4.4 Results Analysis of Comparison Experiments

It can be observed that the Accuracy (%) and Macro-F1 (%) score of the model ASAI-DT on the three public data are higher than the baseline model. It shows that the model in this paper has a certain effect in dealing with aspect-level sentiment analysis.

4.5 Number of Att-GCN Layers

In this paper, the influence of the number of layers of Att-GCN is analyzed and tested on three datasets Lap14, Rest14 and Twitter shown in Fig. 6.

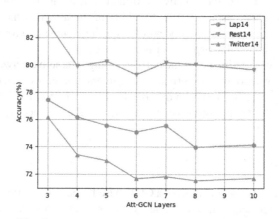

Fig. 6. Att-GCN layer number test on three datasets

As shown in Fig. 6, the Att-GCN layer number corresponds to the experimental settings from 3 through 10 layers. As the layer number has been increased, the accuracy of the ASAI-DT model will be decreased. Among them, the model accuracy is highest when the number of Att-GCN layers is set up as 3. These results demonstrates that the increasing of Att-GCN layers number will lead to the over-fitting of the proposed ASAI-DT model.

Table 4. Ablation experiment of ASAI-DT model

Models	Twitter Datasets		Lap14 Datasets		Rest14 Datasets	
	Accuracy(%)	Macro-F1(%)	Accuracy(%)	Macro-F1(%)	Accuracy(%)	Macro-F1(%)
w/o aspect-attention	73.84	72.39	75.86	71.24	80.36	71.36
w/o dependency subtree	73.41	71.37	75.54	72.20	79.73	69.19
w/o interactive network	74.42	72.99	76.18	71.9	81.07	71.99
ASAI-DT	**76.16**	**74.74**	**77.43**	**73.46**	**83.04**	**76.33**

4.6 Ablation Experiment of ASAI-DT Model

The ablation experimental results of the ASAI-DT are shown in Table 4. The paper first removes aspect-level attention sub-model (w/o aspect-attention). After removing the aspect-attention between Att-GCN, the accuracies correspondingly have been dropped by 2.32%, 1.57% and 2.68% on the three datasets about Twitter, Lap14 and Rest14, respectively. When the paper removes the corresponding dependency subtree structure (w/o dependency subtree), and the accuracy correspondingly decreases by 2.75%, 1.89% and 3.31% respectively. Because aspect-attention is integrated with the Att-GCN between the dependent subtree, the corresponding aspect-attention is also removed as well as the dependent subtree structure is removed.

From the above analysis, the aspect-attention is critical to the Att-GCN layer, which can strengthen the connection of words related to the aspect words. Meanwhile, according to the structure of sub-tree, the effective information acquisition is concentrated on the local information of the aspect words. In addition, the attention mechanism and the use of dependent sub-trees in the process of acquiring syntactic information make the influence of data unrelated to aspect words also reduced.

For the ablation experiments of the corresponding interactive network (w/o interactive network), after removing the corresponding interaction network between Att-GCN, the corresponding accuracy has been dropped by 1.74%, 1.25% and 1.97% on the three datasets respectively. It demonstrates that integration of the sentence information benefits the interactive network.

5 Conclusion

The proposed ASAI-DT model utilizes a GCN-based aspect word attention mechanism, sub-dependency tree, and a interactive network to solve the ABSA tasks. In particular, the dependencies captured by the attention mechanism and the syntactic information provided by the DT can accurately capture the relationship between aspects and their contexts. Therefore, both the semantic information conveyed by the sentence and the obtained syntactic information are effectively extracted and conveyed to classify the sentiment of words. The results of experiments demonstrate that the ASAI-DT outperforms popular baselines on public datasets.

Acknowledgment. This work is supported by National Natural Science Foundation of China (61902116).

References

1. Zhang, C., et al.: Aspect-based sentiment classification with aspect-specific graph convolutional networks (2019)
2. Sun, K., et al.: Aspect-level sentiment analysis via convolution over dependency tree. In: Conference on Empirical Methods in Natural Language Processing and the 9th International Joint Conference on Natural Language Processing (EMNLP-IJCNLP) (2019)
3. Kipf, T.N., Welling, M.: Semi-supervised classification with graph convolutional networks (2016)
4. Vaswani, A., et al.: Attention is all you need. arXiv (2017)
5. Wang, R., et al.: Deep & cross network for ad click predictions. In: ADKDD'17. ACM (2017)
6. Dong, L., et al.: Adaptive recursive neural network for target-dependent twitter sentiment classification. In: Proceedings of the 52nd Annual Meeting of the Association for Computational Linguistics (volume 2: Short Papers) (2014)
7. Li, Y., Sun, X., Wang, M.: Embedding extra knowledge and A dependency tree based on A graph attention network for aspect-based sentiment analysis. In: 2021 International Joint Conference on Neural Networks (IJCNN). IEEE (2021)
8. Zhang, J., Sun, X., Li, Y.: Mining syntactic relationships via recursion and wandering on A dependency tree for aspect-based sentiment analysis. In: 2022 International Joint Conference on Neural Networks (IJCNN). IEEE (2022)
9. Yu, W., et al.: Aspect-level sentiment analysis based on graph attention fusion networks. In: 2022 IEEE 24th Int Conf on High Performance Computing and Communications; 8th International Conference on Data Science and Systems; 20th International Conference on Smart City; 8th International Conference on Dependability in Sensor, Cloud and Big Data Systems and Application (HPCC/DSS/SmartCity/DependSys). IEEE (2022)
10. Huo, Y., Jiang, D., Sahli, H.: Aspect-based sentiment analysis with weighted relational graph attention network. In: Companion Publication of the 2021 International Conference on Multimodal Interaction (2021)
11. Li, R., et al.: Dual graph convolutional networks for aspect-based sentiment analysis. In: The 59th Annual Meeting of the Association for Computational Linguistics and the 11th International Joint Conference on Natural Language Processing (Volume 1: Long Papers) (2021)
12. Jiang, T., et al.: Aspect-based sentiment analysis with dependency relation weighted graph attention. Information **14**(3), 185 (2023)
13. Pang, S., Yan, Z., Huang, W., Tang, B., Dai, A., Xue, Y.: Highway-based local graph convolution network for aspect based sentiment analysis. In: Wang, L., Feng, Y., Hong, Y., He, R. (eds.) NLPCC 2021. LNCS (LNAI), vol. 13028, pp. 544–556. Springer, Cham (2021). https://doi.org/10.1007/978-3-030-88480-2_43
14. Wang, Y., et al.: Aspect-based sentiment analysis with dependency relation graph convolutional network. In: 2022 International Conference on Asian Language Processing (IALP). IEEE (2022)
15. Pontiki, M., et al.: Semeval-2014 task 4: aspect based sentiment analysis. In: SemEval, vol. 2014, p. 27 (2014)
16. Pennington, J., et al.: Glove: global vectors for word representation. In: Conference on Empirical Methods in Natural Language Processing (2014)
17. Kingma, D., Ba, J.: Adam: a method for stochastic optimization. Comput. Sci. (2014)
18. Zhou, J., et al.: SK-GCN: modeling syntax and knowledge via graph convolutional network for aspect-level sentiment classification. Knowl.-Based Syst. **205**, 106292 (2020)

19. Zhang, M., Qian, T.: Convolution over hierarchical syntactic and lexical graphs for aspect level sentiment analysis. In: Empirical Methods in Natural Language Processing Association for Computational Linguistics (2020)
20. Chen, C., et al.: Inducing target-specific latent structures for aspect sentiment classification. In: Empirical Methods in Natural Language Processing Association for Computational Linguistics (2020)

Efficient Collaboration via Interaction Information in Multi-agent System

Meilong Shi[1], Quan Liu[1,2(✉)], and Zhigang Huang[1]

[1] School of Computer and Technology, Soochow University, Suzhou 215006, Jiangsu, China
quanliu@suda.edu.cn
[2] Provincial Key Laboratory for Computer Information Processing Technology, Soochow University, Suzhou 215006, Jiangsu, China

Abstract. Cooperative multi-agent reinforcement learning (CMARL) has shown promise in solving real-world scenarios. The interaction information between agents contains rich global information, which is easily neglected after perceiving other agents' behavior. To tackle this problem, we propose Collaboration Interaction Information Modelling via Hypergraph (CIIMH), which first perceives the behavior of other agents by mutual information optimization and constructs the dynamic interaction information via hypergraph. Perceived behavioral features of other agents are further aggregated in the hypergraph convolutional network to obtain interaction information. We compare our method with three existing baselines on StarCraft II micromanagement tasks (SMAC), Level-based Foraging (LBF), and Hallway. Empirical results show that our method outperforms baseline methods on all maps.

Keywords: Multi-Agent reinforcement learning · Collaboration Modelling · Interaction Information · Hypergraph

1 Introduction

Real-world reinforcement learning tasks often involve multiple agents and are formulated as Cooperative Multi-Agent Reinforcement Learning (CMARL), such as autonomous driving [3,7], traffic light control [9,22] and online visual game [17,24]. CMARL aims to learn efficient policies that control multiple agents and maximize the cumulative reward from the given task.

The mainstream CMARL architecture is centralized training with decentralized execution (CTDE), in which each agent can only observe its own local information during the decentralized execution phase. Therefore, perceiving other agents' behavior is critical to modeling global collaboration in the centralized

Supported by National Natural Science Foundation of China (61772355, 61702055, 61876217, 62176175). Project Funded by the Priority Academic Program Development of Jiangsu Higher Education Institutions (PAPD).

training phase. However, the interaction information between agents contains
rich global information, which is easily overlooked after perceiving other agents'
behavior.

To capture interaction information, we propose Collaboration Interaction
Information Modelling via Hypergraph (CIIMH). Each CIIMH agent first learns
to perceive other agents' behavior and further captures rich interaction information by graph neural network (GNN). However, there is no prior knowledge
of agents' interaction to construct ordinary graphs. Moreover, the interaction
between agents changes dynamically over time, which is difficult for ordinary
graphs to learn. To this end, we consider HyperGraph Convolution Network
(HGCN) [6], whose GNN is based on hypergraphs. Different from the ordinary
full-connected edges in Fig. 1(b), one hyperedge in Fig. 1(c) can connect multiple vertices, which naturally represents the complex and dynamic relationship
between agents in Fig. 1(a).

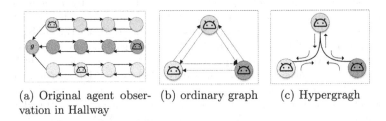

(a) Original agent obser-   (b) ordinary graph   (c) Hypergragh
vation in Hallway

Fig. 1. The illustration of original agent observation in Hallway, ordinary graph, and
hypergraph. The interaction information in (a) requires six ordinary edges, but only
one hyperedge.

In summary, the contributions of this paper are as follows:

(1) We propose a novel CMARL method is proposed, which learns the interaction information in the multi-agent system, and then leverages the learned
information to the cooperation modeling of agents.
(2) We designed various model variants and provided ablation studies, demonstrating that interaction information learning by CIIMH can improve the
performance of the model.
(3) We elaborate experiments and the results illustrate that our proposed
method has an average improvement of 22.49%, which proves CIIMH is
superior to the baselines by utilizing the interaction information.

2   Background

2.1   Dec-POMDP

In this paper, we consider formulating a fully CMARL task as Dec-POMDP
[12] $\langle I, S, A, P, R, O, \Omega, n, \gamma \rangle$. Here, $I \equiv \{1, 2, \ldots, n\}$ is the agent set with n

agents, $s \in S$ is the state of the environment. At each timestep, each agent i chooses an action from its action space A^i, generating a joint action $\boldsymbol{a} \in \prod_{i=1}^{n} A^i$, receiving a shared reward $r = R(s, \boldsymbol{a})$, and transiting to the next state s' by the transition probability $P(s'|s, \boldsymbol{a})$. Additionally, each agent i observes the state s by its own local partial observation $o_i \in \Omega$ according to the individual observation function $O(s, i)$. The action-observation history of each agent i forms its trajectory $\tau_i \in T \equiv (\Omega \times A)$. The policy of agent i is denoted as $\pi^i(a^i|o^i)$: $T \times A^i \longmapsto [0, 1]$. Dec-POMDP maximizes the expected cumulative discounted reward $E_{\boldsymbol{\tau} \sim \pi}[\sum_{t=0}^{\infty} \gamma^t r_t|s_0]$, where $\boldsymbol{\pi}$ is the joint policy, $\boldsymbol{\tau}$ is the joint action-observation trajectory and $\gamma \in [0, 1)$ is the discount factor.

2.2   Centralized Training with Decentralized Execution

Centralized Training with Decentralized Execution (CTDE) is commonly used in CMARL tasks. All agents can access the joint action-observation trajectory during the centralized training phase. During the decentralized execution phase, each agent is only able to make decisions based on its local observations. One mainstream approach to implementing CTDE is value decomposition. Value decomposition decomposes the total Q value Q_{tot} into individual Q values Q_i of each agent i. This idea was first implemented on VDN [16], which assumes the joint value comes from the sum of the individual values. On its basis, QMIX [14] leverages a monotone network to realize the nonlinear mapping from the individual value function to the joint value function. QPLEX [19] proposed to decompose the Q value into a non-positive advantage function, and MAVEN is an exploration value decomposition method [11].

2.3   Hypergraph Convolution Network

A hypergraph [4] can be defined as $\mathcal{G} = (\mathcal{V}, \mathcal{E})$, where $\mathcal{V} = \{v_1, ..., v_N\}$ denotes the set of vertices, $\mathcal{E} = \{\epsilon_1, ..., \epsilon_M\}$ denotes the set of hyperedges, N and M are the numbers of vertices and hyperedges, respectively. For a vertex $v_i \in \mathcal{V}$ and a hyperedge $\epsilon_j \in \mathcal{E}$, the binary incidence matrix value $h_{i,j}$ denotes if v_i is connected by ϵ_j. A diagonal matrix $\mathbf{W} \in \mathbb{R}^{M \times M}$ consists of the non-negative weight w_ϵ of each hyperedge. Besides, the degrees of vertices and hyperedges are defined as $d(v_i) = \sum_{\epsilon_j \in \mathcal{E}} w_{\epsilon_j} h_{i,j}$ and $d(\epsilon_j) = \sum_{v_i \in V} h_{i,j}$ respectively, which in turn form vertex diagonal degree matrix $\mathbf{D_v}$ and hyperedge diagonal degree matrix $\mathbf{D_e}$ respectively.

To reduce the computational complexity, we follow [6], which derives spectral convolution as

$$\mathbf{x}^{(l+1)} = \mathbf{D_v}^{-1/2} \mathbf{H} \mathbf{W} \mathbf{D_e}^{-1} \mathbf{H}^\top \mathbf{D_v}^{-1/2} \mathbf{x}^{(l)} \mathbf{P}^{(l)}, \tag{1}$$

where $\mathbf{W}$ and $\mathbf{P}$ denotes the learnable weight matrix. There have been studies using hypergraphs for CMARL, HGCN-MIX [2] using hypergraph to implement value decomposition, HGAC [23] leverages hypergraph in Actor-Critic method. As a value-based method, our method differs from them in that we incorporate hypergraphs into the learning of interaction information after perceiving agents.

3   CIIMH

This section introduces the CIIMH training process, and its overall process is shown in Fig. 2. For decentralized execution, we first use the GRU unit [5] to preprocess the original observation o_i^t and obtain the current trajectory τ_i^t. The agent network (Fig. 2 (b)) of each CIIMH agent contains an agent perceiving network (Fig. 2 (a)) and an interaction learning network (Fig. 2 (c)). The CIIMH agent perceives the behavior of other agents and samples n perception variations. Learned perception variations are used to combine interaction information to generate a global representation. This is ultimately used to obtain a Q value via an MLP.

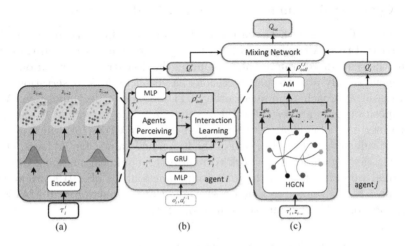

Fig. 2. The overall architecture of CIIMH. (a) Agent perceiving (b) Agent Network (C) Interaction Learning

3.1   Agent Perceiving

We hope that the agent perceiving module (Fig. 2 (a)) can perceive other agents' behavior by its local trajectory τ_i^t. To this end, n multivariate Gaussian distributions are used to sample n corresponding perception variations $\{z_{i\to1}^t, z_{i\to2}^t \cdots , z_{i\to n}^t\}$, where each perception variation $z_{i\to j}^t$ represent the perception from i to j, and the means and variances of Gaussian distributions are encoded by the current trajectory τ_i^t in parallel

$$\left(\mu^{t,i\to j}, \sigma^{t,i\to j}\right) = MLP\left(\tau_i^t\right) \quad z_{i\to j}^t \sim \mathcal{N}\left(\mu^{t,i\to j}, \sigma^{t,i\to j}\right), \quad (2)$$

where t is the current moment, and for agent $j \in \{1, \ldots, n\}$, $\mu^{t,i\to j}$ and $\sigma^{t,i\to j}$ are the mean and variance of the Gaussian distribution encoded by τ_i^t, respectively.

We desire the learned perception variation $z_{i\to j}^t$ should perceive the action selection of each specific agent j. Hence, the goal of the agent perception is to maximize the mutual information between the action a_j^t taken by the corresponding agent j and the perceptual variation $z_{i \to j}^t$ of Gaussian distributions conditioned on agent i's trajectory. Unfortunately, it is difficult to optimize mutual information objectives directly. To optimize the mutual information objective, **Theorem** 1 is derived from the information bottleneck [1].

Theorem 1. *For any agent* $i, j \in \{1, \ldots, n\}$, *the mutual information* $I\left(z_{i \to j}^t, a_j^t \mid \tau_i^t\right)$ *has an evidence lower bound*

$$I\left(z_{i \to j}^t, a_j^t \mid \tau_i^t\right) \geq \mathbb{E}_{\mathcal{D}}\left[-D_{\mathrm{KL}}\left(p\left(z_{i \to j}^t \mid \tau_i^t\right) \| q_\xi\left(z_{i \to j}^t \mid \tau_i^t, a_j^t\right)\right)\right], \quad (3)$$

where $z_{i \to j}^t$ *denotes* i's *perception variation to* j, a_j^t *is the action* j *takes*, q_ξ *is the variational approximator, all the distribution* p *and* q_ξ *are sampled from the replay buffer* $\mathcal{D}$.

Proof. According to the definition of mutual information, we have

$$I\left(z_{i \to j}^t, a_j^t \mid \tau_i^t\right)$$

$$= \mathbb{E}_{z_{i \to j}^t, a_j^t, \tau_i^t}\left[\log \frac{p\left(z_{i \to j}^t \mid a_j^t, \tau_i^t\right)}{p\left(z_{i \to j}^t \mid \tau_i^t\right)}\right]$$

$$= \mathbb{E}_{z_{i \to j}^t, a_j^t, \tau_i^t}\left[\log \frac{q_\xi\left(z_{i \to j}^t \mid a_j^t, \tau_i^t\right)}{p\left(z_{i \to j}^t \mid \tau_i^t\right)}\right] \quad (4)$$

$$+ \mathbb{E}_{a_j', \tau_i^t}\left[D_{\mathrm{KL}}\left(p\left(z_{i \to j}^t \mid a_j^t, \tau_i^t\right) \| q_\xi\left(z_{i \to j}^t \mid a_j^t, \tau_i^t\right)\right)\right]$$

$$\geq \mathbb{E}_{z_{i \to j}^t, d_j^t, \tau_i^t}\left[\log \frac{q_\xi\left(z_{i \to j}^t \mid a_j^t, \tau_i^t\right)}{p\left(z_{i \to j}^t \mid \tau_i^t\right)}\right]$$

$$= \mathbb{E}_{\mathcal{D}}\left[-D_{\mathrm{KL}}\left(p\left(z_{i \to j}^t \mid \tau_i^t\right) \| q_\xi\left(z_{i \to j}^t \mid \tau_i^t, a_j^t\right)\right)\right],$$

proof completed.

According to **Theorem** 1, the evidence lower bound of is obtained, so we derive the loss function of the agent perceiving module as

$$\mathcal{L}_{perc} = \sum_{i \neq j} \mathbb{E}_{\mathcal{D}}\left[D_{\mathrm{KL}}\left(p\left(z_{i \to j}^t \mid \tau_i^t\right) \mid q_\xi\left(z_{i \to j}^t \mid \tau_i^t, a_j^t\right)\right)\right]. \quad (5)$$

3.2   Interaction Learning

Although the learned perception variations can perceive the behavior of other agents, we hope them could further capture rich interaction information, rather than directly inputting them to the Q network. To this end, we construct a hypergraph, where the nodes and hyperedges represent agents and the interaction relationship between agents, respectively. And the perception variations are then fed to the HGCN for learning the interaction information on the constructed hypergraph.

Constructing Hypergraph. Since interaction information in multi-agent tasks varies over time, we dynamically construct an incidence matrix for the hypergraph. To represent different levels of interaction, we use a real-value incidence matrix to describe the relationship between agents. First, we construct the initial vertex features ρ_v^i in the hypergraph using local trajectories τ_i of each agent, which is defined as

$$\rho_v^i = Softmax(MLP(\tau_i)). \tag{6}$$

In order to efficiently learn interaction information between agents, we use cosine similarity to represent the interaction between the features of two agents

$$h_{i,j}^{sim} = \begin{cases} Softmax(cosine(\rho_v^i, \rho_v^j)) & \text{if } i \neq j \\ 1 & \text{if } i = j \end{cases}. \tag{7}$$

Besides, we define a diagonal matrix to prevent the incidence matrix from changing too quickly. Its value is the average of $h_{i,j}^{sim}$ of the current agent i.

$$h_i^{avg} = \frac{\sum_{j=1}^{m}(h_{i,j}^{sim})}{m}, \tag{8}$$

where h_i^{avg} is updated with an interval of time T, and we keep the incidence matrix from changing too quickly by keeping h_i^{avg} constant over the period. Therefore, the incidence matrix H is denoted as

$$H = \underbrace{\begin{bmatrix} h_{1,1}^{sim} & \cdots, & h_{1,m}^{sim} \\ h_{2,1}^{sim} & \cdots, & h_{2,m}^{sim} \\ \vdots & \cdots, & \vdots \\ h_{n,1}^{sim} & \cdots, & h_{n,m}^{sim} \end{bmatrix}}_{similarity} \underbrace{\begin{bmatrix} h_1^{avg} & \cdots, & 0 \\ 0 & \cdots, & 0 \\ \vdots & \cdots, & \vdots \\ 0 & \cdots, & h_n^{avg} \end{bmatrix}}_{average}. \tag{9}$$

HGCN. After the construction of the hypergraph, HGCN can be used to extract interaction information from the hypergraph. Recall the HGCN formula in Eqt. 1, each perception variation $z_{i\to j}^t$ is fed to a two-layer HGCN with hypergraph H

$$z_{i\to 1}^{glo,t}, z_{i\to 2}^{glo,t} \cdots, z_{i\to n}^{glo,t} = HGCN(H, HGCN(H, z_{i\to 1}^t, z_{i\to 2}^t \cdots, z_{i\to n}^t)). \tag{10}$$

After that, the interaction information is aggregated by hyperedge. In other words, global perception variations $z_{i\to j}^{glo,t}$ contain both agents' local and interaction information. To better integrate local features, we utilize the attention mechanism [18]

$$\begin{cases} q_i = W_q\rho_v^i \\ k_{i,j} = W_k z_{i\to j}^{glo,t} \\ v_{i,j} = W_v z_{i\to j}^{glo,t} \end{cases}, \tag{11}$$

where the query q_i is computed from the local vertex features ρ_v^i, key and value are computed from the global perception variation $z_{i \to j}^{glo,t}$, and W_q, W_k, W_v are the attention parameters.

Finally, we combine all the global perception variations of i to get the ultimate collaboration variation ρ_{coll}^i

$$\rho_{coll}^i = \sum_{j \neq i} f_{att}(q_i, k_{i,j}, v_{i,j}), \tag{12}$$

where f_{att} is the attention function in [18]. And we get the agent i's Q value $Q(\tau^i, \rho_{coll}^i, \cdot)$.

3.3   Overall Optimization Objective

As a value decomposition method based on the CTDE paradigm, the proposed model can be trained by minimizing the following TD loss:

$$\mathcal{L}_{td} = \frac{1}{2} \left(y_{td} - Q_{tot}(\boldsymbol{\tau}, \rho_{coll}, \boldsymbol{a}) \right)^2, \tag{13}$$

where the TD target $y_{td} = r + \gamma \max_{\boldsymbol{a}'} Q_{tot}(\boldsymbol{\tau}', \rho_{coll}, \boldsymbol{a}')$, we leverage QMIX [14] as the value decomposition method to get the total Q value Q_{tot} by individual Q value $Q_i, i \in \{1, \ldots, n\}$, and QMIX can also be replaced by other mixing networks. Combining $\mathcal{L}_{perc}$ and $\mathcal{L}_{td}$ losses, we get the overall target is defined as

$$\mathcal{L}_{overall} = \mathcal{L}_{td} + \lambda_{perc} \mathcal{L}_{perc}, \tag{14}$$

where λ_{perc} is the adjustable factor for $\mathcal{L}_{perc}$.

4   Experiment

4.1   Environments and Settings

Environments. The three experimental environments used in this paper are briefly described as follows:

SMAC [15]: SMAC contains a series of StarCraft II games, where agents controlled by our algorithm fight against enemy agents controlled by the built-in game AI. The game ends when all the agents on either side are killed, and the game is won only when the enemy agents are all killed.

Hallway [20]: Multiple agents are randomly initialized in different hallways and can only observe their current location. The game is considered to be won only when all agents reach the destination simultaneously. To increase difficulty, we set three hallway lengths: 2, 6, and 10.

LBF [13]: LBF is a foraging grid game, where each agent and food is initialized with a level value. The agents combine for foraging, and the group level is the sum of the levels of the agents in the group. The group can forage for smaller levels of food. In this experiment, four agents are set to forage in a grid environment of 10×10 and 16×16, the food is set to two portions, and the observation range is set to two grids.

Table 1. The experimental parameter in this paper.

Parameter Name	Value	Parameter Name	Value
optimizer	RMSprop	$\epsilon - greedy$ start value	1
replay buffer	5000	$\epsilon - greedy$ end value	0.05
batch	32	average matrix update interval	20
learning rate	0.0005	$\mathcal{L}_{perc}$ discount factor	0.20

Settings. To demonstrate the validity of CIIMH, we choose several common CTDE algorithms including QMIX [14], QPLEX [19], MAVEN [11], as baseline methods. To ensure fairness, the parameters used in all the algorithms in this experiment are consistent with the standard library PyMARL [15]. We independently run each algorithm with five different seeds and test the winning rate every 10,000 steps during the training process. The specific parameter values are shown in Table 1, where the $\epsilon - greedy$ strategy decays from 1 to 0.05 at a rate of 0.00002 per time step.

4.2 Experimental Results and Analysis

Comparative Experiment. To validate the effect, we conduct the comparative experiment on SMAC and Hallway. According to Fig. 3, CIIMH achieves the best performance in all six maps.

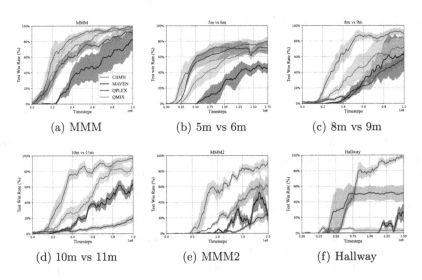

(a) MMM   (b) 5m vs 6m   (c) 8m vs 9m

(d) 10m vs 11m   (e) MMM2   (f) Hallway

Fig. 3. Comparisons with Baselines on SMAC and Hallway

It is worth noting that the relative improvement of CIIMH is not clear on the hard task *5m vs 6m*, and QPLEX can achieve more than 60% of the winning rate within 0.75 million timesteps. We point out that this is because the task contains only five marines and the number of enemies is similar to the number of teammates. In this case, the five agents might win by taking opposing actions. For example, each agent attacks an enemy in a different direction. Therefore, the *5m vs 6m* agents only need weak collaboration, so interaction information is less meditative.

Table 2. Comparative experiment result on SMAC and Hallway in Last 0.1 million timesteps

Map	Algorithm					
	MMM	5 m vs 6 m	8 m vs 9 m	10 m vs 11 m	MMM2	Hallway
CIIMH/%	**98.01**	**78.91**	**91.91**	**95.02**	**87.94**	**97.01**
MAVEN/%	75.46	47.54	60.32	54.83	31.46	23.55
QMIX/%	96.04	66.50	70.86	80.96	63.10	3.34
QPLEX/%	91.50	71.95	52.84	16.35	22.98	50.14

In other SMAC maps, CIIMH's advantages are revealed. Table 2 exhibits the average test win rate of the last 0.1 million time steps in each experiment. CIIMH has achieved a winning rate of more than 95% on the hard maps *MMM* and *10m vs 11m*, 87.94% on the super hard map *MMM2*. The final test win rate is better than the three baseline methods.

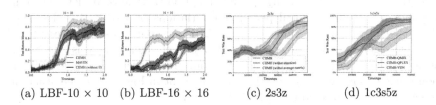

(a) LBF-10 × 10   (b) LBF-16 × 16   (c) 2s3z   (d) 1c3s5z

Fig. 4. Ablation experiment results

Ablation Experiment. To demonstrate the effect of each component of CIIMH, a variety of model variants have been developed to conduct the ablation studies. For the interaction information (II) module, we only remove the II module of CIIMH and directly input perception variations into the Q network, aiming to observe CIIMH performance after eliminating II. The ablation experiment shows that after removing the II, the CIIMH is less effective and unstable. Specifically, LBF-16 × 16 (Fig. 4 (a)) decreased by 29.07%, and relatively easily LBF-10 × 10 (Fig. 4 (b)) by 5.82%, which prove the necessity of II for CIIMH. For the other components in CIIMH, including the attention mechanism and the average matrix, we removed them to observe the effect on $2s3z$ (Fig. 4 (c)). As expected, the algorithm's performance degrades. Furthermore, we explored the effect of different mixing networks on CIIMH in $1c3s5z$ (Fig. 4 (d)), and the results demonstrated that QMIX performs better than other mixing networks.

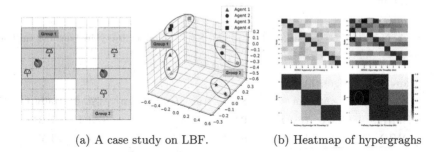

(a) A case study on LBF.   (b) Heatmap of hypergraghs

Fig. 5. Visualization Research. (a) Each agent's perception of all other three agent's on LBF. The blue and green parts of the figure represent the different groups. (b) Heatmap of incidence matrices of MMM2 (top) and Hallway (below), where hyperedge of 0 means the agent is dead. (Color figure online)

Visualization Research. To illustrate how CIIMH components contribute to cooperation, we evaluate CIIMH in multiple tests. Figure 5(a) shows a test timestep with its corresponding global perception variations. We leverage dimensionality reduction to global perception variations by t-SNE [10]. The results show that agents in the same group of LBF are more closely represented, for example, agent 1 and 4, agent 2 and 3. For each agent, there is one in-group agent in the observation range and two out-group agents not in the observation range. Thus the representation of each agent in Fig. 5(a) to the other three agents is also correspondingly divided into 2:1.

Moreover, we investigate the effect of cosine similarity on CIIMH hypergraphs using a visualization analysis. As illustrated in Fig. 5(b), each hyperedge is initialized randomly in the MMM2 and Hallway scenes, and we record the hypergraphs during training. As we expected, the hypergraph successfully extracted information about the interactions between the agents, e.g. hyperedges 2 and 4 of agent 0 in $MMM2$ and hyperedge 1 of agent 0 in Hallway.

5 Conclusion

In this paper, we propose CIIMH, which leverages hypergraphs to learn interaction information in multi-agent systems. The key insight of this paper is that an effective collaboration model requires agents' interaction information. We have designed a series of comparative experiments on SMAC and Hallway, demonstrating the effectiveness of CIIMH. In addition, the ablation experiment shows that the algorithm's performance is significantly reduced without interaction information. This proves the necessity of interaction information. In future research, we will continue to pay attention to the interaction feature extraction problem under the Decentralized Training with Decentralized Execution (DTDE) architecture [21]. In addition, the experiments in this paper mainly consider the scenario where the number of agents is less than 20, the interaction feature extraction problem in large-scale multi-agent reinforcement learning tasks [8] is also worthy of study, and we will further explore this issue in future research.

References

1. Alemi, A.A., Fischer, I., Dillon, J.V., Murphy, K.: Deep variational information bottleneck. arXiv preprint arXiv:1612.00410 (2016)
2. Bai, Y., Gong, C., Zhang, B., Fan, G., Hou, X., Lu, Y.: Cooperative multi-agent reinforcement learning with hypergraph convolution. In: 2022 International Joint Conference on Neural Networks (IJCNN), pp. 1–8. IEEE (2022)
3. Bhalla, S., Ganapathi Subramanian, S., Crowley, M.: Deep multi agent reinforcement learning for autonomous driving. In: Goutte, C., Zhu, X. (eds.) Canadian AI 2020. LNCS (LNAI), vol. 12109, pp. 67–78. Springer, Cham (2020). https://doi.org/10.1007/978-3-030-47358-7_7
4. Bretto, A.: Hypergraph Theory. An Introduction. Mathematical Engineering, Springer, Cham (2013). https://doi.org/10.1007/978-3-319-00080-0
5. Cho, K., et al.: Learning phrase representations using RNN encoder-decoder for statistical machine translation. arXiv preprint arXiv:1406.1078 (2014)
6. Feng, Y., You, H., Zhang, Z., Ji, R., Gao, Y.: Hypergraph neural networks. In: Proceedings of the AAAI Conference on Artificial Intelligence, vol. 33, pp. 3558–3565 (2019)
7. Kiran, B.R., et al.: Deep reinforcement learning for autonomous driving: a survey. IEEE Trans. Intell. Transp. Syst. (2021)
8. Li, J., Yu, T.: Large-scale multi-agent deep reinforcement learning-based coordination strategy for energy optimization and control of proton exchange membrane fuel cell. Sustain. Energy Technol. Assess. **48**, 101568 (2021)
9. Luo, Y.C., Tsai, C.W.: Multi-agent reinforcement learning based on two-step neighborhood experience for traffic light control. In: Proceedings of the 2021 ACM International Conference on Intelligent Computing and its Emerging Applications, pp. 28–33 (2021)
10. Van der Maaten, L., Hinton, G.: Visualizing data using T-SNE. J. Mach. Learn. Res. **9**(11), 2579–2605 (2008)
11. Mahajan, A., Rashid, T., Samvelyan, M., Whiteson, S.: Maven: multi-agent variational exploration. In: Advances in Neural Information Processing Systems, vol. 32 (2019)
12. Oliehoek, F.A., Amato, C.: A concise introduction to decentralized pomdps (2015)
13. Papoudakis, G., Christianos, F., Schäfer, L., Albrecht, S.V.: Benchmarking multi-agent deep reinforcement learning algorithms in cooperative tasks. arXiv preprint arXiv:2006.07869 (2020)

14. Rashid, T., Samvelyan, M., Witt, C.S., Farquhar, G., Foerster, J., Whiteson, S.: Qmix: monotonic value function factorisation for deep multi-agent reinforcement learning. In: International Conference on Machine Learning, pp. 4292–4301 (2018)

15. Samvelyan, M., et al.: The starcraft multi-agent challenge. arXiv preprint arXiv:1902.04043 (2019)

16. Sunehag, P., et al.: Value-decomposition networks for cooperative multi-agent learning. arXiv preprint arXiv:1706.05296 (2017)

17. Tufano, R., Scalabrino, S., Pascarella, L., Aghajani, E., Oliveto, R., Bavota, G.: Using reinforcement learning for load testing of video games. In: Proceedings of the 44th International Conference on Software Engineering, pp. 2303–2314 (2022)

18. Vaswani, A., et al.: Attention is all you need. In: Advances in Neural Information Processing Systems, vol. 30 (2017)

19. Wang, J., Ren, Z., Liu, T., Yu, Y., Zhang, C.: Qplex: duplex dueling multi-agent q-learning. In: International Conference on Learning Representations (ICLR) (2021)

20. Wang, T., Wang, J., Zheng, C., Zhang, C.: Learning nearly decomposable value functions via communication minimization. arXiv preprint arXiv:1910.05366 (2019)

21. Wen, G., Fu, J., Dai, P., Zhou, J.: DTDE: a new cooperative multi-agent reinforcement learning framework. Innovation $2(4)$ (2021)

22. Wu, T., et al.: Multi-agent deep reinforcement learning for urban traffic light control in vehicular networks. IEEE Trans. Veh. Technol. $69(8)$, 8243–8256 (2020)

23. Zhang, B., Bai, Y., Xu, Z., Li, D., Fan, G.: Efficient policy generation in multi-agent systems via hypergraph neural network. In: Tanveer, M., Agarwal, S., Ozawa, S., Ekbal, A., Jatowt, A. (eds.) ICONIP 2022. LNCS, vol. 13624, pp. 219–230. Springer, Cham (2022)

24. Zhang, R., Zong, Q., Zhang, X., Dou, L., Tian, B.: Game of drones: multi-UAV pursuit-evasion game with online motion planning by deep reinforcement learning. IEEE Trans. Neural Networks Learn. Syst. $\mathbf{10}$, 7900–7909 (2022)

A Deep Graph Matching-Based Method for Trajectory Association in Vessel Traffic Surveillance

Yuchen Lu[1,2], Xiangkai Zhang[1,2], Xu Yang[1,2,3(✉)], Pin Lv[1(✉)], Liguo Sun[1], Ryan Wen Liu[4], and Yisheng Lv[1]

[1] Institute of Automation, Chinese Academy of Sciences, Beijing, China
{luyuchen2021,zhangxiangkai2023,xu.yang,pin.lv}@ia.ac.cn
[2] University of Chinese Academy of Sciences, Beijing, China
[3] Centre for Artificial Intelligence and Robotics, Hong Kong Institute of Science and Innovation, Chinese Academy of Sciences, Beijing, China
[4] Hubei Key Laboratory of Inland Shipping Technology, School of Navigation, Wuhan University of Technology, Wuhan, China

Abstract. Vessel traffic surveillance in inland waterways extensively relies on the Automatic Identification Syst em (AIS) and video cameras. While video data only captures the visual appearance of vessels, AIS data serves as a valuable source of vessel identity and motion information, such as position, speed, and heading. To gain a comprehensive understanding of the behavior and motion of known-identity vessels, it is necessary to fuse the AIS-based and video-based trajectories. An important step in this fusion is to obtain the correspondence between moving targets by trajectory association. Thus, we focus solely on trajectory association in this work and propose a trajectory association method based on deep graph matching. We formulate trajectory association as a graph matching problem and introduce an attention-based flexible context aggregation mechanism to exploit the semantic features of trajectories. Compared to traditional methods that rely on manually designed features, our approach captures complex patterns and correlations within trajectories through end-to-end training. The introduced dustbin mechanism can effectively handle outliers during matching. Experimental results on synthetic and real-world datasets demonstrate the exceptional performance of our method in terms of trajectory association accuracy and robustness.

Keywords: Vessel traffic surveillance · Automatic Identification System · Trajectory association · Deep graph matching

1 Introduction

As an economic and efficient mode of transportation, inland waterway has gradually gained recognition. However, the presence of obstacles such as bridges,

B. Luo et al. (Eds.): ICONIP 2023, LNCS 14448, pp. 413–424, 2024.
https://doi.org/10.1007/978-981-99-8082-6_32

tunnels, and locks in the narrow and high-density inland waterways poses significant challenges to maritime management and monitoring. Thus, it is necessary to further improve the intelligence of the waterway monitoring system.

To achieve this objective, the Automatic Identification System (AIS) and video cameras have been widely used in inland waterways. AIS data can provide vessel identity, type, and motion information, such as position, speed, and heading. In contrast, video cameras provide the visual appearance of vessels but lack vital information regarding vessel identity and motion. Consequently, numerous efforts are focused on fusing the AIS data and video data to further improve vessel traffic surveillance [1,5,10,12].

In fact, AIS is unable to provide real-time data due to its low temporal resolution, whereas video cameras have the advantages of intuitiveness and real-time capabilities. This results in problems such as time delay, information lag, and random outliers. Therefore, most current methods that only consider the absolute position of trajectories and rely on manually designed features encounter difficulty in ensuring effective results. However, compared with absolute position, relative position of trajectories is likely to be more important. Specifically, the implied semantic information of trajectories, such as the target's motion pattern, and the interaction behavior between moving targets (relative position, collision, evasion, etc.) plays a critical role in the context understanding of trajectory association.

Based on the above circumstances, we propose a trajectory association method based on deep graph matching. Specifically, we suggest using a unique graph structure, called target trajectory graph, to represent the trajectories set to be matched. Next, we redefine trajectory association as a graph matching problem. Finally, we determine the correspondence of the trajectories using the deep graph matching method. Our main contributions are summarized as follows:

- The proposed target trajectory graph implies trajectory semantic information, signifying its potential for enhancing context understanding and the accuracy of trajectory association.
- We have developed a comprehensive trajectory association framework based on deep graph matching, consisting of an attention [14] mechanism and a dustbin mechanism proposed by [13], which can effectively extract the deep trajectory semantic information from target trajectory graphs while eliminating outliers.
- Experiments on real and synthetic datasets demonstrate the accuracy and robustness of the proposed method in handling outliers.

2   Related Works

2.1   AIS and Video Data Fusion

The fusion of AIS data and video data is crucial for promoting the traffic situational awareness of maritime transportation systems. For instance, work [1] proposed a method for tracking a single vessel by combining AIS and video data, which enabled the camera to focus on the vessel based on its AIS-derived position and applied the Kalman filter to ensure smooth tracking. However, this method can hardly provide accurate identification of each vessel when multiple vessels are present in the field of view.

Consequently, the subsequent researchers have shifted their attention to the fusion of information from multiple vessels. For example, work [10] proposed a fusion method by estimating the vessels' length, which is effective on shore-based visual sensors. Nevertheless, significant errors may arise in the fusion results when the relative azimuth angle between the vessel's course over ground (COG) and the camera bearing is small.

Recently, some researchers have considered using bipartite graph matching algorithms to solve fusion problems. The work [12] proposed using a manually-designed feature similarity metric to calculate the similarity between AIS-based and video-based trajectory points, and using the Hungarian [8] algorithm to obtain the optimal fusion result. However, in this work, the difference in temporal resolution between AIS and video data leads to unreliable position of trajectory points, which affects the fusion result. Moreover, the manually designed similarity metric shows poor generalization performance. Another work [5] proposed applying dynamic time warping (DTW) [11] to calculate the similarity between trajectory sequence data and also utilizing the Hungarian algorithm to obtain fusion results. To the best of our knowledge, this is the first and only work that treats a trajectory as a whole for fusion. In general, most existing AIS and video data fusion methods are not robust for complex scenes and are unable to handle outliers well.

2.2   Deep Graph Matching

Graph matching can be expressed as a Quadratic Assignment Problem (QAP) [9], which is an NP-complete problem [6] and difficult to be solved. Classical methods mainly adopt different heuristic methods, like random walk [2], graduated assignment algorithm [4] and so on. However, these optimization-based methods face the challenge of performance saturation. Therefore, learning-based methods have received more and more attention in recent years.

Since the seminal work [18], deep neural networks (DNNs) have emerged as a promising paradigm for solving a wide range of combinatorial problems, including graph matching [16]. In deep graph matching methods, a general line of research [15,17] embeds structural information into node features and learns similarity measurement through graph neural networks (GNNs). Graph matching is degraded to a linear assignment problem, which can be optimally solved through the Sinkhorn network [3]. SuperGlue [13] is a representative method in this line, which has successfully applied graph matching to the field of image matching.

SuperGlue consists of two main components, the graph embedding layer and the optimal matching layer. In particular, the graph embedding layer transforms the features of each node to a matching descriptor. The transformation process employs an attentional graph neural network to transfer information between nodes within and across graphs via self-attention and cross-attention mechanisms respectively. The optimal matching layer outputs the assignment matrix by matching the matching descriptors, which is formulated by the optimal transport problem and solved by the Sinkhorn algorithm. Specifically, a dustbin mechanism is introduced as an effective means of handling outliers and partial assignments, with a learnable parameter linked to the number of outliers instead of the empirical settings used in previous works.

3   Preliminaries

3.1   Target Trajectory Graph

In order to consider contextual information, we define a target trajectory graph to represent the set of trajectories to be matched. This definition is based on a weighted undirected graph.

Specifically, given a set of trajectories of size N, a corresponding target trajectory graph with N vertices is denoted as G^t. Each node n_i of G^t represents a target. The attribute of n_i corresponds to the trajectory information of the target, denoted as $\mathbf{t}_i = \{(x_1, y_1), (x_2, y_2), ..., (x_p, y_p), ..., (x_P, y_P)\}$, where $1 \leq i \leq N$, $1 \leq p \leq P$. P is the total number of trajectory points, which are arranged in chronological order. (x_p, y_p) represents the position of the p-th trajectory point, where x_p, y_p respectively represent the horizontal and vertical coordinates.

3.2   Target Trajectory Graph Matching

Graph matching is a fundamental problem in pattern recognition that aims to find correspondences among nodes in two or more graphs while maintaining consistency in their edges. Leveraging the proposed target trajectory graph, we formulate trajectory association between AIS data and video data as a pairwise graph matching problem, called target trajectory graph matching. Due to the presence of outliers in AIS and video data in practice, the formulation here is based on target trajectory graphs with outliers.

Given a AIS-based target trajectory graph G_a^t and a video-based target trajectory graph G_v^t, we respectively use $V_a = \{n_1^a, n_2^a, ..., n_{N_a}^a\}$ and $V_v = \{n_1^v, n_2^v, ..., n_{N_v}^v\}$ to denote the node sets of G_a^t and G_v^t, where N_a and N_v are the set sizes. Outliers may exist in either G_a^t or G_v^t. Target trajectory graph matching is defined by identifying a set of node assignments between G_a^t and G_v^t, which can be mathematically represented by an assignment matrix $\mathbf{P} \in \{0,1\}^{N_a \times N_v}$. If $\mathbf{P}_{ij} = 1$, it means assigning the node n_i^a in G_a^t to node n_j^v in G_a^t. $\mathbf{P}$ can be defined as follows:

$$\mathbf{P} \in \mathcal{D} := \left\{ \mathbf{P} \mid \mathbf{P}_{ij} \in \{0,1\}, \sum_{i=1}^{N_a} \mathbf{P}_{ij} \leq 1, \sum_{j=1}^{N_v} \mathbf{P}_{ij} \leq 1, \right\}, \tag{1}$$

where the constraints $\sum_{i=1}^{N_a} \mathbf{P}_{ij} \leq 1$ and $\sum_{j=1}^{N_v} \mathbf{P}_{ij} \leq 1$ correspond to the one-to-one correspondence assumption, which means that one node in G_a^t can be assigned to at most one node in G_v^t.

In this study, target trajectory graph matching is designed to maximize the affinity score by identifying the optimal assignment matrix $\mathbf{P}^*$. Based on a given affinity matrix $\mathbf{K} \in \mathbb{R}^{N_a \times N_v}$, it is essentially a linear programming problem in the following form:

$$\mathbf{P}^* = \arg\max_{\mathbf{P} \in \mathcal{D}} \mathrm{tr}\left(\mathbf{K}^{\top}\mathbf{P}\right),\tag{2}$$

where $\mathrm{tr}(\cdot)$ is the matrix trace operation.

4 The Proposed Method

The proposed method for obtaining the assignment matrix from two target trajectory graphs mainly derives from [13]. It consists of two parts, which are illustrated in Fig. 1 and explained in the following subsections. The first part obtains a matching descriptor for each node in target trajectory graphs. The second part computes the affinity matrix from matching descriptors and then obtains the assignment matrix.

4.1 Attentional Graph Neural Network

The temporal resolution of AIS data is low, and thus AIS cannot provide accurate and real-time ship position information. In contrast, video data provides a higher temporal resolution and more reliable and detailed ship position information. This stark difference makes it challenging to solve the trajectory association problem by solely relying on position information of trajectory points. Instead, successful solutions demand the consideration of both the relative position of target trajectories and the semantic information derived from them, including the behavior and movement patterns of the targets. This aspect holds paramount importance.

In the target trajectory graph, each node represents one target. For one target, integrating the features of surrounding targets can intuitively enhance its distinctiveness and uniqueness in a single target trajectory graph. Analyzing the spatial relationship between the target and other visible targets, especially those with unique motion or position features, is helpful. For instance, a stationary vessel is very likely to be a reliable reference for position.

On the other hand, knowledge of targets in the second target trajectory graph can help to eliminate ambiguities through the comparison of potential matches. In practice, when attempting to match a given target, humans may engage in a back-and-forth process of evaluating potential matching targets, analyzing each target, and seeking contextual cues that aid in distinguishing the true match from other similar targets. This implies an iterative process that can effectively direct attention to specific locations. Therefore, we design the graph embedding layer as an attentional graph neural network.

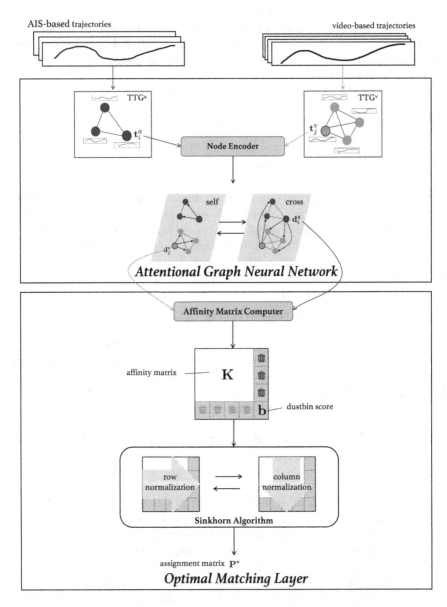

Fig. 1. Framework of the proposed method. It is made up of two major parts: the attentional graph neural network (Sect. 4.1), and the optimal matching layer (Sect. 4.2).

Node Encoder. The initial node feature $\mathbf{t}_i$ for each node i is encoded as follows:

$$\mathbf{l}_i = Enc(\mathbf{t}_i), \tag{3}$$

where Enc represents the encoding network layer and $Enc(\mathbf{t}_i)$ transforms the initial node feature $\mathbf{t}_i$ to a lower-dimensional vector $\mathbf{l}_i$. To address the variable length of the trajectory sequence data, we use an Long Short-Term Memory(LSTM) network [7] for Enc. The LSTM network is well-suited for handling sequence data with differing lengths and can capture long-term temporal dependencies in the trajectories.

Attentional Aggregation. Each vector $\mathbf{l}_i$ communicates with the others and can be further embedded through the attention mechanism, which aggregates information from other vectors. Specifically, each vector $\mathbf{l}_i$ is first linearly transformed into three ones termed by query, key, and value as follows:

$$\begin{bmatrix} \mathbf{q}_i \\ \mathbf{k}_i \\ \mathbf{v}_i \end{bmatrix} = \begin{bmatrix} \mathbf{W}^q \\ \mathbf{W}^k \\ \mathbf{W}^v \end{bmatrix} \mathbf{l}_i + \begin{bmatrix} \mathbf{b}^q \\ \mathbf{b}^k \\ \mathbf{b}^v \end{bmatrix}, \tag{4}$$

where $\mathbf{W}^q$, $\mathbf{W}^k$, $\mathbf{W}^v$, $\mathbf{b}^q$, $\mathbf{b}^k$, $\mathbf{b}^v$ are the projection parameters to be learned. After aggregation through the attention mechanism, the output vector $\mathbf{z}_i$ is obtained as follows:

$$\mathbf{z}_i = \sum_{j \in \mathcal{N}(i)} p_{ij} \mathbf{v}_j, \tag{5}$$

where $\mathcal{N}(i)$ denotes the set of neighbors that pass information with node i. In the case of intradomain message propagation, $\mathcal{N}(i)$ denotes all nodes of the graph except for i. For interdomain message propagation, $\mathcal{N}(i)$ denotes all nodes of the other graph. The probability p_{ij} measures the strength of the connection between node i and its neighbor j, and is defined as follows:

$$p_{ij} = \text{Softmax}_j \left(\mathbf{q}_i^\top \mathbf{k}_j \right). \tag{6}$$

By implementing T times attention in parallel on vetor $\mathbf{z}_i$, a multi-head attention mechanism can be achieved to enhance feature expression capability. Then, further linear transformation is used to merge various outputs to obtain the final matching descriptor:

$$\mathbf{d}_i = \mathbf{W} \left[\mathbf{z}_i^1 \| \mathbf{z}_i^2 \| \cdots \| \mathbf{z}_i^T \right], \tag{7}$$

where $\mathbf{z}_i$ is the single-head attention output given by equation(5), $[\cdot \| \cdot]$ denotes the concatenation operation, and W is the linear transformation matrix to be learned. The matching descriptor $\mathbf{d}_i$ is actually obtained by alternately utilizing attention modules within and between graphs, which facilitate mutual communication between node features.

4.2   Optimal Matching Layer

The second major block of our method is the optimal matching layer, which computes a partial assignment matrix from the affinity matrix $\mathbf{K}$. The affinity matrix $\mathbf{K}$ is obtained by the similarity of matching descriptors:

$$\mathbf{K}_{i,j} = \langle \mathbf{d}_i^a, \mathbf{d}_j^v \rangle, \forall (i,j) \in \{1, ..., N_a\} \times \{1, ..., N_v\}, \tag{8}$$

where $\langle \cdot, \cdot \rangle$ is the inner product.

Then, a dustbin mechanism is introduced to expand each node set to explicitly assign unmatched outliers to their corresponding sets. The expanded affinity matrix is modified by adding elements corresponding to these outliers:

$$\overline{\mathbf{K}}_{i,N_v+1} = \overline{\mathbf{K}}_{N_a+1,j} = \overline{\mathbf{K}}_{N_a+1,N_v+1} = \alpha \in \mathbb{R}, \tag{9}$$

where α is a learnable parameter called dustbin score. Then, the Sinkhorn algorithm iteratively normalizes $\exp(\overline{\mathbf{K}})$ along rows and columns, similar to a softmax operation. After k iterations, the assignment matrix $\mathbf{P}^*$ is obtained by discarding the dustbin.

Both the graph neural network and the optimal matching layer are differentiable, which enables backpropagation. The loss function employed in the network training is given by:

$$\text{Loss} = - \sum_{(i,j) \in \mathcal{M}^I} \log \overline{\mathbf{P}}_{i,j} - \sum_{i \in \mathcal{M}^{OA}} \log \overline{\mathbf{P}}_{i,N_v+1} - \sum_{j \in \mathcal{M}^{OV}} \log \overline{\mathbf{P}}_{N_a+1,j} \tag{10}$$

where $\mathcal{M}^I$ denotes the inlier set in the ground truth. $\mathcal{M}^{OA}$ and $\mathcal{M}^{OV}$ respectively denote the outlier set of different modes in the ground truth.

5   Experiments

The FVessel dataset [5], which was constructed based on the Wuhan segment of the Yangtze River scene, is the benchmark dataset used in our experiments. It consists of 26 videos and the corresponding AIS data. The videos were captured by the HIKVISION DS-2DC4423IW-D dome camera in different locations, such as bridges and riversides, and under varying weather conditions, including sunny, cloudy, and low-light. The AIS data was captured by a Saiyang AIS9000-08 Class-B AIS receiver. We conduct extensive experiments on the FVessel dataset in order to quantitatively evaluate the effectiveness of our proposed method. In addition, a running time analysis is performed to confirm the practicality of our approach.

5.1   Comparison Experiments on the FVessel Dataset

The comparison methods consist of the DeepSORVF [5] method denoted by DeepSORVF and the proposed method denoted by OUR. OUR limits the maximum length of the trajectory to 600 and sets the terminating iteration numbers for the Sinkhorn method to 20.

Table 1. Matching accuracy, recall and runtime on the FVessel dataset.

Video Index	Acc (%) DeepSORVF	Acc (%) OUR	Rec (%) DeepSORVF	Rec (%) OUR	Runtime (s) DeepSORVF	Runtime (s) OUR
video-01	40.00	50.00	50.00	25.00	1.16	0.04
video-02	57.14	75.00	66.67	50.00	2.29	0.04
video-03	100.00	100.00	100.00	80.00	1.79	0.03
video-04	100.00	100.00	100.00	100.00	0.12	0.03
video-05	80.00	80.00	80.00	100.00	0.61	0.03
video-06	100.00	100.00	100.00	100.00	0.59	0.02
video-07	100.00	100.00	100.00	100.00	0.06	0.02
video-08	60.00	100.00	60.00	80.00	0.47	0.03
video-09	100.00	50.00	100.00	100.00	0.04	0.02
video-10	50.00	50.00	100.00	100.00	0.37	0.02
video-11	66.67	100.00	66.67	100.00	0.16	0.02
video-12	100.00	100.00	100.00	100.00	0.33	0.03
video-13	66.67	100.00	66.67	100.00	0.91	0.03
video-14	100.00	100.00	75.00	66.67	0.17	0.02
video-15	100.00	100.00	100.00	100.00	0.49	0.02
video-16	100.00	50.00	100.00	100.00	0.04	0.02
video-17	100.00	100.00	100.00	100.00	0.19	0.03
video-18	0.00	83.33	0.00	83.33	2.32	0.04
video-19	100.00	100.00	100.00	100.00	0.12	0.02
video-20	100.00	100.00	100.00	100.00	0.26	0.02
video-21	100.00	100.00	100.00	100.00	1.18	0.03
video-22	50.00	100.00	100.00	100.00	0.05	0.03
video-23	100.00	50.00	100.00	50.00	0.65	0.04
video-24	50.00	100.00	33.33	100.00	0.20	0.04
video-25	57.14	85.71	50.00	85.71	1.51	0.05
video-26	100.00	75.00	80.00	75.00	0.52	0.10
Average	79.91	**86.50**	79.30	**88.30**	0.64	**0.03**

Table 2. Matching accuracy and recall on the synthetic dataset.

Outlier#	Acc (%) DeepSORVF	Acc (%) OUR	Rec (%) DeepSORVF	Rec (%) OUR
0	73.02	**89.58**	73.70	**85.42**
1	50.41	**88.52**	66.76	**80.42**
2	39.54	**83.33**	64.54	**81.08**
3	29.58	**73.33**	59.07	**66.14**

Specifically, we extract video-based trajectories based on vessel tracking ground truth labeled in FVessel, and obtain AIS-based trajectories by pre-processing AIS data with the method proposed in [5]. We use these two types of trajectories as input and evaluate the association results as well as the runtime of two methods separately.

For DeepSORVF, the similarity matrix between trajectories is calculated by the DTW algorithm, followed by the Hungarian algorithm to identify the corresponding matching results. For OUR, due to the limited size of the dataset, leave-one-out cross-validation is employed to conduct experiments.

As there exist outliers, two criteria, namely matching accuracy and matching recall, are used. These criteria are defined as follows:

$$\text{Acc} = \frac{|\mathcal{M}^{CI}|}{|\mathcal{M}^{I*}|}, \tag{11}$$

$$\text{Rec} = \frac{|\mathcal{M}^{CI}|}{|\mathcal{M}^{I}|}, \tag{12}$$

where $\mathcal{M}^{CI}$ denotes the correct inlier set in the model output results, $\mathcal{M}^{I*}$ denotes the inlier set in the model output results, and $\mathcal{M}^{I}$ denotes the inlier set in the ground truth.

The comparison result of two criteria above and the runtime is shown in Table 1. It can be observed that the performance of the proposed method surpasses that of DeepSORVF. Remarkably, OUR achieves high accuracy and recall at a faster rate compared to DeepSORVF, adhering to the real-time specifications for marine monitoring.

5.2 Experiments on the Synthetic Dataset

In order to further evaluate the ability of the proposed method to handle outliers, we construct synthetic datasets with different amounts of outliers based on the FVessel dataset. As the number of trajectories within each sample is limited to a maximum of 10, we set the number of outliers to 1, 2, and 3 respectively. An outlier is generated by randomly selecting one trajectory and applying a random translation to it.

For OUR, we divide the FVessel dataset into two parts, with 17 for training and the remaining 9 for testing. Due to the limited number of training samples, we use data augmentation methods to expand the training set. For DeepSORVF, we only record its performance on the test set.

The experimental results with different amounts of outliers are shown in Table 2. With an increased number of outliers, the performance of both methods generally decreases. Nevertheless, as the interference of outliers increases, our method demonstrates the capability to handle outliers with robustness, as indicated through the experimental results.

6   Conclusion

This paper introduces a deep graph matching-based method for trajectory association in Vessel Traffic Surveillance, handling outliers and enhancing the data fusion performance. Our proposed learnable trajectory association method replaces the previous handcrafted heuristics with a powerful neural model. Experiments on real and synthetic datasets demonstrate the effectiveness and robustness of the proposed method.

Acknowledgements. This work is supported partly by National Key R&D Program of China (grant 2020AAA0108902), partly by National Natural Science Foundation (NSFC) of China (grants 61973301, 61972020), and partly by Youth Innovation Promotion Association CAS.

References

1. Chen, J., Hu, Q., Zhao, R., Guojun, P., Yang, C.: Tracking a vessel by combining video and ais reports. In: 2008 Second International Conference on Future Generation Communication and Networking, vol. 2, pp. 374–378. IEEE (2008)
2. Cho, M., Lee, J., Lee, K.M.: Reweighted random walks for graph matching. In: Daniilidis, K., Maragos, P., Paragios, N. (eds.) ECCV 2010. LNCS, vol. 6315, pp. 492–505. Springer, Heidelberg (2010). https://doi.org/10.1007/978-3-642-15555-0_36
3. Cuturi, M.: Sinkhorn distances: lightspeed computation of optimal transport. In: Advances in Neural Information Processing Systems, vol. 26 (2013)
4. Gold, S., Rangarajan, A.: A graduated assignment algorithm for graph matching. IEEE Trans. Pattern Anal. Mach. Intell. **18**(4), 377–388 (1996)
5. Guo, Y., Liu, R.W., Qu, J., Lu, Y., Zhu, F., Lv, Y.: Asynchronous trajectory matching-based multimodal maritime data fusion for vessel traffic surveillance in inland waterways. IEEE Trans. Intell. Transp. Syst. (2023)
6. Hartmanis, J.: Computers and intractability: a guide to the theory of np-completeness (michael r. garey and david s. johnson). Siam Rev. **24**(1), 90 (1982)
7. Hochreiter, S., Schmidhuber, J.: Long short-term memory. Neural Comput. **9**(8), 1735–1780 (1997)
8. Kuhn, H.W.: The hungarian method for the assignment problem. Naval Res. Logist. Quart. **2**(1–2), 83–97 (1955)
9. Loiola, E.M., De Abreu, N.M.M., Boaventura-Netto, P.O., Hahn, P., Querido, T.: A survey for the quadratic assignment problem. Eur. J. Oper. Res. **176**(2), 657–690 (2007)
10. Lu, Y., et al.: Fusion of camera-based vessel detection and AIS for maritime surveillance. In: 2021 26th International Conference on Automation and Computing (ICAC), pp. 1–6. IEEE (2021)
11. Müller, M.: Dynamic time warping. In: Information Retrieval for Music and Motion, pp. 69–84. Springer, Heidelberg (2007). https://doi.org/10.1007/978-3-540-74048-3_4
12. Qu, J., Liu, R.W., Guo, Y., Lu, Y., Su, J., Li, P.: Improving maritime traffic surveillance in inland waterways using the robust fusion of AIS and visual data. Ocean Eng. **275**, 114198 (2023)

13. Sarlin, P.E., DeTone, D., Malisiewicz, T., Rabinovich, A.: Superglue: learning feature matching with graph neural networks. In: Proceedings of the IEEE/CVF Conference on Computer Vision and Pattern Recognition, pp. 4938–4947 (2020)
14. Vaswani, A., et al.: Attention is all you need. In: Advances in Neural Information Processing Systems, vol. 30 (2017)
15. Wang, R., Yan, J., Yang, X.: Learning combinatorial embedding networks for deep graph matching. In: Proceedings of the IEEE/CVF International Conference on Computer Vision, pp. 3056–3065 (2019)
16. Yan, J., Yang, S., Hancock, E.R.: Learning for graph matching and related combinatorial optimization problems. In: Proceedings of the Twenty-Ninth International Joint Conference on Artificial Intelligence, IJCAI-20. International Joint Conferences on Artificial Intelligence Organization, pp. 4988–4996 (2020)
17. Yu, T., Wang, R., Yan, J., Li, B.: Learning deep graph matching with channel-independent embedding and hungarian attention. In: International Conference on Learning Representations (2020)
18. Zanfir, A., Sminchisescu, C.: Deep learning of graph matching. In: Proceedings of the IEEE Conference on Computer Vision and Pattern Recognition, pp. 2684–2693 (2018)

Few-Shot Anomaly Detection in Text with Deviation Learning

Anindya Sundar Das[1]([✉])(ID), Aravind Ajay[2](ID), Sriparna Saha[2](ID),
and Monowar Bhuyan[1](ID)

[1] Department of Computing Science, Umeå University, Umeå 90781, Sweden
{aninsdas,monowar}@cs.umu.se
[2] Department of Computer Science Engineering, Indian Institute of Technology
Patna, Patna, India

Abstract. Most current methods for detecting anomalies in text concentrate on constructing models solely relying on unlabeled data. These models operate on the presumption that no labeled anomalous examples are available, which prevents them from utilizing prior knowledge of anomalies that are typically present in small numbers in many real-world applications. Furthermore, these models prioritize learning feature embeddings rather than optimizing anomaly scores directly, which could lead to *suboptimal anomaly scoring* and *inefficient use of data* during the learning process. In this paper, we introduce FATE, a deep few-shot learning-based framework that leverages limited anomaly examples and learns anomaly scores explicitly in an end-to-end method using deviation learning. In this approach, the anomaly scores of normal examples are adjusted to closely resemble reference scores obtained from a prior distribution. Conversely, anomaly samples are forced to have anomalous scores that considerably deviate from the reference score in the upper tail of the prior. Additionally, our model is optimized to learn the distinct behavior of anomalies by utilizing a multi-head self-attention layer and multiple instance learning approaches. Comprehensive experiments on several benchmark datasets demonstrate that our proposed approach attains a new level of state-of-the-art performance (Our code is available at https://github.com/arav1ndajay/fate/).

Keywords: Anomaly detection · Natural language processing · Few-shot learning · Text anomaly · Deviation learning

1 Introduction

Anomaly detection (AD) is defined as the process of identifying unusual data points or events that deviate notably from the majority and do not adhere to expected normal behavior. Although the research on anomaly detection in the text domain is not particularly extensive, it has numerous relevant applications in Natural Language Processing (NLP) [3,9,14,25,28]. In general, anomaly detection faces two formidable obstacles. First, anomalies are dissimilar to each other

B. Luo et al. (Eds.): ICONIP 2023, LNCS 14448, pp. 425–438, 2024.
https://doi.org/10.1007/978-981-99-8082-6_33

and frequently exhibit distinct behavior. Additionally, anomalies are rare data events, which means they make up only a tiny fraction of the normal dataset. Acquiring large-scale anomalous data with accurate labels is highly expensive and difficult; hence, it is exceedingly difficult to train supervised models for anomaly detection. Existing deep anomaly detection methods in text deal with these challenges by predominantly focusing on unsupervised learning, popularly known as one-class classification [22,28] or by employing self-supervised learning [20] in which detection models are trained solely on normal data. These models often identify noise and unimportant data points as anomalies [1], as they do not have any previous understanding of what constitutes anomalous data. This results in higher rates of false positives and false negatives [8,27]. Nevertheless, these deep methods learn the intermediate representation of normal features independently from the anomaly detection techniques using Generative Adversarial Networks (GANs) [4,30] or autoencoders [29]. As a result, the representations generated may not be adequate for anomaly detection.

In order to counter these challenges, we propose a transformer based **Few-shot Anomaly detection in TE**xt with deviation learning (**FATE**) framework in which we leverage a few labeled anomalous instances to train an anomaly-aware detection model, infusing a prior understanding of anomalousness into the learning objective. This setup is possible since only a small number of anomalous data points are required, which can be obtained either from a detection system that has already been established and validated by human experts or directly identified and reported by human users. In this framework, we use a *multi-head self-attention* layer [17] to jointly learn a vector representation of the input text and an anomaly score function. We optimize our model using a *deviation loss* [24] based objective, where the anomaly scores of normal instances are pushed to closely match a reference score, which is calculated as the mean of a set of samples drawn from a known probability distribution (prior). Meanwhile, the scores for anomalous instances are forced to have statistically significant deviations from the normal reference score. FATE needs a tiny number of labeled anomalies for training, only 0.01%–0.07% of all anomalies in each dataset and just 0.008%–0.2% of total available training data.

The **key contributions** of our work are listed below: i) We propose a new transformer-based framework for text anomaly detection based on few-shot learning that can directly learn anomaly scores end-to-end. Unlike existing approaches that focus on latent space representation learning, our model explicitly optimizes the objective of learning anomaly scores. To the best of our knowledge, this is the first effort to utilize a limited number of labeled anomalous instances to learn anomaly scores for text data. ii) Our proposed model employs various techniques, including multi-head self-attention, multiple-instance learning, Gaussian probability distribution and Z-score-based deviation loss to learn anomaly scores that are well-optimized and generalizable. We illustrate that our approach surpasses competing approaches by a considerable margin, especially in terms of sample efficiency and its capability to handle data contamination. iii) We conduct extensive evaluations to determine the performance benchmarks of FATE on three publicly accessible datasets. iv) Our study shows that FATE attains a new state-of-the-art in detecting anomalies in text data, as supported by extensive empirical results from three benchmark text datasets.

2 Related Work

Although there is a limited amount of literature on anomaly detection in text data, some important studies still exist. Our method is related to works from anomaly detection in text and few-shot anomaly detection.

Over the past few years, deep anomaly detection for images has attracted much attention with current works [4, 11, 35, 37] showcasing optimistic, cutting-edge outcomes. A few methods for detecting text anomalies involve pre-trained word embeddings, representation learning, or deep neural networks. In one of the earlier studies [19], autoencoders are used for representation learning for document classification. Most of the recent works treat the task of anomaly detection in text data as a problem of detecting topical intrusions, which entails considering samples from one domain as inliers or normal instances. In contrast, a few samples from different disciplines are considered as outliers or anomalies. In one such study, CVDD [28] leverages pre-trained word embeddings and a self-attention mechanism to learn sentence representations by capturing various semantic contexts and context vectors representing different themes. The network then detects anomalies in sentences and phrases concerning the themes in an unlabeled text corpus. Recent works [2, 6, 18, 36] use self-supervised learning to differentiate between types of transformations applied to normal data and learn features of normality. Token-wise perplexity is frequently used as the anomaly score for AD tasks. The DATE method [20] utilizes transformer self-supervision to learn a probability score indicating token-wise anomalousness, which is then averaged over the sequence length to obtain the sentence-level anomaly score for AD task.

The approaches mentioned above depend entirely on normal data to learn features of normality and they lack in utilizing the small number of labeled anomaly data points that are readily available. Therefore, the dearth of prior knowledge of anomalies often leads to ineffective discrimination by the models [1, 8, 27]. Few-shot anomaly detection addresses the problem by utilizing a small number of labeled anomalous samples and therefore incorporating the prior knowledge of anomaly into the model. Injection of few-labeled anomalous samples into a belief propagation algorithm enhances the performance of anomalous node detection in graphs, as demonstrated in works such as [21] and [33]. These techniques emphasize learning the feature space as the first step, which is then used to calculate the anomaly score. However, in contrast, Deviation Networks are used in [24] and [23] to efficiently leverage a small amount of labeled data for end-to-end learning of anomaly scores for both multivariate and image data, respectively. Multiple Instance Learning (MIL) [23, 32, 34] has also been investigated for anomaly detection in both image and videos in weakly supervised learning settings [23, 32, 34]. In this work, we introduce a transformer-based framework that leverages deviation learning and MIL methods in a novel and efficient way to learn scores reflective of text anomalousness.

3 FATE: The Proposed Approach

In this section, we explain the problem statement, recount the proposed FATE framework, and outline its various components.

3.1 Problem Statement

The objective of our FATE model is to employ a limited number of labeled anomalous samples along with a vast amount of normal data to train an anomaly-aware detection model that can explicitly learn anomaly scores. Given a training dataset $\mathcal{X} = \{x_1, x_2, ..., x_I, x_{I+1}, x_{I+2}, ..., x_{I+O}\}$ consisting of I normal data samples (inlier) $\mathcal{X}_\mathcal{I} = \{x_1, x_2, ..., x_I\}$, a very small set of O labeled anomalies (outlier) $\mathcal{X}_\mathcal{O} = \{x_{I+1}, x_{I+2}, ..., x_{I+O}\}$ with $I \gg O$ (O is much smaller than I), which yields some insights into actual anomalies, we aim to learn an anomaly scorer $\psi_K : \mathcal{X} \rightarrow \mathbb{R}$ which can assign scores to data instances based on whether they are anomalous (outlier) or normal (inlier). The objective is to ensure $\psi_K(x_i)$ lies close to the mean anomaly scores of inlier data samples, defined by the sample mean from a prior distribution $\mathcal{N}(\mu; \sigma)$, and $\psi_K(x_j)$ deviates significantly from the mean anomaly score and lies in the upper tail of the distribution, i.e., $\psi_K(x_j) > \psi_K(x_i)$, where $x_i \in \mathcal{X}_\mathcal{I}$ and $x_j \in \mathcal{X}_\mathcal{O}$.

3.2 The Proposed Framework

The general architecture of our proposed FATE model is shown in Fig. 1. It comprises two major components - an Anomaly Score Generator Network and a Reference Score Generator. The Anomaly Score Generator accepts an input text, a sequence of words of finite length, and processes it to produce a scalar output, i.e., anomaly score. The Reference Score Generator provides a reference for the anomaly score of normal samples.

Anomaly Score Generator Network. The main components of the Anomaly Score Generator Network are *Sentence Encoder*, *Multi-Head Self-Attention (MHSA)* layer [17] and *MIL-driven top-K scorer*.

- **Sentence Encoder.** Let $x_i = (x_{i1}, x_{i2}, ..., x_{iN})$ be an input text or token sequence comprising of N words. The Sentence Encoder is a BERT-based [5,26] encoder that converts the input sequence x_i into a sequence of contextualized representation $h(x_i) = (h_{i1}, h_{i2}, ..., h_{iN}) \in \mathbb{R}^{d \times N}$, which is d-dimensional vector embedding representations of the N tokens in x_i.

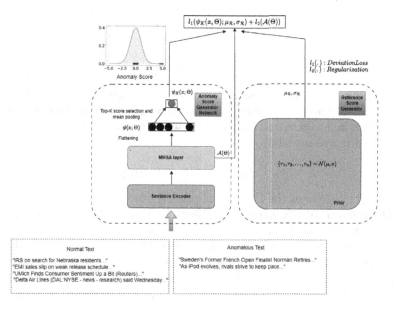

Fig. 1. FATE: a proposed framework. The anomaly scoring network $\psi(x; \theta)$ is parameterized by Θ, comprising the Sentence Encoder and Multi-Head Self-Attention (MHSA) layer. The attention matrix is denoted by $\mathcal{A}(\Theta)$. $\mu_{\mathcal{R}}$ and $\sigma_{\mathcal{R}}$, are the mean and standard deviations of n-number of normal samples, respectively, which are defined by a prior probability distribution $\mathcal{N}(\mu; \sigma)$. ψ_K denotes a top-K MIL-based anomaly scoring function. The loss $l_1(.)$ is deviation loss that ensures normal samples have anomaly scores close to $\mu_{\mathcal{R}}$, while scores for anomalous objects deviate significantly from $\mu_{\mathcal{R}}$. Loss $l_2(.)$ is the regularization term to enforce the attention heads to be nearly orthogonal.

- **Multi-Head Self-Attention.** The MHSA layer accepts $h(x_i)$ as input and transforms it into vector embeddings $\psi(x_i; \Theta)$ of fixed-length m; each of the m vectors represents a set of distinct anomaly scores, capturing multiple aspects and context of anomalousness in the text sequence, x_i. Formally, the MHSA layer computes an attention matrix $A = (a_1, a_2, ..., a_m) \in (0, 1)^{N \times m}$ from embedding representation $h(x_i) \in \mathbb{R}^{d \times N}$ as follows:

$$A = softmax(tanh(h(x_i)^{\mathsf{T}} \Theta_1) \Theta_2) \qquad (1)$$

where $\Theta_1 \in \mathbb{R}^{d \times r_a}$ and $\Theta_2 \in \mathbb{R}^{r_a \times m}$ are learnable weight matrices. The complexity of the MHSA module is determined by intermediate dimensionality r_a. Here the *softmax* activation is applied on each column to ensure the resultant column vector a_j of A is normalized. The m vectors $a_1, a_2, ..., a_m$ are known as attention heads, and each head assigns relative importance to each token in the sequence.

The self-attention matrix A is applied on embedding representation $h(x_i)$, which yields a fixed-length score matrix $S = (s_1, s_2, ..., s_m) \in \mathbb{R}^{d \times m}$ comprising of m anomaly score vectors. Every column vector s_k of S, that represents

d anomaly scores, is a convex combination of embedding vectors, $h_{i1}, h_{i2}, ..., h_{iN}$.

$$S = h(x_i)A \tag{2}$$

To ensure independence between score vectors, eliminate redundant information and ensure that each score vector emphasizes on distinct aspects of semantic anomalousness, an orthogonality constraint called $MHSA\_loss$ is included in the model objective.

$$l_2(x_i; \Theta) = ||A^\mathsf{T} A - I||_F^2 \tag{3}$$

where $||.||_F^2$ denotes the Frobenius norm [7] of a matrix.

- **MIL-driven top-K scorer.** In Multiple-Instance-Learning (MIL) [23,32, 34], each data point is represented as a bag of multiple instances in different feature subspaces resulting in more generalized representations. The score matrix S represents sets of orthogonal score vectors. The anomaly scoring vector $\psi(x_i; \Theta) \in \mathbb{R}^{\hat{N}}$, in which each input text is represented as a set of $\hat{N}$ distinct anomaly scores (multiple instances) is obtained after flattening S (Fig. 1). We then select $L(x_i)$, a set of K instances in $\psi(x_i; \Theta)$ with the highest anomaly scores. The overall MIL-driven top-K anomaly score of the input text is defined as:

$$\psi_K(x_i; \Theta) = \frac{1}{K} \sum_{x_{ij} \in L(x_i)} \psi(x_{ij}; \Theta) \tag{4}$$

where $|L(x_i)| = K$, $\psi(x_i; \Theta) = (\psi(x_{i1}; \Theta), \psi(x_{i2}; \Theta), ..., \psi(x_{i\hat{N}}; \Theta))$ and $\hat{N} = d * m$.

Reference Score Generator. Once the anomaly scores have been obtained using $\psi_K(x_i; \Theta)$, the network output is augmented with a reference score $\mu_\mathcal{R} \in \mathbb{R}$ that provides prior knowledge to aid in the optimization process. This reference score is derived from the average anomaly scores of a set of n randomly chosen normal instances, denoted as $\mathcal{R}$. To define the prior, we opt for a Gaussian distribution since [12] has provided extensive evidence that the Gaussian distribution is an excellent fit for anomaly scores across various datasets. We define a reference score based on Gaussian probability distribution:

$$\mu_\mathcal{R} = \frac{1}{n} \sum_{j=1}^{n} r_i \tag{5}$$

where $r_i \sim \mathcal{N}(\mu, \sigma)$, denotes anomaly score of randomly selected i-th normal sample drawn from a Normal distribution and $\mathcal{R} = \{r_1, r_2, ..., r_n\}$ is prior-driven anomaly score set.

Deviation Learning. The deviation is determined as a Z-score with respect to the selected prior distribution:

$$Z_{dev}(x_i; \Theta) = \frac{\psi_K(x_i; \Theta) - \mu_{\mathcal{R}}}{\sigma_{\mathcal{R}}} \tag{6}$$

where $\sigma_{\mathcal{R}}$ is sample standard deviation of the set of reference scores $\mathcal{R} = \{r_1, r_2, ..., r_n\}$. The model learns the deviation between normality and anomalousness by optimizing a deviation-based contrastive loss [10], as given by Eq. 7.

$$l_1(\psi_K(x_i; \Theta), \mu_{\mathcal{R}}, \sigma_{\mathcal{R}}) = (1 - y_i)|Z_{dev}(x_i; \Theta)| + y_i \max(0, \alpha - Z_{dev}(x_i; \Theta)) \tag{7}$$

where $y_i = 0$ if $x_i \in \mathcal{X}_{\mathcal{I}}$ (normal sample set), and $y_i = 1$ if $x_i \in \mathcal{X}_{\mathcal{O}}$ (anomalous sample set). The hyperparameter α in FATE is similar to the confidence interval parameter in Z-scores. By using this deviation loss, FATE aims to minimize the difference between anomaly scores of normal examples and $\mu_{\mathcal{R}}$ while also ensuring that there is a minimum deviation of α between the anomaly scores of anomalous samples and $\mu_{\mathcal{R}}$.

FATE optimizes both the deviation loss (Eq. 7) and the orthogonality constraint (Eq. 3) objectives during the training process. The overall loss function is formulated as follows:

$$l_{\mathcal{FATE}}(x_i; \Theta) = l_1(\psi_K(x_i; \Theta), \mu_{\mathcal{R}}, \sigma_{\mathcal{R}}) + l_2(x_i; \Theta) \tag{8}$$

The procedure for training FATE is described in Algorithm 1. FATE uses the optimized network ψ_K during the testing phase to generate anomaly scores for each testing instance. The samples are ordered according to their anomaly scores, and those with the highest anomaly scores are identified as outliers. The Gaussian prior employed in generating the reference score makes our anomaly score inherently interpretable.

4   Implementation Details

We will first present here the datasets' specifics and subsequently provide information regarding the methodology used in the experimentation.

4.1   Datasets

To assess how well FATE performs in relation to the standard baseline methods, we evaluate it on three benchmark datasets: 20Newsgroups [13], AG News [38], and Reuters-21578 [15]. We preprocess them based on the methods suggested in [20,28]. This includes converting all text to lowercase, removing punctuation marks and numbers, and eliminating stopwords. To prepare the training and testing splits for our few-shot setup, we use the following method for each dataset: Firstly, we create the inlier data (normal samples) by selecting examples from only one label of the train split of the corresponding dataset. Then we construct the outlier training set comprising a very small number of samples (5 to 40) by

---

Algorithm 1. FATE: Training with few labeled anomalies

Input : $\mathcal{X}$: set of all training samples, $\mathcal{X}_{\mathcal{I}}$: set of normal samples, $\mathcal{X}_{\mathcal{O}}$: set of anomalous samples, $\mathcal{X} = \mathcal{X}_{\mathcal{I}} \cup \mathcal{X}_{\mathcal{O}}$ and $\emptyset = \mathcal{X}_{\mathcal{I}} \cap \mathcal{X}_{\mathcal{O}}$

Output: $\psi_K : \mathcal{X} \rightarrow \mathbb{R}$: Anomaly score generator

1: Initialize parameters in Θ randomly.
2: **for** $e = 1 \rightarrow number\_of\_epochs$ **do**
3:    **for** $b = 1 \rightarrow number\_of\_batches$ **do**
4:        $\mathcal{X}_v \leftarrow$ Select a random set of v data instances consisting of an equal number of examples from both $\mathcal{X}_{\mathcal{I}}$ and $\mathcal{X}_{\mathcal{O}}$.
5:        Generate a set of n anomaly scores $\mathcal{R}$ by sampling randomly from a normal distribution. $\mathcal{N}(\mu, \sigma)$
6:        Calculate the mean $\mu_{\mathcal{R}}$ and standard deviation $\sigma_{\mathcal{R}}$ from sample score set $\mathcal{R}$.
7:        Compute $deviation\_loss_{batch} = \frac{1}{b} \sum_{x \in \mathcal{X}_v} l_1(x, \mu_{\mathcal{R}}, \sigma_{\mathcal{R}}; \Theta)$.
8:        Compute $MHSA\_loss_{batch} = \frac{1}{b} \sum_{x \in \mathcal{X}_v} l_2(x; \Theta)$.
9:        Calculate total loss $loss_{batch} = deviation\_loss_{batch} + MHSA\_loss_{batch}$.
10:       Perform parameter optimization using Adam optimizer on $loss_{batch}$ Θ.
11:    **end for**
12: **end for**
13: return ψ_K.

---

randomly sampling an equal number of instances from each outlier class (other labels in the train split). When it comes to testing, we create the inlier data by selecting examples solely from one label of the test split of the corresponding dataset while using data with other labels in the test split as outlier test examples.

- The **20Newsgroups** [13] dataset consists of about 20,000 newsgroup documents grouped into 20 distinct topical subcategories. To conduct our experiments, we adhere to the approach outlined in [20] and [28] by only selecting articles from six primary classes: **computer, recreation, science, miscellaneous, politics,** and **religion**.
- The **AG News** [38] dataset consists of articles gathered from diverse news sources assembled to classify topics.
- **Reuters-21578** [15] is a well-known dataset in natural language processing that contains a collection of news articles published on Reuter's newswire in 1987. Following [28], we conduct experiments on 7 selected categories: **earn, acq, crude, trade, money-fx, interest,** and **ship**.

4.2 Model Training and Hyperparameters

We adhere to the process illustrated in Fig. 1 to train the FATE network. We used PyTorch 1.12.1[2] framework for all our experiments. For the sentence encoder, we fine-tune an SBERT model[3] and obtain the embeddings from the Encoder layer

---

[2] https://pytorch.org/
[3] https://www.sbert.net/.

(before the pooling takes place). We employ a maximum sequence length of 128 and a batch size of 16. In the self-attention layer, we fix $r_a = 150$ and $m = 5$ for all experiments. Using an Adam optimizer with the learning rate set to 1e−6, we train our model for 5 epochs on AG News. To account for the smaller dataset sizes of 20Newsgroups and Reuters-21578, we trained our model for 50 epochs and 40 epochs, respectively. In some cases where the number of inlier sentences in the training data is lesser than 500 (e.g., some subsets in Reuters-21578), we assign the number of epochs to 80. K in the top-K scorer has been set to 10%, and the number of few-shot anomalies in each training has been set to 10. We have opted for the standard normal distribution $\mathcal{N}(0, 1)$ as prior for reference score. The number of normal reference samples n (Eq. 5) has been set to 5000. We have selected a confidence interval $\alpha = 5$ (Eq. 7).

4.3   Baselines and Evaluation Metrics

We compare our proposed method FATE with existing state-of-the-art text AD models such as OCSVM[4] [31], CVDD (See footnote 4) [28] and DATE[5] [20].

In our experiments, we utilize the commonly used performance metrics for detection, which are the Area Under Receiver Operating Characteristic Curve (AUROC) and Area Under Precision-Recall Curve (AUPRC). AUROC metric describes the true positive versus false positive ROC curve, where AUPRC characterizes compact representation of precision-recall curve. Higher values for both AUROC and AUPRC correspond to better performance.

5   Results and Analysis

In this section, we will compare our model's performance against other state-of-the-art methods and then comprehensively analyse our proposed methodology from various perspectives to gain a profound understanding of its capabilities.

Main Results. The outcomes of our proposed technique and those of previous state-of-the-art approaches on all datasets are shown in Table 1. The AUROC values show that our method outperforms the top-performing DATE model by 3.85 and 3.2% points on 20Newsgroups and AG News, respectively. In addition, our proposed technique surpasses CVDD by a significant margin of 17.4 points and 10.1 points on 20Newsgroups and AG News, respectively. Furthermore, for the Reuters-21578 dataset splits, our approach outperforms the best-performing CVDD method by 2.9 points and DATE model by 7.8 points. This clearly illustrates the efficacy of our proposed method.

---

[4] OCSVM and CVDD implementations are taken from official CVDD work: https://github.com/lukasruff/CVDD-PyTorch.

[5] The experimentation is conducted using the official DATE published code and setup: https://github.com/bit-ml/date

Table 1. AUROC (in %) and AUPRC (in %) results of all the baselines and proposed models on 20Newsgroups, AG News, and Reuters-21578 are reported below. The reported values are the mean results obtained from multiple runs.

Datasets	Inlier	OCSVM		CVDD		DATE		FATE	
		AUROC	AUPRC	AUROC	AUPRC	AUROC	AUPRC	AUROC	AUPRC
20 News	comp	78.0	88.5	74.0	85.3	92.1	96.8	**92.7**	**97.1**
	rec	70.0	87.3	60.6	81.8	83.4	93.5	**89.7**	**96.6**
	sci	64.2	86.0	58.2	82.4	69.7	**89.9**	**73.7**	89.5
	misc	62.1	96.2	75.7	97.3	86.0	97.3	**89.2**	**99.3**
	pol	76.1	93.5	71.5	91.8	81.9	95.8	**89.5**	**97.0**
	rel	78.9	95.0	78.1	94.1	86.1	96.8	**87.5**	**97.4**
	average	71.6	91.1	69.7	88.8	83.2	95.2	**87.1**	**96.2**
AG News	business	79.9	93.0	84.0	93.2	90.0	94.7	**90.8**	**96.1**
	sci	80.7	91.2	79.0	88.0	84.0	92.9	**89.5**	**96.7**
	sports	92.4	97.3	89.9	95.4	95.9	97.8	**99.2**	**99.7**
	world	83.2	91.9	79.6	88.5	90.1	95.0	**93.3**	**96.5**
	average	84.0	93.4	83.1	91.3	90.0	95.1	**93.2**	**97.3**
Reuters-21578	earn	91.1	87.2	93.9	88.7	97.4	97.5	**97.6**	**98.0**
	acq	93.1	95.7	92.7	95.2	93.5	95.0	**98.5**	**99.0**
	crude	92.4	99.1	**97.3**	99.3	78.0	97.7	95.8	**99.6**
	trade	99.0	**99.9**	**99.3**	**99.9**	95.0	99.7	97.8	**99.9**
	money-fx	88.6	99.3	82.5	99.0	88.3	98.7	**94.2**	**99.7**
	interest	97.1	**99.9**	95.9	99.8	93.0	98.8	**97.4**	**99.9**
	ship	91.2	99.7	96.1	99.8	77.7	98.9	**96.5**	**99.9**
	average	93.2	97.3	94.0	97.4	89.0	98.0	**96.8**	**99.4**

Ablation Study. To validate the contributions of different components in the proposed approach, here we introduce four variants for ablation study: i) ***FATE without Top-K:*** We remove the Top-K MIL layer and obtain anomaly score by averaging the outputs of the MHSA layer. ii) ***FATE without MHSA:*** In this variant, the MHSA layer is removed, and the output from the sentence encoder is directly fed to the Top-K layer for anomaly score computation iii) ***FATE with BCELoss:*** We derive this variant by replacing deviation loss with BCE loss iv) ***FATE with FocalLoss:*** In this implementation, focal loss [16] substitutes deviation loss.

Table 2 presents the results of the ablation study. We observe that removing the top-K layer significantly affects the model's performance on some datasets. Removal of the MHSA layer impacts the model's performance to some extent; moreover, it adversely impacts the overall stability of the model. We also found that the FATE variant with deviation loss performs much better on average than its BCE loss and Focal loss counterparts.

We also investigate the top-K MIL module by varying the value of K. As shown in Fig. 3, the FATE model generally achieves improved performance as K increases from 1% to 10%. However, it is observed that the performance starts to decline on some datasets beyond this value of K. We find K=10% works well on most of the datasets.

Sample Efficiency. To determine the optimal number of labeled anomalies needed to train FATE, we vary the number of labeled outliers from 5 to 40 in

Table 2. AUROC (in %) performance of the ablation study

Datasets	Inlier	FATE without Top-K	FATE without MHSA	FATE with BCELoss	FATE with FocalLoss	FATE
20 News	comp	81.1	90.0	**92.7**	91.6	**92.7**
	rec	84.6	86.3	83.2	82.9	**89.7**
	sci	67.2	70.2	71.4	72.5	**73.7**
	misc	84.2	87.3	89.0	88.4	**89.2**
	pol	84.6	84.0	88.4	84.9	**89.5**
	rel	85.1	86.1	87.4	84.5	**87.5**
AG News	business	84.7	88.7	87.6	84.7	**90.8**
	sci	78.0	88.6	84.7	87.1	**89.5**
	sports	94.5	97.5	99.0	98.8	**99.2**
	world	71.0	92.9	**93.5**	92.1	93.3
Reuters-21578	earn	89.3	96.8	98.1	96.9	**97.6**
	acq	97.7	97.8	98.3	98.1	**98.5**
	crude	85.5	94.1	92.4	90.9	**95.8**
	trade	**98.3**	96.4	98.2	97.8	97.8
	money-fx	**97.2**	92.9	94.1	92.3	94.2
	interest	96.2	96.1	**99.5**	98.8	97.4
	ship	94.2	95.6	**98.2**	91.1	96.5

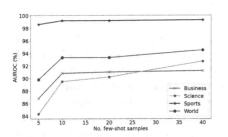

Fig. 2. Sample efficiency: AUROC vs. no of labeled outlier samples for few-shot learning on AG News

Fig. 3. Sensitivity: AUROC performance with respect to Top-K (%)

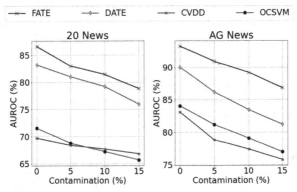

Fig. 4. AUROC vs. different anomaly contamination (%) in inlier training data. The experiments are conducted on 20 News and AG News datasets.

the training dataset. In contrast, the test data is remains unaltered. As shown in Fig. 2, the FATE model achieves remarkable performance with only 10–20 labeled outlier samples, which is a very small fraction (0.03–0.06% on AG News) of the available inlier training samples. This suggests that the FATE model is capable of efficiently utilizing labeled samples. Nonetheless, we also observe that the model's performance is affected when there are no representative samples in the outlier set from one of the outlier classes, highlighting the limitation of FATE in managing previously unseen classes.

Robustness. We further investigate the robustness of FATE in scenarios where the normal samples are contaminated with outliers. In this setup, anomalies are injected into normal training data as contamination. We compare our model's performance with other baselines. As shown in Fig. 4, all methods experience a decline in performance as the contamination rate increases from 0% to 15%. However, our proposed method is impressively resilient, as it outperforms the best-performing DATE model by 5.6% on AG News and by 2.8% on 20 News even when the contamination rate is as high as 15%.

6   Concluding Remarks

In this paper, we proposed FATE, a new few-shot transformer-based framework for detecting anomalies in text that uses a sentence encoder and multi-head self-attention to learn anomaly scores end-to-end. This is done by approximating the anomaly scores of normal samples from a prior distribution and then using a deviation loss that relies on Z-scores to push the anomaly scores of outliers further away from the prior mean. Evaluation on three benchmark datasets demonstrated that FATE significantly outperforms state-of-the-art methods, even in the presence of significant data contamination, and has a high level of efficacy in using labeled anomalous samples, delivering exceptional results even with minimal examples. In our future work, we aim to investigate the effectiveness of FATE in detecting anomalies in text data when dealing with unseen classes and also aim to investigate the interpretability of FATE to help explain its decision-making process. This will help make it more transparent and understandable for end-users.

Acknowledgements. This work was partially supported by the Wallenberg AI, Autonomous Systems and Software Program (WASP) funded by Knut and Alice Wallenberg Foundation.

References

1. Aggarwal, C.C., Aggarwal, C.C.: Supervised outlier detection. Outlier Anal. 219–248 (2017)
2. Arora, U., Huang, W., He, H.: Types of out-of-distribution texts and how to detect them. arXiv preprint arXiv:2109.06827 (2021)

3. Crawford, M., Khoshgoftaar, T.M., Prusa, J.D., Richter, A.N., Al Najada, H.: Survey of review spam detection using machine learning techniques. J. Big Data **2**(1), 1–24 (2015)

4. Deecke, L., Vandermeulen, R., Ruff, L., Mandt, S., Kloft, M.: Image anomaly detection with generative adversarial networks. In: Berlingerio, M., Bonchi, F., Gärtner, T., Hurley, N., Ifrim, G. (eds.) ECML PKDD 2018, Part I. LNCS (LNAI), vol. 11051, pp. 3–17. Springer, Cham (2019). https://doi.org/10.1007/978-3-030-10925-7_1

5. Devlin, J., Chang, M.W., Lee, K., Toutanova, K.: Bert: pre-training of deep bidirectional transformers for language understanding. arXiv preprint arXiv:1810.04805 (2018)

6. Gangal, V., Arora, A., Einolghozati, A., Gupta, S.: Likelihood ratios and generative classifiers for unsupervised out-of-domain detection in task oriented dialog. In: Proceedings of the AAAI Conference on Artificial Intelligence, vol. 34, pp. 7764–7771 (2020)

7. Golub, G.H., Van Loan, C.F.: Matrix computations. JHU Press, Baltimore (2013)

8. Görnitz, N., Kloft, M., Rieck, K., Brefeld, U.: Toward supervised anomaly detection. J. Artif. Intell. Res. **46**, 235–262 (2013)

9. Guthrie, D., Guthrie, L., Allison, B., Wilks, Y.: Unsupervised anomaly detection. In: IJCAI, pp. 1624–1628 (2007)

10. Hadsell, R., Chopra, S., LeCun, Y.: Dimensionality reduction by learning an invariant mapping. In: 2006 IEEE Computer Society Conference on Computer Vision and Pattern Recognition (CVPR 2006), vol. 2, pp. 1735–1742. IEEE (2006)

11. Hendrycks, D., Mazeika, M., Dietterich, T.: Deep anomaly detection with outlier exposure. arXiv preprint arXiv:1812.04606 (2018)

12. Kriegel, H.P., Kroger, P., Schubert, E., Zimek, A.: Interpreting and unifying outlier scores. In: Proceedings of the 2011 SIAM International Conference on Data Mining, pp. 13–24. SIAM (2011)

13. Lang, K.: Newsweeder: learning to filter netnews. In: Machine Learning Proceedings 1995, pp. 331–339. Elsevier (1995)

14. Lee, N., Bang, Y., Madotto, A., Khabsa, M., Fung, P.: Towards few-shot fact-checking via perplexity. arXiv preprint arXiv:2103.09535 (2021)

15. Lewis, D.D.: Reuters-21578 text categorization test collection, distribution 1.0 (1997)

16. Lin, T.Y., Goyal, P., Girshick, R., He, K., Dollár, P.: Focal loss for dense object detection. In: Proceedings of the IEEE International Conference on Computer Vision, pp. 2980–2988 (2017)

17. Lin, Z., et al.: A structured self-attentive sentence embedding. arXiv preprint arXiv:1703.03130 (2017)

18. Mai, K.T., Davies, T., Griffin, L.D.: Self-supervised losses for one-class textual anomaly detection. arXiv preprint arXiv:2204.05695 (2022)

19. Manevitz, L., Yousef, M.: One-class document classification via neural networks. Neurocomputing **70**(7–9), 1466–1481 (2007)

20. Manolache, A., Brad, F., Burceanu, E.: Date: detecting anomalies in text via self-supervision of transformers. arXiv preprint arXiv:2104.05591 (2021)

21. McGlohon, M., Bay, S., Anderle, M.G., Steier, D.M., Faloutsos, C.: Snare: a link analytic system for graph labeling and risk detection. In: Proceedings of the 15th ACM SIGKDD International Conference on Knowledge Discovery and Data Mining, pp. 1265–1274 (2009)

22. Moya, M.M., Koch, M.W., Hostetler, L.D.: One-class classifier networks for target recognition applications. NASA STI/Recon Technical Report N **93**, 24043 (1993)

23. Pang, G., Ding, C., Shen, C., Hengel, A.V.D.: Explainable deep few-shot anomaly detection with deviation networks. arXiv preprint arXiv:2108.00462 (2021)
24. Pang, G., Shen, C., van den Hengel, A.: Deep anomaly detection with deviation networks. In: Proceedings of the 25th ACM SIGKDD International Conference on Knowledge Discovery & Data Mining, pp. 353–362 (2019)
25. Peng, J., Feldman, A., Vylomova, E.: Classifying idiomatic and literal expressions using topic models and intensity of emotions. arXiv preprint arXiv:1802.09961 (2018)
26. Reimers, N., Gurevych, I.: Sentence-Bert: sentence embeddings using SIAMESE Bert-networks. arXiv preprint arXiv:1908.10084 (2019)
27. Ruff, L., et al.: Deep semi-supervised anomaly detection. arXiv preprint arXiv:1906.02694 (2019)
28. Ruff, L., Zemlyanskiy, Y., Vandermeulen, R., Schnake, T., Kloft, M.: Self-attentive, multi-context one-class classification for unsupervised anomaly detection on text. In: Proceedings of the 57th Annual Meeting of the Association for Computational Linguistics, pp. 4061–4071 (2019)
29. Sakurada, M., Yairi, T.: Anomaly detection using autoencoders with nonlinear dimensionality reduction. In: Proceedings of the MLSDA 2014 2nd Workshop on Machine Learning for Sensory Data Analysis, pp. 4–11 (2014)
30. Schlegl, T., Seeböck, P., Waldstein, S.M., Schmidt-Erfurth, U., Langs, G.: Unsupervised anomaly detection with generative adversarial networks to Guide Marker discovery. In: Niethammer, M., et al. (eds.) IPMI 2017. LNCS, vol. 10265, pp. 146–157. Springer, Cham (2017). https://doi.org/10.1007/978-3-319-59050-9_12
31. Schölkopf, B., Platt, J.C., Shawe-Taylor, J., Smola, A.J., Williamson, R.C.: Estimating the support of a high-dimensional distribution. Neural Comput. **13**(7), 1443–1471 (2001)
32. Sultani, W., Chen, C., Shah, M.: Real-world anomaly detection in surveillance videos. In: Proceedings of the IEEE Conference on Computer Vision and Pattern Recognition, pp. 6479–6488 (2018)
33. Tamersoy, A., Roundy, K., Chau, D.H.: Guilt by association: large scale malware detection by mining file-relation graphs. In: Proceedings of the 20th ACM SIGKDD International Conference on Knowledge Discovery and Data Mining, pp. 1524–1533 (2014)
34. Tian, Y., Pang, G., Chen, Y., Singh, R., Verjans, J.W., Carneiro, G.: Weakly-supervised video anomaly detection with robust temporal feature magnitude learning. In: Proceedings of the IEEE/CVF International Conference on Computer Vision, pp. 4975–4986 (2021)
35. Wang, M., Shao, Y., Lin, H., Hu, W., Liu, B.: CMG: a class-mixed generation approach to out-of-distribution detection. In: Proceedings of ECML/PKDD-2022 (2022)
36. Wu, Q., Jiang, H., Yin, H., Karlsson, B.F., Lin, C.Y.: Multi-level knowledge distillation for out-of-distribution detection in text. arXiv preprint arXiv:2211.11300 (2022)
37. Zhang, S., et al.: Label-assisted memory autoencoder for unsupervised out-of-distribution detection. In: Oliver, N., Pérez-Cruz, F., Kramer, S., Read, J., Lozano, J.A. (eds.) ECML PKDD 2021. LNCS (LNAI), vol. 12977, pp. 795–810. Springer, Cham (2021). https://doi.org/10.1007/978-3-030-86523-8_48
38. Zhang, X., Zhao, J., LeCun, Y.: Character-level convolutional networks for text classification. In: Advances in Neural Information Processing Systems, vol. 28 (2015)

MOC: Multi-modal Sentiment Analysis via Optimal Transport and Contrastive Interactions

Yi Li[1,2], Qingmeng Zhu[1,2], Hao He[2(✉)], Ziyin Gu[1,2], and Changwen Zheng[2]

[1] University of Chinese Academy of Sciences, Beijing, China
liyi212@mails.ucas.ac.cn
[2] Institute of Software Chinese Academy of Sciences, Beijing, China
{qingmeng,hehao21,ziyin2020,changwen}@iscas.ac.cn

Abstract. Multi-modal sentiment analysis (MSA) aims to utilize information from various modalities to improve the classification of emotions. Most existing studies employ attention mechanisms for modality fusion, overlooking the heterogeneity of different modalities. To address this issue, we propose an approach that leverages optimal transport for modality alignment and fusion, specifically focusing on distributional alignment. However, solely relying on the optimal transport module may result in a deficiency of intra-modal and inter-sample interactions. To tackle this deficiency, we introduce a double-modal contrastive learning module. Specifically, we propose a model **MOC** (*M*ulti-modal sentiment analysis via *O*ptimal transport and *C*ontrastive interactions), which integrates optimal transport and contrastive learning. Through empirical comparisons on three established multi-modal sentiment analysis datasets, we demonstrate that our approach achieves state-of-the-art performance. Additionally, we conduct extended ablation studies to validate the effectiveness of each proposed module.

Keywords: Multi-modal sentiment analysis · Optimal transport · Contrastive learning

1 Introduction

With the rapid development of social media platforms such as Twitter and WeChat, it has become increasingly popular to share personal life updates in the form of images and text on these apps. As a result, there has been a significant increase in the amount of multimedia data, leading to a growing interest in the analysis of sentiment in multi-modal data. This has given rise to the field of multi-modal sentiment analysis. The objective of this task is to determine the emotional polarity by leveraging features from various modalities. In multi-modal sentiment analysis, information from different modalities is often

---

Y. Li and Q. Zhu—Equal contribution.

B. Luo et al. (Eds.): ICONIP 2023, LNCS 14448, pp. 439–451, 2024.
https://doi.org/10.1007/978-981-99-8082-6_34

complementary [15,25]. For instance, as depicted in Fig. 1(a), BERT classifies the emotion of the text as "Neutral". However, when the text is combined with the accompanying image, our MOC is able to capture the true sentiment as "Positive".

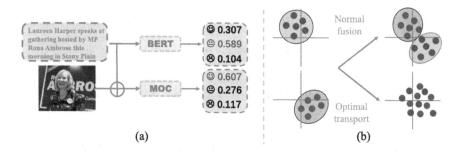

(a)   (b)

Fig. 1. Subfigure (a) demonstrates the inherent complementarity of information derived from diverse modalities. Subfigure (b) accentuates the disparity between fusion techniques employing normal fusion and optimal transport.

A fundamental model in Multi-modal Sentiment Analysis (MSA) is tasked with mapping embeddings from various modalities into a shared latent space to capture informative feature information. In existing research, self-supervised learning [32] is commonly employed to acquire comprehensive representations. For instance, Self-MM [25] achieves informative representations through self-supervised multitask learning. Han Wei [8] introduces the information entropy theory, aiming to maximize the mutual information between different modalities. In addition to self-supervised methods, attention mechanisms have gained widespread usage in MSA models. CLMLF [36] concatenates the feature vectors of images and texts to form a unified representation, which is then passed through the Transformer encoder to leverage its inherent attention mechanism. PMR [14] employs cross-attention to facilitate downstream tasks on multi-modal sequences, effectively reinforcing the interplay between the target modality and the original modality via attention mechanisms. In summary, the key to successful MSA lies in finding effective approaches to fuse feature vectors from different modalities.

However, the aforementioned existing methods fail to consider the inherent heterogeneity among different modalities. To address this issue, we propose the utilization of the optimal transport method to explicitly align and fuse multiple modalities, taking into account the perspective of distributional alignment, which effectively eliminates the heterogeneity. Specifically, the fusion process of multi-modal features is formulated as an optimal transport (OT) problem. By minimizing the wasserstein distance between different distributions, the embeddings of different modalities are transported to a common aligned space using the optimal transport plan, thereby overcoming spatial heterogeneity. The distinction between fusion using optimal transport and conventional methods is illustrated in Fig. 1.

Nevertheless, relying solely on the optimal transport module may result in a lack of intra-modal and inter-sample interactions. Therefore, we incorporate double-modal contrastive learning. Within this module, two data augmentation methods are introduced to construct positive pairs. Additionally, the contrastive learning paradigm facilitates the acquisition of modality-specific discriminative knowledge, further enhancing the representation learning process [12].

In this paper, we build a novel MSA model and the main contributions are as follows:

(1) We address the issue of heterogeneity among different modalities by proposing an optimal transport module for conducting modality alignment and fusion. Notably, this is the first attempt to leverage the optimal transport technique to align and fuse image-text modalities for sentiment analysis, thereby mitigating the heterogeneity problem.
(2) In order to facilitate intra-modality and inter-sample interactions, we introduce a novel contrastive learning module, which bypasses the need for attention mechanisms.
(3) We conduct experiments on three public datasets and the experimental results show that the proposed MOC achieves superior results to the existing competitive approaches (including the state-of-the-art).

2 Related Work

We divide this section into three subsections: (1) Multi-modal sentiment analysis; (2) Optimal transport; (3) Contrastive learning.

Multi-modal Sentiment Analysis: In recent years, there has been significant progress in multi-modal sentiment analysis. CLMLF [36] conducts sentiment analysis by using contrastive learning between image modality and text modality. MultiSentiNet [38] and HSAN [37] simply concatenate feature vectors of different modalities to obtain the joint multi-modal features. Yang [27] proposes the MVAN, which focuses on interaction between modalities by multi-view learning. Co-MN-Hop6 [21] is a co-memory network for iteratively modeling the interactions between multiple modalities. MGNNS [23] utilizes multi-channel graph neural networks with sentiment-awareness for image-text sentiment detection.

However, it is worth noting that the existing models mentioned above commonly overlook the inherent heterogeneity present in different modalities. However, our proposed MOC addresses this limitation from the perspective of distributional alignment.

Optimal Transport: In recent years, optimal transport is gaining more and more attention for its potential in machine learning. Arjovski et al. [1] transform the GAN (Generative Adversarial Network) problem into an OT theory problem. In [17], Solomon et al. propose an approximation algorithm for optimal transport distances between different geometric domains. In [11], the authors develop a fast framework, referred to as WEGL (Wasserstein Embedding for Graph Learning), by which we can embed graphs in a vector space. Optimal transport also plays an important role in domain adaptation. Flamary et al. [4] propose a model to align the representations between source and target domains by a regularized unsupervised optimal transportation. [20] proposes a novel framework for machine translation by using optimal transport.

Despite the significant progress that optimal transport has made in various fields, its application in sentiment analysis remains largely unexplored. In this work, we make a pioneering effort by leveraging optimal transport to tackle the task of MSA.

Contrastive Learning: Contrastive learning has become a popular reached field in representation learning [13,28,33,36]. In both fields of natural language processing (NLP) and computer vision (CV), plenty of models based on contrastive learning are proposed. ConSERT [22], SimCSE [7], CLEAR [19] interpolate contrastive learning in NLP. As for CV, MoCo [9], SimCLR [2], SimSiam [3] are proposed. Indeed, contrastive learning has also emerged as a prominent technique in the multi-modal field, and its potential for enhancing the quality of learned representations has attracted increasing attention. Notable works such as [10,16,26] have demonstrated the efficacy of contrastive learning in improving representation quality in multi-modal tasks.

Nevertheless, in the realm of contrastive learning, the problem of false positive samples has been a persistent challenge. In our work, we tackle this issue by leveraging the label associated with each pair of samples.

3   Preliminary of Optimal Transport

Suppose that we have two sets of points : $X = (x^1, \cdots, x^n)$ and $Y = (y^1, \cdots, y^m)$. We calculate the probability measures $\mu = \sum_{i=1}^{n} \alpha_i \delta(x^i)$ and $v = \sum_{j=1}^{m} \beta_j \delta(y^j)$ for two sets firstly, where the weights $\alpha_i = \frac{1}{n}, \beta_j = \frac{1}{m}$ can be considered as probability simplexs, here $\delta(\cdot)$ represents the Dirac (unit mass) function. Subsequently, let C represent the cost matrix , the element C_{ij} is the cost of moving point x^i to y^j. Hence the optimal transport between μ and v can be formulated as the following formulation:

$$OT(\mu, \nu; \mathbf{C}) := \min_{T \in R_+^{(n \times m)} s.t. T\mathbf{1}_m = \alpha, T^\top \mathbf{1}_n = \beta} \langle \boldsymbol{T}, \boldsymbol{C} \rangle. \tag{1}$$

where $\langle T, C \rangle := tr(T^T C) = \sum_{ij} T_{ij} C_{ij}$.

Wasserstein Distance: We can consider the cost matrix C with p-norm, where the element C_{ij} is defined as $||x^i, y^j||_p^p$. In this sense, the p-norm wasserstein distance can be formulated as: $\mathcal{W}_p(\mu, \nu) := OT(\mu, \nu; || \cdot ||_p^p)^{\frac{1}{p}}$. In this paper, we set $p = 2$.

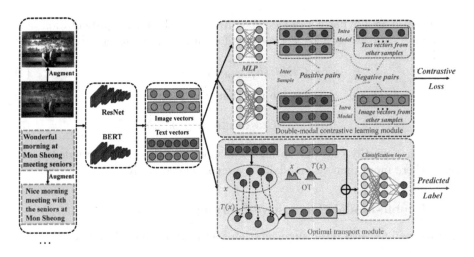

Fig. 2. Overview of our MOC model with two modalities.

4   Method

In this section, we present a comprehensive exposition of our MOC, the overall architecture is depicted in Fig. 2.

4.1   Optimal Transport

Given an image $\mathcal{I}$ and its corresponding sentence $\mathcal{T}$. We initially obtain their embeddings h_I and h_T using ResNet and BERT models respectively. The prior research [18] emphasizes that text encompasses more task-specific features compared to the image modality. Consequently, we consider the text space as the target space and the image space as the original space. We show the process of transferring embedding in image space into text space in the following:

(1) Calculate the measures μ, ν for features of two modalities h_T, h_I;
(2) Seek an optimal transport plan T from μ to ν by **sinkhorn** algorithm [5], and the element T_{ij} means the probability to transport from the i-th feature dimension of h_I to the j-th dimension of h_T;
(3) Utilize T to transfer the h_I into new embedding $\hat{h}_I$ [34]:

$$\hat{h}_I = \mathbf{diag}(1/v)(T^T + \Delta_T)h_I, \tag{2}$$

Δ_T is a learnable parameter.
(4) After obtaining $\hat{h}_I$, the fused vector h_{fusion} can be acquired by:

$$h_{fusion} = \min_{h}[\alpha\mathcal{W}_2(\hat{h}_I, h) + (1-\alpha)\mathcal{W}_2(h_T, h)], \tag{3}$$

where α is a hyper-parameter. The h_{fusion} is the final representation of $\mathcal{I}$ and $\mathcal{T}$, and we employ a two-layer forward network to get the predicted label:

$$\hat{y} = GELU(h_{fusion}W_1 + b_1)W_2 + b_2, \tag{4}$$

where $GELU$ is an activation function. The loss function in this module is:

$$\mathcal{L}_{op} = \text{Cross-Entropy}(\hat{y}, y_{true}), \tag{5}$$

It is noteworthy that our optimal transport module can be extended to the cases where more than two modalities are considered easily.

4.2   Double-Modal Contrastive Learning

In this module, two data augmentation methods are introduced to construct positive samples.

We employ RandAugment [35] for image data and back-translation [39] for text data. RandAugment is a straightforward augmentation technique that uniformly samples from a predefined set of augmentation transformations. This method ensures the augmentation process encompasses a diverse range of transformations for the image data. Initially, the original text expressed in language E, undergoes translation into language C. Subsequently, the translated text in language C is translated back into language E, yielding the augmented text. This iterative translation process facilitates the generation of augmented text samples with increased variability and complexity.

For an embedding h^i_m of a minibatch $\{h^1_m, h^2_m \cdots, h^N_m\}(m \in \{I, T\})$, $\hat{h}^i_I, \hat{h}^i_T$ are obtained by aforementioned augmentation methods. And $\{h^i_I, \hat{h}^i_I\}, \{h^i_T, \hat{h}^i_T\}$, $\{h^i_I, \hat{h}^i_T\}, \{h^i_T, \hat{h}^i_I\}$ are construct as positive pairs. In contrast to the self-supervised setting, negative pairs in our approach consist of pairs $\{h^i_m, h^j_m\}$, where $j \in \Gamma_i$ and Γ_i represents the set of samples with labels that are not equal to the anchor label. In this context, negative pairs are defined based on the disparity in labels between anchor sample and samples in Γ_i. This distinction allows us to formulate our contrastive loss as follows:

$$\mathcal{L}_c = -\sum_i^N log \frac{\sum\limits_{m_1,m_2\in\{I,T\}} e^{sim(h^i_{m_1}, \hat{h}^i_{m_2})/\tau}}{\sum\limits_{m_1,m_2\in\{I,T\}} e^{sim(h^i_{m_1}, \hat{h}^i_{m_2})/\tau} + \sum\limits_{j\in\Gamma_i}\sum\limits_{m\in\{I,T\}} e^{sim(h^i_m, h^j_m))/\tau}}, \tag{6}$$

where $sim(\cdot)$ denotes cosine similarity, τ is temperature coefficient and we set it to 0.07 in our paper.

In the practical training process, the loss function is:

$$\mathcal{L} = \mathcal{L}_{op} + \beta\mathcal{L}_c, \tag{7}$$

β is the hyper-parameter to balance the influence of $\mathcal{L}_{op}, \mathcal{L}_c$.

5   Experiments

5.1   Datasets and Implementation Details

We perform our experiments on three extensively utilized public datasets: MVSA-Single, MVSA-Multiple, and HFM. Each dataset is split into three portions, with a ratio of 8:1:1 for train, validation, and test respectively. The MVSA-Single and MVSA-Multiple datasets consist of three sentiment classes, while the HFM dataset is a binary classification dataset. The statistical details of these datasets are presented in Table 1.

To evaluate the performance of our model, we adopt accuracy and F1 scores as the evaluation metrics. These metrics provide comprehensive insights into accuracy and balance between precision and recall of our model's predictions. We utilize the Pytorch to conduct our experiments with a single GPU. We use the BERT-base to encode the data of text and ResNet-50 for image. The batch size is set to 32 for three datasets, we use AdamW optimizer and learning rate is set to 2e-5.

Table 1. Details of three datasets.

Datasets	Train	Test	Val	Total
MVSA-single	3611	450	450	4511
MVSA-multiple	13624	1700	1700	17024
HFM	19816	2410	2409	24635

5.2   Baselines

In order to provide a comprehensive comparison, we have chosen both well-known multi-modal and uni-modal models to be compared with our proposed model. Specifically, our multi-modal baselines consist of CLMLF [36], MultiSentiNet [38], HSAN [37], Co-MN-Hop6 [21], and MGNNS [23]. These models are widely recognized in the field of multi-modal sentiment analysis and serve as benchmarks for our evaluation.

Additionally, we have selected a set of uni-modal models for comparison. These include BERT [6], CNN [29], Bi-LSTM [31], ResNet [30], and OSDA [24] for uni-modal sentiment analysis.

5.3   Overall Results

The experimental results on the MVSA-single, MVSA-multiple, and HFM datasets are presented in Table 2 and Table 3. In our analysis, we categorize the models into three groups: Text, Image, and Multi-modal. Based on the results, we can make the following observations: (1) Our proposed method surpasses

Table 2. The results on MVSA-single and MVSA-multiple datasets. The best results are in **bold**.

Modality	Model	MVSA-single		MVSA-multiple	
		Acc	F1	Acc	F1
Text	CNN	0.6819	0.5590	0.6564	0.5766
	BiLSTM	0.7012	0.6506	0.6790	0.6790
	BERT	0.7111	0.6970	0.6759	0.6624
	TGNN	0.7034	0.6594	0.6967	0.6180
Image	Resnet-50	0.6467	0.6155	0.6188	0.6098
	OSDA	0.6675	0.6651	0.6662	0.6623
Multi-modal	MultiSentiNet	0.6984	0.6984	0.6886	0.6811
	HSAN	0.6988	0.6690	0.6796	0.6776
	Co-MN-Hop6	0.7051	0.7001	0.6892	0.6883
	MGNNS	0.7377	0.7270	0.7249	0.6934
	CLMLF	0.7533	0.7346	0.7200	0.6983
Multi-modal	MOC	**0.7610**	**0.7426**	**0.7279**	**0.7073**

Table 3. The results on HFM dataset.

Modality	Model	HFM	
		Acc	F1
Text	CNN	0.8003	0.7532
	BiLSTM	0.8190	0.7753
	BERT	0.8389	0.8326
Image	ResNet-50	0.7277	0.7138
	ResNet-101	0.7248	0.7122
Multi-modal	D&R Net	0.8402	0.8060
	CLMLF	0.8543	0.8487
Multi-modal	MOC	**0.8607**	**0.8547**

the performance of existing competitive methods on all three datasets. This indicates the effectiveness and superiority of our approach in addressing the multi-modal sentiment analysis task. (2) The multi-modal models achieve superior performance compared to the uni-modal models. This finding highlights the advantage of leveraging multiple modalities to extract richer and complementary information for conducting downstream sentiment analysis tasks.

By achieving better performance than existing methods and demonstrating the advantage of leveraging multiple modalities, our approach showcases its potential in capturing and leveraging diverse sources of information to improve the accuracy and effectiveness of sentiment analysis tasks.

Table 4. The ablation study on MVSA-single, MVSA-multiple and HFM datasets.

Model	HFM		MVSA-single		MVSA-multiple	
	Acc	F1	Acc	F1	Acc	F1
MOC	**0.8607**	**0.8547**	**0.7610**	**0.7426**	**0.7279**	**0.7073**
W/O OT&C	0.8456	0.8317	0.7289	0.7159	0.7047	0.6798
W/O OT	0.8517	0.8421	0.7394	0.7267	0.7167	0.6878
W/O C	0.8543	0.8474	0.7488	0.7377	0.7201	0.6987

5.4   Ablation Study

In order to investigate the contributions of different components in our MOC model, we conduct an ablation study as presented in Table 4. Specifically, we train and evaluate several variants of the model:

W/O OT&C: This variant simply concatenates the feature vectors from the image and text modalities without utilizing the optimal transport and contrastive learning modules.

W/O OT: This variant includes the contrastive learning module but excludes the optimal transport module.

W/O C: This variant includes the optimal transport module but excludes the contrastive learning module.

The results show that the performance of the model drops when any of the modules is removed, compared to the complete MOC model. This suggests that both optimal transport and contrastive learning modules contribute to the overall performance of multi-modal sentiment analysis. Furthermore, the removal of the optimal transport module has a greater impact on performance, indicating that reducing the heterogeneity between modalities is crucial for effective multi-modal fusion.

Overall, the ablation study confirms the importance of both the optimal transport and contrastive learning modules in our MOC model, highlighting their roles in aligning modalities and facilitating interaction between them for improved sentiment analysis performance.

6   Analysis

In this section, further experiments are conducted to explore the deep properties of MOC.

6.1   Hyper-parameter Heatmap

Our model incorporates two hyperparameters, namely α and β. The value of α determines the wasserstein center of text and image embeddings, while β controls the balance between the impact of $\mathcal{L}_{op}$ and $\mathcal{L}_c$. We search for suitable values of α and β within the range $\{0.1, \cdots, 0.9\}$ and $\{0.1, 0.5, 0.9\}$ respectively. To examine the influence of α and β, we conduct experiments on the MVSA-single dataset, and results are presented in Fig. 3. As depicted, the optimal performance is achieved when $\alpha = 0.2$ and $\beta = 0.1$. Notably, this combination also yields the best results on the other two datasets.

6.2   Visualization

In this section, we present the results of the visualization experiment conducted on the HFM dataset. The purpose of this experiment is to visualize the feature vectors before the classification layer of the model. To achieve this, we employ the TSNE dimensionality reduction algorithm, which projects high-dimensional vectors into a 2-dimensional space while preserving the distribution characteristics.

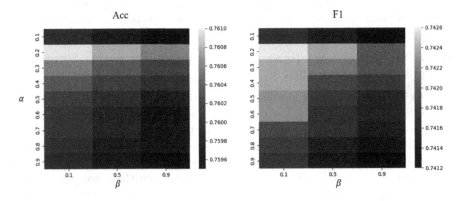

Fig. 3. The impact of α, β on MVSA-single dataset.

In Fig. 4, we provide a scatter plot visualization of the feature vectors. It can be observed that our proposed MOC model effectively separates the two clusters of scatters, indicating a clear distinction between the sentiment categories. On the other hand, the feature vectors produced by the BERT model show a higher degree of overlap and intermingling between the two clusters. This suggests that the BERT model struggles to capture the distinct characteristics of different sentiment categories.

Overall, the visualization results demonstrate the superiority of our MOC model in effectively separating and distinguishing sentiment clusters compared to the BERT model. This further validates the effectiveness of our proposed approach in multi-modal sentiment analysis.

Negative

Positive

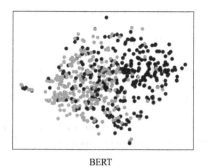 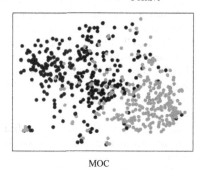

BERT   MOC

Fig. 4. The visualization of positive and negative instances on HFM dataset.

7   Conclusion

In this paper, we propose a novel MOC for the multi-modal sentiment analysis task, which aims to eliminate the heterogeneity of different modalities and simultaneously address the scarcity of modality interactions. Specifically, MOC leverages optimal transport to move image embeddings to the text space and utilizes the wasserstein barycenter to fuse feature vectors from different modalities. Additionally, contrastive learning is introduced to enhance the encoder's ability to learn modality-specific information and facilitate inter-sample interactions. To the best of our knowledge, this is the first work that utilizes optimal transport to improve the performance of multi-modal sentiment analysis (MSA). Extensive experimental results on three prevalent multi-modal sentiment analysis datasets demonstrate that MOC achieves superior results compared to existing competitive approaches.

Acknowledgements. This work is supported by the National Key R&D Program of China (2022YFC3103800) and National Natural Science Foundation of China (62101552).

References

1. Arjovsky, M., Chintala, S., Bottou, L.: Wasserstein GAN (2017)
2. Chen, T., Kornblith, S., Norouzi, M., Hinton, G.E.: A simple framework for contrastive learning of visual representations. In: Proceedings of the 37th International Conference on Machine Learning, ICML 2020, 13–18 July 2020, Virtual Event (2020)
3. Chen, X., He, K.: Exploring simple siamese representation learning. In: IEEE Conference on Computer Vision and Pattern Recognition, CVPR 2021, virtual, 19–25 June 2021, pp. 15750–15758. Computer Vision Foundation/IEEE (2021)
4. Courty, N., Flamary, R., Tuia, D., Rakotomamonjy, A.: Optimal transport for domain adaptation. IEEE Trans. Knowl. Data Eng. (2021)

5. Cuturi, M.: Sinkhorn distances: lightspeed computation of optimal transport. In: Burges, C.J.C., Bottou, L., Ghahramani, Z., Weinberger, K.Q. (eds.) Advances in Neural Information Processing Systems 26: 27th Annual Conference on Neural Information Processing Systems (2013)

6. Devlin, J., Chang, M., Lee, K., Toutanova, K.: BERT: pre-training of deep bidirectional transformers for language understanding. In: NAACL-HLT (2019)

7. Gao, T., Yao, X., Chen, D.: Simcse: simple contrastive learning of sentence embeddings. In: Moens, M., Huang, X., Specia, L., Yih, S.W. (eds.) Proceedings of the 2021 Conference on Empirical Methods in Natural Language Processing, EMNLP 2021 (2021)

8. Han, W., Chen, H., Poria, S.: Improving multimodal fusion with hierarchical mutual information maximization for multimodal sentiment analysis. In: EMNLP (2021)

9. He, K., Fan, H., Wu, Y., Xie, S., Girshick, R.B.: Momentum contrast for unsupervised visual representation learning. In: CVPR (2020)

10. Huang, P., Patrick, M., Hu, J., Neubig, G., Metze, F., Hauptmann, A.: Multilingual multimodal pre-training for zero-shot cross-lingual transfer of vision-language models. In: NAACL-HLT (2021)

11. Kolouri, S., Naderializadeh, N., Rohde, G.K., Hoffmann, H.: Wasserstein embedding for graph learning. In: ICLR (2021)

12. Li, J., et al.: Metamask: revisiting dimensional confounder for self-supervised learning. In: ICML (2022)

13. Liu, X., et al.: Self-supervised learning: generative or contrastive. CoRR (2021)

14. Lv, F., Chen, X., Huang, Y., Duan, L., Lin, G.: Progressive modality reinforcement for human multimodal emotion recognition from unaligned multimodal sequences. In: CVPR (2021)

15. Ngiam, J., Khosla, A., Kim, M., Nam, J., Lee, H., Ng, A.Y.: Multimodal deep learning. In: Getoor, L., Scheffer, T. (eds.) Proceedings of the 28th International Conference on Machine Learning, ICML (2011)

16. Radford, A., et al.: Learning transferable visual models from natural language supervision. In: ICML (2021)

17. Solomon, J., et al.: Convolutional Wasserstein distances: efficient optimal transportation on geometric domains. ACM Trans. Graph. **34**(4), 66:1–66:11 (2015)

18. Tsai, Y.H., Bai, S., Liang, P.P., Kolter, J.Z., Morency, L., Salakhutdinov, R.: Multimodal transformer for unaligned multimodal language sequences. In: Korhonen, A., Traum, D.R., Màrquez, L. (eds.) Proceedings of the 57th Conference of the Association for Computational Linguistics, ACL (2019)

19. Wu, Z., Wang, S., Gu, J., Khabsa, M., Sun, F., Ma, H.: CLEAR: contrastive learning for sentence representation. CoRR abs/2012.15466 (2020)

20. Xu, J., Zhou, H., Gan, C., Zheng, Z., Li, L.: Vocabulary learning via optimal transport for neural machine translation. In: ACL/IJCNLP (2021)

21. Xu, N., Mao, W., Chen, G.: A co-memory network for multimodal sentiment analysis. In: SIGIR (2018)

22. Yan, Y., Li, R., Wang, S., Zhang, F., Wu, W., Xu, W.: Consert: a contrastive framework for self-supervised sentence representation transfer. In: Zong, C., Xia, F., Li, W., Navigli, R. (eds.) ACL/IJCNLP (2021)

23. Yang, X.: Multimodal sentiment detection based on multi-channel graph neural networks. In: ACL (2021)

24. Yang, X., Feng, S., Wang, D., Zhang, Y.: Image-text multimodal emotion classification via multi-view attentional network. IEEE Trans. Multim. **23**, 4014–4026 (2021)

25. Yu, W., Xu, H., Yuan, Z., Wu, J.: Learning modality-specific representations with self-supervised multi-task learning for multimodal sentiment analysis. In: AAAI (2021)

26. Yuan, X., et al.: Multimodal contrastive training for visual representation learning. In: CVPR (2021)

27. Zadeh, A., Liang, P.P., Poria, S., Vij, P., Cambria, E., Morency, L.: Multi-attention recurrent network for human communication comprehension. In: AAAI (2018)

28. Li, J., Qiang, W., Zheng, C., Su, B., Xiong, H.: Metaug: contrastive learning via meta feature augmentation. In: International Conference on Machine Learning, pp. 12964–12978. PMLR (2022)

29. Chen, Y.: Convolutional neural network for sentence classification. Master's thesis, University of Waterloo (2015)

30. He, K., Zhang, X., Ren, S., Sun, J.: Deep residual learning for image recognition. In: Proceedings of the IEEE Conference on Computer Vision and Pattern Recognition, pp. 770–778 (2016)

31. Zhou, P., et al.: Attention-based bidirectional long short-term memory networks for relation classification. In: Proceedings of the 54th Annual Meeting of the Association for Computational Linguistics (Volume 2: Short Papers), pp. 207–212 (2016)

32. Li, J., et al.: Modeling multiple views via implicitly preserving global consistency and local complementarity. IEEE Trans. Knowl. Data Eng. (2022)

33. Qiang, W., Li, J., Zheng, C., Su, B., Xiong, H.: Interventional contrastive learning with meta semantic regularizer. In: International Conference on Machine Learning, pp. 18018–18030. PMLR (2022)

34. Cao, Z., Xu, Q., Yang, Z., He, Y., Cao, X., Huang, Q.: Otkge: multi-modal knowledge graph embeddings via optimal transport. Adv. Neural. Inf. Process. Syst. **35**, 39090–39102 (2022)

35. Cubuk, E.D., Zoph, B., Shlens, J., Le, Q.V.: Randaugment: practical automated data augmentation with a reduced search space. In: Proceedings of the IEEE/CVF Conference on Computer Vision and Pattern Recognition Workshops, pp. 702–703 (2020)

36. Li, Z., Xu, B., Zhu, C., Zhao, T.: CLMLF: a contrastive learning and multi-layer fusion method for multimodal sentiment detection. arXiv preprint arXiv:2204.05515 (2022)

37. Xu, N.: Analyzing multimodal public sentiment based on hierarchical semantic attentional network. In: 2017 IEEE International Conference on Intelligence and Security Informatics (ISI), pp. 152–154. IEEE (2017)

38. Xu, N., Mao, W.: Multisentinet: a deep semantic network for multimodal sentiment analysis. In: Proceedings of the 2017 ACM on Conference on Information and Knowledge Management, pp. 2399–2402 (2017)

39. Sennrich, R., Haddow, B., Birch, A.: Improving neural machine translation models with monolingual data. arXiv preprint arXiv:1511.06709 (2015)

Two-Phase Semantic Retrieval for Explainable Multi-Hop Question Answering

Qin Wang, Jianzhou Feng[✉], Ganlin Xu, and Lei Huang

School of Information Science and Engineering, Yanshan University, Qinhuangdao,
China
fjzwxh@ysu.edu.cn

Abstract. Explainable Multi-Hop Question Answering (MHQA)
requires an ability to reason explicitly across facts to arrive at the answer.
The majority of multi-hop reasoning methods concentrate on semantic
similarity to obtain the next hops or act as entity-centric inference. However, approaches that ignore the rationales required for problems can easily lead to blindness in reasoning. In this paper, we propose a two-**Phase**
text **R**etrieval method with an entity **M**ask mechanism (**PRM**), which
focuses on the rationale from global semantics along with entity consideration. Specifically, it consists of two components: 1) The **rationale-
aware retriever** is pre-trained via a dual encoder framework with an
entity mask mechanism. The learned representations of hypotheses and
facts are utilized to obtain top K candidate core facts by a sentence-level
dense retrieval. 2) The **entity-aware validator** determines the reachability of hypotheses and core facts with an entity granularity sparse
matrix. Our experiments on three public datasets in the scientific domain
(i.e., OpenbookQA, Worldtree, and ARC-Challenge) demonstrate that
the proposed model has achieved remarkable performance over the existing methods.

Keywords: Question answering · Multi-hop · Explainable · Entity
mask mechanism

1 Introduction

MHQA is the task of integrating two or more pieces of facts from external
resources to address a particular problem [29]. Compared to single-hop QA,
a primary challenge of multi-hop reasoning tasks is that the answers often do
not exist in an isolated fact but instead need to be inferred from several related
facts in an external corpus. Explainable MHQA requires the model to output the
correct reasoning chain (or some equivalent explanation representations) along
with the correct answer.

Recent work in MHQA includes generating reasoning chains using large-
scale sequence-to-sequence (Seq2Seq) models [11,23] and retrieval-based reasoning methods. Among former approaches, the results are dominated by the

B. Luo et al. (Eds.): ICONIP 2023, LNCS 14448, pp. 452–465, 2024.
https://doi.org/10.1007/978-981-99-8082-6_35

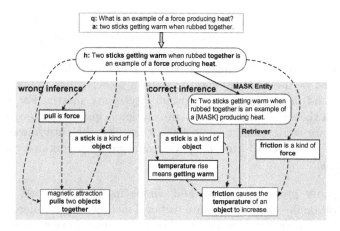

Fig. 1. Inference results using entity tracking (**left**) and our solution (**right**)

internal probability distribution which is still invisible and uncontrollable. In contrast, retrieval-based models can guarantee faithful facts and perform better at readability and scalability. In this mode, the retriever-reader approaches [3,9] cannot reflect the reasoning process explicitly, which reduces the comprehensibility of answers. A large quantity of studies emphasizes the significant role of entities in multi-hop processes. The methods based on graph structures [1,15], employing entities to identify inference paths [12,14] for Knowledge Base Question Answering (KBQA) or utilizing knowledge bases with unstructured texts [6,16] mostly interest in entity information to obtain the next hop. Besides, some methods generate reasoning paths by enjoying the benefit of semantic similarities [24,26]. However, in these approaches, the rationales of the problem are inevitably ignored, which limits the capabilities of reasoning and makes the process unreliable. In the multi-hop iterative retrieval process, they will even bring a continuous accumulation of errors, culminating in a catastrophic decline in model performance. Most retrieval-augmented methods emphasize the significance of entities in textual knowledge bases [1,7,21]. Existing work is also passionate about evaluating the evidence that holds the answer entity [8], but practically, few studies show a direct relationship between the retriever performance and the fact holding the answer entity. Therefore, we believe that the rationales can alleviate the entity-centric inference to some extent.

As shown in Fig. 1, a hypothesis h is generated by a question q and a candidate answer a. In the left part, the entity-centric strategy exhibits the process of generating incorrect reasoning chains that lead to a wrong conclusion. The two explanatory sentences of "*pull*" and "*object*" are erroneously introduced from the hypothesis by entity tracking, which attributes to the final consequence. Therefore, the lack of rationale comes with obvious limitations.

We address the above issues by presenting the guidance of rationales, which can largely restrict the search space to a small range. To this end, we propose a two-stage retrieval method merging both the rationale- and entity-aware approaches. Firstly, we form hypotheses by concatenating the question and each option [26]. Secondly, a rationale-aware retriever will get the top K facts with the highest score under the dual encoder framework with entity mask mechanism. Thirdly, we use an entity-aware validator to determine the reachability of hypotheses and facts via an entity granularity sparse matrix. For example, as illustrated in the right part of Fig. 1, our solution acquires the core fact through rationale-aware retriever first (solid line), then uses the entity-aware validator to verify the core fact (dotted line). Overall, our work contains the following contributions:

(1) This paper proposes a novel and simple model framework, which combines rationale- and entity-aware retrieval methods in the inference process. Therefore, adequate attention could be paid to the global semantics between hypotheses and facts.
(2) The proposed model explains the reasoning process explicitly with faithful facts, which facilitates model analyzing and debugging.
(3) The experimental results on three benchmark datasets in the scientific domain reveal that the model proposed has a remarkable improvement.

2   Related Work

Large-scale pre-training models have achieved great success in various NLP tasks, but they are still incompetent in MHQA systems that relate to language understanding and knowledge retrieval tasks. With the equipment of semantic understanding abilities of large language models, some researchers begin to perform MHQA tasks on unstructured text data directly. Some recent datasets with complete annotated explanations [10,27] and partial inference chains [17] have been provided to break through the bottleneck of multi-hop reasoning.

Open-domain MHQA mainly includes two methods: augmenting retrieval skills and generating reasoning paths. The way to augment retrieval is to strengthen the ability of coarse-grained filtering of open-domain corpus based on the information extraction model, e.g. AutoROCC [28], FID [9]. To alleviate multi-hop problems, researchers have successively proposed subsequent iterative retrieval methods, such as MUPPET [7], PullNet [21] and GoldEn [19]. Generating reasoning paths obtain the answers according to the hops of reasoning chains. Recent approaches include incorporating knowledge into the model first and proceeding with answer generation [18,31] and the method that frames question answering as an abductive reasoning problem, constructing plausible explanations for each choice and then selecting the candidate with the best explanation as the final answer [24,26]. Large-scale Language Models (LLMs) learn context from a few examples without updating model parameters. [2,11,23] use prompts to generate reasoning chains. But the reasoning chains are not reliable due to the inherent limitations of generative models.

3   Methodology

3.1   Problem Definition

we need to obtain the correct answer a from options o. The hypotheses $h = \{h_1, h_2, ..., h_M\}$ are formed by combining q with each option. The approach proposed by [5] can handle questions in the form of interrogative sentences. For other questions, the answer options will be filled directly into the underlined part of the question sentence. In addition, an external fact corpus $\mathcal{K}$ containing an abundance of core facts and grounding facts is also required. We need to retrieve all of the valid facts that can support the hypothesis within $\mathcal{K}$.

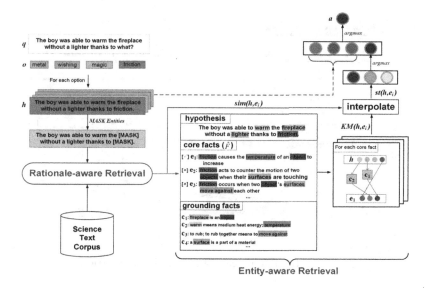

Fig. 2. Overall framework of our proposed model. Note that we have omitted the subscripts of some symbols for better understanding.

3.2   Overview

We assume that each problem requires a core fact as the rationale [17], while other facts needed are called grounding facts. As shown in Fig. 2, there are two stages in PRM: rationale-aware dense retrieval and entity-aware sparse retrieval. Term-based sparse retrieval (BM25 [20]) and dense retrieval in representation space (DPR [13]) are two main retrieval methods. In rationale-aware dense retrieval, we use encoders to convert hypotheses and facts into dense representation in vector space, and the hypotheses will retrieve relevant facts using cosine similarity. In entity-aware sparse retrieval, sentences will be transformed into a sparse representation of entity granularity, and then entity matching is employed to complete the retrieval process.

3.3   Rationale-Aware Dense Retrieval

This paper assumes that the core fact for solving a certain problem comes from abstract domain facts, which has a low similarity with the hypothesis. In previous entity-centric reasoning processes, overlooking the global semantics between sentences makes inferring process aimless. Therefore, we propose a rationale-aware global dense retriever. As shown in Fig. 3, we pretrain the retriever. Formally, we first mask a certain number of entities in hypotheses. The masking mechanism is to prevent entities from dominating the mapping process. To maintain the relative integrity of hypothesis semantics, the entities are hidden with a probability of φ.

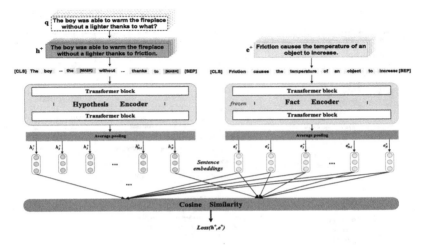

Fig. 3. The training process of rationale-aware retrieval.

Secondly, the hypothesis and the corresponding core fact are inputted into the dual-encoding framework respectively. The mean token vector of sentences is used as the dense representation. Thirdly, we adopt the idea of [13] to train the golden hypothesis h^+ of each question and its corresponding core fact e^+.

The training data is $D = <h_i^+, e_i^+>, i \in [1, N]$, where N is the number of sample pairs. h_i^+ indicates the correct hypothesis of the question. e_i^+ indicates the supporting core fact corresponding to h_i^+ which is also viewed as the positive relevant fact. The core facts corresponding to other hypotheses in the same batch of training are regarded as negative relevant facts $e_{i,j}^-$. We define $sim(\cdot)$ as the cosine similarity function, $E_H(\cdot)$ as the encoder that generates hypothesis dense representation, $E_F(\cdot)$ as the encoder that generates fact dense representation. m is the batch size. Then optimize the loss function as the negative log-likelihood of the positive facts:

$$Loss(h_i^+, e_i^+) = -log\left(\frac{exp\left(sim(h_i^+, e_i^+)\right)}{exp\left(sim(h_i^+, e_i^+)\right) + \sum_{j=1}^{m} exp\left(sim(h_i^+, e_{i,j}^-)\right)}\right) \quad (1)$$

$$sim(h, e) = E_H(h)^T \cdot E_F(e) \tag{2}$$

Each hypothesis from the same question should mask the same entity location. After pre-training, we use cosine similarity to adopt top K candidate core facts $\widehat{F}_{i,j}$ for each hypothesis $h_{i,j}$ from question q_i, i.e., $\widehat{F}_{i,j} = top_k\{sim(h_{i,j}, e_l), l \in [1, L]\}$ where L is the size of fact corpus. Then, we form the core fact pool for each question. The top K core facts are chosen to compose the core fact pool $\widehat{F}_i$, i.e., $\widehat{F}_i = top_k\{e_j, e_j \in [\widehat{F}_{i,1} \cup \widehat{F}_{i,2} \cup ... \cup \widehat{F}_{i,M}]\}$, where M is the option size of question.

3.4 Entity-Aware Sparse Retrieval

This module aims to get the entity matching score between hypotheses and core facts in the core fact pool $\widehat{F}_i$ with no parameters.

Concepts of Entity. Entities are nouns or noun phrases mentioned frequently in the grounding facts. Specifically, we use the spaCy toolkit to prepossess all sentences in the grounding fact base and then extract frequent noun chunks within them as entities to build the vocabulary $\mathcal{C}$.

Entity Linking. Entities occupy an important position in the reasoning process. The link between entities is defined as whether the entities are reachable, and we only consider entities in particular relationships (such as "kind of", "part of", "example of", and "similar to") as reachable to one another. We connect the same entity nodes in the grounding fact base, and determine whether there is a reachable path between any two entities through the *Breadth-First-Search* algorithm. After that, the entity linking relationship graph $\mathcal{P}$ can be constructed.

Sparse Retrieval. The entity-aware retrieval is added to filter the correct supporting core fact. We believe that although there is no common entity between the hypothesis and the core fact when many jump steps are involved, an accessible path between them exists most of the time. See the dotted line flow on the right side of Fig. 1. In addition, we regard the entities that appear in all the hypotheses of a question as ordinary entities, and give a comparatively low weight to them. Other entities in the hypothesis that do not appear at the same time are treated as special entities and given higher weight. The following formulas calculate the reachability scores between entities of the core fact and the hypothesis:

$$se(e_{c_i}, h_{c_j}) = I(c_i, c_j) \cdot \frac{S(c_j, T_h)}{|path(i, j)| + \epsilon} \tag{3}$$

$$S(c_j, T_h) = \begin{cases} 1 & if \ c_j \in T_h \\ \beta & if \ c_j \notin T_h \end{cases} \tag{4}$$

where $I(u, v)$ is an indicator function indicating whether there is a reachable path from u to v in $\mathcal{P}$. $|path(i, j)|$ represents the path length from $entity_i$ to $entity_j$, and a hyperparameter ϵ with a small value is added to prevent division by zero. T_h is the special entity set of hypothesis h, e_{c_i} is the i-th entity in the core fact, and h_{c_j} is the j-th entity in the hypothesis.

After the preceding procedures, we use the *Kuhn-Munkres* algorithm to calculate the entity-aware retrieval scores between hypotheses and core facts. Finally, taking the joint impacts of both rationale-aware and entity-aware retrieval into account, we set γ as their scaling hyperparameter and rate the confidence between core facts and hypotheses:

$$st(h, e) = \gamma \cdot KM(h, e) + (1 - \gamma) \cdot sim(h, e) \tag{5}$$

where $KM(h, e)$ represents the entity matching score using *Kuhn-Munkres* algorithm with entity reachability scores computed in Eq. 3.

3.5 Training Objective

We fine-tune our rationale-aware retriever on downstream tasks. The core fact $\widehat{e}_i$ with the highest confidence score is picked to support the hypothesis, i.e., $\widehat{e}_i = argmax\{st(h_i, e), e \in \widehat{F}\}$, and then select the hypothesis with the highest score among $st(h_i, \widehat{e}_i)$ as the final answer $\widehat{a} = argmax\{st(h_i, \widehat{e}_i), i = [1, M]\}$. Finally, we use the cross-entropy loss function to fine-tune it. During fine-tuning, only the parameters on the linear layer and hypothesis side encoder are updated during iteration. Noteworthy, for each iteration, only the core fact with the highest score participates in the update process due to the limitation of *argmax*. To make full use of the core fact pool, a cross-entropy for the overall retrieval is added to let the overall relevance between spurious hypotheses and the core fact pool be farther than the correct hypothesis. The final training goal is:

$$\mathcal{L} = \frac{1}{N} \sum_N \sum_{i=1}^{M} [y_i log(p_i) + y_i log(avg(\widehat{F}_i))] \tag{6}$$

where $y = \{y_1, y_2, ..., y_M\}$ is the one-hot representation of answer label. p_i represents the probability that the *i-th* choice is the correct answer, and $avg(\widehat{F}_i)$ reflects the average score of all core facts corresponding to the *i-th* hypothesis in the core fact pool.

4 Experiment

4.1 Datasets

We experiment on three challenging multiple-choice datasets to demonstrate our proposed method: OpenBookQA [17], ARC-Challenge [4], and Worldtree [10]. They are all scientific QA tasks. Specifically, Worldtree requires 6 facts on average of each answer (up to 20) [32]. Compared to HotpotQA (almost 2 hops) [29], they involve larger reasoning networks. The details for training/dev/testing splitting method are referred to Table 1.

Table 1. Number of questions in each QA dataset

	Train	Dev	Test	Corpus
Worldtree	1,545	484	1,199	9,725
ARC-Challenge	1,119	299	1,172	14,621,856
OpenBookQA	4,957	500	500	6,492

4.2 Baselines

The compared baselines for our approach are listed as follows: (1) **PathNet** [14] structures an inference path between question and candidate answers, and then encodes the path to get answers. (2) **MRC-Strategies** [22] adds three strategies to handle MRC tasks. (3) **AutoROCC** [28] adopts an unsupervised way to freely combine the retrieved passages to form a reason set. (4) **ExplanationLP** [24] divides facts into grounding facts and abstract facts, and uses Bayesian based on graph and linear programming to improve performance. (5) **GPT-3** [2] is one of the most advanced language models with excellent language generation and context understanding abilities. (6) **N-XKT** [31] converts scientific texts into neural network representations, then makes it implicitly used in downstream question answering tasks. (7) **CB-ANLI** [26] adopts retrieve-reuse-revise framework and derives the answer in combination with similar cases, which has certain interpretability. (8) **KiC** [18] is a semi-parametric language model with six types of knowledge in the external memory, which adaptively retrieves the most helpful pieces of knowledge. (9) **OPT** [30] is a large language model that predicts answers using zero-shot and few-shot prompting learning.

4.3 Settings

We use Bert (bert-base, 768 dimensions, 12 layers) as our encoder. In the rationale-aware pre-training stage, the learning rate is 1e-5 and the batch size is 8. According to the statistics provided by OpenBookQA, there are Avg. question tokens of 11.46 and Avg. science fact tokens of 9.28. Therefore, it is reasonable to set the maximum length of the hypothesis and fact to 200. For downstream specific question answering tasks, the initial learning rate is 5e-6 and the batch size is 4. It has been trained for 10 epochs on OpenBookQA task. On both Worldtree and ARC-Challenge, they are trained for 15 epochs. We also analyzed the scale hyperparameter γ. When the value of γ is 0.4, OpenBookQA can achieve the best results; and when it is 0.6, Worldtree and ARC-Challenge will get the best benefit. In our pilot experiments, the best value of β is 0.2 and φ is 0.3.

4.4 Result Analysis

ARC-Challenge & OpenBookQA. Table 2 shows the performance of our model and several comparative models. Experiments are implemented with the core fact pool size of 10 and 20 respectively. Three fact corpora are used on ARC-Challenge. When using ARC Corpus, we use BM25 [20] to get the top 50 facts of each hypothesis to form a small-scale fact base. As can be seen from Table 2, firstly, on OpenBookQA, when $K=10$, our method improves by 1.40% compared with KiC, which uses T5 as the backbone and incorporates about 290GB of triplet knowledge (much more than the external knowledge we use). Secondly, when using both OpenBookQA and Worldtree as the external fact corpora, ARC-Challenge achieves the best result among all compared models, which demonstrates the effectiveness of our model. Finally, PRM on ARC-Challenge and OpenbookQA is still valid compared to OPT-30B[1] and GPT-3[2] which have an extremely large number of parameters. Note that although the large language models improve the performance of QA tasks with the increasing parameters, the hallucinations in the explanation cast doubt on the generated facts.

Worldtree. BM25 and Sentence-BERT are deemed as the representatives for sparse and dense retrieval, respectively. We also fine-tune on slightly larger models like BERT-large and RoBERTa-large. As shown in Table 3, our model delivers a considerable improvement in inferring accuracy, and the best results are obtained when the core fact pool capacity is 20 (increased by 4.68%, 4.61% and 3.30% respectively).

Table 2. The accuracy of ARC-Challenge and OpenBookQA test sets. *External KB* indicates the external fact corpus referenced in ARC-Challenge: 1. ARC Corpus; 2. OpenBook; 3. Worldtree [27]. All baseline results come from published papers.

Model	OpenBookQA	ARC-Challenge	External KB
IR BM25 (K = 10)	27.20	24.49	3
AutoROCC	-	41.24	1
Reading Strategies+GPT	55.20	40.70	1
ExplanationLP	-	32.16	1
CB-ANLI	-	36.77	1
N-XKT	56.56	38.09	3
KiC	58.20	-	-
OPT-30B	-	31.10	-
GPT-3	57.60	-	-
PRM (K = 10)	**59.60**	-	-
PRM (K = 20)	57.80	**42.27**	2, 3
PRM (K = 20)	-	40.19	3
PRM (K = 20)	-	36.86	1

---

[1] OPT-30B has 30B parameters, and the accuracy on ARC-Challenge is under the zero-shot setting.

[2] GPT-3 has 175B parameters, and the accuracy on OpenBookQA is under the zero-shot setting from [25].

Table 3. the accuracy on worldtree test set

Model	Overall	Challenge	Easy
BM25(K = 20)	39.45	30.41	43.54
Sentence-BERT(K = 20)	43.14	31.96	48.20
PathNet	41.50	36.42	43.32
BERT-large	46.19	31.96	52.96
RoBERTa-large	50.20	35.05	57.04
PRM(K = 20)	**54.88**	**41.03**	**60.34**
PRM(K = 30)	53.54	39.21	59.43

4.5 Ablation

Rationale-Aware Pre-training. In the process of pre-training, we anticipate that the model can obtain a convergent mapping from hypothesis to core fact and has the ability to retrieve core facts. Therefore some ablation experiments have been conducted to demonstrate this presume. *Explanation* indicates whether the core fact pool contains golden core fact. We perform the analysis on the *test-set* of OpenBookQA, and Fig. 4 shows the change in the accuracy of answer and explanation at different periods of pre-training. As can be observed, as the number of pre-training steps rises, the explanation accuracy increases rapidly, along with the sluggish increase in answer accuracy. The result reveals that rationale-aware pre-training module is effective, which enables the model to retrieve accurate core facts even when certain entities are hidden and can further improve the accuracy of task.

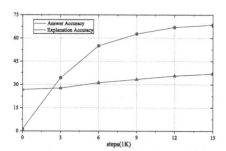
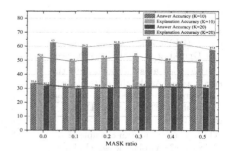

Fig. 4. The accuracy answers and explanation on OpenBookQA.

Fig. 5. The effect of MASK ratios on OpenBookQA.

Entity Masking Ratios. To further investigate the impact of entity masking strategy on rationale-aware retrieval module, we conduct a comparison experiment on the entity masking ratios on OpenBookQA. As shown in Fig. 5, the ratio of masked entities has little influence on the accuracy of final answers. Firstly,

With the ratio increasing, the explanation accuracy descends since the mapping of hypothesis to core fact relies on entities at small scales. Then, the importance of entities gradually decreases, which causes the decline of explanation accuracy. After the entities drop to a threshold (close to 0.3), contextual semantics play a critical role in the mapping process. The explanation accuracy peaks at 53%/65% (0.4%/2.0% higher than without any entity masking), which also indicates the efficiency of our entity masking strategy. Finally, the impact of serious semantic loss makes the explanation accuracy continually reduced.

4.6 Explainability

We investigate the specific explainability of our proposed method, as well as its impact on downstream tasks. We evaluate the dev-set of Worldtree ($K = 20$). To evaluate the explainability, the facts predicted will be considered accurate if they are part of the golden explanations. As shown in Table 4, there are a series of qualitative examples to illustrate the relationship between explainability and reasoning ability. In the first example, the correct answer is derived from an accurate explanation, and the number of explanations is relatively limited. When there is only one fact in the golden explanation set, this situation occurs for a total of 83.33% of correct answers. When the number of golden explanations is under 6, the accurate explanations is 68.14%. The second row depicts the situation when incorrect results are produced from accurate explanations. We

Table 4. Example of predicted explanations and answers. The <u>underlined</u> choices represent the correct answers. The **bold** choices represent the predicted answers. *Accurate* indicates whether the predicted explanations are part of the golden explanations. *Correct* indicates whether predicted answers are the golden answers.

Question	Explanation	Accurate	Correct
Which of these is most flexible? (A) Broom handle (B) Wooden ruler **(C) Drinking straw** (D) Sewing needle	(1) a drinking straw is flexible	Y	Y
A student left a bar of chocolate in the sun on a hot day. As the chocolate melted, which property changed? **(A) its mass** (B) its shape (C) its weight (D) its <u>composition</u>	(1) melting means matter; a substance changes from a solid into a liquid by increasing heat energy; (2) chocolate is a kind of solid usually	Y	N
Which description is an example of camouflage? (A) plant leaves closing up when touched (B) butterflies migrating south for the winter **(C) green insects sitting on green <u>leaves</u>** (D) earthworms moving through soil	(1) an example of camouflage is an organism looking like leaves	N	Y
What do all vertebrate animals have in common? **(A) They all have fur** (B) They all have backbones (C) They all are warm-blooded (D) They all have young that look like the adults	(1) a mammal usually has fur; (2) fur is often part of an animal	N	N

think it happened when the explanations obtained are not sufficient. The example of rows 3 and 4 show two scenarios in which inaccurate explanations occur. The issue maybe is that too many facts contained in the golden explanation sets makes the accurate explanations hard to find.

Table 5. Examples of the difference between predicted explanations and golden explanations.

Question	Predicted Explanations	Golden Explanations	Accurate
If you bounce a rubber ball on the floor, it goes up and then comes down. What causes the ball to come down? (A) magnetism *(B) gravity* (C) electricity (D) friction	(1) gravity accelerates an object while that object falls by pulling on it; (2) a ball is a kind of object	(1) gravity; gravitational force causes objects that have mass; substances to be pulled down; to fall on a planet; (2) a ball is a kind of object; (3) come down is similar to falling	Y
When notebook paper is folded to make an airplane, what physical property of the paper changes? (**A**) **Mass** (B) Weight (C) Shape (D) Smell	(1) as the shape of an object changes, the mass of that object will stay the same; (2) mass is a kind of physical property	(1) folding an object causes that object to change shape; (2) a paper is a kind of object; (3) shape is a kind of physical property	N

In addition, the existing researches reveal that automatic evaluation algorithms always underestimate the explainability of models. As shown in Table 5, although the predicted explanations are different from the golden explanations, we still believe they are reasonable for answers. Apparently, literal explanations like the second row will be rationally regarded as spurious explanations.

5 Conclusion

This paper presents a novel method for abstract scientific QA. Firstly, domain facts are divided into core facts and grounding facts, and then the mapping relationship between correct hypotheses covered entities and core facts is learned by the pre-trained model. Secondly, an entity-aware retrieval module is added, which is a nonparametric sparse matrix from the grounding fact corpus in the field. Finally, the inference model is obtained through fine-tuning. We use experiments to prove the effectiveness of diluting the impact of entities and paying attention to semantics in reasoning, and also make contributions to the interpretability to some degree. In future work, we will further research how to integrate context semantics and entities more naturally in the reasoning process, so that the multi-hop reasoning process can be more accurate and clearer.

Acknowledgments. This work is supported by the National Natural Science Foundation of China (62172352), the Central leading local science and Technology Development Fund Project (No. 226Z0305G), Project of Hebei Key Laboratory of Software

Engineering (22567637H), the Natural Science Foundation of Hebei Province (F20222 03028) and Program for Top 100 Innovative Talents in Colleges and Universities of Hebei Province (CXZZSS2023038).

References

1. Asai, A., Hashimoto, K., Hajishirzi, H., Socher, R., Xiong, C.: Learning to retrieve reasoning paths over wikipedia graph for question answering. arXiv preprint arXiv:1911.10470 (2019)
2. Brown, T.B., et al.: Language models are few-shot learners (2020)
3. Chen, D., Fisch, A., Weston, J., Bordes, A.: Reading wikipedia to answer open-domain questions. arXiv preprint arXiv:1704.00051 (2017)
4. Clark, P., et al.: Think you have solved question answering? Try arc, the AI2 reasoning challenge. arXiv preprint arXiv:1803.05457 (2018)
5. Demszky, D., Guu, K., Liang, P.: Transforming question answering datasets into natural language inference datasets. arXiv preprint arXiv:1809.02922 (2018)
6. Dhingra, B., Jin, Q., Yang, Z., Cohen, W.W., Salakhutdinov, R.: Neural models for reasoning over multiple mentions using coreference. arXiv preprint arXiv:1804.05922 (2018)
7. Feldman, Y., El-Yaniv, R.: Multi-hop paragraph retrieval for open-domain question answering. arXiv preprint arXiv:1906.06606 (2019)
8. Izacard, G., Grave, E.: Distilling knowledge from reader to retriever for question answering. arXiv preprint arXiv:2012.04584 (2020)
9. Izacard, G., Grave, E.: Leveraging passage retrieval with generative models for open domain question answering. arXiv preprint arXiv:2007.01282 (2020)
10. Jansen, P.A., Wainwright, E., Marmorstein, S., Morrison, C.T.: Worldtree: a corpus of explanation graphs for elementary science questions supporting multi-hop inference. arXiv preprint arXiv:1802.03052 (2018)
11. Wei, J., et al.: Chain of thought prompting elicits reasoning in large language models. arXiv, abs/2201.11903 (2022)
12. Jiang, Y., Joshi, N., Chen, Y.C., Bansal, M.: Explore, propose, and assemble: an interpretable model for multi-hop reading comprehension. arXiv preprint arXiv:1906.05210 (2019)
13. Karpukhin, V., et al.: Dense passage retrieval for open-domain question answering. arXiv preprint arXiv:2004.04906 (2020)
14. Kundu, S., Khot, T., Sabharwal, A., Clark, P.: Exploiting explicit paths for multi-hop reading comprehension. arXiv preprint arXiv:1811.01127 (2018)
15. Lan, Y., Jiang, J.: Query graph generation for answering multi-hop complex questions from knowledge bases. Association for Computational Linguistics (2020)
16. Lin, B.Y., Sun, H., Dhingra, B., Zaheer, M., Ren, X., Cohen, W.W.: Differentiable open-ended commonsense reasoning. arXiv preprint arXiv:2010.14439 (2020)
17. Mihaylov, T., Clark, P., Khot, T., Sabharwal, A.: Can a suit of armor conduct electricity? A new dataset for open book question answering. arXiv preprint arXiv:1809.02789 (2018)
18. Pan, X., Yao, W., Zhang, H., Yu, D., Yu, D., Chen, J.: Knowledge-in-context: towards knowledgeable semi-parametric language models. In: The Eleventh International Conference on Learning Representations (2023)
19. Qi, P., Lin, X., Mehr, L., Wang, Z., Manning, C.D.: Answering complex open-domain questions through iterative query generation. arXiv preprint arXiv:1910.07000 (2019)

20. Robertson, S., Zaragoza, H., et al.: The probabilistic relevance framework: BM25 and beyond. Found. Trends® Inf. Retrieval **3**(4), 333–389 (2009)
21. Sun, H., Bedrax-Weiss, T., Cohen, W.W.: Pullnet: open domain question answering with iterative retrieval on knowledge bases and text. arXiv preprint arXiv:1904.09537 (2019)
22. Sun, K., Yu, D., Yu, D., Cardie, C.: Improving machine reading comprehension with general reading strategies. arXiv preprint arXiv:1810.13441 (2018)
23. Kojima, T., Gu, S.S., Reid, M., Matsuo, Y., Iwasawa, Y.: Large language models are zero-shot reasoners. arXiv, abs/2205.11916 (2022)
24. Thayaparan, M., Valentino, M., Freitas, A.: Explanationlp: abductive reasoning for explainable science question answering. arXiv preprint arXiv:2010.13128 (2020)
25. Touvron, H., et al.: Llama: open and efficient foundation language models (2023)
26. Valentino, M., Thayaparan, M., Freitas, A.: Case-based abductive natural language inference. arXiv e-prints pp. arXiv-2009 (2020)
27. Xie, Z., Thiem, S., Martin, J., Wainwright, E., Marmorstein, S., Jansen, P.: Worldtree V2: a corpus of science-domain structured explanations and inference patterns supporting multi-hop inference. In: Proceedings of the 12th Language Resources and Evaluation Conference, pp. 5456–5473 (2020)
28. Yadav, V., Bethard, S., Surdeanu, M.: Quick and (not so) dirty: unsupervised selection of justification sentences for multi-hop question answering. arXiv preprint arXiv:1911.07176 (2019)
29. Yang, Z., et al.: Hotpotqa: a dataset for diverse, explainable multi-hop question answering. arXiv preprint arXiv:1809.09600 (2018)
30. Zhang, S., et al.: OPT: open pre-trained transformer language models. arXiv preprint arXiv:2205.01068 (2022)
31. Zhou, Z., Valentino, M., Landers, D., Freitas, A.: Encoding explanatory knowledge for zero-shot science question answering. arXiv preprint arXiv:2105.05737 (2021)
32. Jansen, P., Balasubramanian, N., Surdeanu, M., Clark, P.: What's in an explanation? characterizing knowledge and inference requirements for elementary science exams. In: Proceedings of COLING 2016, the 26th International Conference on Computational Linguistics: Technical Papers, pp. 2956–2965 (2016)

Efficient Spiking Neural Architecture Search with Mixed Neuron Models and Variable Thresholds

Zaipeng Xie[1,2]([✉]), Ziang Liu[2], Peng Chen[2], and Jianan Zhang[2]

[1] Key Laboratory of Water Big Data Technology of Ministry of Water Resources,
Hohai University, Nanjing, China
[2] College of Computer and Information, Hohai University, Nanjing, China
{zaipengxie,ziangliu,pengchen,jianan_zhang}@hhu.edu.cn

Abstract. Spiking Neural Networks (SNNs) are emerging as energy-efficient alternatives to artificial neural networks (ANNs) due to their event-driven computation and effective processing of temporal information. While Neural Architecture Search (NAS) has been extensively used to optimize neural network structures, its application to SNNs remains limited. Existing studies often overlook the temporal differences in information propagation between ANNs and SNNs. Instead, they focus on shared structures such as convolutional, recurrent, or pooling modules. This work introduces a novel neural architecture search framework, MixedSNN, explicitly designed for SNNs. Inspired by the human brain, MixedSNN incorporates a novel search space called SSP, which explores the impact of utilizing Mixed spiking neurons and Variable thresholds on SNN performance. Additionally, we propose a training-free evaluation strategy called Period-Based Spike Evaluation (PBSE), which leverages spike activation patterns to incorporate temporal features in SNNs. The performance of SNN architectures obtained through MixedSNN is evaluated on three datasets, including CIFAR-10, CIFAR-100, and CIFAR-10-DVS. Results demonstrate that MixedSNN can achieve state-of-the-art performance with significantly lower timesteps.

Keywords: Spiking Neural Network · Neural Architecture Search ·
Brain-inspired Computing · Mixed Neuron Models

1 Introduction

Spiking Neural Networks (SNNs) [19] have rapidly ascended as a potent field of research, simulating the intricate behavior of biological neurons through a highly congruent model. Setting themselves apart from traditional neural networks, SNNs exploit discrete spikes or pulses to represent neuron firing, thereby promoting superior efficiency and biological realism. This unique approach has

Supported by The Belt and Road Special Foundation of the State Key Laboratory of Hydrology-Water Resources and Hydraulic Engineering under Grant 2021490811.

B. Luo et al. (Eds.): ICONIP 2023, LNCS 14448, pp. 466–481, 2024.
https://doi.org/10.1007/978-981-99-8082-6_36

attracted substantial interest across a diverse range of applications. SNNs have demonstrated a capacity to serve as energy-efficient alternatives to conventional Artificial Neural Networks (ANNs).

The recent trend in SNN research involves adapting architectures originally designed for ANNs, such as VGG-Net [25] and ResNet [9]. Although many studies [4,14,16] have utilized Neural Architecture Search (NAS) to optimize network structures in ANNs, there remains limited exploration of such techniques for SNNs [10,22]. Notably, these studies frequently employ NAS search spaces and algorithms developed for ANNs, while neglecting the distinctive temporal differences in information propagation between ANNs and SNNs. Spiking neurons asynchronously transmit information via sparse and binary spikes over multiple time steps. Given the fundamental differences between SNN and ANN, substantial attention has been dedicated to the optimization of SNNs by exploring efficient spiking neuron models [1,6,31]. Similar to those in the human brain, spiking neurons often utilize varied activation patterns and thresholds to enhance neural circuit functionality. However, the performance implications of integrating mixed activation patterns and thresholds necessitate deeper investigation. Additionally, the consideration of how temporal information propagation impacts network architecture design frequently remains overlooked.

This work introduces MixedSNN, a novel framework for searching Spiking Neural Network (SNN) architectures. MixedSNN investigates Mixed Neuron types and Variable Thresholds within spiking layers to design high-performance SNNs. We utilize a dedicated search space, SSP, to explore diverse spiking neuron types and thresholds to enhance efficiency. Effective navigation within SSP requires rapid network performance evaluation. To address this, Mixed-SNN employs a training-free approach named Period-Based Spike Evaluation (PBSE) to capture temporal information propagation in SNNs. Furthermore, we incorporate PBSE scores as fitness metrics in a niche genetic algorithm [15] for innovative SNN architecture discovery. Experimental results demonstrate that the resulting SNN architecture discovered by MixedSNN outperforms state-of-the-art algorithms on the CIFAR10-DVS dataset. Moreover, for architectures with similar network depths, our model exhibits considerable improvements on CIFAR-10 and CIFAR-100 datasets. Our contributions can be summarized as follows:

- We propose a novel Neural Architecture Search paradigm, MixedSNN, that utilizes mixed spiking neurons and flexible thresholds to enhance the performance of Spiking Neural Networks in classification tasks.
- MixedSNN implements a unique training-free evaluation approach, known as Period-Based Spike Evaluation, which takes into account the temporal information of spiking neuron outputs. This approach provides potential avenues for assessing the performance of a spiking neural network without the need for comprehensive training.
- We evaluate the performance of MixedSNN by comparing it with several state-of-the-art algorithms. Experimental results demonstrate that the SNN architecture realized through MixedSNN excels in classification tasks on the

CIFAR10-DVS dataset. Furthermore, in comparison with network architectures of comparable depths, the SNN architecture exhibits significant performance improvements across a wide range of datasets, as evidenced by enhancements on the CIFAR-10 and CIFAR-100 datasets.

2   Related Work

Spiking Neural Networks (SNNs) [27], owing to their bio-inspired approach, capabilities for temporal processing, energy efficiency, and advancements in supporting hardware, have garnered significant attention. However, the integration of spiking neurons, characterized by their discrete spiking nature and complex temporal dynamics [1,6,31], with backpropagation presents a notable challenge, distinctly separating them from traditional artificial neural networks (ANN).

Achieving high SNN performance is challenging due to non-differentiable backpropagation. Spiking neurons' binary signaling contrasts with continuous activations in ANNs, hindering traditional gradient-based training. The conversion from ANNs to SNNs offers a significant strategy [8,12,24,28] to address the non-differentiability challenge in SNNs. Guo et al. [8] propose a combined ANN-to-SNN framework utilizing weight factorization and self-distillation. Wang et al. [28] introduce a technique combining soft-reset and parameter normalization to preserve membrane potential. Although the converted SNNs can achieve accuracies comparable to ANNs, they often rely on pre-trained ANNs and may require substantial timesteps, resulting in energy-inefficient spike generation.

In contrast to the ANN-to-SNN conversion, recent years have witnessed the emergence of direct training with surrogate gradient approaches [3,18,21,23,29–31,34]. These methods have gained prominence due to their superior performance and reduced timestep requirements. For example, Deng et al. [3] propose a temporal efficient training approach that optimizes each timestep's output rather than just the final integrated output. Zheng et al. [34] introduce a threshold-dependent batch normalization (tdBN) method for training deep SNNs with spatio-temporal backpropagation, mitigating gradient issues and balancing neuron firing rates. Thus, adopting a direct training approach offers a path to obtaining energy-efficient SNNs.

The weight-sharing [16,22] and cell-based [4,10,14,33] approaches are popular techniques for automatic neural network discovery. The weight-sharing strategy employs a supernet encompassing subnetworks, efficiently searching architectures through parameter and weight sharing. For example, Liu et al. [16] propose a differentiable method for architecture search, departing from traditional evolution or reinforcement learning methods. Na et al. [22] investigate a spike-aware architecture search to optimize SNNs regarding accuracy and spike count. The cell-based approach represents neural networks using repeated cell structures. Dong et al. [4] introduce NAS-Bench-201, an extension of NAS-Bench-101 [33], serving as a unified benchmark for NAS algorithms. Li et al. [14] propose T2IGAN, a text-to-image synthesis network architecture discovered through NAS. Kim [10] presents a NAS approach incorporating temporal feedback connections for SNN design. However, when applying NAS to SNNs, the focus often lies on adapting an equivalent ANN architecture, resulting in limited research addressing the distinct structures of SNNs.

As datasets and neural network architectures become more intricate, mitigating the computational burden of NAS methods is increasingly vital. Recent investigations [2,10,17,20,26,32] have explored training-free evaluation techniques predicting neural network performance without actual training. For instance, Chen et al. [2] merge linear regions theory and neural tangent kernel (NTK) for architecture assessment. Sun et al. [26] apply the maximum entropy principle for evaluation. Xu et al. [32] propose a Gradient Kernel-based method, while Mellor et al. [20] introduce a network kernel matrix approach based on linear regions theory. Lopes et al. [17] present an evaluation using the kernel matrix. Furthermore, Kim et al. [10] extend the kernel matrix approach to capture the temporal characteristics of SNNs.

While training-free evaluation approaches reduce computational costs for architecture search, limited attention has been given to the temporal characteristics of SNNs. To address this gap, we introduce a novel Training-Free method that employs the kernel matrix, specifically focusing on capturing the temporal characteristics of SNNs.

3 Preliminaries

Spiking neurons in SNNs accumulate synaptic inputs for membrane potential updates. In contrast, neurons in Artificial Neural Networks (ANNs) produce continuous and differentiable output. When the membrane potential surpasses a threshold, spiking neurons generate discrete spikes, followed by a reset. Various neuron models differ in their biological realism and computational efficiency. The Integrate-and-Fire (IF) model [1] is simple, with spikes occurring after potential accumulation reaches a threshold. The charge equations for IF models are:

$$H_t = V_{t-1} + X_t, \tag{1}$$

where H_t is the membrane potential values before a spike is fired, X_t symbolizes the presynaptic inputs at time t, and V_t is the membrane potential value. The Leaky Integrate-and-Fire (LIF) [7] model's charge equation is as follows:

$$H_t = V_{t-1} + \frac{1}{\tau} \cdot [X_t - (V_{t-1} - V_{reset})], \tag{2}$$

where τ is a membrane time constant controlling the decay of membrane potential and presynaptic inputs, and V_{reset} signifies the reset value for the membrane potential post-activation. The Parametric Leaky Integrate-and-Fire (PLIF) spiking neuron model [6] learns both the synaptic weights and τ:

$$H_t = V_{t-1} + \frac{1}{\tau} \cdot [X_t - (V_{t-1} - V_{reset})], \tag{3}$$

where $1/\tau = Sigmoid(\omega)$ and ω is a trainable parameter. The Quadratic Integrate-and-Fire (QIF) model [7] introduces a quadratic term:

$$H_t = V_{t-1} + \frac{1}{\tau} \cdot [X_t + a_0 \cdot (V_{t-1} - V_{rest}) \cdot (V_{t-1} - V_c)], \tag{4}$$

where V_c represents critical voltage. The Exponential Integrate-and-Fire (EIF) model [7] has an exponential term for rapid potential escalation:

$$H_t = V_{t-1} + \frac{1}{\tau} \cdot \left[X_t + V_{rest} - V_{t-1} + \Delta_T \cdot e^{\frac{(V_{t-1} - \theta_{rh})}{\Delta_T}} \right], \tag{5}$$

where Δ_T is the sharpness parameter and θ_{rh} is the rheobase threshold.

The selection of an appropriate neuron model for a spiking neural network depends on the specific requirements and the balance between computational efficiency and biological fidelity.

4   Methodology

Several neuron models can be used to construct mixed neuron models that allow for a more comprehensive representation of neural behavior and functionality within the network. By incorporating diverse neuron types, the SNN can benefit from their respective strengths, such as computational efficiency, biological realism, and capturing various dynamics. This approach enables the network to achieve a balance between computational efficiency and biological fidelity while leveraging the unique properties of different neuron models, ultimately enhancing the SNN's capabilities, flexibility, and ability to handle complex tasks.

Neural architecture search methods often prioritize shared structures between ANNs and SNNs. They frequently transfer evaluation strategies from ANNs to SNNs, neglecting the unique neuron model traits. We propose MixedSNN, a NAS approach that tackles these issues. MixedSNN establishes a search space for diverse spiking neuron models and membrane potential thresholds. It employs a training-free evaluation strategy, PBSE, considering spiking neurons' temporal characteristics. This strategy integrates within MixedSNN's framework, including a dedicated search algorithm for optimal PBSE-based designs. Figure 1 illustrates our proposed MixedSNN strategy.

Fig. 1. Diagram of our MixedSNN strategy. It creates a search space SSP, explores with the search algorithm, evaluates using PBSE, and outputs the best architecture.

4.1   Search Space

This study introduces a designated search space, SSP, encompassing distinct spiking neuron layers. Each layer can choose a spiking neuron model and its corresponding membrane potential threshold. The set of spiking neuron models, SN, in SSP includes IF, LIF, PLIF, QIF, and EIF models as defined:

$$SN \triangleq \{IF, LIF, PLIF, QIF, EIF\}. \tag{6}$$

Using a step size of 0.1 for the membrane potential threshold, we define the threshold set $SV \triangleq \{0.1, 0.2, \cdots, 1.9, 2.0\}$.

Let NSP_i be the Cartesian product of SN and SV for each spiking neuron layer. Thus, NSP_i can be expressed as $NSP_i = SN \times SV$. The search space, SSP, is computed by forming the Cartesian product of each NSP_i, corresponding to a specific SNN layer. Hence, SSP can be expressed as:

$$SSP = NSP_1 \times NSP_2 \cdots \times NSP_N. \tag{7}$$

4.2   Evaluation Strategy

We propose a novel evaluation strategy, Period-Based Spike Evaluation (PBSE), built on the Homogeneous Poisson Point Process, to capture spiking neuron temporal characteristics more effectively. While Mellor et al. [20] and Lopes et al. [17] focus on assessing network potential through neural layer activations, Kim et al. [10] examine spiking neuron output per timestep. However, these approaches overlook the crucial temporal dynamics in spiking neural networks.

We leverage the Homogeneous Poisson Process (HPP) to effectively model consistent random event occurrence scenarios within defined time intervals. Our utilization of HPP in this study focuses on characterizing the activation of spiking neurons across timesteps, providing a more adept capture of temporal intricacies in Spiking Neural Networks. We hypothesize that spiking neuron outputs within each layer conform to an HPP distribution, i.e., $n_{ik}^l(t) \sim \text{Poisson}(\lambda_{ik}^l(t))$, where $n_{ik}^l(t)$ denotes the activation state of the i-th neuron in the l-th spiking neural layer upon introducing the k-th data point from the batch data. These distributions are presumed mutually independent, and we represent t as a finite set, $t \in \{1, 2, \ldots, \text{TimeStep}\}$, with TimeStep as the maximum time steps.

To accurately gauge the variations in the outputs of data points within the layer, we estimate the expectation of the HPP in $[0, TimeStep]$ using the number of neuron activations. Then, the expectation of $n_{ik}^l(t)$ can be given by

$$E\left(n_{ik}^l(T)\right) = \lambda_{ik}^l T. \tag{8}$$

Subsequently, we use maximum likelihood estimation to ascertain the parameter $\hat{\lambda}$, an estimation of λ for the HPP. This is done under the condition that $T = TimeStep$. The process can be represented as follows:

$$\hat{\lambda}_{ik}^l = \frac{E\left(n_{ik}^l(T)\right)}{T}. \tag{9}$$

Therefore, the activation of the spiking neuron layer within the network can be characterized by the estimated parameter $\hat{\lambda}$ for each spiking neuron. We denote the activation vector at the l-th layer for the k-th data point in the input batch as c_k^l, while N_l represents the number of spiking neurons in the l-th layer. The definition of c_k^l is as follows:

$$c_k^l = \{\hat{\lambda}_{1k}^l, \hat{\lambda}_{2k}^l, \cdots, \hat{\lambda}_{N_l k}^l\}. \tag{10}$$

Upon computing the activation vector of the l-th layer for the k-th data point in the batch, we employ Manhattan distance as a metric to quantify the disparity in the activations of different data within the spiking neuron layer:

$$d_M\left(c_k^l, c_j^l\right) = \sum_{i=1}^{N_l} \left|\hat{\lambda}_{ik}^l - \hat{\lambda}_{ij}^l\right|. \tag{11}$$

Finally, we compute K_M^l across the complete timestep. This computation can be expressed as follows:

$$K_M^l = \begin{pmatrix} N_l - d_M\left(c_1^l, c_1^l\right) & \cdots & N_l - d_M\left(c_1^l, c_{N_b}^l\right) \\ \vdots & \ddots & \vdots \\ N_l - d_M\left(c_{N_b}^l, c_1^l\right) & \cdots & N_l - d_M\left(c_{N_b}^l, c_{N_b}^l\right) \end{pmatrix}, \tag{12}$$

where N_b denotes the batch size used in the search process. The evaluation metric s for the architecture candidate is then computed as follows:

$$s = log\left(det\left|\sum_{l=1}^{N_c} K_M^l\right|\right), \tag{13}$$

where N_c represents the number of spiking neuron layers.

4.3   Search Algorithm

MixedSNN employs the niche genetic algorithm [15] to navigate the search space SSP. This algorithm adeptly balances exploration and exploitation, upholding diversity within the population through fitness sharing and crowding distance mechanisms. These mechanisms preserve diversity and prevent premature convergence. By doing so, the algorithm facilitates comprehensive exploration of the solution space, leading to improved convergence towards high-quality solutions. This approach suits optimization of intricate problems requiring diversity preservation. In MixedSNN's framework, we select the top 10 spiking neural network architectures for training and to identify the optimal neural architecture. Algorithm 1 delineates the search algorithm employed in MixedSNN, with individuals represented within the population denoted as P_i and defined as:

$$P_i = \{Neuron, V_{\text{threshold}}\}. \tag{14}$$

Algorithm 1: MixedSNN Algorithm

Input: Population number N, Elite population number N_{ne}, Evolutionary Generations T, Cross probability P_c, Mutation probability P_m, Ratio Factor n_e, Fitness Function $PBSE(P(t))$, Manhattan $Manhadun(F_i)$, Penalty Value P_V, Architecture Population $P(t)$

Output: Spiking Neural Architecture

1 for t *in* 1 to T do
2   for i *in* 1 to N do
3          $p_i, p_j \in P(t)$;
4          $F \leftarrow PBSE(P(t))$;
5          $NP(t) \leftarrow$ Top N_{ne} Individuals;
6          $p_c \leftarrow random(0,1), p_m \leftarrow random(0,1)$;
7   if $p_c \leq P_c$ then
8              $Cross(p_i, p_j)$;
9   if $p_m \leq P_m$ then
10             $Mutation(p_i)$;
11   Calculate σ_n, σ_{th};
12     $(D_{neuron}, D_{threshold}) \leftarrow (\sigma_n/n_e, \sigma_{th}/n_e)$;
13   for j *in* 1 to $N + N_{ne}$ do
14         $p_j, p_k \in P(t) + NP(t)$;
15   if $|Manhadun(p_j) - Manhadun(p_k)| \leq (D_{neuron}, D_{threshold})$ then
16   if $F_i > F_j$ then
17                 $F_j \leftarrow F_j - P_V$;
18   else $F_i \leftarrow F_i - P_V$
19   Find a Spiking Neural Architecture;

The process of updating the niche radius can be described by

$$(D_{neuron}, D_{threshold}) = (\sigma_n/n_e, \sigma_{th}/n_e). \tag{15}$$

Here, D_{neuron} and $D_{threshold}$ designate the niche radius for neuron types and threshold coding. σ_n and σ_{th} represent the standard deviation of the Manhattan distance between the neuron coding and threshold coding. Additionally, n_e corresponds to the ratio factor. When the Manhattan distance, encoded by the neuron gene and the threshold gene of two individuals, falls below the neuron and threshold niche radius, it indicates that the individuals are in an overcrowded state. In such scenarios, a fitness penalty is imposed on individuals with lower fitness scores to influence crowding selection. The incorporation of the crowding mechanism enhances global search capabilities, fostering increased diversity within the population.

5 Experiments

5.1 Experimental Setups

Experiments are performed using the SNN implementation developed with the SpikingJelly [5] Python library. Our SNNs' performance is assessed on three datasets: CIFAR10 [11], CIFAR100 [11], and CIFAR10-DVS [13]. CIFAR10 contains 60,000 color images across 10 classes, while CIFAR100 provides 100 fine-grained classes. These datasets are established benchmarks, evaluating SNNs on intricate image classification and object recognition tasks. CIFAR10-DVS converts CIFAR10 static images into a spatio-temporal event-stream representation, enabling dynamic visual information benchmarking for real-time processing. These datasets promote fair SNN architecture and learning method comparisons, advancing effective SNN model development.

Six hyperparameters are utilized in our experiments. The first hyperparameter, denoted as τ, determines the membrane time constant, controlling the decay rate of the membrane potential and presynaptic inputs. We set τ to a value of 3/4 for moderate decay. Another hyperparameter, N_b, is assigned a value of 100, representing the number of training batches. Table 1 provides an overview of these hyperparameters, which significantly influence the training dynamics and can impact the models' performance and convergence. Careful selection and tuning of these hyperparameters are crucial for achieving optimal results when training SNNs on the given datasets.

Table 1. Hyperparameters for training our SNNs on three datasets.

Datasets	Learning Rate	Schedual	Epoch	Batch Size
CIFAR10	0.1	CosLR	1000	32
CIFAR100	0.1	CosLR	2000	128
CIFAR10-DVS	0.1	CosLR	1000	32

SNN Architectures. In our study, the experimental SNN adopts a VGG5-like architecture. The architecture is improved by incorporating a TDBN structure [34] instead of the conventional batch normalization module. Figure 2(a) shows the adapted VGG5 architecture, utilizing spiking neurons for image classification. It includes five convolutional layers comprising three normal cells and two linear cells. The normal cell (Fig. 2(b)) produces spikes based on input and internal state, while the linear cell (Fig. 2(c)) provides weighted sum outputs. The network's last layer is a softmax layer for estimating class probabilities.

The MixNeuron layer integrates specific spiking neuron types in each cell, forming the MixedVGG5 structure (Fig. 3). This integration of mixed neuron models within the MixNeuron layer is a pivotal element of MixedSNN, enhancing the neural network's performance and capabilities. The combinations of mixed neurons and thresholds are denoted as mixed neuron $MN = [EIF, EIF, IF, LIF, QIF]$ and variable threshold $VT = [0.5, 0.5, 0.3, 1.5, 1.1]$.

5.2 Performance Evaluation

Table 2 presents a performance comparison between the obtained spiking neural network architecture, MixedVGG5, and baseline SNN models on three benchmark datasets. Remarkably, MixedVGG5 achieves comparable performance to previous works despite its significantly shallow network and the use of small timesteps. Specifically, on the CIFAR10-DVS dataset, MixedVGG5 achieves a classification accuracy of 78.55% without using data augmentation methods, surpassing the state-of-the-art results. When compared to models such as ResNet19 [34], AutoSNN [22], VGGSNN (VGG11) [3], and VGG11 [21], our MixedVGG5 architecture exhibits significantly improved accuracy with a simple structure.

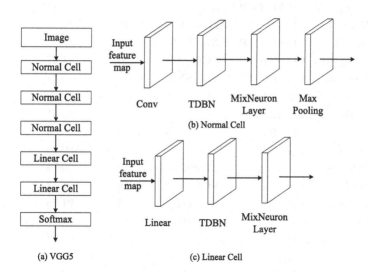

Fig. 2. The experimental SNN architecture. (a) VGG5-like SNN for image classification. (b) Normal cell structure. (c) Linear cell structure.

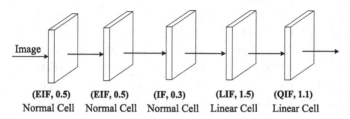

Fig. 3. The obtained spiking neural network architecture (MixedVGG5) resulting from the neural architecture search using MixedSNN

Furthermore, the experiments demonstrate that when considering networks with similar structures, MixedVGG5 achieves competitive classification accuracy on CIFAR-10 and CIFAR-100 datasets. For the CIFAR10 dataset, compared to Rathi [24], Wu [29], and Wu [30], MixedVGG5 demonstrates improvements of 3.94%, 1.81%, 0.80%, and 0.32%, respectively, with a timestep value of 4. Similarly, for the CIFAR100 dataset, compared to Lu (VGG15) [18] and Rathi (ResNet20) [23], MixedVGG5 with its shallow architecture improves the performance by 1.25% and 0.38%, respectively.

Table 2. Comparison of classification accuracy (%) achieved by MixedVGG5 and baseline SNNs on the CIFAR10, CIFAR100, and CIFAR10-DVS datasets.

Paper	Datasets	Architecture	Timesteps	Accuracy(%)
Wu [30]	CIFAR10-DVS	CIFARNet	12	60.5
Wu [29]	CIFAR10-DVS	SpikingCNN	8	65.59
Zheng [34]	CIFAR10-DVS	ResNet19	10	67.80
Wu [31]	CIFAR10-DVS	LIAF-Net	10	70.40
Na [22]	CIFAR10-DVS	AutoSNN	8	72.50
Meng [21]	CIFAR10-DVS	VGG-11	20	77.27
Deng [3]	CIFAR10-DVS	VGGSNN	10	77.33
Ours	CIFAR10-DVS	**MixedVGG5**	10	**78.55**
Rathi [24]	CIFAR10	VGG5	75	86.91
Wu [29]	CIFAR10	CIFARNet	8	89.04
Rathi [23]	CIFAR10	VGG6	10	90.05
Wu [30]	CIFAR10	CIFARNet	12	90.53
Rathi [24]	CIFAR10	VGG16	100	91.13
Rathi [23]	CIFAR10	ResNet20	5	91.78
kundu [12]	CIFAR10	VGG16	100	91.29
Zheng [34]	CIFAR10	ResNet19	2	92.34
Na [22]	CIFAR10	AutoSNN	8	93.15
Kim [10]	CIFAR10	SNASNet-Bw	8	93.64
Ours	CIFAR10	**MixedVGG5**	4	**90.85**
Lu [18]	CIFAR100	VGG15	62	63.20
Rathi [23]	CIFAR100	ResNet20	5	64.07
kundu [12]	CIFAR100	VGG11	120	64.98
Rathi [24]	CIFAR100	VGG11	125	65.52
Na [22]	CIFAR100	AutoSNN	8	69.16
Kim [10]	CIFAR100	SNASNet-Bw	5	73.04
Ours	CIFAR100	**MixedVGG5**	6	**64.45**

The MixedSNN framework applied to VGG5 has achieved state-of-the-art results on the CIFAR10-DVS dataset for event-based vision tasks. On the standard CIFAR10 and CIFAR100 image datasets, MixedVGG5 demonstrates competitive performance compared to other shallow spiking networks. However, its accuracy lags slightly behind a few deeper spiking models on these image datasets. Applying the MixedSNN methodology to deeper spiking architectures could help close this performance gap while maintaining efficient event-driven processing.

5.3 Ablation Experiments

Ablation on Validity of the Search Space in MixedSNN. We assessed 24 spiking neural network architectures to confirm the efficacy of the SSP search space. These architectures encompass mixed neurons (MN) and individual neurons ($SN = IF, LIF, PLIF, QIF, EIF$), coupled with variable thresholds (VT) and three common unified thresholds ($UV = 0.3, 0.5, 0.7$). Experiments were performed on CIFAR10 and CIFAR100 datasets using a timestep of 4, and on the CIFAR10-DVS dataset with a timestep of 10. The results are depicted in Fig. 4, highlighting the three highest accuracy bars.

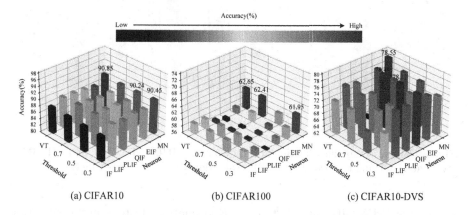

Fig. 4. Comparison of classification accuracy on CIFAR10, CIFAR100, and CIFAR10-DVS datasets using mixed neurons (MN) versus individual neuron types and variable thresholds (VT) versus fixed unified thresholds.

In CIFAR10 and CIFAR100 image classification experiments, we observe minimal impact on performance when varying threshold settings but keeping neuron combinations fixed. However, altering neuron combinations significantly affects accuracy with a consistent threshold. Using the mixed neurons (MN) architecture yields superior performance across different thresholds. In contrast, for the CIFAR10-DVS event-based task, changing thresholds greatly impacts accuracy despite fixed neuron combinations. The variable threshold (VT) neuron is not consistently optimal across neuron combinations. The quadratic integrate-and-fire (QIF) neuron performs well in CIFAR10-DVS across thresholds. The MN and VT architecture achieves top accuracy across all three datasets.

These results validate the MixedSNN framework, showing the network architecture's neuron types and thresholds significantly influence performance. This affirms the value of exploring diverse neuron behaviors and thresholds, spotlighting the importance of the proposed search space.

Ablation on Validity of the Evaluation Strategy in MixedSNN. To validate the effectiveness of the proposed PBSE evaluation strategy, we performed experiments comparing PBSE with the SAHA (Sparsity-Aware Hamming Distance) method proposed in [10]. The experimental results are described in Fig. 5. In the experiments on static datasets (CIFAR10, CIFAR100), as shown in Figs. 5(a) and (b), we evaluate architectures from both search strategies across timesteps of 2, 4, and 6. On CIFAR10, both strategies achieve similar accuracy, with our architecture reaching 90.93% at timestep 6. Accuracy increases overall with larger timesteps.

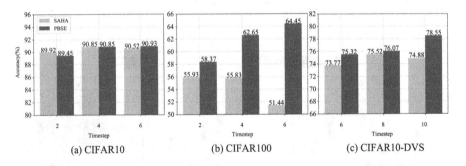

Fig. 5. Comparison of classification accuracy on CIFAR10, CIFAR100, and CIFAR10-DVS datasets using the proposed PBSE evaluation versus the SAHA method.

For the CIFAR100 dataset, the PBSE evaluation strategy performs better than SAHA. Our architecture's accuracy rises steadily with timestep, achieving 64.45% at timestep 6. The accuracy of our architecture at timestep = 2 and timestep = 6 is 58.37% and 64.45%, which are both higher than the best accuracy of SAHA. The gap between strategies is more pronounced on CIFAR100's task.

For the CIFAR10-DVS dataset, as shown in Figs. 5(c). Our architecture surpasses several state-of-the-art deep spiking networks. At timestep 6, our accuracy is 1.55% higher than the comparison. Our model achieves state-of-the-art 78.55% accuracy at timestep 10, exceeding the comparison by 3.03%. Our architecture attains better accuracy with fewer timesteps, validating the proposed PBSE evaluation approach.

6 Conclusion

This study presents MixedSNN, a dedicated neural architecture search framework specifically designed for spiking neural networks. While SNNs offer energy-efficient computation and effective temporal information processing, the optimization of SNNs through NAS has been relatively unexplored. MixedSNN addresses this gap by leveraging the unique characteristics of SNNs and introducing a novel search space called SSP, which investigates the impact of mixed spiking neurons and variable thresholds on network performance. Furthermore, a training-free evaluation strategy, known as Period-Based Spike Evaluation (PBSE), is proposed to capture the temporal features of SNNs by considering spike activation patterns. Experimental evaluations on CIFAR10, CIFAR100, and CIFAR10-DVS datasets demonstrate that MixedSNN achieves state-of-the-art performance with significantly reduced timesteps, demonstrating the effectiveness of the framework in optimizing SNN architectures. This study paves the way for further research and advancements in the design and optimization of SNNs using NAS techniques.

References

1. Abbott, L.F.: Lapicque's introduction of the integrate-and-fire model neuron (1907). Brain Res. Bull. **50**(5–6), 303–304 (1999)
2. Chen, W., Gong, X., Wang, Z.: Neural architecture search on ImageNet in four GPU hours: a theoretically inspired perspective. In: International Conference on Learning Representations (2021)
3. Deng, S., Li, Y., Zhang, S., Gu, S.: Temporal efficient training of spiking neural network via gradient re-weighting. In: International Conference on Learning Representations (2022)
4. Dong, X., Yang, Y.: NAS-BENCH-201: extending the scope of reproducible neural architecture search. In: 8th International Conference on Learning Representations, (ICLR), Addis Ababa, Ethiopia, 26–30 April 2020 (2020)
5. Fang, W., Chen, Y., Ding, J., Chen, D., Yu, Z., et al.: SpikingJelly (2020). https://github.com/fangwei123456/spikingjelly. Accessed 20 June 2023
6. Fang, W., Yu, Z., et al.: Incorporating learnable membrane time constant to enhance learning of spiking neural networks. In: Proceedings of the IEEE/CVF International Conference on Computer Vision, pp. 2661–2671 (2021)
7. Gerstner, W., Kistler, et al.: Neuronal Dynamics: From Single Neurons to Networks and Models of Cognition. Cambridge University Press (2014)
8. Guo, Y., Peng, W., Chen, Y., et al.: Joint A-SNN: joint training of artificial and spiking neural networks via self-distillation and weight factorization. Pattern Recogn. **142**, 109639 (2023)
9. He, K., Zhang, X., et al.: Deep residual learning for image recognition. In: Proceedings of the IEEE Conference on Computer Vision and Pattern Recognition, pp. 770–778 (2016)
10. Kim, Y., Li, Y., Park, H., et al.: Neural architecture search for spiking neural networks. In: Avidan, S., Brostow, G., Cissé, M., Farinella, G.M., Hassner, T. (eds.) Computer Vision-ECCV 2022: 17th European Conference, Tel Aviv, Israel, 23–27 October 2022, Proceedings, Part XXIV, pp. 36–56. Springer, Cham (2022). https://doi.org/10.1007/978-3-031-20053-3_3

11. Krizhevsky, A., Hinton, G.: Learning multiple layers of features from tiny images. Technical report 0, University of Toronto, Toronto, Ontario (2009)
12. Kundu, S., Datta, G., Pedram, M., Beerel, P.A.: Spike-thrift: towards energy-efficient deep spiking neural networks by limiting spiking activity via attention-guided compression. In: Proceedings of the IEEE/CVF Winter Conference on Applications of Computer Vision, pp. 3953–3962 (2021)
13. Li, H., Liu, H., Ji, X., et al.: CIFAR10-DVS: an event-stream dataset for object classification. Front. Neurosci. **11**, 309 (2017)
14. Li, W., Wen, S., Shi, K., Yang, Y., Huang, T.: Neural architecture search with a lightweight transformer for text-to-image synthesis. IEEE Trans. Netw. Sci. Eng. **9**(3), 1567–1576 (2022)
15. Li, X., Epitropakis, M.G., Deb, K., et al.: Seeking multiple solutions: an updated survey on niching methods and their applications. IEEE Trans. Evol. Comput. **21**(4), 518–538 (2016)
16. Liu, H., Simonyan, K., Yang, Y.: DARTS: differentiable architecture search. In: International Conference on Learning Representations (2019)
17. Lopes, V., Alirezazadeh, S., Alexandre, L.A.: EPE-NAS: efficient performance estimation without training for neural architecture search. In: Farkaš, I., Masulli, P., Otte, S., Wermter, S. (eds.) ICANN 2021. LNCS, vol. 12895, pp. 552–563. Springer, Cham (2021). https://doi.org/10.1007/978-3-030-86383-8_44
18. Lu, S., Sengupta, A.: Exploring the connection between binary and spiking neural networks. Front. Neurosci. **14**, 535 (2020)
19. Maass, W.: Networks of spiking neurons: the third generation of neural network models. Neural Netw. **10**(9), 1659–1671 (1997)
20. Mellor, J., Turner, J., et al.: Neural architecture search without training. In: International Conference on Machine Learning, pp. 7588–7598. PMLR (2021)
21. Meng, Q., Xiao, M., Yan, S., et al.: Training high-performance low-latency spiking neural networks by differentiation on spike representation. In: Proceedings of the IEEE/CVF Conference on Computer Vision and Pattern Recognition, pp. 12444–12453 (2022)
22. Na, B., Mok, J., Park, S., et al.: AutoSNN: towards energy-efficient spiking neural networks. In: International Conference on Machine Learning, pp. 16253–16269. PMLR (2022)
23. Rathi, N., Roy, K.: DIET-SNN: a low-latency spiking neural network with direct input encoding and leakage and threshold optimization. IEEE Trans. Neural Netw. Learn. Syst. (2021)
24. Rathi, N., Srinivasan, G., Panda, P., Roy, K.: Enabling deep spiking neural networks with hybrid conversion and spike timing dependent backpropagation. In: International Conference on Learning Representations (2020)
25. Simonyan, K., Zisserman, A.: Very deep convolutional networks for large-scale image recognition. In: 3rd International Conference on Learning Representations, (ICLR), San Diego, CA, USA, 7–9 May 2015 (2015)
26. Sun, Z., Lin, M., Sun, et al.: MAE-DET: revisiting maximum entropy principle in zero-shot NAS for efficient object detection. In: International Conference on Machine Learning, pp. 20810–20826. PMLR (2022)
27. Taherkhani, A., Belatreche, A., Li, Y., et al.: A review of learning in biologically plausible spiking neural networks. Neural Netw. **122**, 253–272 (2020)
28. Wang, Y., Zhang, M., Chen, Y., Qu, H.: Signed neuron with memory: towards simple, accurate and high-efficient ANN-SNN conversion. In: International Joint Conference on Artificial Intelligence (2022)

29. Wu, J., Chua, Y., Zhang, M., et al.: A tandem learning rule for effective training and rapid inference of deep spiking neural networks. IEEE Trans. Neural Netw. Learn. Syst. (2021)
30. Wu, Y., Deng, L., Li, G., Zhu, J., Xie, Y., Shi, L.: Direct training for spiking neural networks: faster, larger, better. In: Proceedings of the AAAI Conference on Artificial Intelligence, vol. 33, pp. 1311–1318 (2019)
31. Wu, Z., Zhang, H., et al.: LIAF-Net: leaky integrate and analog fire network for lightweight and efficient spatiotemporal information processing. IEEE Trans. Neural Netw. Learn. Syst. 33(11), 6249–6262 (2021)
32. Xu, J., Zhao, L., Lin, J., et al.: KNAS: green neural architecture search. In: International Conference on Machine Learning, pp. 11613–11625. PMLR (2021)
33. Ying, C., Klein, A., Christiansen, E., Real, E., Murphy, K., Hutter, F.: NAS-Bench-101: towards reproducible neural architecture search. In: International Conference on Machine Learning, pp. 7105–7114. PMLR (2019)
34. Zheng, H., Wu, Y., Deng, L., et al.: Going deeper with directly-trained larger spiking neural networks. In: Proceedings of the AAAI Conference on Artificial Intelligence, vol. 35, pp. 11062–11070 (2021)

Towards Scalable Feature Selection: An Evolutionary Multitask Algorithm Assisted by Transfer Learning Based Co-surrogate

Liangjiang Lin[1], Zefeng Chen[1(✉)], and Yuren Zhou[2]

[1] School of Artificial Intelligence, Sun Yat-sen University, Zhuhai, China
linlj23@mail2.sysu.edu.cn, chenzef5@mail.sysu.edu.cn
[2] School of Computer Science and Engineering, Sun Yat-sen University, Guangzhou, China
zhouyuren@mail.sysu.edu.cn

Abstract. When faced with large-instance datasets, existing feature selection methods based on evolutionary algorithms still face the challenge of high computational cost. To address this issue, this paper proposes a scalable evolutionary algorithm for feature selection on large-instance datasets, namely, transfer learning based co-surrogate assisted evolutionary multitask algorithm (cosEMT). Firstly, we tackle the feature selection on large-instance datasets via an evolutionary multitasking framework. The co-surrogate models are constructed to measure the similarity between each auxiliary task and main task, and the knowledge transfer between tasks is realized through instance-based transfer learning. Through the numerical relationship between the relative and absolute number of transferable instances, we propose a novel dynamic resource allocation strategy to make more efficient use of limited computational resources and accelerate evolutionary convergence. Meanwhile, an adaptive surrogate model update mechanism is proposed to balance the exploration and exploitation of the base optimizer embedded in the cosEMT framework. Finally, the proposed algorithm is compared with several state-of-the-art feature selection algorithms on twelve large-instance datasets. The experimental results show that the cosEMT framework can obtain significant acceleration in the convergence speed and high-quality solutions. All verify that cosEMT is a highly competitive method for feature selection on large-instance datasets.

Keywords: Feature Selection · Evolutionary Multitasking (EMT) · Surrogate · Transfer Learning · Resource Allocation

1 Introduction

Feature selection (FS) [1], as an important aspect of machine learning, is widely used in many real-world applications. In practice, feature selection methods can be classified into three categories: 1) filter-based; 2) wrapper-based; 3) embedded

B. Luo et al. (Eds.): ICONIP 2023, LNCS 14448, pp. 482–493, 2024.
https://doi.org/10.1007/978-981-99-8082-6_37

methods [2]. Considering that the wrapper-based methods can always achieve better classification accuracy while being applicable to any classifier, our study puts emphasis on the design of efficient wrapper-based methods for FS when the dataset is large (i.e., containing a large number of training instances). Among existing wrapper-based methods, evolutionary algorithms (EAs) are a type of powerful gradient-free optimization methods for FS. However, when faced with large-instance datasets, the repeated evaluations of candidate solutions in an evolved population require masses of computational cost.

At present, several effective algorithms to handle the scalability issue brought by large-instance datasets have been proposed, such as hardware acceleration techniques [4], distributed computing (such as using the MapReduce paradigm) [5] and the multitask evolutionary ML (MEML) algorithm with minions [3]. The former two approaches require advance hardware or plenty of hardware resources to support them so that they can achieve high efficiency. In contrast, the MEML does not depend on plenty of hardware, and it is the first work introducing the framework of evolutionary multitasking (EMT) [14] to deal with the FS on large-instance datasets. It referred to the auxiliary tasks that are assigned with small portions of the original dataset as *minions*, and the effect of minions has been demonstrated in the application of single-objective FS [3] and multiobjective hyper-parameter tuning of deep neural network models [6].

In this work, we continue to employ the idea of minions and concentrate on utilizing the surrogate model with instance-based transfer learning to further improve the efficiency of EAs applied to the FS on large-instance datasets. The main contributions of this work are summarized as follows.

1. For the FS on large-instance datasets, we propose the transfer learning based co-surrogate assisted evolutionary multitask algorithm (named cosEMT for short). In this algorithm, multiple auxiliary tasks with small training datasets created by subsampling are incorporated to effectively reduce the computational cost of the embedded EA and assist the evolutionary process of the main task through instance-based transfer learning.
2. A Gaussian process (GP) surrogate is built for the main task, and we respectively build a co-surrogate GP model between each auxiliary task and the main task. For making full use of computational resources, we use the numerical relationship of the surrogate models, namely, the absolute and relative number of the transferred instances, to measure the similarities between each auxiliary task and the main task, and then propose a novel dynamic computational resource allocation strategy.
3. We adopt an adaptive selection function to manage and update the surrogate models to improve the accuracy of the surrogate models.

The rest of this paper is organized as follows. In Sect. 2, we present the related background in the literature. In Sect. 3, we describe the details of our proposed algorithm. The results of our experiments are presented in Sect. 4. In Sect. 5, we conclude this work and present the potential directions for future research.

2   Preliminaries: Basics of Multitask Optimization

Consider that there are currently K optimization tasks ready to be executed at the same time. The jth task, denoted as T_j, has a search space χ_j and objective function $f_j : \chi_j \to \mathbb{R}$. Each task may also be subjected to several equality and/or inequality constraints.

In this context, the purpose of multitask optimization is to find a set of optimal solutions $\{x_1^*, x_2^*, \cdots, x_K^*\} = argmax\{f_1(x), f_2(x), \cdots, f_K(x)\}$, such that $x_j^* \in \chi_j$ and all constraints of T_j are satisfied. For an FS problem at hand, we can regard it as the main task, and generate several auxiliary tasks through subsampling. The main task and auxiliary tasks share the same search space for multitask optimization so that knowledge can be exchanged between tasks. As illustrated in Fig. 1, we are primarily concerned with solving K tasks, whereas the $T_1, T_2, \cdots, T_{K-1}$ serve as auxiliary tasks to assist the main task T_K.

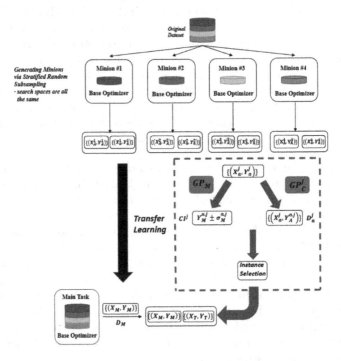

Fig. 1. Note that the computational cost of fitness evaluation on each auxiliary task is smaller than that of the main task, and the evolution of the main task can be assisted by instance-based transfer learning. For the symbols appeared in this figure, their meanings will be explained in Sect. 3.

3   Proposed CosEMT Framework

3.1   Co-surrogate Model for Instance-Based Transfer Learning

In this paper, we use Gaussian processes (GP) [7] as the surrogate model, because they not only provide a prediction for the objective function but also a confidence level of the prediction. Due to the difference in the size of training dataset, there is a time latency in the fitness evaluation of the population between the auxiliary tasks and the main task. To utilize the evaluation latency, we build a GP model for the main task, called GP_M, and respectively build a regression model for each auxiliary task, called co-surrogate model $GP_C^j, \forall j \leq K - 1$, to estimate the underlying relationship between each auxiliary task and the main task. We define $D_A^j = \{(X_A^j, Y_A^j)\}(\forall j \leq K - 1)^1$ as the database of the jth auxiliary task, $D_M = \{(X_M, Y_M)\}$ as the database of the GP_M model, and $D_C^j = \{(X_C^j, Y_C^j)\}(\forall j \leq K-1)$ as the database of the corresponding GP_C^j model, where the difference of population fitness between each auxiliary task and the main task, named $Y_C^j(\forall j \leq K - 1)$, can be calculated as follows:

$$Y_C^j = Y_M - Y_A^j, \forall j \leq K - 1 \tag{1}$$

For the same classification task, the training time complexity of Naive Bayes classifier is $O(n \times f)$, where n is the size of training dataset and f is the number of features. The difference between the auxiliary tasks and the main task lies in the number of the used training dataset n, which depends on the number of auxiliary tasks (according to the generation method of the training dataset for each auxiliary task introduced in Sect. 3.4). Therefore, the evaluation latency between each auxiliary task and the main task is defined as

$$\tau = K - 1 \tag{2}$$

where K is the number of all tasks. Therefore, when all tasks are processed in parallel, the main task trains GP_M by evaluating μ new solutions through a real objective function (will be analyzed in Sect. 3.3), while each auxiliary task can utilize the latency to evaluate additional $\mu \times (\tau - 1)$ solutions, which we define as $\{(X_a^j, Y_a^j)\}, \forall j \leq K - 1$. Therefore, the co-surrogate model GP_C^j can predict the solutions' fitness difference between each auxiliary task and the main task on X_a^j, named $Y_C^{a,j}(\forall j \leq K - 1)$, and then the synthetic values of predicted fitness for the main task on X_a^j are determined as follows:

$$Y_A^{a,j} = Y_C^{a,j} + Y_a^j, \forall j \leq K - 1 \tag{3}$$

Therefore, the auxiliary database for the main task is defined as $D_a^j = \{(X_a^j, Y_A^{a,j})\}, \forall j \leq K - 1$. However, not all instances belonging to D_a^j are suitable for transfer to the database of GP_M. The hypothesis we make here is that

---

[1] The database $\{(X, Y)\}$ is defined as a set, in which each element is represented in the form of (X, Y) where X denotes the decision vector of a solution and Y is the corresponding fitness.

the prediction given by the GP_C^j is unreliable if the predicted value is out of the boundary predicted by GP_M [8]. For the solutions X_a^j, we need to get the corresponding mean fitness value $Y_M^{a,j}$ ($\forall j \leq K - 1$), and the standard deviation $\sigma_M^{a,j}$ ($\forall j \leq K - 1$) through GP_M. Therefore, The lower and upper bounds of the confidence interval are respectively defined as

$$LCI^j = Y_M^{a,j} - \sigma_M^{a,j} \qquad (4)$$

$$UCI^j = Y_M^{a,j} + \sigma_M^{a,j} \qquad (5)$$

where $\forall j \leq K - 1$. For each $(X_a^j, Y_A^{a,j}) \in D_a^j$, only when $Y_A^{a,j}$ is in the range of the confidence interval, the corresponding $(X_a^j, Y_A^{a,j})$ will be considered reliable and stored in the transferable database $D_T^j = \{(X_T^j, Y_T^j)\}, \forall j \leq K - 1$.

3.2 Dynamic Resource Allocation Strategy

A dynamic resource allocation strategy is necessary to better exploit a potentially good fitness function's landscape of the auxiliary tasks to reduce the evaluation cost on the original dataset. Therefore, when the fitness function of an auxiliary task has a larger similarity to the main task, it should be allocated more computational resources. Intuitively, when the number of transferred instances of an auxiliary task occupies a larger proportion in D_T^j, we can consider that the fitness function landscape of the corresponding auxiliary task is more similar to the main task. With this insight, we define the comparison coefficient as

$$a_j = \frac{|D_T^j|}{\sum_{i=1}^{K-1} |D_T^i|}, \forall j \leq K - 1 \qquad (6)$$

where $|\cdot|$ represents the size of a database and $\sum_{j=1}^{K-1} a_j = 1$. However, it is incomplete to only consider the relative proportion of transferred instances' number, because when the number of transferable instances belonging to each auxiliary task is low, more resources should be allocated to the main task. Therefore, the measurement coefficient representing each auxiliary task's own meaningful instance-based transfer to the main task is defined as

$$b_j = \frac{|D_T^j|}{|D_a^j|}, \forall j \leq K - 1 \qquad (7)$$

where $0 \leq b_j \leq 1$. To further clarify, as the value of b_j approaches one, the number of transferable instances for the corresponding auxiliary task increases, which indicates that the auxiliary task performs more meaningful instance-based transfer, so it can be inferred that the auxiliary task has a greater similarity with the main task. Considering the influence of the above two coefficients, we propose the following formulas to execute the dynamic resource allocation strategy:

$$N_j = N_{total} \times a_j \times b_j, \forall j \leq K - 1 \qquad (8)$$

$$N_K = N_{total} - \sum_{j=1}^{K-1} N_j \qquad (9)$$

where N_j and N_K are the size of population assigned to each auxiliary task and the main task, respectively, after the dynamic resource allocation strategy is executed and N_{total} is the total population size across all tasks.

3.3   Mechanism of Surrogate Model Update

It is of paramount importance to select the right candidate solutions to be evaluated using the real fitness function so that the quality of the surrogate models can be improved. Here, new solutions are generated through the population update strategy of the base optimizer embedded in cosEMT framework. At the same time, μ new solutions are selected to be evaluated by the real fitness function of the main task through the following adaptive selection function [9]:

$$ASF(x) = (1 - \gamma) \times (\varphi_x/\varphi_{max}) + \gamma \times (\sigma_x/\sigma_{max}) \qquad (10)$$

The value of γ is set as follow:

$$\gamma = -0.5 \times \cos(\frac{FE}{FE_{max}} \times \pi) + 0.5 \qquad (11)$$

where φ_x and σ_x denote the mean and variance of the predicted fitness values of each new solution x, respectively. φ_{max} and σ_{max} represent the maximum values of φ_x and σ_x across all new solutions, respectively. FE denotes the current number of real fitness function evaluations, and FE_{max} represents the predefined maximum number of real fitness function evaluations. The main motivation behind the adaptive selection function (ASF) is to achieve relatively fast convergence in the early stage by assigning a large weight to the predicted fitness value, while more exploitative search is achieved in the later search stage. Specifically, by calculating the ASF score corresponding to each new solution, the results will be sorted in descending order. The top μ solutions will be evaluated by the true fitness function of the main task and added to the database of GP_M.

3.4   Summary of cosEMT Framework

The pseudocode of the cosEMT framework is presented in Algorithm 1. We assign non-overlapped training data to each auxiliary task by stratified sampling from the original dataset, denoted as $D_{train}^j, \forall j \leq K - 1$ and $D_{train} = \cup_{j=1}^{K-1} D_{train}^j$. At the start, i.e., $t = 0$, we obtain the databases of GP_M and $GP_C^j (\forall j \leq K - 1)$, and allocate equal computational resources to each task $T_j, \forall j \leq K$. At the initial stage of each iteration, we first update the GP surrogate models with the corresponding database. In every generation, we run the base optimizer embedded in cosEMT framework to generate new candidate solutions for each task. Then, the selection fitness of the candidate solutions obtained by the main task will be calculated by the ASF to select μ solutions which will be evaluated by the real fitness function of the main task and added to the database of the GP_M. While the main task performs expensive fitness evaluations, each auxiliary task can perform additional fitness evaluations to generate the auxiliary database. Through the calculation of the synthetic values and the execution of the instance selection strategy, the transferable instances are generated and added to the databases of GP_M and $GP_C^j (\forall j \leq K-1)$. Note that the operations conducted on each task (Lines 3–16) are executed in parallel.

Algorithm 1. Pseudocode of cosEMT

Input: $T = \{T_j\}_{j=1}^K$, $\mathcal{D} = \{D^j = D_{train}^j \cup D_{val}^j\}_{j=1}^K$, FE_{max}, μ, τ, Δt, N_{total}.
Output: Optimized solution(s) from the final population of T_K.
1: Initialization: Initialize the population for each task $N_j = N_{total}/K, \forall j \leq K$. Set $D_M = \{(X_M, Y_M)\}$, $D_C^j = \{(X_C^j, Y_C^j)\}$, $D_T^j = \varnothing$, $D_{Cnew}^j = \varnothing(\forall j \leq K-1)$, $FE = N_{total}$ and $t = 0$.
2: **while** $FE \leq FE_{max}$ **do**
3:    Train GP_M with the database D_M for the main task;
4:    Train GP_C^j with the database D_C^j for each auxiliary task;
5:    Run the base optimizer to generate new candidate solutions (to be evaluated by ASF) for the main task;
6:    Use ASF to determine μ solutions to be evaluated by the real fitness function of the main task T_K. Add the μ new solutions to the database of GP_M;
7:    **for** $j = 1 : K - 1$ **do**
8:       Run the base optimizer to generate and evaluate $\mu \times (\tau - 1)$ new candiate solutions $\{(X_a^j, Y_a^j)\}$ for the jth auxiliary task;
9:       Set $D_a^j = \{(X_a^j, Y_A^{a,j})\}$. Use GP_M to predict the mean fitness values $Y_M^{a,j}$ and the uncertainty $\sigma_M^{a,j}$ on X_a^j; Calculate the confidence interval using Eqs. (4) & (5);
10:       **for** each $(X_a^j, Y_A^{a,j}) \in D_a^j$ **do**
11:          **if** $Y_A^{a,j} \in [LCI^j, UCI^j]$ **then**
12:             The corresponding solution $(X_a^j, Y_A^{a,j})$ is transferable and saved in D_T^j;
13:          **end if**
14:       **end for**
15:       **for** each $(X_T^j, Y_T^j) \in D_T^j$ **do**
16:          Calculate Y_C^j on X_T^j using Eq. (1). Save (X_T^j, Y_C^j) into D_{Cnew}^j;
17:       **end for**
18:    **end for**
19:    **if** $mod(t, \Delta t) \neq 0$ **then**
20:       $D_M \leftarrow D_M, D_C^j \leftarrow D_C^j(\forall j \leq K-1)$;
21:    **else**
22:       $D_M \leftarrow D_M \cup D_T^j(\forall j \leq K-1), D_C^j \leftarrow D_C^j \cup D_{Cnew}^j(\forall j \leq K-1)$;
23:       Execute the dynamic computational resource allocation strategy;
24:       $D_T^j \leftarrow \varnothing, D_{Cnew}^j \leftarrow \varnothing$;
25:    **end if**
26:    Update $FE = FE + \mu, t = t + 1$;
27: **end while**
28: **return** the optimized solutions in D_M

The transfer interval Δt indicates the frequency at which instance-based transfer is conducted across tasks and t denotes the current generation count. When $mod(t, \Delta t) \neq 0$, the training databases of the GP surrogate models are consistent with the previous generation. When $mod(t, \Delta t) = 0$, the similarity between each auxiliary task and the main task will be measured so that the instanced-based transfer learning and the dynamic computational resource allocation strategy will be executed. The process iterates until the number of real fitness evaluations reaches FE_{max}.

Remark 1. Main novelty of our proposed cosEMT over MEML: MEML only realizes the soft knowledge exchange by calculating the mixed probability model, which *does not make full use of the difference of evaluation cost between the auxiliary tasks and the main task, resulting in the waste of computational resources.* However, cosEMT can make full use of the latency of evaluation cost and the numerical relationship obtained by the surrogate models to further capture the similarity between each auxiliary task and the main task, so as to utilize the source-domain instances.

4   Experimental Studies

4.1   Datasets for Experiments

12 datasets (all of their instances are greater than 10000) are selected from UCI data repository [10] and Kaggle [11] to conduct the experiments to verify the superiority of the proposed algorithm. The details of the datasets are listed in Table 1.

Table 1. Details of benchmark datasets.

No.	Dataset	No. of instance	No. of features
1	Electrical Grid Data	10000	14
2	Online Shoppers	12330	18
3	Dry Bean	13611	17
4	MAGIC Telescope	19020	11
5	Letter Recognition	20000	16
6	Avila	20867	10
7	Default of Credit Card	30000	24
8	Bank Marketing	45211	17
9	Secondary Mushroom	61069	21
10	Clickstream Data	165474	14
11	Diabetes Health	253680	22
12	Linkage Comparison	574913	12

4.2   Experimental Configuration

The Naive Bayes method is utilized as the classifier in this paper wherein each dataset is divided into the training set and the validation set with a splitting ratio of 80% and 20%. The rest of the algorithmic settings are specified as follows.

1) Solution Representation: Binary coded.
2) Total population size N_{total}: 100.

3) Parameters for cosEMT:
   a) Number of Auxiliary Tasks: 3;
   b) Number of New Samples for Updating GP μ: 10;
   c) Transfer Interval Δt: 2.
4) Evaluation Criterion: mean classification scores (Mean).

In this paper, several well-known algorithms are compared with cosEMT, which are binary genetic algorithm (BGA), sticky binary PSO (SBPSO) [12], and two variants of a recently proposed effective feature selection framework MEML [3], namely, MEML-BGA and MEML-SBPSO. For a clear demonstration of the superiority of our cosEMT framework, we choose BGA as the base optimizer used within cosEMT, and we denote the resultant as cosEMT-BGA. In our experiments, all comparative algorithms use the same classifier and the evaluation criterion as cosEMT in the evaluation step. The maximum number of real fitness evaluations FE_{max} is set to 4000, acting as the termination condition of each algorithm.

4.3   Results of Comparative Experiments

The results shown in Table 2 are the mean classification scores achieved by cosEMT and comparative algorithms for each comparative dataset over 30 runs. It can be seen from Table 2 that cosEMT-BGA achieves the best performance on most datasets except Letter Recognition, and the MEML framework is ranked second on most datasets under the Wilcoxon signed rank test with a confidence level of 95%, which shows the advancement of multitasking framework for knowledge transfer.

Table 2. Comparison of the mean performance and standard deviation of different comparative algorithms on different datasets. Result in dark and light grey backgrounds indicate the best and second-best results, respectively. Wilcoxon signed rank test with a confidence level of 95% is used. "$\star$" indicates the result is significantly outperformed by cosEMT-BGA. "$\dagger$" indicates the result is not significantly different from cosEMT-BGA. "$\ddagger$" indicates the result is significantly better than cosEMT-BGA.

Dataset	BGA	SBPSO	MEML-BGA	MEML-SBPSO	cosEMT-BGA
	Mean ± Std	Mean ± Std	Mean ± Std	Mean ± Std	Mean ± Std
Electrical Grid Data	74.07 ± 0.02 $\star$	74.21 ± 0.05 $\star$	74.47 ± 0.13 $\dagger$	74.60 ± 0.15 $\dagger$	74.62 ± 0.08
Online Shoppers	89.12 ± 0.98 $\star$	89.10 ± 1.17 $\star$	89.30 ± 0.89 $\dagger$	89.15 ± 0.35 $\star$	89.77 ± 0.65
Dry Bean	91.05 ± 0.11 $\star$	91.25 ± 0.74 $\star$	91.27 ± 0.56 $\star$	91.30 ± 0.16 $\star$	91.97 ± 0.19
MAGIC Telescope	76.85 ± 0.16 $\star$	77.17 ± 0.13 $\star$	77.15 ± 0.39 $\star$	77.17 ± 0.09 $\star$	77.84 ± 0.14
Letter Recognition	64.77 ± 0.22 $\star$	65.56 ± 0.10 $\dagger$	65.55 ± 0.21 $\dagger$	65.68 ± 1.55 $\dagger$	65.58 ± 0.59
Avila	87.04 ± 0.17 $\star$	89.68 ± 0.14 $\star$	89.62 ± 0.62 $\star$	91.74 ± 0.30 $\dagger$	91.92 ± 0.38
Default of Credit Card	81.27 ± 0.75 $\star$	81.29 ± 1.02 $\star$	82.02 ± 0.18 $\dagger$	81.44 ± 0.14 $\star$	82.34 ± 0.33
Bank Marketing	91.05 ± 0.19 $\star$	91.30 ± 0.17 $\star$	91.27 ± 0.52 $\star$	91.25 ± 0.73 $\star$	91.77 ± 0.14
Secondary Mushroom	65.63 ± 0.78 $\star$	66.36 ± 0.19 $\dagger$	66.23 ± 0.45 $\dagger$	66.48 ± 0.49 $\dagger$	66.49 ± 0.62
Clickstream Data	59.19 ± 1.24 $\star$	58.68 ± 0.33 $\star$	59.21 ± 0.49 $\star$	59.36 ± 0.87 $\dagger$	59.89 ± 0.51
Diabetes Health	85.86 ± 0.43 $\star$	85.94 ± 0.31 $\dagger$	85.94 ± 0.17 $\dagger$	86.02 ± 0.60 $\dagger$	86.04 ± 0.38
Linkage Comparison	99.29 ± 0.65 $\star$	99.36 ± 0.36 $\star$	99.44 ± 0.81 $\dagger$	99.42 ± 0.35 $\dagger$	99.67 ± 0.52

The convergence trends of the five comparison algorithms on the Default of Credit Card dataset and the Bank Marketing dataset are depicted in Fig. 2(a) and Fig. 2(b), respectively. We can see that cosEMT, MEML, and SBPSO perform better than the basic EC algorithm BGA on both datasets, and cosEMT obtains the fastest convergence and the optimal solution. We can observe that multitasking frameworks such as cosEMT and MEML have obtained different degrees of improvement compared to the corresponding single-task algorithms, which shows that the auxiliary tasks formed by stratified sampling can effectively reduce the computational cost and help the optimization process of the original large dataset through knowledge transfer. Since auxiliary tasks only produce a fraction of the original computational cost of evaluating the whole dataset, they provide an effective channel to explore the model configurations of the main task. At the same time, cosEMT is superior to MEML both in convergence speed and solution quality on most datasets, which suggests that the multitasking framework with the use of the surrogate model can effectively reduce the cost of fitness evaluation.

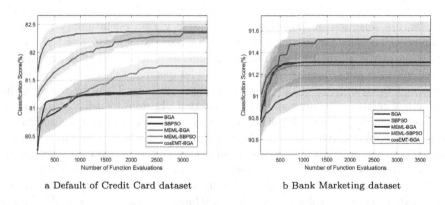

a Default of Credit Card dataset   b Bank Marketing dataset

Fig. 2. Comparison of the convergence trends of the five algorithms on two representative datasets.

In Fig. 3, we can observe that cosEMT-BGA can achieve the same performance level at less computational cost than BGA on each dataset (each dataset is represented by an alphabetical abbreviation). Up to 54.9% reduction in function evaluation calls can be achieved with cosEMT in the evaluation savings. This further demonstrates the potential of the proposed cosEMT framework to reduce the evaluation cost on large-instance datasets.

In practice, using the cosEMT framework can reduce the time required for baseline EAs by several hours or even days, which can greatly improve the efficiency of solving feature selection on large-instance datasets, even other complex optimization problems involving large-instance data in the field of machine learning [13, 15, 16].

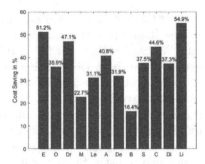

Fig. 3. Computational cost savings by using cosEMT framework to achieve the same performance level as BGA (given 4000 function evaluations) on each dataset.

5   Conclusion

In this paper, we investigate how to efficiently apply evolutionary algorithms to solve the feature selection problem on large-instance datasets. First, we tackle the feature selection on large-instance datasets via an evolutionary multitask framework so that the auxiliary tasks with small dataset can assist the optimization of the original large dataset through instance-based transfer learning when all tasks are processed in parallel. And an instance selection mechanism is proposed to ensure the quality of transferable solutions. Then, by utilizing the numerical relationship between the relative and absolute transferred instances, a novel dynamic resource allocation strategy is proposed to better utilize limited computational resources and an adaptive surrogate model update mechanism is adopted to better balance the exploration and exploitation of the base optimizer.

A series of experiments have been carried out to showcase the influence of each parameter and component, and the efficacy of cosEMT. Compared with several state-of-the-art evolutionary FS algorithms, our proposed cosEMT framework can consume less computational cost to obtain better solutions on most large-instance datasets. In particular, up to 54.9% reduction in function evaluation calls can be achieved while achieving high-quality solutions.

Although our work has achieved good results, there are still several topics for further improvement, for example, the embedding of more powerful optimizers. In addition, expanding the Gaussian process model suitable for high-dimensional search space will be another focus of our future research.

Acknowledgements. This research is supported by the National Natural Science Foundation of China under Grant 62206313 and 62232008.

References

1. Xue, B., Zhang, M., Browne, W.N.: Particle swarm optimization for feature selection in classification: a multi-objective approach. IEEE Trans. Cybern. **43**(6), 1656–1671 (2013)

2. Xue, Y., Xue, B., Zhang, M.: Self-adaptive particle swarm optimization for large-scale feature selection in classification. ACM Trans. Knowl. Discovery Data (TKDD) **13**(5) (2019)

3. Zhang, N., Gupta, A., Chen, Z., Ong, Y.S.: Evolutionary machine learning with minions: a case study in feature selection. IEEE Trans. Evol. Comput. **26**(1), 130–144 (2022)

4. Franco, M.A., Krasnogor, N., Bacardit, J.: Speeding up the evaluation of evolutionary learning systems using GPGPUs. In: Genetic and Evolutionary Computation Conference, GECCO 2010, Proceedings, Portland, Oregon, USA, 7–11 July 2010 (2010)

5. Verma, A., Llorà, X., Goldberg, D.E., Campbell, R.H.: Scaling genetic algorithms using MapReduce. In: International Conference on Intelligent Systems Design and Applications ISDA 2009 (2009)

6. Chen, Z., Gupta, A., Zhou, L., Ong, Y.S.: Scaling multiobjective evolution to large data with minions: a Bayes-informed multitask approach. IEEE Trans. Cybern., 1–14 (2022)

7. Matheron, G.: Principles of geostatistics. Econ. Geol. **58**(8), 1246–1266 (1963)

8. Wang, X., Jin, Y., Schmitt, S., Olhofer, M.: Transfer learning based co-surrogate assisted evolutionary bi-objective optimization for objectives with non-uniform evaluation times. Evol. Comput. **30**(2), 221–251 (2022)

9. Wang, X., Jin, Y., Schmitt, S., Olhofer, M.: An adaptive Bayesian approach to surrogate-assisted evolutionary multi-objective optimization. Inf. Sci. **519**, 317–331 (2020)

10. Bache, K., Lichman, M.: UCI machine learning repository (2013)

11. Kaggle dataset website. https://www.kaggle.com/datasets/alexteboul/diabetes-health-indicators-dataset

12. Nguyen, B.H., Xue, B., Andreae, P.: A novel binary particle swarm optimization algorithm and its applications on knapsack and feature selection problems. Intell. Evol. Syst. (2017)

13. Telikani, A., Tahmassebi, A., Banzhaf, W., Gandomi, A.H.: Evolutionary machine learning: a survey. ACM Comput. Surv. (CSUR) **54**(8), 1–35 (2021)

14. Gupta, A., Ong, Y.S., Feng, L.: Multifactorial evolution: toward evolutionary multitasking. IEEE Trans. Evol. Comput. **20**(3), 343–357 (2016)

15. Li, P., Tang, H., Hao, J., Zheng, Y., Fu, X., Meng, Z.: ERL-RE^2: efficient evolutionary reinforcement learning with shared state representation and individual policy representation. arXiv preprint arXiv:2210.17375 (2022)

16. Li, P., Hao, J., Tang, H., Zheng, Y., Fu, X.: Race: improve multi-agent reinforcement learning with representation asymmetry and collaborative evolution. In: International Conference on Machine Learning, PMLR 2023, pp. 19490–19503 (2023)

CAS-NN: A Robust Cascade Neural Network Without Compromising Clean Accuracy

Zhuohuang Chen[1], Zhimin He[1(⊠)], Yan Zhou[1], Patrick P. K. Chan[2(⊠)], Fei Zhang[3], and Haozhen Situ[4]

[1] School of Electronic and Information Engineering, Foshan University, Foshan 528000, China
`zhmihe@gmail.com`
[2] Shien-Ming Wu School of Intelligent Engineering, South China University of Technology, Guangzhou 510006, China
`patrickchan.scut@gmail.com`
[3] College of Computer and Information Engineering, Henan Normal University, Xinxiang 453007, China
[4] College of Mathematics and Informatics, South China Agricultural University, Guangzhou 510642, China

Abstract. Adversarial training has emerged as a prominent approach for training robust classifiers. However, recent researches indicate that adversarial training inevitably results in a decline in a classifier's accuracy on clean (natural) data. Robustness is at odds with clean accuracy due to the inherent tension between the objectives of adversarial robustness and standard generalization. Training a single classifier that combines high adversarial robustness and high clean accuracy appears to be an insurmountable challenge. This paper proposes a straightforward strategy to bridge the gap between robustness and clean accuracy. Inspired by the idea underlying dynamic neural networks, *i.e.,* adaptive inference, we propose a robust cascade framework that integrates a standard classifier and a robust classifier. The cascade neural network dynamically classifies clean and adversarial samples using distinct classifiers based on the confidence score of each input sample. As deep neural networks suffer from serious overconfident problems on adversarial samples, we propose an effective confidence calibration algorithm for the standard classifier, enabling accurate confidence scores for adversarial samples. The experiments demonstrate that the proposed cascade neural network increases the clean accuracies by 10.1%, 14.67%, and 9.11% compared to the advanced adversarial training (HAT) on CIFAR10, CIFAR100, and Tiny ImageNet while keeping similar robust accuracies.

Keywords: Adversarial training · Adversarial learning

1 Introduction

Deep neural networks (DNNs) are notoriously vulnerable to small and imperceptible perturbations in the input data known as *adversarial examples* [4,12,16].

B. Luo et al. (Eds.): ICONIP 2023, LNCS 14448, pp. 494–505, 2024.
https://doi.org/10.1007/978-981-99-8082-6_38

Adversarial training has been demonstrated to be the most effective approach to improving the accuracy on adversarial examples *robust accuracy* [4]. However, adversarial training leads to an undesirable reduction in natural unperturbed inputs (*clean accuracy*) [13,15]. There is an inherent tension between the goals of adversarial robustness and standard generalization due to the fundamental disparities in the features learned through standard training and robust training [13].

Subsequently, several vanilla adversarial training strategies have been proposed to achieve a better trade-off between clean accuracy and robust accuracy [8,9,14,15]. However, the goals of adversarial robustness are fundamentally at odds with standard generalization [13,15], Consequently, it is challenging and even impossible to train a single classifier to attain high adversarial robustness without compromising clean accuracy. Despite these recent advances, closing the gap between clean accuracy and robust accuracy remains an open challenge.

Dynamic neural network (NN) is an emerging research topic due to its high efficiency and adaptability [6]. It adapts model structures to allocate appropriate computation according to different inputs during inference, thereby reducing redundant computation on simple samples. Inspired by the concept of adaptive inference in dynamic NN, we propose a robust cascade neural network (CAS-NN) to alleviate the gap between clean accuracy and adversarial robustness. Rather than employing a single classifier, CAS-NN cascades two classifiers, *i.e.,* a standard classifier for clean samples and a robust classifier for adversarial samples. After obtaining the confidence score of the standard classifier on the input sample, early exiting occurs if the confidence score surpasses a predefined threshold; otherwise, the robust classifier is activated and makes the final decision. CAS-NN employs confidence-based criteria to determine the adopted classifier. The overconfidence problem of deep neural networks on adversarial samples [4,7] is an obstacle of such decision paradigm as adversarial samples obtain high confidences, leading to early exiting. In this paper, we address this problem by designing a confidence calibration on the standard classifier. The main contributions of this paper are as follows.

- we propose a cascade framework to alleviate the gap between clean accuracy and adversarial robustness. It can dynamically handle clean samples and adversarial samples using a standard model and a robust model.
- we propose an effective calibration method that provides low confidence scores for adversarial samples. The calibrated standard neural network outputs predictive scores that accurately reflect the actual likelihood of correctness.
- experiments on CIFAR10, CIFAR100, and Tiny ImageNet demonstrate that CAS-NN achieves comparable robust accuracy to state-of-the-art adversarial training algorithms while causing minimal or no decline in the clean accuracy.

2   Related Work

A variety of works have attempted to alleviate the gap between clean accuracy and adversarial robustness. An effective strategy is dual-domain training whose loss function integrates both clean and adversarial objectives. Zhang *et al.* [15] characterized the trade-off by decomposing the robust error into the sum of the natural error and the boundary error. They proposed TRADES, which minimized the cross-entropy loss of clean samples and a KL divergence term that encourages smoothness in the neighborhood of samples. Rade *et al.* [8] observed that adversarial training resulted in a large margin along certain adversarial directions, leading to a significant drop in clean accuracy. They extended the objective function of TRADES by incorporating an additional helper loss, calculated using helper examples with proper labels in order to reduce the excessive margin. However, combining the optimization tasks of both clean and adversarial objectives into a single model and enforcing complete matching of the feature distributions between adversarial and clean examples may result in sub-optimal solutions.

More recent attempts to enhance the trade-off between clean accuracy and adversarial robustness have primarily focused on refining adversarial training by early-stopping [9], additional data [10] and weight initialization for training with large perturbations [11]. Despite these advancements, the problem remains far from being solved, with a substantial gap between clean accuracy and robustness observed in practical image datasets.

3   Preliminaries

Given a classification task with c classes, $f_\theta : \mathcal{X} \to \mathcal{Y}$ represent a deep neural network parameterized by θ, where $\mathcal{X} \subseteq \mathbb{R}^d$ and $\mathcal{Y} \subseteq \mathbb{R}^c$ are the input and output space. The predicted label of an input sample x is given by $\arg\max_k f_\theta(x)_k$, where $f_\theta(x)_k$ is the k-th component of $f_\theta(x)$. The confidence score of an input sample x is defined as the maximum softmax prediction probability, *i.e.*, $s(x) = \max_k f_\theta(x)_k$. Adversarial training is formulated as

$$\min_\theta \mathbb{E}_{x \in \mathcal{D}} (\max_{x' \in \mathcal{B}_\epsilon(x)} \ell(f_\theta(x'), y), \tag{1}$$

where $\mathcal{D} = \{(x_i, y_i)\}_{i=1}^n$ is the training set. $\mathcal{B}_\epsilon(x)$ denotes the l_p-norm ball centered at x with radius ϵ, which has been extensively employed in recent studies [7,14,15]. The inner maximization problem aims to generate an adversarial sample x' with respect to the training sample x. Existing adversarial training algorithms approximately optimize the inner maximization problem using a gradient-based iterative solver, *e.g.*, multi-step PGD [7] and CW attacks [2].

Following the terminologies in Refs. [8,15], the *clean* (or *natural*) and *robust* (or *adversarial*) accuracies refer to the test accuracies on clean samples and adversarial samples, respectively.

4   Robust Cascade Neural Network

We propose a robust cascade neural network (CAS-NN) that combines two neural networks (NNs) to achieve both high standard accuracy and robust accuracy.

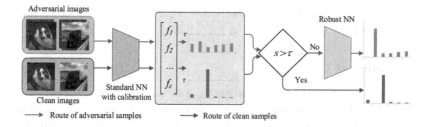

Route of adversarial samples   Route of clean samples

Fig. 1. Schematic of the robust cascade system.

Specifically, the standard NN is used to classify clean samples, while the robust NN is employed to classify adversarial samples. The route is determined by the confidence score of the input sample, which is measured by the standard NN. The confidence score is defined as the maximum softmax prediction probability of the standard NN. The diagram of the robust cascade system is illustrated in Fig. 1. Given a sample x, the standard NN first makes a prediction and outputs the confidence score $s(x) = \max_k f_\theta(x)_k$. If the confidence score surpasses the threshold τ, the input sample is likely to be untainted. Then the inference terminates and the system outputs the prediction. If $s \leq \tau$, the standard NN is not confident of the prediction and the input sample is likely to be contaminated, prompting the activation of the robust NN to deliver the final prediction.

4.1   Confidence Calibration

DNNs suffer from a serious overconfident problem on adversarial samples [4, 7]. CAS-NN cannot solely depend on the confidence score of the standard NN to determine the activation of the robust NN. It is necessary to calibrate the confidence score of the standard NN on adversarial samples.

The confidence score should be indicative of the actual likelihood of correctness. Since the standard NN's predictions on adversarial samples are incorrect, it should output low confidence scores on these samples. Based on the low confidence scores, adversarial samples will be passed to the robust NN for the final decision. Since the confidence score is defined as the maximum softmax prediction probability, a uniform distribution of outputs results in the lowest confidence score. Therefore, we employ the Kullback-Leibler (KL) divergence to encourage the softmax prediction probability of the standard NN on adversarial samples to approximate a uniform distribution. This approach ensures that the standard NN produces the lowest confidence score for adversarial samples. Given a training set $\mathcal{D} = \{(x_i, y_i)\}_{i=1}^n$ with n samples, the loss function used to calibrate the confidence score of standard NN is defined as

$$\mathcal{L}_c = \mathbb{E}_{x \in \mathcal{D}}(\mathrm{KL}(P_u || f_\theta(x^{(s)}) + \mathrm{KL}(P_u || f_\theta(x^{(r)})), \tag{2}$$

where KL denotes the KL divergence, which quantifies the difference between two distributions. $x^{(s)}$ and $x^{(r)}$ refer to the adversarial samples generated by an adversarial attack against the standard NN and the robust NN, respectively.

In this paper, we use the PGD attack [7]. P_u represents a discrete uniform distribution, where $P_u(X = k) = 1/c$, $k \in \{1, 2, ..., c\}$.

4.2   Training of CAS-NN

CAS-NN comprises two neural networks: a standard NN and a robust NN. The robust NN can be trained using any adversarial training algorithm, making our extension compatible with state-of-the-art (SOTA) adversarial training techniques. During the training of a standard NN, confidence regularization should be taken into account, resulting in the loss defined as

$$\mathcal{L} = \mathcal{L}_{std} + \lambda \mathcal{L}_c, \tag{3}$$

where $\mathcal{L}_{std}$ is the original loss term of a standard NN. In this paper, we use the softmax cross entropy, which ensures a high confidence score for the clean sample. λ serves as a parameter that controls the trade-off between the original loss and the confidence regularization term. Although the training of the standard NN involves adversarial samples, we call it a standard NN because it does not incorporate the labels of adversarial samples during training.

Algorithm 1 outlines the pseudo-code for the training process of CAS-NN. We first train a robust NN f_{θ_r} on the training dataset with an adversarial training algorithm. It is important to emphasize that various adversarial training algorithms can be employed to train the robust NN. In this paper, we use a standard adversarial training [7]. The calibrated standard NN f_{θ_s} is trained by stochastic gradient descent. In each epoch, a mini-batch of samples is randomly selected, and their corresponding adversarial samples are generated through PGD attack against f_{θ_s} and f_{θ_r} as shown in Lines 8 and 9. The parameters of the calibrated standard NN f_θ are optimized based on Eq. (3) (Line 10). $\mathcal{B}_\epsilon(x)$ is the closed ball with radius ϵ centered at x, where $\mathcal{B}_\epsilon(x) = \{x' | \parallel x - x' \parallel_\infty \leq \epsilon\}$. $\prod_{\mathcal{B}_\epsilon(x)}$ denotes the projection function that maps the adversarial variant back to the ϵ-ball centered at x.

5   Adaptive Attack Against CAS-NN

Since CAS-NN is composed of two neural networks, the intuitive attack is that we generate adversarial samples targeted on the standard NN or the robust NN. However, neither of these strategies exploits the structure and mechanism of CAS-NN. In this paper, we propose an adaptive attack against CAS-NN in order to better evaluate the robustness of CAS-NN. The detailed process is shown in Algorithm 2. The adaptive attack is a gradient-based method that utilizes the model's gradient to search for the perturbation direction. In each iteration, the gradient information is calculated with respect to different models based on the confidence score of the adversarial sample $s(x')$. Specifically, if the confidence score exceeds the threshold τ, the gradient is calculated based on the robust NN f_{θ_r}; otherwise, the gradient is obtained based on the standard NN f_{θ_s}. $Attack(f_\theta, x')$ represents one-step update of the adversarial sample x'

Algorithm 1. Training of a robust cascade neural network.

Require: $\mathcal{D}=\{(\boldsymbol{x}_i,y_i)\}_{i=1}^n$: training data; m: batch size; η: learning rate; ϵ: attack radius; α: attack step size; λ: the trade-off parameter between the original loss and the confidence regularization term. K: number of attack iterations.

Ensure: the cascade NN with a calibrated standard NN (f_{θ_s}) and a robust NN (f_{θ_r})

1: train a robust NN f_{θ_r} via an adversarial training algorithm
2: randomly initial the parameters $\boldsymbol{\theta}_s$ of the standard NN f_{θ_s}
3: **repeat**
4:     Sample a mini-batch of samples $\mathcal{D}_b = \{\boldsymbol{x}_j, y_j\}_{j=1}^m$ from $\mathcal{D}$
5:     **for** $j = 1, ..., m$ **do**
6:         $\boldsymbol{x}_j^{(s)}$ and $\boldsymbol{x}_j^{(r)}$ are initialized by a uniformly random point in the l_p-norm ball centered at $\boldsymbol{x}$ with radius ϵ
7:         **for** $k = 1, ..., K$ **do**
8:             $\boldsymbol{x}_j^{(s)} = \prod_{\mathcal{B}_\epsilon}(\boldsymbol{x}_j^{(s)} + \alpha\mathrm{sign}(\nabla_{\boldsymbol{x}_j^{(s)}}\ell(f_{\theta_s}(\boldsymbol{x}_j^{(s)}), y_j)))$
9:             $\boldsymbol{x}_j^{(r)} = \prod_{\mathcal{B}_\epsilon}(\boldsymbol{x}_j^{(r)} + \alpha\mathrm{sign}(\nabla_{\boldsymbol{x}_j^{(r)}}\ell(f_{\theta_r}(\boldsymbol{x}_j^{(r)}), y_j)))$
10:    $\boldsymbol{\theta}_s = \boldsymbol{\theta}_s - \eta\nabla_{\boldsymbol{\theta}_s}\mathbb{E}_{(\boldsymbol{x}_j,y_j)\sim\mathcal{D}_b}(\mathrm{CE}(f_{\theta_s}(\boldsymbol{x}_j), y_j) + \lambda\mathrm{KL}(P_u||f_{\theta_s}(\boldsymbol{x}_j^{(s)})) + \lambda\mathrm{KL}(P_u||f_{\theta_s}(\boldsymbol{x}_j^{(r)})))$
11: **until** training completed

using the gradient calculated by the model f_θ. Any gradient-based attack such as PGD, CW, and MM can be utilized to update the adversarial sample.

Algorithm 2. Adaptive attack against the robust cascade neural network.

Require: $(\boldsymbol{x}, y)$: target sample; f_{θ_s}: the calibrated standard NN; f_{θ_r}: the robust NN; τ: the threshold of confidence score; K: number of attack iterations.

Ensure: adversarial sample $\boldsymbol{x}'$

1: $\boldsymbol{x}'$ is initialized by a uniformly random point in the l_p-norm ball centered at $\boldsymbol{x}$
2: **for** $i = 0, ..., K - 1$ **do**
3:     $s(\boldsymbol{x}') = \max_k f_{\theta_s}(\boldsymbol{x}')_k,$
4:     **if** $s(\boldsymbol{x}') > \tau$ **then** $\boldsymbol{x}' \leftarrow Attack(f_{\theta_s}, \boldsymbol{x}')$
5:     **else** $\boldsymbol{x}' \leftarrow Attack(f_{\theta_r}, \boldsymbol{x}')$
    return $\boldsymbol{x}'$

6   Experiments

Datasets. We extensively assess the proposed method on three datasets, *i.e.*, CIFAR10, CIFAR100, and Tiny ImageNet. Both CIFAR-10 and CIFAR-100 contain 50,000 training images and 10,000 test images that are in the size of 32×32. CIFAR-10 covers 10 classes while CIFAR-100 consists of 100 classes. Tiny ImageNet is a more complex dataset, which is a subset of ImageNet. It comprises 200 classes with each class containing 500 training images and 50 validation images that are in the size of 64×64.

Evaluation of Adversarial Robustness. Robust accuracy is evaluated using three adversarial attacks: PGD [7], CW [2], and Minimum-Margin(MM) attacks [3]. PGD is a reliable and computationally efficient method for assessing the robustness of a model. CW exhibits strong attack performance by introducing an alternative loss. Both PGD and CW employ 20 steps. The MM attack searches for the optimal adversarial sample with the minimum margin. It achieves comparable performance to AutoAttack while requiring only 3% of the computational time. The target selection numbers of the MM attack are set to 3 and 9. As CAS-NN comprises two neural networks, we employ three strategies to attack CAS-NN: adversarial samples generated according to the standard (Std) NN, the robust (Rob) NN, and the adaptive (Ada) path in Sect. 5. The attack radius ε is 8/255 for all attacks.

Implementation Details. The calibrated standard NN is trained according to Algorithm 1. The robust NN in CAS-NN is trained with adversarial training [7], where the adversarial samples are generated by PGD20 with l_∞ norm. Both the standard NN and the robust NN share the same architecture and are optimized by an SGD optimizer with 0.9 momentum. The learning rate and weight decay are 0.02 and 5×10^{-4}, respectively. We use NF-ResNet18 [1] as the backbone of the neural network. The thresholds τ of CAS-NN are set to 0.2, 0.1, and 0.1 for the CIFAR10, CIFAR100, and Tiny ImageNet, respectively. The trade-off parameter λ between the original loss and the confidence regularization term is set to 1. The batch size m, attack step size α, and the number of attack iterations K are 256, 2/255 and 10, respectively. Each experiment is independently run five times, and the average result is reported.

6.1   Comparisons with Prominent Adversarial Training Algorithms

Table 1 reports the clean accuracy and robust accuracy of CAS-NN and some prominent adversarial training algorithms. A high clean accuracy and robust accuracy indicate the model can well classify clean samples and adversarial samples, respectively. It can be observed that current adversarial training algorithms exhibit a significant decrease in clean accuracy compared to the standard training method (ST). Specifically, AT, TRADES, and HAT exhibit an average degradation of 14.21%, 13.44%, and 12.54% across three datasets, respectively. However, CAS-NN achieves comparable clean accuracy to ST on CIFAR10 and CIFAR100, with only a decrease of 3.23% on Tiny ImageNet.

We present the robust accuracies of CAS-NN against adversarial attacks based on the standard (Std) NN, robust (Rob) NN, and adaptive (Ada) path. CAS-NN achieves high robust accuracies on adversarial samples generated based on the standard NN. The result is intuitive as the adversarial samples are classified by the robust NN which can well classify adversarial samples generated based on the standard NN. In the case of PGD and CW attacks, CAS-NN has similar robust accuracies on adversarial samples generated based on the robust NN and the adaptive path. However, CAS-NN exhibits 11.71 to 19.96 % lower robust accuracies against the adaptive MM attack compared to the one targeted

Table 1. Clean accuracy (%) and robust accuracy (%) of different robust classifiers on CIFAR10, CIFAR100 and Tiny ImageNet. Robust accuracy is measured by different attacks, including PGD20 [7], CW20 [2], MM3 [3] and MM9 [3]. ST denotes standard training. ↑ suggests that higher values correspond to better performance.

Datasets	Classifiers	Clean acc(↑)	Robust acc (↑) PGD20			CW20			MM3			MM9		
CIFAR10	ST	92.47	0.02			0.01			0.00			0.00		
	AT [7]	81.58	46.73			45.78			41.82			40.85		
	TRADES [15]	80.90	48.10			44.61			42.43			42.33		
	HAT [8]	82.61	46.38			43.22			41.19			41.13		
	CAS-NN	92.71	Std	Rob	Ada	Std	Rob	Ada	Std	Rob	Ada	Std	Rob	Ada
			81.50	46.73	46.73	81.49	45.80	45.78	78.59	55.38	43.67	78.43	54.47	40.54
CIFAR100	ST	73.33	7.44			0.00			0.00			0.00		
	AT [7]	56.30	24.86			23.79			21.32			20.58		
	TRADES [15]	56.83	25.14			21.67			19.29			18.84		
	HAT [8]	57.91	24.57			21.14			19.42			19.38		
	CAS-NN	72.58	Std	Rob	Ada	Std	Rob	Ada	Std	Rob	Ada	Std	Rob	Ada
			56.12	24.86	24.86	56.06	23.81	23.80	50.23	42.52	24.73	49.90	41.83	21.87
Tiny ImageNet	ST	60.56	8.04			0.00			0.00			0.00		
	AT [7]	45.84	18.95			16.72			14.38			13.99		
	TRADES [15]	48.30	18.73			14.42			12.33			11.91		
	HAT [8]	48.22	17.01			13.25			11.54			11.27		
	CAS-NN	57.33	Std	Rob	Ada	Std	Rob	Ada	Std	Rob	Ada	Std	Rob	Ada
			45.75	18.95	18.86	45.40	16.72	16.72	39.10	31.70	16.94	38.75	31.32	15.03

on the robust NN. The adversarial attack with an adaptive path proves to be the most effective among all the datasets. In the subsequent experiments, we only consider the adaptive attack as it is the most effective attack against CAS-NN. CAS-NN exhibits comparable or even higher robust accuracies than AT, TRADES, and HAT across four adversarial attacks (PGD20, CW20, MM3, and MM9). This indicates that CAS-NN can notably increase clean accuracy without compromising robust accuracy.

6.2 Performance of Confidence Calibration

A well-calibrated classifier ensures that the average confidence of a model aligns closely with its actual accuracy. The effectiveness of confidence calibration is evaluated using the Expected Calibration Error (ECE) [5], which measures how closely the average confidence of a model aligns with its actual accuracy. A better-calibrated classifier exhibits a lower ECE value. Table 2 shows the ECE values of the standard NN and the calibrated NN on adversarial samples. The standard NN has very high ECE values (greater than 89%) on adversarial samples. However, after confidence calibration, the average values of ECE decrease to 2.26%, 0.62%, and 1.92% in CIFAR10, CIFAR100, and Tiny ImageNet, respectively. The calibrated NNs achieve ECE values lower than 3.22% across all datasets.

After confidence calibration, the standard NN outputs low confidence scores for adversarial samples, accurately reflecting their actual accuracy. The majority of adversarial samples have confidence scores below the threshold, triggering the

Table 2. Expected Calibration Error (ECE) of the proposed confidence calibration on adversarial samples generated by PGD, CW, MM3, and MM9

Attack	CIFAR10		CIFAR100		Tiny ImageNet	
	Standard NN	Calibrated NN	Standard NN	Calibrated NN	Standard NN	Calibrated NN
PGD	98.15	2.71	89.33	0.75	91.85	1.50
CW	97.22	3.22	97.80	0.75	92.82	1.76
MM3	98.07	1.47	99.04	0.48	92.82	2.41
MM9	98.49	1.63	99.05	0.49	92.20	2.01

Table 3. Percentage of samples (%) that are classified by the robust NN.

Datasets	Clean	PGD	CW	MM3	MM9
CIFAR10	0.00	100.00	99.98	99.93	99.88
CIFAR100	0.00	100.00	99.93	99.83	99.74
Tiny ImageNet	0.87	100.00	99.90	99.90	99.88

activation of the robust classifier to make a final decision for these samples. Table 3 shows the percentage of samples whose confident scores given by the calibrated standard NN fall below the threshold. The majority of adversarial samples generated by PGD, CW, MM3, and MM9 fall below the threshold and are consequently transferred to the robust NN. Almost all clean samples have confidence scores exceeding the threshold, resulting in their classification by the standard NN.

6.3   Margin Along the Initial Adversarial Direction

Previous research [8] demonstrates that adversarial training leads to an excessive increment in the margin along initial adversarial directions to attain robustness, which prevents the neural network from using high discriminative features in those directions, leading to a substantial drop in clean accuracy. As described in Ref. [8], there exists a connection between the augmentation of the margin along initial adversarial directions and the reduction in clean accuracy. In contrast to adversarial training, the calibrated standard NN in CAS-NN does not require precise classification of adversarial samples, and its objective function does not involve the one-hot label vectors of the adversarial sample. Consequently, the confidence calibration has minimal impact on the margins between the training samples and the decision boundary.

Figure 2 shows the margin between clean samples and the decision boundary along initial adversarial directions. The horizontal axes denote the margins between clean samples and the decision boundary of the standard NN while the vertical axes are the margins between these samples and the decision boundaries of the adversarially trained NN (the upper row) and the calibrated standard NN (the bottom row). The standard NN typically exhibits margins ranging from 0 to

2, with an average margin of 1.36. However, after adversarial training, the margins have a substantial increment, resulting in an average margin of 11.56, which is 8.5 times greater than that of the standard NN. For the calibrated standard NN, there is either no or only a slight increment in the margin across different datasets. The average margin across the three datasets is 1.55, suggesting that confidence calibration minimally alters the geometry of the decision boundary, thereby exerting little effect on clean accuracy.

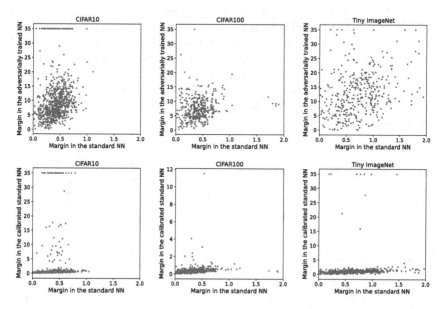

Fig. 2. Margins between clean samples and the decision boundaries of standard NN, adversarially trained NN and calibrated standard NN along initial adversarial directions. Each dot denotes a clean sample.

6.4 Analysis of the Hyper-parameter

Figure 3 illustrates the accuracies of CAS-NN on clean samples and adversarial samples as well as the pass rates for different values of the hyper-parameter τ. We are only interested in the pass rate of adversarial samples that are misclassified by the standard NN. With an increasing threshold, a greater number of samples exhibit confidence scores below the threshold, resulting in an increased pass rate. When the threshold is 1.0, all samples are classified by the robust NN, resulting in CAS-NN functioning as the traditional adversarial training algorithm. We can observe a significant degradation of clean accuracy in this setting. If the threshold is set to 0, all samples are classified by the standard NN, rendering CAS-NN comparable to the classifier with standard training. In this setting, CAS-NN has the highest clean accuracy but the lowest robust accuracy.

For CIFAR10, the confidence scores of the majority of clean samples surpass 0.6, whereas the confidence scores of all adversarial samples fall below 0.2. Despite a small fraction of clean samples being classified by the robust NN with thresholds of 0.6, 0.7, 0.8, or 0.9, the clean accuracy has no degradation as the robust NN accurately classifies these samples. CAS-NN attains identical clean accuracy and robust accuracy when the threshold is within the range of $[0.2, 0.9]$. However in the more challenging tasks such as CIFAR100 and Tiny ImageNet, a higher threshold substantially amplifies the pass rate of clean samples, consequently resulting in a more pronounced decline in clean accuracy. When the thresholds are 0.2, 0.1, and 0.1 for CIFAR10, CIFAR100, and Tiny ImageNet, respectively, the pass rates of adversarial samples reach 1.0, suggesting that the confidence scores of adversarial samples are significantly reduced through confidence calibration. Based on the aforementioned observations, the pass rate can guide the setting of the threshold. Under the premise of ensuring a high pass rate of adversarial samples, the threshold should be set as small as possible to achieve a high clean accuracy.

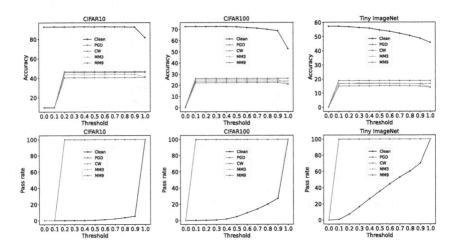

Fig. 3. Accuracies (%) of CAS-NN on clean samples and adversarial samples generated by PGD, CW, MM3, and MM9 as well as the pass rates (%) for different thresholds τ.

7   Conclusion

Considering the inherent contradiction between the adversarial robustness and standard generalization, we propose a novel strategy called CAS-NN, to bridge the gap between robustness and standard accuracy. CAS-NN dynamically classifies clean samples and adversarial samples with different neural networks according to the confidence scores of these samples. In order to mitigate the overconfidence problem in adversarial samples, we designed a confidence calibration algorithm. Experimental results on CIFAR10, CIFAR100, and Tiny ImageNet show that CAS-NN achieves comparable or slightly lower clean accuracy compared to the standard neural network while attaining similar robust accuracy as existing adversarial training algorithms. In future work, we will develop more strategies based on adaptive inference to construct a safety-critical system.

Acknowledgements. This work is supported by Guangdong Basic and Applied Basic Research Foundation (Nos. 2022A1515140116, 2022A1515010101), National Natural Science Foundation of China (Nos. 61972091, 61802061).

References

1. Brock, A., De, S., Smith, S.L.: Characterizing signal propagation to close the performance gap in unnormalized resnets. In: ICLR (2021)
2. Carlini, N., Wagner, D.: Towards evaluating the robustness of neural networks. In: S&P, pp. 39–57 (2017)
3. Gao, R., et al.: Fast and reliable evaluation of adversarial robustness with minimum-margin attack. In: ICML, pp. 7144–7163 (2022)
4. Goodfellow, I.J., Shlens, J., Szegedy, C.: Explaining and harnessing adversarial examples. In: ICLR (2015)
5. Guo, C., Pleiss, G., Sun, Y., Weinberger, K.Q.: On calibration of modern neural networks. In: ICML, pp. 1321–1330 (2017)
6. Han, Y., Huang, G., Song, S., Yang, L., Wang, H., Wang, Y.: Dynamic neural networks: a survey. TPAMI **44**(11), 7436–7456 (2021)
7. Madry, A., Makelov, A., Schmidt, L., Tsipras, D., Vladu, A.: Towards deep learning models resistant to adversarial attacks. In: ICLR (2018)
8. Rade, R., Moosavi-Dezfooli, S.M.: Reducing excessive margin to achieve a better accuracy vs. robustness trade-off. In: ICLR (2022)
9. Rice, L., Wong, E., Kolter, Z.: Overfitting in adversarially robust deep learning. In: ICML, pp. 8093–8104 (2020)
10. Sehwag, V., et al.: Robust learning meets generative models: can proxy distributions improve adversarial robustness? In: ICLR (2022)
11. Shaeiri, A., Nobahari, R., Rohban, M.H.: Towards deep learning models resistant to large perturbations. arXiv:2003.13370 (2020)
12. Szegedy, C., et al.: Intriguing properties of neural networks. In: ICLR (2014)
13. Tsipras, D., Santurkar, S., Engstrom, L., Turner, A., Madry, A.: Robustness may be at odds with accuracy. In: ICLR (2018)
14. Wang, Y., Zou, D., Yi, J., Bailey, J., Ma, X., Gu, Q.: Improving adversarial robustness requires revisiting misclassified examples. In: ICLR (2020)
15. Zhang, H., Yu, Y., Jiao, J., Xing, E.P., Ghaoui, L.E., Jordan, M.I.: Theoretically principled trade-off between robustness and accuracy. In: ICML, pp. 7472–7482 (2019)
16. Zheng, J., Chan, P.P., Chi, H., He, Z.: A concealed poisoning attack to reduce deep neural networks' robustness against adversarial samples. Inf. Sci. **615**, 758–773 (2022)

Multi-scale Information Fusion Combined with Residual Attention for Text Detection

Wenxiu Zhao and Changlei Dongye[✉]

Shandong University of Science and Technology, Qingdao, China
dycl@sdust.edu.cn

Abstract. Driven by deep learning and neural networks, text detection technology has made further developments. Due to the complexity and diversity of scene text, detecting text of arbitrary shapes has become a challenging task. Previous segmentation-based text detection methods can hardly solve the problem of missed detection in complexity scene text detection. In this paper, we propose a text detection model that combines residual attention with a multi-scale information fusion structure to effectively capture text information in natural scenes and avoid text omission. Specifically, the multi-scale information fusion structure extracts text features from different levels to achieve better text localisation and facilitate the fusion of text information. At the same time, residual attention is combined with features from high-resolution images to enhance the contextual information of the text and avoid text omission. Finally, text instances are obtained by a binarisation method. The proposed model is very helpful for text detection in complex scenes. Experiments conducted on three public benchmark datasets show that the model achieves state-of-the-art performance.

Keywords: text detection · deep learning · residual attention · multi-scale information fusion

1 Introduction

Text detection is one of the most challenging computer vision tasks, with potential applications in areas such as autonomous driving, scene understanding, and machine translation. With the continuous development of object detection and instance segmentation algorithms, significant progress has also been made in the field of text detection. Recent research has shifted its focus from traditional horizontal or multi-directional text detection to more challenging scene text detection with arbitrary shapes. Compared to multi-directional text detection, scene text detection with arbitrary shapes has a broader application scope. However, significant challenges are still posed by arbitrary shape text detection due to the diversity in text shape, size, angle, and complex background.

Compared to text detection algorithms based on manually labeled features, deep learning-based text detection algorithms have the ability to automatically

B. Luo et al. (Eds.): ICONIP 2023, LNCS 14448, pp. 506–518, 2024.
https://doi.org/10.1007/978-981-99-8082-6_39

extract deep semantic features from images, making them more capable of generalization and becoming the mainstream algorithm for text detection. Currently, text detection methods can be classified into two major categories: segmentation-based text detection algorithms and regression-based text detection algorithms.

Segmentation-based text detection models typically combine pixel-level predictions with post-processing algorithms to obtain bounding boxes. Zhang et al. detected multi-directional text using a semantic segmentation algorithm [1], while Liao et al. proposed a novel architecture capable of detecting and recognizing text instances of arbitrary shapes [2]. PixelLink [3] has been used to detect small text by connecting pixels to form text boxes using pixel-level connectivity. PAN [4]uses pixel clustering algorithms to predict pixel clusters in the center of text cores and uses clustering algorithms in post-processing to reconstruct text instances. PSENet [5] is based on the breadth-first search algorithm (BFS) and uses a gradually increasing scale expansion algorithm to successfully predict adjacent text instances. CRAFT [6] implementation allows for the localization of individual character regions, which are then connected to form a single text instance. DBNet [7] conducts adaptive binarization on each pixel and utilises network learning to acquire the binarization threshold. The network output directly obtains the text box, which streamlines the post-processing process. PRPN [8] employs a directional pooling progressive region prediction network to predict the probability distribution of text regions. TextPMs [9] uses the sigmoid alpha function to describe the probability distribution of text region pixels and aggregates text instances using region growth.

To filter potential boxes, deep convolutional networks were employed as strong predictors in some early vertex regression-based methods [10]. A text box was formed by connecting the different tiny text fragments identified by CTPN [11] with RNN to form a larger text section. TextBoxes [12], an end-to-end rapid scene text detector, only used standard non-maximum suppression as post-processing. In 2018, the authors improved the algorithm to identify text in any direction [13]. DMPNet [14] was a brand-new convolutional neural network-based method for identifying text with smaller quadrilaterals. EAST [15] predicted the values of each pixel and regressed the distance of each pixel to the four sides using a U-shaped structure. DeRPN [16] was a candidate area extraction technique based on dimension decomposition. It could decouple the target's breadth and height to split the detection dimension, minimizing the effect of target form changes on detection. TextBPN [17] suggested an adaptive boundary model that iteratively deformed text areas using graph neural networks and recurrent neural networks. DRRG [18] initially divided text instances into text components, and then utilized graph convolutional networks to deduce the relationships between text components and adjacent components.

However, these methods do not take into account the complexity and large text aspect ratios of text images resulting in text omissions (See Fig. 1). To address this issue, we propose a novel text detection network that combines residual attention with a multi-scale information fusion structure. By leveraging these components, our model effectively captures comprehensive textual information in natural scenes and mitigates text omissions. The main contributions of this paper are summarized as follows:

1. We introduce a text detection model that integrates residual attention with a multi-scale information fusion structure. Experiments demonstrate that this new method significantly enhances text detection in complex scenes on three public benchmark datasets.
2. The multi-scale information fusion structure plays a crucial role in our proposed model by effectively extracting text features from various levels within an image. This fusion of information leads to improved text localization and facilitates the integration of textual details.
3. To overcome text omissions in challenging scenarios, we designed a residual attention module. By leveraging high-resolution image features, this module captures regions of interest and enhances the contextual information of the text, effectively reducing text omissions.

Fig. 1. Some examples of complex scenes and large text aspect ratios.

2 Method

2.1 Network Structure

The proposed network model is shown in Fig. 2. The model consists of four parts, including the CNN backbone, multi-scale feature fusion structure, residual attention module, and binary output head. The extraction and fusion network is composed of ResNet and a multi-scale feature fusion structure. The multi-scale feature fusion structure combines feature maps from various levels through channel concatenation and pixel-wise addition to acquire the spatial information of text. Before pixel-wise addition, the residual attention module is embedded in the output branch with the highest resolution to enhance text features. Finally, the signal binary map is obtained by predicting the text region using a binarization formula.

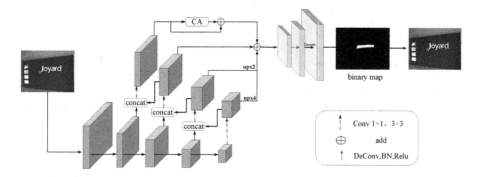

Fig. 2. Overview of the proposed network structure.

2.2   Multi-scale Information Fusion Structure

The semantic information contained in different levels of semantic feature maps varies. The low-level large-scale feature map is used to locate the position information of the text, while the high-level small-scale feature map focuses on the boundary information of the text. Therefore, how to fuse different levels of semantic information to obtain the complete text area becomes a key issue. A multi-scale information fusion structure is proposed to combine different feature maps and generate feature maps containing rich text information. In the multi-scale information fusion structure, as shown in Fig. 3, there are two operations: feature expansion and feature fusion. Feature expansion concatenates two same-sized feature maps to generate a deeper feature map. For text detection tasks, this approach ensures that different text feature information is fully retained in the channels, avoiding missed text features. The up-sampling and element-wise addition operations are used to achieve feature fusion of the expanded feature map, which is beneficial for better classifying text information in the feature map. $C2$ to $C5$ represent the feature maps extracted by the network, and $P2$ to $P5$ represent the information after feature expansion.

2.3   Residual Attention Module

Text images in complex scenes are difficult to distinguish between foreground and background information, and most of the text has a large aspect ratio, which makes it difficult to locate text information in images. Therefore, we predicted a weight for the highest resolution output layer to obtain useful details in the high-resolution feature map. Specifically, we have introduced a residual attention module and integrated it into the output branch of the highest resolution after feature expansion. The structure of the residual attention module is shown in Fig. 4. $C^{AVG} \in R^{C \times 1 \times 1}$ is obtained by calculating the global average pooling (AvgPool) and the shared fully connected layer of the feature map $F \in R^{C \times H \times W}$, while $C^{MAX} \in R^{C \times 1 \times 1}$ is obtained by calculating the global maximum pooling (MaxPool) and the shared fully connected layer of the feature

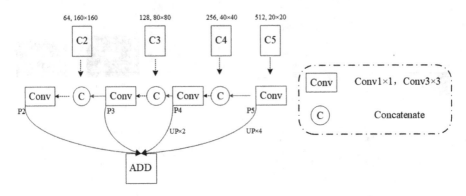

Fig. 3. Multi-scale information structure.

map $F \in R^{C \times H \times W}$. Then, the weight vectors of these two channels are merged and weighted to obtain the channel attention vector $C \in R^{C \times 1 \times 1}$. The channel attention vector $C \in R^{C \times 1 \times 1}$ is multiplied pixel-wise with $F \in R^{C \times H \times W}$, and the resulting product is added to $F \in R^{C \times H \times W}$ to obtain the final output $F' \in R^{C \times H \times W}$. The formula for the residual attention module is shown below as Eq. (1):

$$F' = \left(\sigma\left(MLP\left(C^{AVG}\right) + MLP\left(C^{AVG}\right)\right) \times F\right) + F \tag{1}$$

where σ represents the sigmoid function, MLP represents the multilayer perceptron, containing 1×1 convolution and Relu functions.

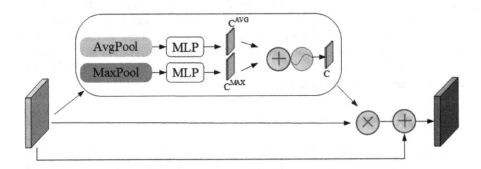

Fig. 4. Residual attention module.

2.4 Loss Function

The loss function L of the model is mainly divided into two parts, L_c and L_d. L_c uses the binary cross-entropy loss function, and L_d uses Dice loss. The weight coefficient α was set to 5. In order to balance positive and negative samples and prevent negative samples from affecting the gradient, we used hard negative mining for both BCE loss and Dice loss. The formula is as follows:

$$L = L_d + \alpha \times L_c \tag{2}$$

$$L_d = 1 - 2\frac{\sum_{i \in S_l} y_i \times x_i}{\sum_{i \in S_l} |y_i| + \sum_{i \in S_l} |x_i| + \varepsilon} \tag{3}$$

$$L_c = \sum_{i \in S_l} (y_i \log x_i + (1 - y_i) \log (1 - x_i)) \tag{4}$$

where x_i is the predicted value and y_i indicates the true value, S_l is a sampled dataset with a ratio of positive and negative samples of 1:3.

3   Analysis of Experimental Results

3.1   Dataset

MSRA-TD500 [19]contains 500 natural images, the training set contains 300 images randomly selected from the original dataset, and the remaining 200 images constitute the testset.

ICDAR 2015 [20] has been publicly collected and published as a benchmark for evaluating text detection algorithms to track the latest advances in the field of natural image Chinese this inspection, especially the detection of arbitrary text. ICDAR 2015 contains 1000 training images and 500 test images, a total of 4500 words, and provides corresponding real labels, text labels are labeled by word position.

MLT-2017 [21] is a dataset containing multiple language types, which is used to verify the robustness of the model, especially for different types of text detection tasks. It contains 7200 training set data and 1800 test set data.

3.2   Experimental Details

The model used ResNet50 as the feature extraction network. To verify the practicality of multiscale information fusion in lightweight networks, ResNet18 was also used for experimental analysis. Due to the distinct characteristics of different datasets, deformable convolution [22] was employed on ResNet18 and ResNet50 networks on the MSRA-TD500 dataset. The learning rate was set to 0.00175, and the batch_size was set to 8. During training, the images were cropped to 640×640 and data augmentation methods such as random cropping and random rotation were applied.

We primarily employed inference speed testing as the key metric to assess the computational efficiency of our model. During the inference phase, we resized input images, adjusting the height to suit each dataset. The model was deployed on an RTX3060 GPU and an Intel(R) Core(TM) i5-10400F CPU. For the inference speed test, we used a batch size of 1.

3.3   Ablation Experiments and Analysis

We conducted an ablation study to assess the effectiveness of the proposed components. Table 1 demonstrate the ablation study results of the model using ResNet18 as the feature extraction network on the MSRA-TD500 datasets. The baseline model data was obtained by re-implementing the DB algorithm. As shown by the quantitative results in Table 1, even with the baseline model, good performance can still be achieved.

Effectiveness of multi-scale information fusion structure. Compared to the baseline model, the Baseline_MF model uses a multi-scale information fusion structure in the feature fusion stage, which can fully utilize multi-scale semantic information. This module can better predict text features and improve model performance.

Effectiveness of residual attention module. The Baseline_RA model adds a residual attention module on top of the baseline model. The data in Table 1 show that a simple residual attention module can improve overall performance metrics.

Table 1. Ablation results of the model on the MSRA-TD500 dataset. (Unit: %).

Method	Precision	Recall	F-measure
Baseline	90.4	76.3	82.8
Baseline_MF	86.5	80.4	83.3
Baseline_RA	88.0	78.7	83.6
Ours	90.7	79.2	84.5

3.4   Comparison with State-of-the-Arts

Table 2 shows the results of the quantitative evaluation of the three datasets and a comparison with other SOTA models. The evaluation results of the model on the MLT-2017 dataset. For testing we set the height to 736. Using the ResNet18-based model, precision, recall, and F-measure improved by 1.5%, 0.9%, and 1.2%, respectively, compared to DBNet. Compared to PAN[5], precision improved by 3.4%, while recall decreased by 5.1% and F-measure decreased by 0.5%. Using the ResNet50-based model, precision improved by 4.9%, while recall decreased by 0.9%, and F-measure improved by 1.3% compared to DBNet. Based on the quantitative results, the detection performance of the proposed model is significantly better than previous models. It can be observed from the table that even using the shallow ResNet18, our detection results outperform the use of the deep ResNet50 and the PSE with complex post-processing. In terms of frames per second (FPS), we can observe that our model is faster than previous methods on the MLT-2017 dataset while ensuring better accuracy.

Table 2. Comparison of this model with other SOTA models. (Unit:%) (The best results are highlighted in red, with the second-best in blue.)

	MLT-2017				ICDAR 2015				MSRA-TD500			
	Precision	Recall	F-measure	FPS	Precision	Recall	F-measure	FPS	Precision	Recall	F-measure	FPS
PAN[5]	80	69.8	73.4	–	–	–	–	–	–	–	–	–
PSE[6]	73.8	68.2	70.9	30	86.9	84.5	85.6	1.6	–	–	–	–
TextSnake[8]	–	–	–	–	84.9	80.4	82.6	1.1	83.2	73.9	78.3	1.1
PRPN[10]	–	–	–	–	88.5	83.7	86.0	–	85.2	84.9	84.7	–
TextPMs[11]	81.7	61.5	72.5	–	88.5	84.3	86.1	–	90.9	85.3	86.7	8.3
TextBoxes++[14]	–	–	–	–	87.2	76.7	81.7	11.6	84.8	79.6	80.6	–
EAST[16]	–	–	–	–	83.6	73.5	78.2	13.2	–	–	–	–
TextBPN[18]	–	–	–	–	90.4	84.0	87.7	–	86.0	84.5	85.2	13
DBNet-resnet18	81.9	63.8	71.7	41	87.7	77.5	82.3	47	90.4	76.3	82.8	62
DBNet-resnet50	83.1	67.9	74.7	19	91.3	80.3	85.4	12	91.5	79.2	84.9	32
DBNet++-resnet18	–	–	–	–	90.1	77.2	83.1	44	90.1	77.2	83.1	55
DBNet++-resnet50	–	–	–	–	90.9	83.9	87.3	10	91.5	83.3	87.3	29
ours-resnet18	83.4	64.7	72.9	44	87.6	79.6	83.4	46	90.7	79.2	84.5	65
ours-resnet50	88.0	67.0	76.0	22	90.2	82.2	86.2	14	91.7	82.4	86.9	30

The evaluation results of the model on the ICDAR2015 dataset. We set the image height to 736 when testing with the ResNet18-based model, and we adjusted the image height to 1152 when testing with the ResNet50-based model. DBNet++ is a model that was built on top of the DBNet by adding an adaptive feature fusion module that uses spatial attention mechanism to learn text information. The following experimental results demonstrate the performance of the model based on different backbone networks (ResNet18, ResNet50). Compared to DBNet, the model's performance has been enhanced by utilizing both backbone networks. While this resulted in a slight reduction in frames per second (FPS), it led to a notable increase in overall accuracy. Compared with DBNet, the model's performance has been improved using both backbone networks. Compared with the SOTA model DBNet++, our model performs well with ResNet18 as the feature extraction network. The model based on ResNet50 as the feature extraction network performs better than PRPN with an increase in precision of 1.7%, a decrease in recall of 1.5%, and an increase in F-measure of 0.2%. According to the quantitative experimental results, due to the complexity of the ICDAR2015 dataset, which includes dense text and partially blurred small targets, the detection difficulty is high, and the model's advantages are not significant.

The evaluation results of our study on the MSRA-TD500 dataset. For testing we set the height to 736. Compared with DBNet, using the ResNet18-based model improved precision by 0.3%, recall by 2.9%, and F-measure by 1.7%. However, using the ResNet50-based model, precision decreased by 0.8%, recall increased by 2.0%, and F-measure increased by 0.6%. Compared with the recent method TextBPN, using the ResNet50-based model improved precision by 4.3%, recall decreased by 3.3%, and F-measure increased by 0.5%. Importantly, our ResNet50-based model achieved this improved performance while maintaining a superior computing speed compared to TextBPN, highlighting its efficiency. It is worth noting that the MSRA-TD500 dataset contains text of multiple orientations and arbitrary sizes, and our model demonstrated good performance in handling such diverse text scenes.

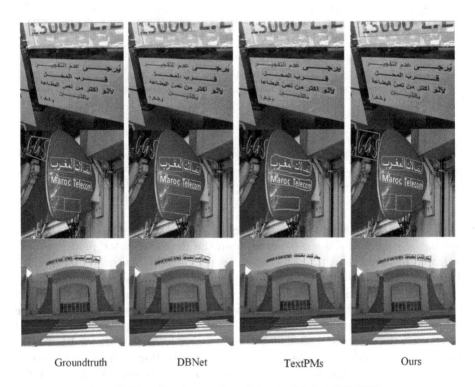

Groundtruth   DBNet   TextPMs   Ours

Fig. 5. Visualization results of the model on MLT-2017.

3.5   Analysis of Visual Results

This section presents the visual results of our model on three different datasets, demonstrating its excellent performance in detecting various types of texts. Figure 5 shows the visualization results of our model on the MLT-2017 dataset. Although this dataset contains complex backgrounds that may affect the detection results, our proposed model exhibits lower miss rates and avoids text fusion compared to DBNet (as shown in the third row of Fig. 5), resulting in visual results that are closer to the ground truth.

Figure 6 shows the visualization results of our model on the ICDAR2015 dataset, which contains many blurry and highlighted texts, making the detection task more challenging. Compared with the latest algorithm TextPMs, our proposed model exhibits lower miss rates (as shown in the first and second rows of Fig. 6), and benefits from the residual attention module that helps to better distinguish between fuzzy background and text information, resulting in more accurate detection results.

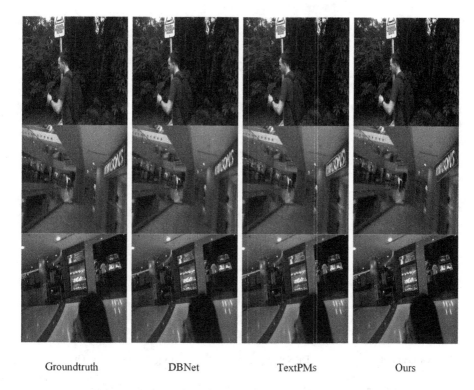

| Groundtruth | DBNet | TextPMs | Ours |

Fig. 6. Visualization results of the model on ICDAR2015.

Figure 7 shows the visualization results of our model on the MSRA-TD500 dataset, which contains texts of various orientations and sizes that may be easily overlooked (as shown in the first row of Fig. 7). Despite this difficulty, our model still exhibits relatively accurate detection results, demonstrating its effectiveness in detecting different types of text objects.

3.6   Limitations

We have also undertaken an exploration of the limitations inherent in our current model. One such limitation is the model's susceptibility to missed detections when dealing with densely-packed text instances. Additionally, smaller text instances, as well as those rich in characters, have proven to be particularly challenging and can lead to instances of missed detection. Another limitation arises when handling text instances with irregular angles or skewed orientations, potentially affecting the model's ability to accurately detect text in such cases. These limitations are in line with the challenges commonly encountered by segmentation-based scene text detectors.

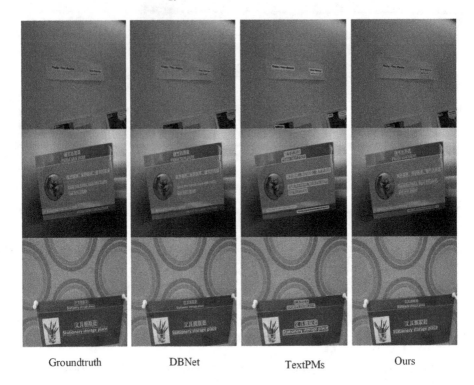

Groundtruth   DBNet   TextPMs   Ours

Fig. 7. Visualization results of the model on MSRA-TD500.

4   Conclusion

This paper presents a text detection model that combines residual attention and multi-scale information fusion to address the issue of missing detections in natural scenes. Specifically, the multi-scale information fusion structure can integrate semantic information from different levels to enhance text information. The residual attention mechanism is embedded in the highest resolution output branch to assist in locating text information and focus the model more on text features.

To verify the effectiveness of this model, the paper evaluates it on three commonly used text detection datasets and compares it with other existing text detection models. Experimental results show that the proposed model achieves excellent detection performance on multiple datasets, especially in some challenging scenarios. Therefore, this text detection model proposed in this paper can provide an effective solution for practical text detection problems. In the future, we will continue to focus on the detection of dense text instances as well as angular irregular text instances.

References

1. Zhang, Z., Zhang, C., Shen, W., Yao, C., Liu, W., Bai, X.: Multi-oriented text detection with fully convolutional networks. In: Proceedings of the IEEE Conference on Computer Vision and Pattern Recognition, pp. 4159–4167 (2016)
2. Lyu, P., Liao, M., Yao, C., Wu, W., Bai, X.: Mask textspotter: an end-to-end trainable neural network for spotting text with arbitrary shapes. In: Proceedings of the European Conference on Computer Vision (ECCV), pp. 67–83 (2018)
3. Deng, D., Liu, H., Li, X., Cai, D.: Pixellink: detecting scene text via instance segmentation. In: Proceedings of the AAAI Conference on Artificial Intelligence, vol. 32 (2018)
4. Huang, Z., Zhong, Z., Sun, L., Huo, Q.: Mask R-CNN with pyramid attention network for scene text detection. In: 2019 IEEE Winter Conference on Applications of Computer Vision (WACV), pp. 764–772. IEEE (2019)
5. Wang, W., et al.: Shape robust text detection with progressive scale expansion network. In: Proceedings of the IEEE/CVF Conference on Computer Vision and Pattern Recognition, pp. 9336–9345 (2019)
6. Baek, Y., Lee, B., Han, D., Yun, S., Lee, H.: Character region awareness for text detection. In: Proceedings of the IEEE/CVF Conference on Computer Vision and Pattern Recognition, pp. 9365–9374 (2019)
7. Liao, M., Wan, Z., Yao, C., Chen, K., Bai, X.: Real-time scene text detection with differentiable binarization. In: Proceedings of the AAAI Conference on Artificial Intelligence, vol. 34, pp. 11474–11481 (2020)
8. Zhong, Y., Cheng, X., Chen, T., Zhang, J., Zhou, Z., Huang, G.: PRPN: progressive region prediction network for natural scene text detection. Knowl.-Based Syst. **236**, 107767 (2022)
9. Zhang, S.-X., Zhu, X., Chen, L., Hou, J.-B., Yin, X.-C.: Arbitrary shape text detection via segmentation with probability maps. IEEE Trans. Pattern Anal. Mach. Intell. **45**(3), 2736–2750 (2022)
10. Huang, W., Qiao, Yu., Tang, X.: Robust scene text detection with convolution neural network induced MSER trees. In: Fleet, D., Pajdla, T., Schiele, B., Tuytelaars, T. (eds.) ECCV 2014. LNCS, vol. 8692, pp. 497–511. Springer, Cham (2014). https://doi.org/10.1007/978-3-319-10593-2_33
11. Tian, Z., Huang, W., He, T., He, P., Qiao, Yu.: Detecting text in natural image with connectionist text proposal network. In: Leibe, B., Matas, J., Sebe, N., Welling, M. (eds.) ECCV 2016. LNCS, vol. 9912, pp. 56–72. Springer, Cham (2016). https://doi.org/10.1007/978-3-319-46484-8_4
12. Liao, M., Shi, B., Bai, X., Wang, X., Liu, W.: Textboxes: a fast text detector with a single deep neural network. In: Proceedings of the AAAI Conference on Artificial Intelligence, vol. 31 (2017)
13. Liao, M., Shi, B., Bai, X.: Textboxes++: a single-shot oriented scene text detector. IEEE Trans. Image Process. **27**(8), 3676–3690 (2018)
14. Liu, Y., Jin, L.: Deep matching prior network: Toward tighter multi-oriented text detection. In: Proceedings of the IEEE Conference on Computer Vision and Pattern Recognition, pp. 1962–1969 (2017)
15. Zhou, X., et al.: East: an efficient and accurate scene text detector. In: Proceedings of the IEEE Conference on Computer Vision and Pattern Recognition, pp. 5551–5560 (2017)
16. Xie, L., Liu, Y., Jin, L., Xie, Z.: Derpn: taking a further step toward more general object detection. In: Proceedings of the AAAI Conference on Artificial Intelligence, vol. 33, pp. 9046–9053 (2019)

17. Zhang, S.-X., Zhu, X., Yang, C., Wang, H., Yin, X.-C.: Adaptive boundary proposal network for arbitrary shape text detection. In: Proceedings of the IEEE/CVF International Conference on Computer Vision, pp. 1305–1314 (2021)
18. Zhang, S.-X., et al.: Deep relational reasoning graph network for arbitrary shape text detection. In: Proceedings of the IEEE/CVF Conference on Computer Vision and Pattern Recognition, pp. 9699–9708 (2020)
19. Yao, C., Bai, X., Liu, W., Ma, Y., Tu, Z.: Detecting texts of arbitrary orientations in natural images. In: 2012 IEEE Conference on Computer Vision and Pattern Recognition, pp. 1083–1090. IEEE (2012)
20. Karatzas, D., et al.: ICDAR 2015 competition on robust reading. In: 2015 13th International Conference on Document Analysis and Recognition (ICDAR), pp. 1156–1160. IEEE (2015)
21. Nayef, N., et al.: ICDAR 2017 robust reading challenge on multi-lingual scene text detection and script identification-RRC-MLT. In: 2017 14th IAPR International Conference on Document Analysis and Recognition (ICDAR), vol. 1, pp. 1454–1459. IEEE (2017)
22. Zhu, X., Hu, H., Lin, S., Dai, J.: Deformable convnets V2: more deformable, better results. In: Proceedings of the IEEE/CVF Conference on Computer Vision and Pattern Recognition, pp. 9308–9316 (2019)

Encrypted-SNN: A Privacy-Preserving Method for Converting Artificial Neural Networks to Spiking Neural Networks

Xiwen Luo[1], Qiang Fu[1(✉)], Sheng Qin[1], and Kaiyang Wang[2]

[1] Guangxi Key Lab of Brain-inspired Computing and Intelligent Chips, School of Electronic and Information Engineering, Guangxi Normal University, Guilin, China
qiangfu@gxnu.edu.cn
[2] International Joint Audit Institute, Nanjing Audit University, Nanjing, China

Abstract. The transformation from Artificial Neural Networks (ANNs) to Spiking Neural Networks (SNNs) presents a formidable challenge, particularly in terms of preserving privacy to safeguard sensitive data during the conversion process. In response to these privacy concerns, a novel Encrypted-SNN approach is proposed for the ANN-SNN conversion. By incorporating noise into the gradients of both ANNs and SNNs, privacy protection without compromising network performance can be enhanced. The proposed method is tested using popular datasets including CIFAR10, MNIST, and Fashion MNIST, achieving respective accuracies of 88.1%, 99.3%, and 93.0% respectively. The influence of three distinct privacy budgets ($\epsilon = 0.5$, 1.0, and 1.6) on the accuracy of the model are also discussed. Experimental results demonstrate that the Encrypted-SNN approach effectively optimizes the balance between privacy and performance. This has practical implications for data privacy protection and contributes to the enhancement of security and privacy within SNNs.

Keywords: Artificial Neural Networks · Spiking Neural Networks · ANN-SNN conversion · Privacy Protection

1 Introduction

Artificial Neural Networks (ANNs) [1] and Spiking Neural Networks (SNNs) [2] are two prominent paradigms in the field of neural networks. ANNs have been widely used in various applications [3–5] due to their ability to approximate complex nonlinear functions and perform tasks such as classification and regression. In contrast to the continuous activation of neurons in ANNs, biological neurons utilize discrete spikes to process and transmit information, where spike time and spike rate are important. SNNs [2,6–8] are biologically more realistic than ANNs, with advantages such as low power consumption, fast reasoning, and event-driven information processing. Given the inherent benefits presented by both ANNs and

B. Luo et al. (Eds.): ICONIP 2023, LNCS 14448, pp. 519–530, 2024.
https://doi.org/10.1007/978-981-99-8082-6_40

SNNs [9,10], the conversion of ANNs to SNNs has attracted considerable attention in research and development efforts [11–15]. Recent studies have shown that SNN can be effectively constructed by approximating the activation value of ANN neurons with the firing rate of spike neurons [11,12,16,17] and [18,19] under the guidance of rate coding. This work not only simplifies the training process of the spike-based learning method described above but also enables SNN to achieve the best-reported results in many challenging tasks.

However, privacy issues in ANNs and SNNs have also attracted significant attention. Previous works have explored data privacy in the conversion from ANN to SNN [20–24]. The problem addressed by research [23] is how to balance privacy protection and energy efficiency in neural systems. The primary objective of the proposed method is to create energy-efficient SNNs from pre-trained ANN models while preserving the confidentiality of sensitive information present on the dataset. The approach of [24] combines the principles of differential privacy protection and SNNs, which allows the differentially private SNNs to achieve a powerful balance between preserving privacy and ensuring high accuracy. These existing methods exhibit certain limitations as they only focus on addressing privacy issues during the conversion of ANN [21,24–26], SNN [24], and from ANN to SNN [23], and do not show that SNNs can be converted in the case of encrypted ANN models, which has the potential to enhance privacy protection significantly.

This work aims to investigate the privacy issue associated with the conversion from ANN to SNN and proposes an approach to address this problem. The proposed method takes into account both privacy protection and model accuracy in both the encrypted ANN and the converted SNN. It achieves a balance between privacy and utility in the converted Encrypted-SNN. There are three significant contributions to this work. Firstly, by incorporating gradient perturbation twice during the conversion from ANNs to SNNs, we effectively address the privacy concern in the ANN to SNN conversion, thereby safeguarding sensitive information throughout the conversion process. Secondly, we convert the encrypted ANN into an SNN, and experiments demonstrate the feasibility of this conversion method. On the Fashion MNIST dataset, the test accuracy of the SNN surpasses that of the encrypted VGG16 network by three percent. Experiments are also carried out to evaluate the proposed Encrypted-SNN model on the CIFAR10 [27], MNIST [28], and Fashion MNIST [29] public datasets, and the accuracy effects under three different privacy budgets ($\epsilon = 0.5$, 1.0, and 1.6) are given. The experimental results show that Encrypted-SNN provides strong privacy protection while ensuring high performance.

2   Background

This section offers an introduction to SNNs, the concept of privacy protection, and the significance of privacy preservation in SNNs.

2.1   Spiking Neural Networks

To build efficient neuromorphic systems, various SNNs training techniques have been proposed. Alternative gradient learning techniques bypass the nondifferentiable problem of the Leak-Integrate-and-Fire (LIF) neurons by defining an approximate backward gradient function [30–32]. The ANN-SNN conversion technique uses weight or threshold balancing to convert pre-trained ANN to SNN, replacing ReLU with LIF activation [11–15]. Compared with other methods, the SNN generated by the conversion has a precision comparable to that of similar ANN on various tasks. The conversion process of this work involves replacing BatchNorm2d layers with Identity layers, ReLU layers with hardtanh layers, and Linear layers with bias-free Linear layers.

2.2   Privacy Protection

If two datasets differ by only one data instance, they are considered adjacent. In the context of (ϵ, δ)-differential privacy, we adopt the definition proposed by Dwork et al [33]. According to this definition, a mechanism M satisfies (ϵ, δ)-differential privacy for any pair of adjacent datasets D and D', as well as any event E. The probability of observing event E in the output of mechanism M on dataset D' is at most e^{ϵ} times the probability of observing event E in the output of mechanism M on dataset D, with the inclusion of an additional term δ, i.e.

$$P[M(D) \in E] \leq e^{\epsilon} \cdot P[M(D^{'}) \in E] + \delta, \tag{1}$$

This notion of privacy analysis is based on the framework of the Gaussian Differential Privacy (GDP) introduced by Dong et al [34]. In GDP, a mechanism M is said to provide μ-GDP if the distribution of the mechanism's output on dataset D is statistically indistinguishable from the distribution of a Gaussian noise $N(\mu, 1)$.

Privacy protection plays a crucial role in the realm of SNNs, and studying methods for safeguarding privacy within SNNs holds immense significance. The DPSNN [24] approach has been proposed to combine differential privacy and SNNs, aiming to achieve both accuracy and privacy in the model. By considering factors such as gradient accuracy, neuron type, and time window, DPSNN greatly contributes to privacy preservation. Another noteworthy contribution is the private SNN [23] model, which addresses the challenge of data and class leakage. It ensures robust privacy protection while maintaining high accuracy and low power consumption. These advancements underscore the importance of privacy-preserving SNN training and emphasize the necessity for privacy-preserving techniques in SNNs. However, both approaches overlook the encryption of ANNs during the conversion of ANNs to SNNs, potentially compromising privacy in SNNs. This work demonstrates that by introducing gradient perturbations twice during the conversion process, privacy preservation in SNNs can be effectively enhanced.

3   Encrypted-SNN

In this section, we provide an overview of the primary methods employed in this study. Firstly, we introduce the technique of gradient perturbation during the conversion process from ANN to SNN. Secondly, we describe the specific approach utilized for the conversion from ANN to SNN.

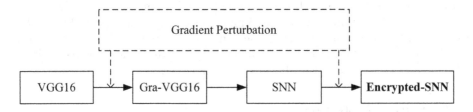

Fig. 1. The implementation flow chart of the Encrypted-SNN.

3.1   Privacy-Preserving Methods for ANN and SNN

Introducing gradient perturbation plays a crucial role in enhancing privacy and reducing information leakage during the propagation of ANN and SNN models. A proven method is used to crop the gradient and inject Gaussian noise into the gradient, namely gradient perturbation. Figure 1 shows two gradient perturbations during the conversion from ANN to SNN. The first time is that the gradient disturbance is carried out when the VGG16 network becomes a Gra-VGG16 network, and the network is encrypted. The second time is to perform gradient perturbation on the converted SNN to form an Encrypted-SNN. By introducing gradient perturbation in the conversion and training process, sensitive information is effectively protected, so that the network has strong privacy protection. The global sensitivity of the model parameters is computed. The sensitivity quantifies the maximum change in the model's output resulting from a small change in the input data. To calculate sensitivity, the gradients of the loss function with respect to the model parameters should be calculated by

$$G = \nabla_\theta L(f(x; \theta), y),   \tag{2}$$

where G denotes the gradients, and ∇_θ denotes the derivative with respect to θ. L denotes the loss function and $f(x; \theta)$ denotes the model. Moreover, x and y denote the inputs and outputs respectively. Then, the $L2$ norm can be calculated by

$$N = \sqrt{\sum_{i=1}^{n}(G_i^2)},   \tag{3}$$

where N denotes the norm, and $\sum_{i=1}^{n}$ denotes the sum from i to n, G_i denotes the ith element of the gradient vector, and norm denotes the norm of the gradient vector. The global sensitivity is then obtained by summing up all these norms. The global sensitivity refers to the measure of the model's sensitivity to changes in the input data, indicating how much the model's predictions can vary. The noise standard deviation is calculated by using the global sensitivity, i.e.

$$\sigma = \frac{\varphi}{2 * \chi * \gamma}, \tag{4}$$

where φ represents the computed global sensitivity, χ denotes the size of the training set, and γ acts as a scaling factor. Thirdly, random noise tensors are generated, preserving the same shape as the model parameters. The noise tensors follow a Gaussian distribution with zero mean and standard deviation σ. Subsequently, the generated noise tensors are added to the gradients of the model parameters. To ensure stability, the gradients are clipped to prevent them from exploding. Finally, the backpropagation and optimization steps proceed using the perturbed gradients.

3.2 ANN-SNN Conversion

This work presents a method to convert the VGG16 network into an SNN. The conversion process involves replacing BatchNorm2d layers with Identity layers, ReLU layers with hardtanh layers, and Linear layers with bias-free Linear layers. These operations are performed to make the neural network more suitable for use in SNNs. While BatchNorm2d and ReLU layers are not necessary for SNNs, Linear layers require bias removal.

4 Experiments and Results

This section provides the proposed Encrypted-SNN, as well as the experiments that were verified on CIFAR10, MNIST, and Fashion-MNIST datasets of encrypted VGG16 and SNNs before conversion, and compares the test accuracy of advanced methods in existing works. The training and testing experiments of encrypted SNNs under three privacy budgets ($\epsilon = 0.5$, 1.0, and 1.6) on public datasets are discussed.

4.1 Experiments Setup

In this paper, the PyTorch [35] library is used to realize the conversion of ANNs to SNNs. GPU Nvidia 3090 is used to accelerate the calculation process. The batch size is set to 64 and the learning rate is set to 0.01 and 0.001. The SGD optimizer is used to optimize the model. The proposed method is validated through experiments conducted on three widely-used public datasets, i.e. CIFAR10 [27], MNIST [28], and Fashion MNIST [29]. For the conversion method, the method of gradient add noise clipping is used to encrypt the ANN, and then the encrypted ANN is converted into an SNN, and then the method of gradient add noise clipping is used to encrypt the SNN, which strengthens the privacy protection in the process of ANN to SNN conversion. For the amount of added noise, the impact of accuracy under three different privacy budgets of $\epsilon = 0.5$, 1.0, and 1.6 is discussed, and the global sensitivity is calculated to determine the optimal value of added noise.

4.2   Experiments Results

Table 1. Performance comparisons of different models on different datasets.

Models	CIFAR10	CIFAR100	MNIST	Fashion-MNIST
Gra-VGG16	86.1	60.1	99.1	90.1
SNN	87.3	63.8	99.1	93.1
Encrypted-SNN	**88.1**	**63.0**	**99.3**	**93.0**

Table 1 indicates that the test accuracy of both the SNN derived from the privacy-preserving VGG16 network with the addition of gradient noise and clipping, and the encrypted SNN, slightly surpasses that of the VGG16 network with privacy protection. On the CIFAR10 and CIFAR100 datasets, the test accuracy has been enhanced from 86.1% and 60.1% in the Gra-VGG16 network, to 88.1% and 63.0% respectively in the Encrypted-SNN. On the MNIST dataset, the test accuracy of the three neural networks shows minimal variation. On the Fashion-MNIST dataset, the test accuracy of the Gra-VGG16 network is 90.1% and the test accuracy of the Encrypted-SNN is 93.0%. Nevertheless, it should be noted that the test accuracy of the SNN model reaches 93.1%, slightly surpassing the accuracy of the Encrypted-SNN model. This difference can be attributed to the complexity of the Fashion-MNIST dataset. Compared to the relatively MNIST dataset, the Fashion-MNIST dataset is more susceptible to noise and exhibits greater sensitivity to privacy-enhancing measures. Table 1 illustrates that the privacy protection function of the SNNs is enhanced when the ANNs with privacy protection function are converted to privacy-preserving SNNs. The test accuracy of the SNNs is slightly improved.

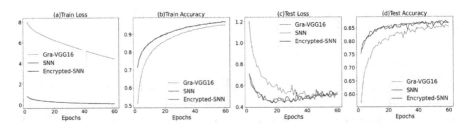

Fig. 2. Loss and accuracy curves of training and test on the CIFAR10 dataset.

The training and test loss of the three differential networks on the CIFAR10 dataset is shown in Fig. 2(a) and (c). The training and test accuracy of the three differential networks on CIFAR10 is shown in Fig. 2(b) and (d) respectively. Both the Gra-VGG16 and the Encrypted-SNN can ensure high accuracy and shorter time steps. The Encrypted-SNN network demonstrates a 2% improvement in test accuracy on the CIFAR10 dataset compared to the Gra-VGG16 network. When the epoch is twenty, the Encrypted-SNN reaches 85.59%, while the neural

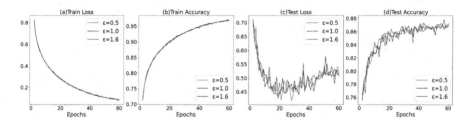

Fig. 3. Loss and accuracy curves of training and test using different epsilon on the CIFAR10 dataset.

network Gra-VGG16 has a test accuracy of 80.30%. The results show that the proposed Encrypted-SNN achieves a good balance between performance and privacy.

As shown in Fig. 3(a) and (c) is the training and test loss of the different epsilon on CIFAR10. (b) and (d) is the training and test accuracy of the different epsilon on CIFAR10. Moreover, when the privacy budget is increased from 0.5 to 1.6, there is little difference in the performance of the neural network as the training period increases. The Encrypted-SNN exhibits relatively stable accuracy as privacy protection strengthens. Encrypted-SNN has strong privacy protection while ensuring high accuracy, and achieves 88.1% test accuracy on the CIFAR10 dataset, which is very close to the method [24]. The test accuracy of the proposed method is not much different from that of the VGG16.

Table 2. Classification accuracy on CIFAR10, CIFAR100.

Approaches	CIFAR10	CIFAR100
VGG16 [36]	89.7	63.4
VGG16 [16]	91.5	62.7
VGG16 [14]	91.4	-
VGG16 [11]	90.9	-
PrivateSNN [23]	89.2	62.3
This work	88.1	63.2

Surprisingly, we find no significant performance loss for Encrypted-SNN. Table 2 shows the performance of the reference ANN, namely VGG16, and previous conversion methods using training data. The representative state-of-the-art conversion methods [11,14,16,23,36] for comparison. The results show that the proposed Encrypted-SNN does not suffer much performance degradation on all datasets. This implies that the gradient noise clipping from the ANN model contains sufficient information to successfully facilitate the conversion to SNNs.

As shown in Fig. 4(a) and (c) is the training and test loss of the three differential networks on the MNIST. (b) and (d) is the training and test accuracy of the

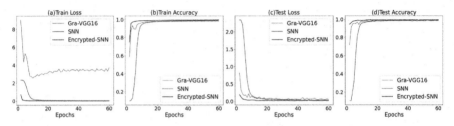

Fig. 4. Loss and accuracy curves of training and test dataset on the MNIST dataset.

three differential networks on MNIST. The Encrypted-SNN demonstrates high performance with fewer time steps on both the training and test datasets. This observation highlights the efficiency and effectiveness of the proposed method in achieving accurate results. When the epoch is three, the test accuracy reaches 96.09%. The test accuracy of the Encrypted-SNN on the MNIST dataset is final 99.47%. The Encrypted-SNN achieves a combination of privacy preservation and performance.

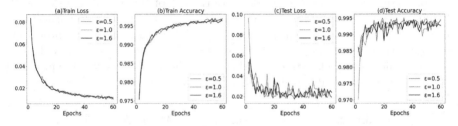

Fig. 5. Loss and accuracy curves of training and test dataset using different epsilon on the MNIST dataset.

Table 3. Performance comparisons of different studies for the Encrypted-SNN on different datasets with different epsilons ($\epsilon = 0.5, 1.0$, and 1.6).

Methods	ϵ	MNIST	Fashion-MNIST
DPSNN [24]	1.6	97.96	87.31
	1.0	98.36	87.93
	0.5	98.63	88.37
Encrypted-SNN **(This work)**	1.6	**99.27**	**92.83**
	1.0	**99.43**	**92.89**
	0.5	**99.47**	**93.01**

As shown in Fig. 5(a) and (c) is the training and test loss of the different epsilon on MNIST. (b) and (d) is the training and test accuracy of the different

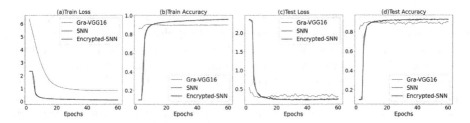

Fig. 6. Loss and accuracy curves of training and test dataset on the Fashion-MNIST dataset. (a) and (c) is the training and test loss curves, moreover (b) and (d) is the training and test accuracy curves of the three different models on the Fashion-MNIST dataset.

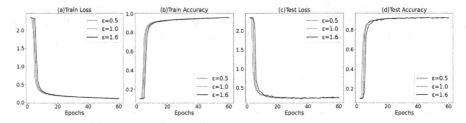

Fig. 7. Loss and accuracy curves of training and test dataset using different epsilon on the Fashion-MNIST dataset.

epsilon on MNIST. Figure 6 shows that the Encrypted-SNN has little difference in training accuracy and testing accuracy under the three different models, which reflects that the Encrypted-SNN has high privacy protection and good performance. Table 3 presents a comparison of the test accuracy between the method proposed in this study and the method described in [24], under varying privacy budgets. It can be seen that the proposed method performs better than the approach of [24] for different privacy budgets on MNIST and Fashion-MNIST datasets. It is 0.84% higher than the research of [24], and 4.64% higher on the more complex Fashion-MNIST dataset. Figure 6 shows that Encrypted-SNN performs better than the Gra-VGG16 network. Figure 7 shows the experiments of the proposed method on the Fashion-MNIST dataset under three privacy budgets. When the privacy protection is enhanced ($\epsilon = 1.6$), the test accuracy of the proposed method requires slightly more iterations than that of the Gra-VGG16 and SNN to achieve the same performance. However, the final test accuracy is comparable to Gra-VGG16 and SNNs.

5   Conclusion

A novel and comprehensive encryption method, i.e. Encrypted-SNN, is proposed to protect privacy during the process of ANN to SNN conversion. It can safeguard private information effectively. To investigate the influence of the privacy budget on performance, three distinct privacy budgets are considered. This approach facilitates superior privacy protection without sacrificing accuracy. Our work involves calculating the global sensitivity, gradient clipping, and adding noise to encrypt the ANNs. The same process is also applied to encrypt the subsequently converted SNN. Experimental results demonstrate that these techniques significantly enhance the privacy protection of SNNs. The results indicate that the Encrypted-SNN method successfully strikes a delicate balance between privacy preservation and performance. Future work will focus on exploring aspects such as the impact of incorporating Laplacian and other types of noise on the security and performance of SNNs.

Acknowledgement. This research is supported by the National Natural Science Foundation of China under Grant 61976063, the Guangxi Natural Science Foundation under Grant 2022GXNSFFA035028, the research fund of Guangxi Normal University under Grant 2021JC006, the AI+Education research project of Guangxi Humanities Society Science Development Research Center under Grant ZXZJ202205.

References

1. Tavanaei, A., Ghodrati, M., Kheradpisheh, S.R., Masquelier, T., Maida, A.: Deep learning in spiking neural networks. Neural Netw. **111**, 47–63 (2019)
2. Pfeiffer, M., Pfeil, T.: Deep learning with spiking neurons: opportunities and challenges. Front. Neurosci. **12**, 774 (2018)
3. Liu, J., Qin, S., Luo, Y., Wang, Y., Yang, S.: Intelligent traffic light control by exploring strategies in an optimised space of deep Q-learning. IEEE Trans. Veh. Technol. **71**(6), 5960–5970 (2022)
4. Liu, J., Sun, T., Luo, Y., Yang, S., Cao, Y., Zhai, J.: Echo state network optimization using binary grey wolf algorithm. Neurocomputing **385**, 310–318 (2020)
5. Deng, M., Ma, H., Liu, L., Qiu, T., Lu, Y., Suen, C.Y.: Scriptnet: a two stream CNN for script identification in camera-based document images. In: Tanveer, M., Agarwal, S., Ozawa, S., Ekbal, A., Jatowt, A. (eds.) ICONIP 2022. CCIS, vol. 1793, pp. 14–25. Springer, Cham (2022). https://doi.org/10.1007/978-981-99-1645-0_2
6. Wu, J., et al.: Progressive tandem learning for pattern recognition with deep spiking neural networks. IEEE Trans. Pattern Anal. Mach. Intell. **44**(11), 7824–7840 (2021)
7. Liu, J., et al.: Exploring self-repair in a coupled spiking astrocyte neural network. IEEE Trans. Neural Netw. Learn. Syst. **30**(3), 865–875 (2018)
8. Fu, Q., Dong, H.: An ensemble unsupervised spiking neural network for objective recognition. Neurocomputing **419**, 47–58 (2021)
9. Xiao, X., Chen, X., Kang, Z., Guo, S., Wang, L.: A spatio-temporal event data augmentation method for dynamic vision sensor. In: Tanveer, M., Agarwal, S., Ozawa, S., Ekbal, A., Jatowt, A. (eds.) ICONIP 2022. CCIS, vol. 1793, pp. 422–433. Springer, Cham (2022). https://doi.org/10.1007/978-981-99-1645-0_35
10. Liu, J., Huang, X., Huang, Y., Luo, Y., Yang, S.: Multi-objective spiking neural network hardware mapping based on immune genetic algorithm. In: Tetko, I.V., Krková, V., Karpov, P., Theis, F. (eds.) ICANN 2019. LNCS, vol. 11727, pp. 745–757. Springer, Cham (2019). https://doi.org/10.1007/978-3-030-30487-4_58
11. Rueckauer, B., Lungu, I.-A., Hu, Y., Pfeiffer, M., Liu, S.-C.: Conversion of continuous-valued deep networks to efficient event-driven networks for image classification. Front. Neurosci. **11**, 682 (2017)

12. Diehl, P.U., Neil, D., Binas, J., Cook, M., Liu, S.C., Pfeiffer, M.: Fast-classifying, high-accuracy spiking deep networks through weight and threshold balancing. In: 2015 International Joint Conference on Neural Networks (IJCNN), pp. 1–8. IEEE (2015)
13. Roy, K., Jaiswal, A., Panda, P.: Towards spike-based machine intelligence with neuromorphic computing. Nature **575**(7784), 607–617 (2019)
14. Han, B., Srinivasan, G., Roy, K.: RMP-SNN: residual membrane potential neuron for enabling deeper high-accuracy and low-latency spiking neural network. In: Proceedings of the IEEE/CVF Conference on Computer Vision and Pattern Recognition, pp. 13 558–13 567 (2020)
15. Li, Y., Deng, S., Dong, X., Gong, R., Gu, S.: A free lunch from ANN: towards efficient, accurate spiking neural networks calibration. In: International Conference on Machine Learning, pp. 6316–6325. PMLR (2021)
16. Sengupta, A., Ye, Y., Wang, R., Liu, C., Roy, K.: Going deeper in spiking neural networks: VGG and residual architectures. Front. Neurosci. **13**, 95 (2019)
17. Kim, S., Park, S., Na, B., Yoon, S.: Spiking-yolo: spiking neural network for energy-efficient object detection. In: Proceedings of the AAAI Conference on Artificial Intelligence, vol. 34, no. 07, pp. 11270–11277 (2020)
18. Hu, Y., Tang, H., Pan, G.: Spiking deep residual networks. IEEE Trans. Neural Netw. Learn. Syst. (2021)
19. Li, Y., Deng, S., Dong, X., Gu, S.: Converting artificial neural networks to spiking neural networks via parameter calibration, arXiv preprint arXiv:2205.10121 (2022)
20. Liu, J., Zhang, S., Luo, Y., Cao, L.: Machine learning-based similarity attacks for chaos-based cryptosystems. IEEE Trans. Emerg. Top. Comput. **10**(2), 824–837 (2020)
21. Abadi, M., et al.: Deep learning with differential privacy. In: Proceedings of the 2016 ACM SIGSAC Conference on Computer and Communications Security, pp. 308–318 (2016)
22. Ren, M., Kornblith, S., Liao, R., Hinton, G.: Scaling forward gradient with local losses, arXiv preprint arXiv:2210.03310 (2022)
23. Kim, Y., Venkatesha, Y., Panda, P.: Privatesnn: privacy-preserving spiking neural networks. In: Proceedings of the AAAI Conference on Artificial Intelligence, vol. 36, no. 1, pp. 1192–1200 (2022)
24. Wang, J., Zhao, D., Shen, G., Zhang, Q., Zeng, Y.: DPSNN: a differentially private spiking neural network, arXiv preprint arXiv:2205.12718 (2022)
25. Cui, Y., Xu, J., Lian, M.: Differential privacy machine learning based on attention residual networks. In: 2023 15th International Conference on Computer Research and Development (ICCRD), pp. 21–26. IEEE (2023)
26. Pasdar, A., Hassanzadeh, T., Lee, Y.C., Mans, B.: ANN-assisted multi-cloud scheduling recommender. In: Yang, H., Pasupa, K., Leung, A.C.-S., Kwok, J.T., Chan, J.H., King, I. (eds.) ICONIP 2020. CCIS, vol. 1332, pp. 737–745. Springer, Cham (2020). https://doi.org/10.1007/978-3-030-63820-7_84
27. Krizhevsky, A., Hinton, G., et al.: Learning multiple layers of features from tiny images (2009)
28. LeCun, Y., Bottou, L., Bengio, Y., Haffner, P.: Gradient-based learning applied to document recognition. Proc. IEEE **86**(11), 2278–2324 (1998)
29. Xiao, H., Rasul, K., Vollgraf, R.: Fashion-MNIST: a novel image dataset for benchmarking machine learning algorithms, arXiv preprint arXiv:1708.07747 (2017)
30. Lee, J.H., Delbruck, T., Pfeiffer, M.: Training deep spiking neural networks using backpropagation. Front. Neurosci. **10**, 508 (2016)

31. Neftci, E.O., Mostafa, H., Zenke, F.: Surrogate gradient learning in spiking neural networks: bringing the power of gradient-based optimization to spiking neural networks. IEEE Signal Process. Mag. **36**(6), 51–63 (2019)

32. Lee, C., Sarwar, S.S., Panda, P., Srinivasan, G., Roy, K.: Enabling spike-based backpropagation for training deep neural network architectures. Front. Neurosci. 119 (2020)

33. Dwork, C., McSherry, F., Nissim, K., Smith, A.: Calibrating noise to sensitivity in private data analysis. In: Halevi, S., Rabin, T. (eds.) TCC 2006. LNCS, vol. 3876, pp. 265–284. Springer, Heidelberg (2006). https://doi.org/10.1007/11681878_14

34. Dong, J., Roth, A., Su, W.J.: Gaussian differential privacy. J. R. Stat. Soc. Ser. B Stat. Methodol. **84**(1), 3–37 (2022)

35. Paszke, A., et al.: Automatic differentiation in pytorch (2017)

36. Zambrano, D., Nusselder, R., Scholte, H.S., Bohté, S.M.: Sparse computation in adaptive spiking neural networks. Front. Neurosci. **12**, 987 (2019)

PoShapley-BCFL: A Fair and Robust Decentralized Federated Learning Based on Blockchain and the Proof of Shapley-Value

Ziwen Cheng[1], Yi Liu[1], Chao Wu[2](✉), Yongqi Pan[1], Liushun Zhao[3],
and Cheng Zhu[1](✉)

[1] National University of Defense Technology, Changsha, China
zhucheng@nudt.edu.cn
[2] Zhejiang University, Hangzhou, China
chao.wu@zju.edu.cn
[3] Xidian University, Xian, China

Abstract. Recently, blockchain-based Federated learning (BCFL) has emerged as a promising technology for promoting data sharing in the Internet of Things (IoT) without relying on a central authority, while ensuring data privacy, security, and traceability. However, it remains challenging to design an decentralized and appropriate incentive scheme that should promise a fair and efficient contribution evaluation for participants while defending against low-quality data attacks. Although Shapley-Value (SV) methods have been widely adopted in FL due to their ability to quantify individuals' contributions, they rely on a central server for calculation and incur high computational costs, making it impractical for decentralized and large-scale BCFL scenarios. In this paper, we designed and evaluated PoShapley-BCFL, a new blockchain-based FL approach to accommodate both contribution evaluation and defense against inferior data attacks. Specifically, we proposed PoShapley, a Shapley-value-enabled blockchain consensus protocol tailored to support a fair and efficient contribution assessment in PoShapley-BCFL. It mimics the Proof-of-Work mechanism that allows all participants to compute contributions in parallel based on an improved lightweight SV approach. Following using the PoShapley protocol, we further designed a fair-robust aggregation rule to improve the robustness of PoShapley-BCFL when facing inferior data attacks. Extensive experimental results validate the accuracy and efficiency of PoShapley in terms of distance

Supported by the Strategic Priority Research Program of the Chinese Academy of Sciences (XDA19090105), National Key Research and Development Project of China (2021ZD0110505), National Natural Science Foundation of China (U19B2042), the Zhejiang Provincial Key Research and Development Project (2023C01043 and 2022C03106), the University Synergy Innovation Program of Anhui Province (GXXT-2021-004), Academy Of Social Governance Zhejiang University, Fundamental Research Funds for the Central Universities (226-2022-00064), Hunan Province Graduate Innovation Project Fund (Grant No. CX20200042).

B. Luo et al. (Eds.): ICONIP 2023, LNCS 14448, pp. 531–549, 2024.
https://doi.org/10.1007/978-981-99-8082-6_41

and time cost, and also demonstrate the robustness of our designed PoShapley-BCFL.

Keywords: Blockchain · Proof of Shapley · Federated learning · Contribution Evaluation

1 Introduction

Nowadays, the proliferation of the Internet of Things (IoT) has led to massive data being generated from various sources. As an emerging distributed learning paradigm, federated learning (FL) [30] has been regarded as a promising solution to promote these data sharing and collaboratively training. However, the conflicts between its centralized framework and the increasing scalability of IoT seriously impede its applications in data sharing. Under this situation, the advent of blockchain-based federated learning (BCFL) [4] has ameliorated the shortcomings. Blockchain emerges as a decentralized ledger technology with the potential to revolutionize the distributed learning paradigm from a centralized design to a decentralized point-to-point collaboration paradigm [8,29]. Specifically, BCFL works on a P2P communication network, removing the need for centralized servers [10]. In such settings, FL participants are customarily treated as equal blockchain nodes with extensive functions, such as performing local training, recording related transactions, and then making the leader who wins the consensus competition complete the aggregation steps [17]. In this way, BCFL mitigates the concerns about a single point of failure and scalability [9] while also enhancing the protection for data privacy, ownership, and security [14]. To maintain these advantages of BCFL in the long term and better serve IoT data sharing, an efficient, fair and robust incentive mechanism which could always attract participants with high-quality data is critical. Since BCFL-enabled IoT data-sharing tasks are performed on specific tasks in complex networks without inspecting the original data, one of the most direct and effective incentive approaches is to evaluate participants' performance in the global model without third parties' assistance and reward them accordingly. Unfortunately, it is still absent.

The Shapley Value (SV) method [1,28] has received much attention due to abilities to quantify the contribution of individuals within a group under the Cooperative Game Theory. It calculates the marginal contributions of each participant in all possible subset consortiums to which it belongs and assigns a weighted sum of marginal contributions as total contribution value to each participant [21], thus, ensuring fairness. This method is also commonly used in FL to evaluate model utility [16,18]. However, the original calculation procedures of SV often incur exponential time concerning the number of participants n, which is not always suited to practical scenarios involving tremendous FL participants, let alone performing a contribution evaluation for participants under the decentralized settings of BCFL. Fairness cannot be guaranteed if the participants upload their self-contribution evaluation results because of self-interest assumptions.

Given the aforementioned dilemmas, a feasible solution for individual performance-based BCFL contribution evaluation is that participants jointly and simultaneously run the calculation of a lightweight Shapley value and mutually oversee without central servers. Fortunately, this idea naturally coincides with the role of blockchain in enabling trusted collaboration with a trusted central authority omitting. More importantly, the consensus mechanism, as one of the critical components of blockchain, defines how different participants collaboratively work to maintain the blockchain networks [2]. With this motivation in mind, we have comprehensively considered combining the blockchain consensus mechanism and the Shapley value, and proposed PoShapley-BCFL, a novel decentralized FL framework with fair and robust contribution evaluation-based incentive designs. Our contributions can be summarized as follows:

- We first developed PoShapley-BCFL, a new combination of federated learning and blockchain technology that guarantees a fair and robust learning process. Our approach achieves this goal by extending a blockchain-consensus-enabled SV calculation procedure and an SV-enabled aggregation procedure into the typical FL framework.
- We proposed a Shapley-value-enabled blockchain protocol, PoShapley, tailored to support the assessment of contributions in decentralized federated learning processes. Our proposed protocol mimics the Proof-of-Work mechanism and allows all participants to compute a monte-carlo-sampling enabled lightweight Shapley value algorithm in parallel until an agreement is achieved, resulting in a more efficient, trustworthy and fair evaluation process.
- Following using the PoShapley protocol, we also developed the fair-robust aggregation method. This method includes a smart-contract-driven client selection process and a Shapley-value-based aggregation process. Specifically, we assigned weights to selected clients based on the ratio of their shapley values, which automatically differentiate low-quality participants and improve model performance when facing inferior data attacks.

The remainder of this paper is organized as follows. Section 2 provided related work and preliminaries. We proposed our PoShapley-BCFL in Sect. 3 with the detailed descriptions of the PoShapley consensus protocol and the SV-based aggregation method. After that, we move to experiments in Sect. 4 to demonstrate the performance of our work. Finally, conclusions are summarized in Sect. 5.

2 Related Work and Preliminaries

2.1 Related Work

Incentive Mechanism in BCFL. Existing incentive approaches in BCFL can be broadly categorized into three types: game-based method [14,20,26,27], auction-based method [5,11,13], and reputation-based method [3,12]. Game-based incentives focus on maximizing FL participants' utilities based on Stackelberg games [14], contract-based games [26] or Bayesian game [20,27]. Auction-based incentives usually reward FL participants with aims of keeping individual rationality and incentive compatibility [11,13], which are usually adopted in FL-enabled data trading systems [5]. Reputation mechanism was introduced in [12] and [3] to promote honest participation in BCFL to earn higher reputation value in blockchain networks. Generally speaking, these value-driven schemes rely on sophisticated utility functions, pricing strategies, or reputation models, which are important in motivating honest participants, ensuring fair compensation, and preventing malicious attacks. However, they often overlook the evaluation of the model itself and typically have high complexity, making them difficult to apply to large-scale and dynamic IoT environments.

Shapley-Value-Based Incentive Mechanism in FL. Recently, Shapley-value-based contribution evaluation stems from cooperative game theory and has been the focus of research due to its remarkable features of fairness [16]. However, the original SV calculation incurs high computational costs, making it challenging to implement in practice. Various approaches were proposed to reduce the time complexity in FL, including methods that aim to decrease the number of permutations sampling for SV calculation, such as Monte-Carlo (MC) sampling-enabled SV methods [6,24]. Other techniques involve using coalition models to minimize individual redundant re-executions, as demonstrated by the Group-SV protocol [18], or training FL sub-models instead of starting from scratch, as in Truncation Gradient Shapley [16,25]. In some works investigating SV methods in BCFL, the paper [15] designs a PoSap protocol to properly reward coins to data owners. The work [22] introduced three Shapley-value-based revenue distribution models for blockchain-enabled data sharing. However, these works did not provide implementations, and thus the feasibility of the proposed scheme is not clear. To this end, our research expanded on the findings of existing works and proposed a consensus mechanism that uses proof of Shapley value to optimize fair and robust decentralized FL. Besides, we provide implementations and experiments to illustrate the feasibility and performances of our work.

2.2   Preliminaries

This paper considers the common Horizontal Federated Learning framework, in which FL members with different samples share the same feature space. For the convenience of presentation, we consider a collaborative learning task with N data owners (i.e., FL participants), each with a private local dataset $\mathcal{D}_i$. During each round t, each participant i downloads the global model w^t and trains on local dataset $\mathcal{D}_i$ for multiple local epochs to get a local model w_i^{t+1}. Then, the local updates and global aggregation can be performed as follows:

$$\Delta_i^{t+1} = w^t - w_i^{t+1}.$$

$$w^{t+1} = w^t + \sum_{i=1}^{N} \frac{|\mathcal{D}_i|}{\sum_{i=1}^{N} |\mathcal{D}_i|} \Delta_i^{t+1}. \tag{1}$$

The original Shapley value is a solution concept from cooperative game theory, which can be defined as:

$$\phi_i = \frac{1}{N} \sum_{S \subseteq I \setminus \{i\}} \frac{1}{\binom{N-1}{|S|}} \left[U\left(S \cup \{i\}\right) - U\left(S\right) \right]. \tag{2}$$

where S denotes the subset of participants from N, $U(\cdot)$ is the utility function, which can be assumed as any form in FL settings, such as accuracy, loss and F1 scores.

3 The Algorithm Design for PoShapley-BCFL

In this section, we proposed a novel blockchain-based serverless federated learning named PoShapley-BCFL. It is expected to effectively complete the contribution evaluation of all participants during the iteration of decentralized collaborative training while also being able to prevent attacks from inferior data sources. We reorganized the entire process of a typical FL and divided PoShapley-BCFL into six procedures. Figure 1 explains the interactions among these procedures, and Algorithm 1 demonstrates the pseudo-codes of each procedure to reveal the details. Initially, the data requester releases some parameters as inputs for PoShapley-BCFL, including an initialized global w^0, evaluation function $U(w)$, number of FL participants N, total training round T and Mining-success criteria ρ (i.e., the error threshold for consensus judgement). After successfully recruiting N participants, PoShapley-BCFL begins to operate.

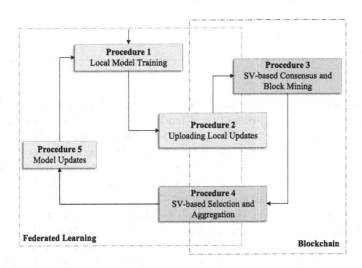

Fig. 1. The modules redesign of proposed PoShapley-BCFL

Algorithm 1: PoShalpey-BCFL Algorithm

---

input : initial FL model w^0, evaluation function $U(w)$, N FL participants,
　　　　　Total training round T, Mining-success criteria ρ
output: final FL model w^T, SVs for all rounds for N participants

1　**for** *each round $t = 0, 1, 2, \cdots, T-1$* **do**
2　　**for** *each participant i in parallel* **do**
3　　　　Procedure Local model training($\mathcal{D}_i$)
4　　　　　$\Delta_i^{t+1} \leftarrow$ Local Training $(\mathcal{D}_i; w^t)$
5　　　　Procedure Upload local updates(i, Δ_i^{t+1})
6　　　　　$Tx_{t+1,i} = \left\{ \Delta_i^{t+1}, hash\left(\Delta_i^{t+1}\right), ID, timestamp \right\}_{Sig_i}$
7　　　　　$Tx_{t+1,i} \rightarrow$ upload to BC
8　　**end**
9　　$SV^t = \{0, 0, \cdots, 0\} \leftarrow$ initialize SV value list at t
10　**while** *Mining-success criteria ρ not met* **do**
11　　　Procedure SV-based consensus and Block mining(w^t, SV^t)
12　　　**for** *each participant k in parallel* **do**
13　　　　$\left| \ \{\mathbb{B}_t, SV^t, M_t\} = \texttt{PoShapley}(w^t, U(w), \left\{\Delta_i^{t+1}\right\}, SV^t)\right.$
14　　　**end**
15　　**end**
16　　Procedure SV-based Selection and Aggregation(SV^t)
17　　**for** M_t **do**
18　　　　$S_a^t \leftarrow \texttt{SV-based-selection-SC}(SV^t, key_{M_t})$
19　　　　$\Delta^{t+1} = \texttt{SV-based Aggregation}(\left\{\Delta_i^{t+1}\right\}, SV^t)$
20　　　　broadcast to BC and to all participants
21　　**end**
22　　**for** *each participant i in parallel* **do**
23　　　　Procedure Models Updates(SV^t)
24　　　　$w^{t+1} = w^t - \eta \Delta^{t+1}$
25　　**end**
26 **end**

In Algorithm 1, Lines 2–8 show the local model training process in **Procedure 1** and uploading process of local model updates in **Procedure 2**. Specifically, in each round t, every FL participant i independently obtains a global model w^t and trains it based on local dataset $\mathcal{D}_i$. After some local training iterations, the local model updates Δ_i^{t+1} are generated and transmitted to blockchain as transactions $Tx_{t+1,i}$, in which the pair of Δ_i^{t+1} and $hash\left(\Delta_i^{t+1}\right)$ (Line 6) can ensure no tampering during transmission. Then, after all participants finish procedure 2, a smart contract deployed on the blockchain triggers the release of an SV list with an initial value of zeros (Line 9), which drives the running of **Procedure 3**. In Lines 10–15, the SV-based consensus procedure begins execution. In this procedure, every FL participant continues to perform the *PoShapley* algorithm (see details in Sect. 3.1) to compute each participant's contributions in this training round as long as mining-success criteria ρ is satisfied. After that, the SV list, one of the outputs of *PoShapley* algorithm, is further used in

Procedure 4 (Line 16–20). A winner of the PoShapley competition at round t adapts a fair and robust SV-based aggregation approach (see details in Sect. 3.2) to obtain new global model updates, which are then broadcasted to all participants for next training round. Lines 22–25 indicate that every participant performs **Procedure 5** to update a new global model and then restart **Procedure 1**.

3.1 The Designs of PoShapley Algorithm

Given the significance of efficiency and attack resistance in contribution evaluation for BCFL systems, we proposed a novel blockchain consensus mechanism named PoShapley. This tailored mechanism facilitates an efficient, fair and robust SV calculation in BCFL, where no central trusted authority exists to evaluate SV utility. The basic concept underlying PoShapley mimics that of PoW [19], which replaces meaningless mathematical puzzles with an improved lightweight SV calculation problem. We present the pseudo-code in Algorithm 2. The initial model w^t represents the initialized global model at training round $t + 1$, which also serves as a benchmark for evaluating the training performance. The utility function $U(w)$ can have multiple forms, including accuracy, loss, recall rate, and F1. An illustration of a completed PoShapley loop is presented below.

Before entering the iterative calculation, a participant k should first perform the preparations according to Line 2–4, such as initializes the SV calculation times as $m_k = 1$, computes the utility value of the global model w^t as $v_0^{m_k}$ (i.e., $v_0^{m_k} = U(w^t)$), constructs an initialized permutation of received model as L_t, and initializes an SV list as ϕ with all values of 0. Next, at each iterative times m_k , the participant k performs a Monte Carlo sampling [16] on permutation L_t to build a list $\pi_{m_k}^k$. By scanning through the $\pi_{m_k}^k$ from the first entity to the last, the jth FL participant's marginal model contribution can be estimated by participant k following the principle of (3), and then be accumulated into the average Shapley value $V_{m_k}^{t,k}$. We show the calculation steps after disassembling Eq. (3) in Lines 6–12.

$$\Delta v_j = \mathbb{E}\left[U\left(s \cup \{m_k^k[j]\}\right) - U(s)\right]$$

$$= \mathbb{E}\left[U\left(w^t + \sum_{p \in s \cup \{m_k^k[j]\}} \frac{|\mathcal{D}_p|}{\sum_{p \in s \cup \{m_k^k[j]\}} |\mathcal{D}_p|} \Delta_p^{t+1}\right)\right.$$

$$\left. -U\left(w^t + \sum_{p \in s} \frac{|\mathcal{D}_p|}{\sum_{p \in s} |\mathcal{D}_p|} \Delta_p^{t+1}\right)\right]. \tag{3}$$

Here, $s = m_k^k[1 : (j-1)]$. And according to index order within SV^t, $V_{m_k}^{t,k}$ is then reordered and denoted as $sv_{m_k}^{t,k}$ (Line 13–14). The $sv_{m_k}^{t,k}$ is then used as input for invoking a *Judgement-Smart-Contract* (Line 15), which provides a global signal J_k that indicates whether the loop should be terminated.

Algorithm 2: PoShapley Algorithm

---

input : initial FL model w^t, evaluation function $U(w)$, participants' model updates $\{\Delta_i^{t+1}, \cdots, \Delta_n^{t+1}\}$, initial SV^t for all participants, Mining-success criteria ρ

output: $SV^t = \{\phi_i^{t+1}, \cdots, \phi_n^{t+1}\}$ for all participants, new block $\mathbb{B}_t$, the winner M_t

---

1 **for** *each participant k in parallel* **do**
2   initialize
3     $m_k = 1; v_0^{m_k} = U\left(w^t\right)$
4     $L_t = \left\{\Delta_i^{t+1}, \cdots, \Delta_n^{t+1}\right\}; \phi = \{0, 0, \cdots, 0\}\, (|\phi| = |L_t|)$
5   **while** *Mining-success criteria not met* **do**
6        $\pi_{m_k}^k \leftarrow$ Monte Carlo sampling permutation of L_t
7   **for** $q = 1, 2, \cdots, \left|\pi_{m_k}^k\right|$ **do**
8           $S = \left\{\pi_{m_k}^k\,[1], \pi_{m_k}^k\,[2], \cdots, \pi_{m_k}^k\,[q]\right\}$
9           $w_S^{t+1} = w^t + \sum_{p \in S} \frac{|\mathcal{D}_p|}{\sum_{p \in S}|\mathcal{D}_p|}\Delta_p^{t+1}$
10          $v_q^{m_k} = U\left(w_S^{t+1}\right)$
11          $\phi_{\pi_{m_k}^k\,[q]} = \frac{1}{m_k}\left((m_k - 1)\,\phi_{\pi_{m_k}^k\,[q]} + v_q^{m_k} - v_{q-1}^{m_k}\right)$
12   **end**
13       $V_{m_k}^{t,k} = \left\{\phi_{\pi_{m_k}^k\,[1]}, \phi_{\pi_{m_k}^k\,[2]}, \cdots, \phi_{\pi_{m_k}^k\,[|\pi_{m_k}^k|]}\right\}$
14       $sv_{m_k}^{t,k} = sort\left\{V_{m_k}^{t,k}\right\}$ by index in SV^t
15       $J_k \leftarrow$ Judgement-SC$(k, sv_{m_k}^{t,k}, \rho)$
16   **if** $J_k == True$ **then**
17          $\mathbb{B}_{k,t} \leftarrow$ generate block(Txs,$sv_{m_k}^{t,k}$)
18   broadcast $\mathbb{B}_{k,t}$ to all participants
19   **if** $verify(\mathbb{B}_{k,t}) == True$ **then**
20   for all participants :
21   stop PoShapley at this round t
22   blockchain add $\mathbb{B}_{k,t}$
23   return SV^t and break
24   **end**
25   **else**
26          $SV^t \leftarrow$ SV-Update-SC$(k, sv_{m_k}^{t,k})$
27          $m_k = m_k + 1$
28   **end**
29   **end**
30 **end**

The automatic judgment operations of this contract are illustrated in Algorithm 3, where the maximum distance between $sv_{m_k}^{t,k}$ and SV^t is calculated. Notably, SV^t refers to the latest average value of all participants' estimated SV, which is updated through smart contracts deployed in blockchain systems in advance (referred in Line 26). When *Judgement-Smart-Contract* returns a *True* value, that is the maximal distance between the average estimated SV and the

---

Algorithm 3: Judgement Smart-Contract

    input : $sv_{m_k}^{t,i}, \rho$
    output: J_i

1 **if** *invoke successful* **then**
2   **for** i *automatically* **do**
3         $\rho_k = \max \left| sv_{m_i}^{t,i} - SV^t \right|$
4   **if** $\rho_i \leq \rho$ **then**
5 | $J_i = True$
6   **else**
7 | $J_i = False$
8   **end**
9   **end**
10   *Return* J_i
11 **end**

Algorithm 4: SV-Update Smart-Contract

    input : $sv_{m_k}^{t,i}$
    output: SV^t

1 **if** *invoke successful* **then**
2   **for** k *automatically* **do**
3   **for** $j = 1, 2, \cdots, n$ **do**
4 | $SV^t[j] = \frac{1}{2}\left(SV^t[j] + sv_{m_k}^t[j]\right)$
5   **end**
6   **end**
7   *Return* SV^t
8 **end**

---

estimated SV from participant k is no greater than threshold ρ, the participant k becomes a candidate responsible for generating a new block $\mathbb{B}_{k,t}$ and broadcasting it (Line 16–18). $\mathbb{B}_{k,t}$ contains all transactions of this training round, and its calculation result $sv_{m_k}^{t,k}$. Line 19–24 show if the verification of $\mathbb{B}_{k,t}$ passes, all participants would stop the PoShapley procedure at this round and append the newest block to the blockchain. Meanwhile, the consensus winner and the final approximated SV results can be acknowledged by all participants. Whereas, if *Judgement-Smart-Contract* returns a *False* value, the participant k should invoke *SV-Update-Smart-Contract* to update the SV^t (Line 26). The automatic updating operations of computing the average value between $sv_{m_k}^{t,k}$ and SV^t are introduced in Algorithm 4. After that, the iteration time m_k is incremented by one, driving participant k to continue the loop at round t.

Algorithm 5: SV-based-selection-Smart-Contract

---

    input : SV^t, $Private_{key}^{M_t}$

    output: S_a^t

1 **if** *invoke successful* **then**

2   verify identity

3   **if** $Public_{key}^{M_t} = f(Private_{key}^{M_t})$ **then**

4   **for** M_t *automatically* **do**

5   **for** $i = 1, 2, \cdots, m$ **do**

6 $v = max(SV^t)$, add the indies of corresponding participant into combination (v, k)

7 $S_a^t = append(S_a^t, (v, k))$

8 SV^t=remove v from SV^t

9   **end**

10   **end**

11   *Return* S_a^t

12   **else**

13   *Return Error*

14   **end**

15 **end**

3.2 Fair and Robust Aggregation

Through leveraging results of the PoShapley protocol, we designed the *SV-based aggregation* (i.e., Line 19 in Algorithm 1) to perform global updates with fairness and inferior data attack tolerance. Specifically, in each round t, after the election of a consensus winner M_t, a *SV-based-selection-Smart-Contract* is triggered by the M_t to generate clients set S_a whose corresponding model updates are to be selected for global model aggregation. The workflows of this contract are illustrated in Algorithm 5.

To ensure security and fairness during the smart contract invocation, we first use the RSA encryption algorithm to verify the identities of M_t (Lines 2–3). Each participant's corresponding public key is submitted and held in the PoShapley-BCFL system when forming the collaborative group. The public key of a winner at each round t can be encapsulated into *SV-based-selection-Smart-Contract* as soon as the consensus process is finished. Only the private key from M_t can pass the verification and triggers the running of automatic selection for S_a, shown in Line 4–11. That is, top m values from the list of SV_t are selected, and their corresponding local updates are accepted for aggregation to get rid of some modifying updates or malicious inferior updates in each global iteration. Here $m \leq N$.

After that, the winner M_t performs the fair and robust aggregation using the formula (4). Unlike the simple average aggregation in (1), we assign aggregation weights based on the shapley value of selected participants and represent the aggregation formula in (4). Finally, the M_t delivers the newest global model updates to blockchain for the next round of training.

$$\Delta^{t+1} = \sum_{k \in S_a^t} \frac{SV_k^t}{\sum_{k \in S_a^t} SV_k^t} \Delta_k^{t+1}. \tag{4}$$

4 Experimental Results and Evaluations

4.1 Experimental Settings

This section introduces the experiment settings, including PoShapley-BCFL components in our experimental setup, dataset settings, evaluation metrics and other learning parameters.

PoShapley-BCFL Components in the Experimental Setup. Figure 2 shows the arrangement components in the experimental setup of our PoShapley-BCFL. We used the Go language (version 1.15.7) and Hyperledger-Fabric-enabled channels, gossip and gRPC protocols to simulate the p2p communications and public ledgers among FL participants. For ease of implementation of PoShapley, the block structure and its chain-based generation process were reprogrammed by Go. Go language was also used to implement the PoShapley-BCFL smart contract, which was deployed to the PoShapley-BCFL blockchain networks using docker. As for the FL participant side, we used Go as the primary programming language, and multithreading settings and goroutine channels were utilized for networking, connections, and coordination among simulated participants. In this process, Pytorch 1.10 from Python (version 3.6) was used as the local training architecture to develop the learning behaviors in FL, which then communicated with some public and trust storage systems (such as IPFS) for model parameter exchange and storage. For the local experiments on the blockchain, the participant invoked the smart contracts through the Go SDK interface and then interacted with the PoShapley-BCFL blockchain network. PoShapley consensus protocol was implemented by the joint of Go and Python and embedded into the PoShapley-BCFL system.

Datasets and Evaluation Metrics. (1) **Datasets**. The dataset used in the experiments is based on the MNIST dataset. To evaluate the proposed algorithms under different FL settings, we designed IID and NIID FL scenarios with 10 participants as follows:

- *IID datasets–Same Size and Same Distribution*: The MNIST dataset, which contains 60000 training samples of ten digits and 10000 testing samples, was evenly divided into ten parts as every participant's local training dataset (i.e., each participant has 6000 training samples and 10000 testing samples).

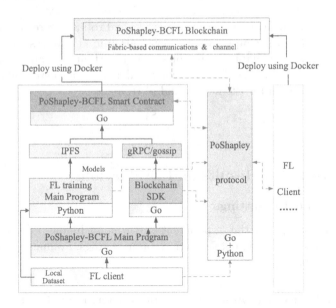

Fig. 2. PoShapley-BCFL components in the experimental setup.

- *NIID datasets*:
  - *NIID-1–Same Size and Different Distributions*: We allocate the same number of MNIST samples for every participant. However, different distributions are set as follows: participant 1 & 2's datasets contain 40% of digits '0' and '1', respectively. The other 8 participants evenly divide the remaining 20% of digits '0' and '1'. Participant 3 & 4's datasets contain 40% of digits '2' and '3', respectively. The other 8 participants evenly divide the remaining 20% of digits '2' and '3'. Similar procedures are applied to the rest of the samples.
  - *NIID-2–Different Sizes and Same Distribution*: We randomly sample from the entire MNIST dataset following pre-defined ratios to achieve NIID-2 settings: The proportions are 5% for participants 1 and 2, respectively; 7.5% for participants 3 and 4, respectively; 10% for participants 5 and 6, respectively; 12.5% for participants 7 and 8, respectively; and 15% for participants 9 and 10, respectively.

(2) **Evaluation metrics**. To comprehensively test the performance of PoShapley, we chose the original shapley algorithm, following the principle of Equation (2), as a benchmark. We also used Adjust-SV from [23] and TMC-SV from [7] as comparison approaches. In addition, inspired by [16], we introduced the following evaluation metrics:

- *Distance metrics*: We used the results of the Original-SV algorithm as a baseline, with distance metric referring to the deviation from the results produced by the Original-SV algorithm. For any participant i, We denote its model contributions calculated by Original-SV in all training rounds as a vector $\phi_i^* = \langle \phi_{i,1}^*, \cdots, \phi_{i,T}^* \rangle$, and the estimated results calculated by any other approach are denoted as $\phi_i = \langle \phi_{i,1}, \cdots, \phi_{i,T} \rangle$. Three distances are introduced as follows.

- *Euclidean Distance*: The Euclidean Distance for any participant i is defined as:

$$ED_i = \sqrt{\sum_{t=1}^{T} \left(\phi_{i,t}^* - \phi_{i,t} \right)^2}. \tag{5}$$

- *Cosine Distance*: The Cosine Distance for any participant i is defined as:

$$CD_i = 1 - \cos\left(\phi_i^*, \phi_i \right). \tag{6}$$

- *Maximum Distance*: The Maximum Distance for any participant i is defined as:

$$MD_i = \max_{t=1}^{T} \left| \phi_i^* - \phi_i \right|. \tag{7}$$

– *Time analysis*: The total time cost of calculating SVs and time complexity is used to evaluate the efficiency of each approach.
– *Accuracy analysis*: The accuracy metrics are used to evaluate the effectiveness of our PoShapley-BCFL with an SV-based aggregation rule, particularly in scenarios where adversarial nodes upload inferior model updates.

Other Learning Parameters. We implemented a MLP neural network architecture as the training model and set learning rate $\eta = 0.01$, total training round $T = 10$, and mining-success criteria $\rho = 0.01$. As for evaluation function $U(w)$, since the *F1 Score* can better measure the performance of the models in most scenarios [22], we use it as the measure of contribution, i.e., $U(w) = F1(w)$.

4.2 Experimental Results Analysis

Firstly, we analyzed the accuracy and time performance of the PoShapley protocol and compared it to state-of-the-art baselines under various FL settings, including both IID and Non-IID (NIID) data silos. Next, we investigated the performance of the PoShapley-BCFL algorithm against inferior data attacks when using the SV-based aggregation.

1) Accuracy analysis of PoShapley: We analyzed the experimental results under the three aforementioned dataset settings. In each case, 10 participants were involved in BCFL with 10 training rounds. And the average distances of all participants' evaluation under different dataset settings were calculated to represent the accuracy performance of PoShapley, shown in Table 1. Under IID data settings, PoShapley achieves the lowest average distance under all three distance metrics, demonstrating that PoShapley achieves the best contribution accuracy. And under Niid-1 settings, the results show that PoShapley still performs with the best accuracy according to the average distances. Notably, the average accuracy gap among the three algorithms is less than that under the IID settings.

Table 1. The Average SV Distance.

Dataset	Distance	Standard deviation of Distance		
		AdjustSV	TMC_SV	Poshapley
IID	$\overline{ED}$	0.0709	0.0921	**0.0558**
	$\overline{CD}$	0.4584	0.4169	**0.3229**
	$\overline{MD}$	0.0464	0.0546	**0.333**
NIID-1	$\overline{ED}$	0.0595	0.0842	**0.0520**
	$\overline{CD}$	0.2877	0.3077	**0.2054**
	$\overline{MD}$	0.0477	0.0639	**0.0401**
NIID-2	$\overline{ED}$	0.0707	0.0617	**0.0456**
	$\overline{CD}$	0.1296	0.1588	**0.1020**
	$\overline{MD}$	0.0560	0.0521	**0.0368**

The results under NIID-2 situations show a similar pattern as in NIID-1, where PoShapley continues outperforming Adjust-SV and TMC-SV approaches regarding average distance. We attribute this advantage to the fact that PoShapley generates more permutations and calculation times of SV due to all participants working in parallel. Moreover, Table 2 compares the standard deviation of SV distance to indicate the stability of different algorithms. PoShapley achieves the slightest standard deviation of SV distances under all metrics and all datasets, making SV approximation more stable and fair for all participants. This advantage is more pronounced under the NIID-1 settings, illustrating that PoShapley is well-suited for these NIID settings.

Table 2. Standard deviation of SV Distance.

Dataset	Distance	Standard deviation of Distance		
		AdjustSV	TMC_SV	Poshapley
IID	Euclidean	0.0244	0.0375	**0.0172**
	consine	0.2391	0.2198	**0.2103**
	max	0.0156	0.0392	**0.011**
NIID-1	Euclidean	0.022	0.0332	**0.0134**
	consine	0.1887	0.2029	**0.0855**
	max	0.0165	0.0217	**0.0087**
NIID-2	Euclidean	0.0314	0.0328	**0.0146**
	consine	0.1569	0.1331	**0.11**
	max	0.0304	0.0229	**0.014**

2) Time cost and complexity analysis: To investigate the time cost of our PoShapley algorithm concerning the number of participants n, we varied n from

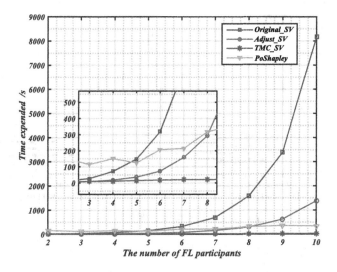

Fig. 3. Time costs with respect to the number of participant.

2 to 10 when performing PoShapley and other compared algorithms under all three dataset settings. The time cost values of each method were determined by calculating their average time under three data settings, and these values were shown in Fig. 3. The original SV method involves training and evaluating additional $2^n - 1$ models, resulting in exponential growth with the number of participants. In contrast, the other three algorithms significantly reduce computational time. It is notable that the runtime of TMC-SV is not significantly affected by an increase in the number of participants. This is because TMC-SV uses the Truncation Monte Carlo policy to drop models with a small marginal utility gain and keeps a low number of models in each round. However, TMC-SV performs poorly in terms of accuracy in our experiments. Adjust-SV uses an approximating algorithm during model reconstruction and outperforms PoShapley in terms of time when $n \leq 7$. However, Adjust-SV still relies on the principle of Eq. (2) when calculating SV, resulting in exponential time consumption as n increases (such as $n \geq 8$ in Fig. 3). As n increases, PoShapley saves even more time than Adjust-SV. In each calculation round of PoShapley, every participant involves in training and evaluating additional n models at each iteration m_k (Line 27 in Algorithm 2). Assuming the maximum m_k among all participants is m and the total training round is T, the number of evaluations of PoShapley is expressed as $\mathcal{O}\left(\sum_{t=1}^{T} mn\right)$, which indicates that the time complexity of PoShapley increases linearly with the number of participants.

3) Accuracy Performance of PoShapley-BCFL: In Fig. 4, we investigated the accuracy performance of PoShapley-BCFL with SV-based aggregation procedure, which is achieved by comparing with a typical weighted aggregation process based on data size (referred to as size-based aggregation). We also set up two

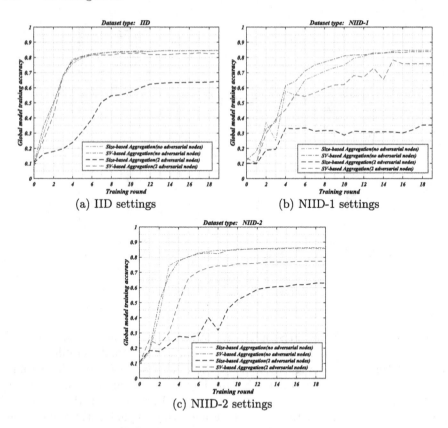

(a) IID settings   (b) NIID-1 settings

(c) NIID-2 settings

Fig. 4. Accuracy convergence performance under various settings.

groups of experimental comparisons: group 1, with 10 regular participants and no adversarial nodes, and group 2, with 8 regular participants and 2 adversarial participants (return randomized parameters). From Fig. 4, we can observe that the proposed SV-based aggregation method achieves almost the same performance in terms of model accuracy under all data settings when there are no adversarial participants. Notably, PoShapley-BCFL exhibits faster convergence than size-based aggregation under the NIID-1 scenario, which can be attributed to the reason that the SV-based approach encourages models with more remarkable contributions to occupy more weight in the aggregated models. Moreover, under all three data settings, PoShapley-BCFL significantly improves accuracy compared to baselines (size-based aggregation) when adversarial participants are attacking the collaborative learning process. Its performance is nearly as close to the settings without adversarial nodes under the IID scenario, and is slightly worse than that without adversarial nodes under the NIID settings, but remarkably outperforms the size-based method. This advantage is attributed to that the SV computing process naturally and automatically detect adversarial workers with lower or no contributions, allocating them with lower or no weights and no longer allowing them to participate in the current aggregation round. The above results confirm that the SV-based aggregation procedure is more robust than size-based methods and that PoShapley-BCFL is more effective when encountering malicious attacks.

5 Conclusion

This paper addresses the challenge of providing fair and robust incentives for blockchain-based decentralized federated learning services that support edge data sharing. We presented insights into designing a new blockchain-based serverless federated learning named PoShapley-BCFL, which has a modular design capable of evaluating model contributions while facilitating robust learning. To meet the lightweight calculation requirements and offer self-assessment in a decentralized setting, we proposed PoShapley. This Shapley-value-enabled blockchain consensus protocol provides fair and efficient contribution evaluation. Based on the results from PoShapley, we further design a fair-robust model aggregation algorithm that can tolerate inferior data attacks. Extensive experiments demonstrated that our proposed methods could promote fair and efficient contribution evaluation during decentralized collaborative learning and improve the final model performance through robust aggregation. For future work, since we observe that the marginal model contribution becomes smaller and smaller during the late stage of model convergence, we plan to study adaptive PoShapley, which can adjust the mining-success threshold during the learning process to prevent degradation of Shapley-value results.

References

1. An, Q., Wen, Y., Ding, T., Li, Y.: Resource sharing and payoff allocation in a three-stage system: integrating network DEA with the Shapley value method. Omega **85**, 16–25 (2019). https://doi.org/10.1016/j.omega.2018.05.008
2. Chen, S., et al.: A blockchain consensus mechanism that uses proof of solution to optimize energy dispatch and trading. Nat. Energy **7**(6), 495–502 (2022). https://doi.org/10.1038/s41560-022-01027-4
3. Chen, Y., et al.: DIM-DS: dynamic incentive model for data sharing in federated learning based on smart contracts and evolutionary game theory. IEEE Internet Things J. (2022). https://doi.org/10.1109/JIOT.2022.3191671
4. Cheng, Z., Pan, Y., Liu, Y., Wang, B., Deng, X., Zhu, C.: Vflchain: blockchain-enabled vertical federated learning for edge network data sharing. In: 2022 IEEE International Conference on Unmanned Systems (ICUS), Guangzhou, China, pp. 606–611. IEEE (2022). https://doi.org/10.1109/ICUS55513.2022.9987097. https://ieeexplore.ieee.org/document/9987097/
5. Fan, S., Zhang, H., Zeng, Y., Cai, W.: Hybrid blockchain-based resource trading system for federated learning in edge computing. IEEE Internet Things J. **8**(4), 2252–2264 (2021). https://doi.org/10.1109/JIOT.2020.3028101
6. Ghorbani, A., Zou, J.: Data Shapley: Equitable Valuation of Data for Machine Learning. No. arXiv:1904.02868 (2019). https://doi.org/10.48550/arXiv.1904.02868. https://arxiv.org/abs/1904.02868. arXiv:1904.02868

7. Ghorbani, A., Zou, J.: Data shapley: equitable valuation of data for machine learning. In: Chaudhuri, K., Salakhutdinov, R. (eds.) Proceedings of the 36th International Conference on Machine Learning. Proceedings of Machine Learning Research, vol. 97, pp. 2242–2251. PMLR (2019). https://proceedings.mlr.press/v97/ghorbani19c.html

8. Hu, D., Chen, J., Zhou, H., Yu, K., Qian, B., Xu, W.: Leveraging blockchain for multi-operator access sharing management in internet of vehicles. IEEE Trans. Veh. Technol. **71**(3), 2774–2787 (2022). https://doi.org/10.1109/TVT.2021.3136364

9. Imteaj, A., Thakker, U., Wang, S., Li, J., Amini, M.H.: A survey on federated learning for resource-constrained IoT devices. IEEE Internet Things J. **9**(1), 1–24 (2022). https://doi.org/10.1109/JIOT.2021.3095077

10. Issa, W., Moustafa, N., Turnbull, B., Sohrabi, N., Tari, Z.: Blockchain-based federated learning for securing internet of things: a comprehensive survey. ACM Comput. Surv. **55**(9), 1–43 (2023). https://doi.org/10.1145/3560816

11. Jiang, L., Zheng, H., Tian, H., Xie, S., Zhang, Y.: Cooperative federated learning and model update verification in blockchain-empowered digital twin edge networks. IEEE Internet Things J. **9**(13), 11154–11167 (2022). https://doi.org/10.1109/JIOT.2021.3126207

12. Kang, J., Xiong, Z., Niyato, D., Zou, Y., Zhang, Y., Guizani, M.: Reliable federated learning for mobile networks. IEEE Wirel. Commun. **27**(2), 72–80 (2020). https://doi.org/10.1109/MWC.001.1900119

13. Li, D., Guo, Q., Yang, C., Yan, H.: Trusted data sharing mechanism based on blockchain and federated learning in space-air-ground integrated networks. Wirel. Commun. Mob. Comput. **2022**, 1–9 (2022). https://doi.org/10.1155/2022/5338876

14. Lin, X., Wu, J., Bashir, A.K., Li, J., Yang, W., Piran, M.J.: Blockchain-based incentive energy-knowledge trading in IoT: joint power transfer and Ai design. IEEE Internet Things J. **9**(16), 14685–14698 (2022). https://doi.org/10.1109/JIOT.2020.3024246

15. Liu, Y., Ai, Z., Sun, S., Zhang, S., Liu, Z., Yu, H.: FedCoin: a peer-to-peer payment system for federated learning. In: Yang, Q., Fan, L., Yu, H. (eds.) Federated Learning. LNCS (LNAI), vol. 12500, pp. 125–138. Springer, Cham (2020). https://doi.org/10.1007/978-3-030-63076-8_9

16. Liu, Z., Chen, Y., Yu, H., Liu, Y., Cui, L.: GTG-shapley: efficient and accurate participant contribution evaluation in federated learning. ACM Trans. Intell. Syst. Technol. **13**(4) (2022). https://doi.org/10.1145/3501811

17. Lo, S.K., et al.: Toward trustworthy AI: blockchain-based architecture design for accountability and fairness of federated learning systems. IEEE Internet Things J. **10**(4), 3276–3284 (2023). https://doi.org/10.1109/JIOT.2022.3144450

18. Ma, S., Cao, Y., Xiong, L.: Transparent contribution evaluation for secure federated learning on blockchain. In: 2021 IEEE 37th International Conference on Data Engineering Workshops (ICDEW), Chania, Greece, pp. 88–91. IEEE (2021). https://doi.org/10.1109/ICDEW53142.2021.00023. https://ieeexplore.ieee.org/document/9438754/

19. Nakamoto, S.: Bitcoin: a peer-to-peer electronic cash system (2009). Cryptography Mailing list at https://metzdowd.com

20. Qinnan, Z., Jianming, Z., Sheng, G., Zehui, X., Qingyang, D., Guirong, P.: Incentive mechanism for federated learning based on blockchain and bayesian game. SCIENTIA SINICA Informationis **52**(6), 971- (2022). https://doi.org/10.1360/SSI-2022-0020. https://www.sciengine.com/publisher/ScienceChinaPress/journal/SCIENTIASINICAInformationis/52/6/10.1360/SSI-2022-0020

21. Shapley, L.S.: A value for n-person games. Contributions to the Theory of Games (1953)
22. Shen, M., Duan, J., Zhu, L., Zhang, J., Du, X., Guizani, M.: Blockchain-based incentives for secure and collaborative data sharing in multiple clouds. IEEE J. Sel. Areas Commun. **38**(6), 1229–1241 (2020). https://doi.org/10.1109/JSAC.2020.2986619
23. Song, T., Tong, Y., Wei, S.: Profit allocation for federated learning. In: 2019 IEEE International Conference on Big Data (Big Data), pp. 2577–2586 (2019). https://doi.org/10.1109/BigData47090.2019.9006327
24. Touati, S., Radjef, M.S., Sais, L.: A Bayesian Monte Carlo method for computing the shapley value: application to weighted voting and bin packing games. Comput. Oper. Res. **125**, 105094 (2021). https://doi-org-s.libyc.nudt.edu.cn:443/10.1016/j.cor.2020.105094. https://www-sciencedirect-com-s.libyc.nudt.edu.cn:443/science/article/pii/S0305054820302112
25. Wang, T., Rausch, J., Zhang, C., Jia, R., Song, D.: A Principled Approach to Data Valuation for Federated Learning. No. arXiv:2009.06192 (2020). https://doi.org/10.48550/arXiv.2009.06192. https://arxiv.org/abs/2009.06192. arXiv:2009.06192
26. Wang, X., Zhao, Y., Qiu, C., Liu, Z., Nie, J., Leung, V.C.M.: Infedge: a blockchain-based incentive mechanism in hierarchical federated learning for end-edge-cloud communications. IEEE J. Sel. Areas Commun. (2022). https://doi.org/10.1109/JSAC.2022.3213323
27. Weng, J., Weng, J., Huang, H., Cai, C., Wang, C.: Fedserving: a federated prediction serving framework based on incentive mechanism. In: IEEE INFOCOM 2021 - IEEE Conference on Computer Communications, pp. 1–10 (2021). https://doi.org/10.1109/INFOCOM42981.2021.9488807
28. Wu, W., et al.: Consortium blockchain-enabled smart ESG reporting platform with token-based incentives for corporate crowdsensing. Comput. Ind. Eng. **172**, 108456 (2022). https://doi.org/10.1016/j.cie.2022.108456
29. Xu, L., Bao, T., Zhu, L.: Blockchain empowered differentially private and auditable data publishing in industrial iot. IEEE Trans. Industr. Inf. **17**(11), 7659–7668 (2021). https://doi.org/10.1109/TII.2020.3045038
30. Yang, Q., Liu, Y., Chen, T., Tong, Y.: Federated machine learning: concept and applications. ACM Trans. Intell. Syst. Technol. **10**(2), 1–19 (2019). https://doi.org/10.1145/3298981

Small-World Echo State Networks for Nonlinear Time-Series Prediction

Shu Mo[2], Kai Hu[3], Weibing Li[2], and Yongping Pan[1($\boxtimes$)]

[1] School of Advanced Manufacturing, Sun Yat-sen University, Shenzhen 518100,
China
panyongp@mail.sysu.edu.cn
[2] School of Computer Science and Engineering, Sun Yat-sen University,
Guangzhou 510006, China
mosh6@mail2.sysu.edu.cn, liwb53@mail.sysu.edu.cn
[3] School of Artificial Intelligence, Sun Yat-sen University, Zhuhai 519000, China
hukai8@mail2.sysu.edu.cn

Abstract. Echo state network (ESN) is a reservoir computing approach
for efficiently training recurrent neural networks. However, it sometimes
suffers from poor performance and robustness due to the non-trainable
reservoir. This paper proposes a novel computational framework for
ESNs to improve prediction performance and robustness. A small-world
network is applied as the reservoir topology, a biologically plausible unsupervised learning method named dual-threshold Bienenstock-Cooper-
Munro learning rule is applied to adjust reservoir weights adaptively,
and a recursive least-squares-based composite learning algorithm is introduced to update readout weights. The proposed method is compared with
several kinds of ESNs on the Mackey-Glass system, a benchmark problem
of nonlinear time-series prediction. Simulation results have shown that
the proposed method not only achieves the best prediction performance
but also exhibits remarkable robustness against noise.

Keywords: Recurrent neural network · Reservoir adaptation ·
Small-world network · Time-series prediction · Composite learning

1 Introduction

Reservoir computing (RC) is an effective learning framework based on a specific
type of recurrent neural networks (RNNs) and has gained considerable attention in recent years [1]. Echo state network (ESN) is one of the pioneering RC
approaches [2] and has been widely studied in the field of neural computation due
to its high learning efficiency and impressive performance in various applications,
e.g., see [3–10]. ESN is derived from RNNs with a pool of large sparsely interconnected neurons (called the reservoir), an input layer feeding external data to
the reservoir, and an output layer weighting reservoir states. The reservoir, consisting of randomly and recurrently connected nonlinear neurons, offers an effective implementation for RNNs. Moreover, the complex dynamics generated by

B. Luo et al. (Eds.): ICONIP 2023, LNCS 14448, pp. 550–560, 2024.
https://doi.org/10.1007/978-981-99-8082-6_42

the reservoir nonlinearly maps input data to spatiotemporal state patterns in a high-dimensional feature space, enabling the linear separation of state vectors of different classes. In this way, ESN effectively avoids gradient explosion/vanishing and reduces computational costs during the training of RNNs.

Although ESNs have demonstrated high performance and flexible implementation in many applications, they usually suffer from poor robustness due to the randomly generated and untrained reservoir. Naturally, there is an open question of choosing the reservoir topology to improve the performance and robustness of ESNs. Deng and Zhang [11] proposed a scale-free highly clustered ESN featured by short characteristic path length, high clustering coefficient, scale-free distribution, and hierarchical and distributed architecture, which enhances the echo state property. Xue et al. [12] developed a decoupled ESN comprising multiple sub-reservoirs interconnected by lateral inhibitory mechanisms, which mitigates the coupling effects among reservoir neurons. Rodan et al. [13] proposed a simple cycle reservoir that minimizes the complexity of the reservoir topology and parameter, which achieves performance comparable to the classical ESN. Cui et al. [14] proposed a mixture ESN with a dynamic reservoir topology based on a mixture of small-world and scale-free topology, exhibiting a broader spectral radius while maintaining a comparable short-term memory capacity to the original ESN. Qiao et al. [15] proposed a growing ESN with multiple sub-reservoirs that adaptively adjust the size, topology, and sparsity of the reservoir and showed better prediction performance and stronger robustness compared to several ESNs with the fixed reservoir size and topology. However, the growing ESN is inherently complex in both the network topology and training process compared to the classical ESN. Kawai et al. [16] constructed a reservoir with a small-world topology that enables ESNs to keep a stable echo state property when the spectral radius is more than a unit. Suarez et al. [17] constructed a neuromorphic reservoir endowed with biologically realistic connection patterns and exhibited good performance and robustness even at the edge of chaos.

While the aforementioned methods primarily concentrate on the topological optimization of ESNs, neural plasticity, a bio-inspired unsupervised adaptation mechanism, represents another approach to enhance the performance and robustness of ESNs. Yusoff et al. [18] employed two synaptic plasticity rules, including anti-Oja [19] and Bienenstock-Cooper-Munro (BCM) [20] learning rules, to tune reservoir weights adaptively, and showed largely improved prediction and classification performance of ESNs. Wang et al. [21] applied a structural plasticity rule proposed in [22] to simultaneously regulate the connection weights and topology within the reservoir, which expanded the memory capacity and overall stability of ESNs. Patane and Xibilia [23] introduced intrinsic plasticity to adapt reservoir parameters in ESNs to estimate some key variables in a sulfur recovery unit of a refinery plant. Morales et al. [24] applied the synaptic and intrinsic plasticity rules to ESNs and analyzed how unsupervised learning through plasticity rules affects the ESN architecture in a way that boosts its performance.

This paper proposes a novel computational framework for ESNs to improve prediction performance and robustness. The proposed approach involves three key steps. First, a small-world network is applied as the reservoir topology, which is less complex and more biologically plausible than the growing ESN; second, a biologically plausible unsupervised learning method named dual-threshold BCM (DTBCM) learning is introduced to tune reservoir weights adaptively; third, a recursive least-squares (RLS)-based composite learning algorithm is introduced to train readout weights, which can guarantee parameter convergence under a relaxed excitation condition named interval excitation.

Throughout this paper, $\mathbb{N}^+$ denotes the set of positive integers, and $\mathbb{R}$, $\mathbb{R}^+$, $\mathbb{R}^k$, and $\mathbb{R}^{n \times m}$ denote the spaces of real numbers, positive real numbers, real k-vectors, and real $n \times m$-matrices, respectively, where $k, n, m \in \mathbb{N}^+$.

2   Small-World Echo State Networks

A block diagram of the proposed small-world ESN is shown in Fig. 1, which consists of three modules: 1) the input layer (K external input neurons), 2) the small-world topology reservoir layer (N reservoir neurons) with internal weights tuned by the DTBCM learning rule in [25], and 3) the output layer (L readout neurons) with readout weights training by RLS-based composite learning algorithm in [10]. In general, the elements of the matrices W_{in}, W, and W_{fb} are randomly generated with certain uniform distributions and then keep fixed, while the only trainable weight matrix W_{out} is initialized to zero. In this study, W is updated by the DTBCM rule. Note that the inputs to reservoir neurons, the reservoir to output neurons, and the output to reservoir neurons are fully connected, while the reservoir neurons are sparsely connected, which indicates that W is a sparse matrix. For the topology shown in Fig. 1, the state and activation of the reservoir are iteratively updated as follows [2]:

$$x(k) = \phi(W_{\text{in}} u(k) + W x(k-1) + W_{\text{fb}} z(k-1)), \tag{1}$$
$$r(k) = (1 - \alpha) r(k-1) + \alpha x(k) \tag{2}$$

where $\phi(\cdot)$ denotes a nonlinear activation function (typically a hyperbolic tangent function) of the reservoir neurons working in an element-wise manner, and $\alpha \in (0, 1]$ is a constant leaky rate used to control the update rate of reservoir activation. In fact, α is data-sensitive, and its choice significantly affects the performance of ESNs. The network output $z(k)$ is a weighted sum of the reservoir activation $r(k)$, which is calculated as follows:

$$z(k) = r(k)^T W_{\text{out}}(k-1). \tag{3}$$

2.1   Topology of the Reservoir Layer

The Watts-Strogatz (WS) small-world network is a network structure between a regular and random network [26]. This network has a large clustering coefficient

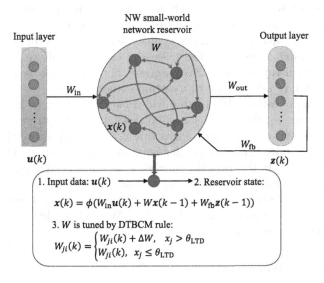

Fig. 1. The computational framework of the proposed small-world ESN, where $W_{\text{in}} \in \mathbb{R}^{N \times K}$, $W \in \mathbb{R}^{N \times N}$, $W_{\text{out}} \in \mathbb{R}^{N \times L}$ and $W_{\text{fb}} \in \mathbb{R}^{N \times L}$ denote input, internal, output, and feedback weight matrices, respectively, and $u(k) \in \mathbb{R}^K$, $x(k) \in \mathbb{R}^N$, $r(k) \in \mathbb{R}^N$ and $z(k) \in \mathbb{R}^L$ denote the network input, reservoir state, reservoir activation, and network output at the time epoch k, respectively. Note that $x(k)$ is driven by $u(k)$, $z(k-1)$ and its past value, and the DTBCM learning rule adaptively tunes W according to the state of presynaptic and postsynaptic neurons.

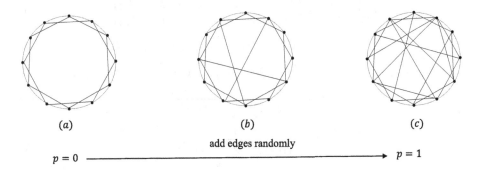

Fig. 2. An illustration of generating an NW small-world network. (a) A regular network consists of 12 nodes, where each node connects to the $K = 2$ nearest neighbors on each side. (b) The network is obtained by adding edges with a certain probability. (c) As the probability of adding edges increases, the network tends to be disordered.

and a small average path length, and the connections may experience disruptions, which are then followed by reconnection with a certain probability p. If $p = 0$, reconnection will not occur, resulting in a regular network; else if $p = 1$, all connections are subject to reconnection, ultimately creating a completely random network. The random reconnection process in the WS small-world network has the potential to disrupt the overall connectivity of the network, thereby hindering the transmission of information to a certain extent.

The Newman-Watts (NW) small-world network provides a slightly modified network structure to address the above limitation [27]. This structure also begins with a regular network but adds edges through randomization with a specific probability p, transforming the regular network into a small-world network, as illustrated in Fig. 2. In this way, the network structure can effectively prevent the occurrence of isolated nodes within the network. This network allows for faster information transmission, where even minor connection changes can dramatically impact network performance. Therefore, the NW small-world network is chosen as the topology of ESNs in this study.

2.2   Reservoir Weights Tuning

Neuroscientific studies have demonstrated that synaptic plasticity can be conducive to stabilizing synaptic connections and adapting to complex internal and external environments [28]. From the perspective of ESNs, synaptic plasticity rules aim to modify the reservoir neuron connection weights based on reservoir states stimulated by network inputs. In this manner, information embedded in input signals can be learned by synaptic plasticity rules. We apply the DTBCM rule to tune the internal weight W of the reservoir, since it is more biologically plausible and robust compared to other synaptic plasticity rules, such as the anti-Oja learning rule. The DTBCM rule is formulated as follows:

$$W_{ji}(k+1) = \begin{cases} W_{ji}(k) + \Delta W_{ji}(k), & x_j(k) > \theta_{\text{LTD}} \\ W_{ji}(k), & x_j(k) \leq \theta_{\text{LTD}} \end{cases}, \tag{4}$$

$$\Delta W_{ji}(k) = \eta \left[x_j(k) \left(x_j(k) - \theta_{\text{LTP}} \right) x_i(k) - \varepsilon W_{ji}(k) \right], \tag{5}$$

$$\theta_{\text{LTP}} = \frac{\sum\limits_{k'=k-h}^{k} x_j^2(k') e^{(k-k')}}{\sum_{k'=k-h}^{k} e^{(k-k')}}, \tag{6}$$

$$\theta_{\text{LTD}} = \rho \sum_{k'=k-h}^{k} x_j(k') \tag{7}$$

where $W_{ji}(k) \in \mathbb{R}$ is the weight between the presynaptic neuron i and the postsynaptic neuron j; $x_i \in \mathbb{R}$ and $x_j \in \mathbb{R}$ denote the states of the presynaptic neuron i and the postsynaptic neuron j, respectively; $\Delta W_{ji}(k) \in \mathbb{R}$ is the weight change; $\varepsilon \in \mathbb{R}^+$ is a time-decay constant that is the same for all synapses; $\theta_{\text{LTP}} \in \mathbb{R}$ is a sliding threshold for long-term potentiation (LTP) to determine the direction of synaptic weight changes, which is a time-weighted average of the squared postsynaptic response x_j over a time interval $h \in \mathbb{R}^+$; the exponential decay term in DTBCM serves as a forgetting factor; $\rho \in \mathbb{R}^+$ is a scaling parameter of postsynaptic neuronal average responses x_j within h; $\theta_{\mathrm{LTD}} \in \mathbb{R}$ acts as a sliding threshold for long-term depression (LTD) to determine the weight change, which is self-adjusted based on neuronal responses and is determined by the average postsynaptic neuronal responses within a specific time interval.

2.3   Readout Weights Training

First-order reduced and controlled error (FORCE) learning is an online training framework for ESNs with a feedback loop from output units to the reservoir [29]. It aims to stabilize complex and potentially chaotic dynamics of the reservoir by keeping a prediction error sufficiently small even from the initial training phase, such that the network output is almost equal to the target output. Nevertheless, FORCE learning does not fully utilize the potential of recurrent connectivity because the degree and form of modifications are limited, and RNNs trained by FORCE learning for complex tasks require more neurons to achieve the same performance as those networks trained by gradient-based methods [30].

FORCE learning that resorts to memory regression extension (MRE) (also termed composite FORCE learning) can enhance the learning speed, stability, and transient performance of the original FORCE learning [10]. The underlying concept involves using a stable filter with memory to create a novel extended regression equation [31]. To present this method, a prediction error between the network output $\mathbf{z}(k)$ and the target output $\mathbf{f}(k)$ is defined by

$$e(k) := r(k)^T \hat{W}_{\mathrm{out}}(k-1) - f(k) \qquad (8)$$

where $\hat{W}_{\mathrm{out}} \in R^{N \times L}$ is the estimate of W_{out}. Multiplying (8) by $\mathbf{r}(k)$, one gets an extended regression equation

$$r(k)e(k) = r(k)r(k)^T \hat{W}_{\mathrm{out}}(k-1) - r(k)f(k) \qquad (9)$$

Applying a stable filter $L(z) = \frac{\lambda}{1-(1-\lambda)z^{-1}}$ to (9), one gets

$$E(k) = \Omega(k)\hat{W}_{\mathrm{out}}(k-1) - Y(k) \qquad (10)$$

with $E(k) := L\left\{r(k)e(k)\right\}$, $Y(k) := L\left\{r(k)f(k)\right\}$, and $\Omega(k) := L\left\{r(k)r^T(k)\right\}$, where $\lambda \in \mathbb{R}^+$ is a filtering constant, and z is a Z-transform operator. Then, the MRE-RLS-based FORCE learning is summarized as follows:

$$\hat{W}_{\mathrm{out}}(k) = \hat{W}_{\mathrm{out}}(k-1) - P(k)(e(k)r(k) + \beta E(k)), \qquad (11)$$

$$P(k) = P(k-1) - \frac{P(k-1)r(k)r(k)^T P(k-1)}{1 + r(k)^T P(k-1)r(k)} \qquad (12)$$

where $P(k) \in R^{N \times N}$ is a symmetric positive definite learning rate matrix with $P(0) = \frac{I}{\alpha}$, in which $I \in R^{N \times N}$ denotes the identity matrix and $\alpha \in \mathbb{R}^+$ is a constant parameter. The constant gain $\beta \in \mathbb{R}^+$ controls the ratio of the prediction error $e(k)$ and the generalized error $E(k)$.

Table 1. Prediction performance of several methods on MGS-17 without noise.

Method	Testing NRMSE$_{84}$	Reservoir size	Sparsity
Growing ESN [15]	1.32e–04	1000	0.0050
Original ESN [32]	1.58e–04	1000	0.0100
Decoupled ESN [12]	1.50e–04	1000	0.0080
Simple cycle reservoir [13]	5.39e–04	1000	0.0010
RLS-FORCE [29]	5.21e–04	500	0.0200
NW-RLS-FORCE	4.24e–04	500	0.0171
DTBCM-RLS-FORCE	5.01e–04	500	0.0200
Composite-RLS-FORCE	1.75e–04	500	0.0200
The proposed method	**8.51e–05**	500	0.0171

3   Numerical Verification

This section introduces a benchmark for nonlinear time-series prediction and presents the corresponding simulation results, where simulations are performed using MATLAB 2022a platform, and the results of each trial are averaged after 100 runs.

The benchmark is the Mackey Glass system (MGS) which has been extensively employed to assess the ability of neural networks to identify complex dynamical systems, where it is formulated as follows [32]:

$$f(k+1) = f(k) + 0.1 \left[\frac{0.2f(k-\tau)}{1 + f(k-\tau)^{10}} - 0.1f(k) \right]. \tag{13}$$

The above MGS has a chaotic attractor when $\tau > 16.8$. Therefore, we set $\tau = 17$ (denoted by MGS-17) and $f(0) = 1.2$.

To evaluate the prediction performance for comparison purposes, we utilize a normalized root mean square error (NRMSE) defined by

$$\text{NRMSE} = \sqrt{\frac{\sum_{i=1}^{n}(f(i) - z(i))^2}{\sum_{i=1}^{n}(f(i) - \bar{f})}} \tag{14}$$

where $z \in \mathbb{R}$ and $f \in \mathbb{R}$ represent the network output and target output, and $\bar{f} \in \mathbb{R}$ refers to the mean of the whole target output.

For comparison, we set up the simulation environment according to [15] and generated 10000 data points according to (13) with 3000, 3000, and 4000 training, validation, and testing data points, respectively. For numerical configuration, set $\alpha = 0.8$, $\beta = 20$, $h = 10$, $\rho = 0.1$, $K = 4$, and $p = 0.0005$. The initial values of the input, reservoir, output, and feedback weight matrices (i.e., W_{in}, W, W_{out}, and W_{fb}) are randomly drawn from a uniform distribution within $[-1, 1]$.

Table 1 shows the average prediction results of the proposed ESN compared WITH several methods, including the original ESN [32], the growing ESN [15],

Table 2. Prediction performance of several methods on MGS-17 with noise.

Method	Testing NRMSE$_{84}$	Reservoir size	Sparsity
Growing ESN [15]	3.14e–02	280	0.018
Original ESN [32]	3.12e–02	400	0.035
Decoupled ESN [12]	3.10e–02	400	0.035
Simple cycle reservoir [13]	3.22e–02	300	0.0033
RLS-FORCE [29]	1.64–03	200	0.0400
NW-RLS-FORCE	9.73e–04	200	0.0340
DTBCM-RLS-FORCE	1.36e–03	200	0.0400
Composite-RLS-FORCE	6.02e–04	200	0.0400
The proposed method	**3.47e–04**	200	0.0340

the decoupled ESN [12], and the simple cycle reservoir [13], over 100 independent simulation trials on the MGS-17. We also independently investigate the impact of the NW small-world network, DTBCM rule, and composite FORCE learning on the performance of the proposed model. For the MGS-17 task, we compare the network prediction outputs with the target sequences of the MGS-17 after 84 steps to obtain a testing NRMSE, in which

$$\text{NRMSE}_{84} = \left(\sum_{i=1}^{100} (z_i(2084) - f_i(2084))^2 / 100\sigma^2 \right)^{\frac{1}{2}},$$

where $z_i(2084)$, $f_i(2084)$, and σ^2 are the network prediction output, target output, and variance of the whole MGS-17 sequence, respectively. It is shown that our approach achieves the highest average testing NRMSE$_{84}$, implying the best prediction performance, even utilizing a smaller network size.

To demonstrate that the proposed ESN is robust to noise, we insert uniform noise into the MGS-17, which is randomly generated from $[-0.01, 0.01]$. Table 2 lists the prediction results of all models in the noise case, which demonstrates the proposed ESN achieving the best prediction performance. The reason why the proposed method demonstrates robustness to noise can be summarized as follows. The small-world network exhibits a high clustering coefficient and short path length, enabling faster information propagation within the network. This facilitates the rapid transmission of useful information to the readout layer, enhancing the performance and robustness of the network [16]. The DTBCM rule uses a dual-threshold mechanism, where the weights are adjusted only when the correlation between pre- and post-synaptic neurons exceeds a certain threshold. This can help filter out noise in the input and thus improve the robustness to noise. The composite FORCE learning utilizes a transfer function to effectively capture the historical values of the activation matrix $r(k)r(k)^T$, which can not only play the role of filtering but also accelerate the convergence of the readout layer. The proposed model utilizes the above three techniques and achieves the best results, albeit at the expense of some computational burden.

4 Conclusion

In this paper, we have proposed a novel computational framework of ESNs to improve prediction performance and robustness. The proposed approach achieves a significantly improved performance and demonstrates robustness to noise in nonlinear time-series prediction compared to previous ESN approaches. This can be attributed to using a small-world network as the reservoir topology, leveraging the DTBCM learning rule to adjust reservoir weights, and training readout weights via the MRE-RLS-based FORCE learning. We would apply the proposed method to tackle more complicated modeling problems in the future.

Acknowledgment. This work was supported in part by the Guangdong Provincial Pearl River Talents Program, China, under Grant No. 2019QN01X154, and in part by the Fundamental Research Funds for the Central Universities, Sun Yat-sen University, China, under Grant No. 23lgzy004.

References

1. Nakajima, K., Fischer, I. (eds.): Reservoir Computing. NCS, Springer, Singapore (2021). https://doi.org/10.1007/978-981-13-1687-6
2. Jaeger, H.: The "echo state" approach to analysing and training recurrent neural networks-with an erratum note. In: German National Research Center for Information Technology GMD Technical Report, Bonn, Germany, vol. 148, no. 34, p. 13 (2001)
3. Park, J., Lee, B., Kang, S., Kim, P.Y., Kim, H.J.: Online learning control of hydraulic excavators based on echo-state networks. IEEE Trans. Autom. Sci. Eng. **14**(1), 249–259 (2016)
4. Schwedersky, B.B., Flesch, R.C.C., Dangui, H.A.S., Iervolino, L.A.: Practical nonlinear model predictive control using an echo state network model. In: International Joint Conference on Neural Networks, Rio de Janeiro, Brazil, 08–13 July 2018, pp. 1–8. IEEE (2018)
5. Jordanou, J.P., Antonelo, E.A., Camponogara, E.: Online learning control with echo state networks of an oil production platform. Eng. Appl. Artif. Intell. **85**(Oct.), 214–228 (2019)
6. Chen, Q., Shi, H., Sun, M.: Echo state network-based backstepping adaptive iterative learning control for strict-feedback systems: an error-tracking approach. IEEE Trans. Cybern. **50**(7), 3009–3022 (2019)
7. Wu, R., Li, Z., Pan, Y.: Adaptive echo state network robot control with guaranteed parameter convergence. In: Liu, X.-J., Nie, Z., Yu, J., Xie, F., Song, R. (eds.) ICIRA 2021. LNCS (LNAI), vol. 13016, pp. 587–595. Springer, Cham (2021). https://doi.org/10.1007/978-3-030-89092-6_53
8. Wu, R., Nakajima, K., Pan, Y.: Performance improvement of FORCE learning for chaotic echo state networks. In: Mantoro, T., Lee, M., Ayu, M.A., Wong, K.W., Hidayanto, A.N. (eds.) ICONIP 2021. LNCS, vol. 13109, pp. 262–272. Springer, Cham (2021). https://doi.org/10.1007/978-3-030-92270-2_23
9. Tanaka, K., Minami, Y., Tokudome, Y., Inoue, K., Kuniyoshi, Y., Nakajima, K.: Continuum-body-pose estimation from partial sensor information using recurrent neural networks. IEEE Robot. Autom. Lett. **7**(4), 11244–11251 (2022)

10. Li, Y., Hu, K., Nakajima, K., Pan, Y.: Composite FORCE learning of chaotic echo state networks for time-series prediction. In: Chinese Control Conference, Heifei, China, 25–27 July 2022, pp. 7355–7360 (2022)

11. Deng, Z., Zhang, Y.: Collective behavior of a small-world recurrent neural system with scale-free distribution. IEEE Trans. Neural Netw. **18**(5), 1364–1375 (2007)

12. Xue, Y., Yang, L., Haykin, S.: Decoupled echo state networks with lateral inhibition. Neural Netw. **20**(3), 365–376 (2007)

13. Rodan, A., Tino, P.: Minimum complexity echo state network. IEEE Trans. Neural Netw. **22**(1), 131–144 (2010)

14. Cui, H., Liu, X., Li, L.: The architecture of dynamic reservoir in the echo state network. Chaos **22**(3), 455 (2012)

15. Qiao, J., Li, F., Han, H., Li, W.: Growing echo-state network with multiple sub-reservoirs. IEEE Trans. Neural Netw. Learn. Syst. **28**(2), 391–404 (2016)

16. Kawai, Y., Park, J., Asada, M.: A small-world topology enhances the echo state property and signal propagation in reservoir computing. Neural Netw. **112**, 15–23 (2019)

17. Suárez, L.E., Richards, B.A., Lajoie, G., Misic, B.: Learning function from structure in neuromorphic networks. Nat. Mach. Intell. **3**(9), 771–786 (2021)

18. Yusoff, M.H., Chrol-Cannon, J., Jin, Y.: Modeling neural plasticity in echo state networks for classification and regression. Inf. Sci. **364–365**, 184–196 (2016)

19. Babinec, Š, Pospíchal, J.: Improving the prediction accuracy of echo state neural networks by anti-oja's learning. In: de Sá, J.M., Alexandre, L.A., Duch, W., Mandic, D. (eds.) ICANN 2007. LNCS, vol. 4668, pp. 19–28. Springer, Heidelberg (2007). https://doi.org/10.1007/978-3-540-74690-4_3

20. Castellani, G., Intrator, N., Shouval, H., Cooper, L.: Solutions of the BCM learning rule in a network of lateral interacting nonlinear neurons. Netw. Comput. Neural Syst. **10**(2), 111 (1999)

21. Wang, X., Jin, Y., Hao, K.: Computational modeling of structural synaptic plasticity in echo state networks. IEEE Trans. Cybern. **52**(10), 11254–11266 (2021)

22. Fauth, M., Wörgötter, F., Tetzlaff, C.: The formation of multi-synaptic connections by the interaction of synaptic and structural plasticity and their functional consequences. PLoS Comput. Biol. **11**(1), e1004031 (2015)

23. Patanè, L., Xibilia, M.G.: Echo-state networks for soft sensor design in an SRU process. Inf. Sci. **566**, 195–214 (2021)

24. Morales, G.B., Mirasso, C.R., Soriano, M.C.: Unveiling the role of plasticity rules in reservoir computing. Neurocomputing **461**, 705–715 (2021)

25. Wang, X., Jin, Y., Du, W., Wang, J.: Evolving dual-threshold Bienenstock-Cooper-Munro learning rules in echo state networks. IEEE Trans. Neural Netw. Learn. Syst., 1–12 (2022). https://doi.org/10.1109/TNNLS.2022.3184004

26. Watts, D.J., Strogatz, S.H.: Collective dynamics of 'small-world' networks. Nature **393**(6684), 440–442 (1998)

27. Newman, M.E., Watts, D.J.: Renormalization group analysis of the small-world network model. Phys. Lett. A **263**(4–6), 341–346 (1999)

28. Benito, E., Barco, A.: Creb's control of intrinsic and synaptic plasticity: implications for creb-dependent memory models. Trends Neuralsci. **33**(5), 230–240 (2010)

29. Sussillo, D., Abbott, L.F.: Generating coherent patterns of activity from chaotic neural networks. Neuron **63**(4), 544–557 (2009)

30. DePasquale, B., Cueva, C.J., Rajan, K., Escola, G.S., Abbott, L.: full-FORCE: a target-based method for training recurrent networks. PLoS ONE **13**(2), e0191527 (2018)
31. Pan, Y., Yu, H.: Composite learning from adaptive dynamic surface control. IEEE Trans. Autom. Control. **61**(9), 2603–2609 (2016)
32. Jaeger, H., Haas, H.: Harnessing nonlinearity: predicting chaotic systems and saving energy in wireless communication. Science **304**(5667), 78–80 (2004)

Preserving Potential Neighbors for Low-Degree Nodes via Reweighting in Link Prediction

Ziwei Li[1,2], Yucan Zhou[1(✉)], Haihui Fan[1], Xiaoyan Gu[1,2], Bo Li[1],
and Dan Meng[1]

[1] Institute of Information Engineering, Chinese Academy of Sciences, Beijing, China
{liziwei,zhouyucan,fanhaihui,guxiaoyan,libo,mengdan}@iie.ac.cn
[2] School of Cyber Security, University of Chinese Academy of Sciences, Beijing,
China

Abstract. Link prediction is an important task for graph data. Methods
based on graph neural networks achieve high accuracy by simultaneously
modeling the node attributes and structure of the observed graph. However, these methods often get worse performance for low-degree nodes.
After theoretical analysis, we find that current link prediction methods
focus more on negative samples for low-degree nodes, which makes it hard
to find potential neighbors for these nodes during inference. In order to
improve the performance on low-degree nodes, we first design a node-
wise score to quantify how seriously the training is biased to negative
samples. Based on the score, we develop a reweighting method called
harmonic weighting(HAW) to help the model preserve potential neighbors for low-degree nodes. Experimental results show that the model
combined with HAW can achieve better performance on most datasets.
By detailedly analyzing the performance on nodes with different degrees,
we find that HAW can preserve more potential neighbors for low-degree
nodes without reducing the performance of other nodes.

Keywords: Link prediction · Sample reweighting · Harmonic
weighting

1 Introduction

Graphs are widely used to model relations among data in the real world, such
as social networks [19], protein interaction networks [24], etc. Link prediction
is a fundamental task on graph data and is widely used in recommendation
systems [17], completing knowledge graphs, and predicting drug interactions
[18]. It focuses on predicting whether there exists an edge between two nodes,
where structure information and attribute information are both significant [21].
However, as these two kinds of information are heterogeneous, it is difficult to
utilize them simultaneously.

Intuitively, Graph Neural Networks(GNNs) can model both structure information and attribute information by enhancing attribute information via graph

B. Luo et al. (Eds.): ICONIP 2023, LNCS 14448, pp. 561–572, 2024.
https://doi.org/10.1007/978-981-99-8082-6_43

structure [7]. Thus, GNN based methods such as GAE [11] and SEAL [22] have achieved better performance on link prediction. Usually, they convert link prediction into a binary classification task, where node pairs with links are regarded as positive samples, and node pairs without links are negative samples. According to this definition, low-degree nodes are related to fewer positive samples and more negative samples. During training, models will pull nodes in each positive sample together and push nodes in each negative sample away. Thus, for low-degree nodes, models are more likely to push their potential neighbors away stronger than pull them together, which makes it hard to preserve potential neighbors for low-degree nodes.

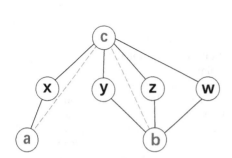

Fig. 1. Potential neighbor (b, c) is easier to predict than potential neighbor (a, c).

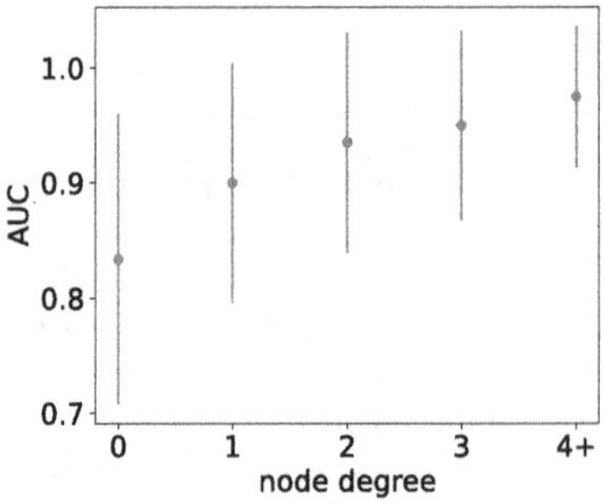

Fig. 2. Link prediction performance on nodes with different degrees on Citeseer.

As shown in Fig. 1, the node pairs (a, c) and (b, c) are both potential neighbors. Node "a" is a low-degree node and is only connected to node "x". The link between (a, x) is a positive sample, and the nonexistent links between node "a" and the other nodes are negative samples. Similarly, the links between (b, y/z/w) and (c, x/y/z/w) are positive samples and the nonexistent links between (b, a/c/x) and (c, a/b) are negative samples. Thus during training, node "c" is pulled close to node "x", "y", "z", and "w", node "a" is pulled close to "x" but pushed away from "c", "y", "z", and "w", node "b" is pulled close to "y", "z", and "w" but pushed away from "c" and "x". As a consequence, low-degree node "a" is pulled close to "c" only once because of the link (a, x) but pushed away from "c" four times because of the nonexistent links (a, c/y/z/w). As a contrast, node "b" is pulled close to "c" three times because of the links (b, y/z/w) but pushed away from "c" twice because of the nonexistent links (b, x/c). Thus, it is harder to preserve potential neighbors of node "a". Figure 2 shows the performance of GAE on nodes of different degrees in Citeseer [6] measured by AUC. Nodes with lower degrees get worse performance, which verifies again that it is harder to predict potential neighbors of low-degree nodes.

In order to preserve potential neighbors for low-degree nodes, an intuitive solution is providing more positive samples or reducing the effect of negative samples [4]. As the number of positive samples is determined by the observed graph, all we can do is reduce the effect of negative samples. Therefore, we design a general loss function weighting method called **harmonic weighting**(HAW), which assigns lower weights to the loss of negative samples. Specifically, we first theoretically analyze why models perform badly on low-degree nodes and quantify how seriously the training is biased to negative samples. Then, we induce a weighting formula to balance the quantified score for each node. Furthermore, we smooth the induced weights with harmonic average to achieve better performance. To show the effectiveness of HAW, we conduct experiments on several datasets by combining HAW with some commonly used link prediction methods.

Contributions. Our main contributions can be summarized as follows. (1) We find that current link prediction methods have difficulty in preserving potential neighbors for low-degree nodes and theoretically analyze the underlying reasons. (2) Based on the theoretical analysis, we derive a non-parametric generalized weighting function called HAW to help models preserve potential neighbors. (3) Experimental results show the effectiveness of HAW on several datasets when combined with different link prediction methods. The code of HAW is publicly available at https://github.com/lzwqbh/HAW.

2 Related Work

Early link prediction methods use heuristics to analyze the connectivity of two nodes and assume that a node pair of higher connectivity is more likely to be linked [1,9,10,23]. Common-neighbor analyzes the connectivity of two nodes by counting the number of shared neighbors between them. It is a first-order heuristic since it only involves the one-hop neighbors of two nodes. Katz [10] scores two nodes with a weighted sum of the number of all the paths between them. It is a high-order heuristic since it requires knowing the entire graph. However, these methods ignore the attribute information of nodes.

By enhancing attribute information via graph structure with graph neural networks, GNN based link prediction methods have achieved promising performance recently, which can be roughly divided into auto-encoder based methods [8,11,16,20] and subgraph based methods [2,3,15,22]. Auto-encoder based methods learn a representation for each node and treat the similarity of two embeddings as the link probability. As the first auto-encoder based method, GAE/VGAE [11] extend auto-encoder/variational auto-encoder into graph scenario by using a GNN as the encoder. ARVGA [16] assumes it is beneficial for models to regularize embeddings. It adds an adversarial training scheme to VGAE to regularize representations to Gaussian distribution. DGAE [20] deepens GAE by stacking multi-layer GNNs to enable GAE to use more structure to enhance attribute information. To avoid the model being over-smoothed, it combines representations at different layers to get the final embedding and uses each embedding to reconstruct node features for regularization. Methods based on subgraphs treat link prediction as a graph classification task. They extract nodes and links around two target nodes to construct a subgraph and use subgraph classification methods to evaluate the connectivity of two nodes which is further used to predict the link probability of two nodes. SEAL [22] proves traditional heuristic methods can be well approximated from subgraph and uses a GNN based graph classification method to evaluate connectivity personalized for each dataset. The graph classification method is composed of a GNN which learns an embedding for each node and a graph pooling layer which induces subgraph embedding from embeddings of nodes in the subgraph. As current pooling layers only consider node embeddings and ignore structure relation between nodes, the current state-of-the-art link prediction algorithm, WALKPOOL [15], designs walk-based pooling to make use of them at the same time. It significantly improves prediction accuracy compared with GAEs and SEAL.

All the GNN based methods mentioned above share the same procedure to select positive samples and negative samples. They treat linked node pairs as positive samples and treat node pairs without links as negative samples. Thus low-degree nodes get more negative samples than positive samples which makes these models hard to preserve potential neighbors of low-degree nodes.

3 Preliminaries

Given a graph $G = (V, E, X)$, $V = \{V_1, ..., V_N\}$ is the set of nodes, N is the number of nodes, E is the set of edges, $e(i, j) \in E$ indicates node "i" and node "j" exist link, and $X = \{X_1, ..., X_N\} \in \mathbb{R}^{N \times F}$ is the feature matrix. The adjacent matrix A is defined below:

$$A(i, j) = \begin{cases} 1 & e(i, j) \in E \\ 0 & otherwise. \end{cases} \tag{1}$$

Given a partially observed graph $G^o = (V, E^o, X)$, $E^o \subset E$, the link prediction method can be expressed as a function $f(E^o, X) = \bar{A}$, which attempts to predict $\bar{A}$ that is close to A. GNN based methods seek to use the observed graph to train GNN models, whose training objective can be expressed as:

$$\arg\min_{\theta} \mathcal{L}(f(E^o, X, \theta), A^o), \tag{2}$$

where θ is the parameter of the GNN model, $\mathcal{L}$ is the loss function that measures the quality of a prediction, and A^o is the adjacent matrix corresponding to E^o. As links are sparse in graphs, to avoid information of observed links vanishing, the loss function is separated into two parts relying on the state in A^o and both calculate the average of them. Still, calculating the prediction score for all node pairs where no link exists between them is impractical. As a consequence link prediction methods sample an affordable number of node pairs from them to train the GNN model. Considering the sampling procedure, we get the loss function that link prediction methods commonly used:

$$L = \frac{1}{|E^o|} \sum_{e(i,j) \in E^o} \mathcal{L}(\bar{A}(i, j), 1) + \frac{1}{|E^-|} \sum_{e^-(i,j) \in E^-} \mathcal{L}(\bar{A}(i, j), 0), \tag{3}$$

where E^- indicates the negative samples of links $E^- \cap E^o = \emptyset$, $|.|$ counts the number of items in a set.

4   Proposed Method

In this section, we first analyze the node-wise unbalance problem from a training perspective theoretically and get a node-wise score that indicates the unbalances between positive and negative samples for every node. Then based on the score we put forward the objective of balancing the bias, and give a feasible solution. In order to weaken the influence on nodes with higher degrees, we combine harmonic average and the thought of Laplace smoothing and put forward HAW.

4.1   Theoretical Analysis

In order to analyze why models perform differently for nodes with different degrees after training, we analyze the difference between nodes in the link prediction loss function. We rewrite the loss function as follows:

$$
L = \frac{\sum_{k=1}^{N} \left(\frac{1}{|E^o|} \sum_{e(i,j) \in E_k^o} \mathcal{L}(\bar{A}(i,j), 1) + \frac{1}{|E^-|} \sum_{e^-(i,j) \in E_k^-} \mathcal{L}(\bar{A}(i,j), 0) \right)}{2}, \quad (4)
$$

where $E_k^o = \{e | e(k,j) \text{ or } e(j,k) \in E^o\}$, $E_k^- = \{e^- | e^-(k,j) \text{ or } e^-(j,k) \in E^-\}$. We called each summing term in the rewriting formula of L as L_k:

$$
L_k \triangleq \frac{1}{|E^o|} \sum_{e(i,j) \in E_k^o} \mathcal{L}(\bar{A}(i,j), 1) + \frac{1}{|E^-|} \sum_{e^-(i,j) \in E_k^-} \mathcal{L}(\bar{A}(i,j), 0). \quad (5)
$$

L_k is highly related to the training of node "k". As discussed before, if a node has more negative samples than positive samples, models will find it harder to preserve its potential neighbors. Thus we define a node-wise unbalanced score:

$$
score_k \triangleq \frac{\frac{|E_k^-|}{|E^-|}}{\frac{|E_k^o|}{|E^o|}} = \frac{|E^o| \cdot |E_k^-|}{2 \cdot \deg(k) \cdot |E^-|}. \quad (6)
$$

The higher the score is, the harder for models to predict neighbors. As negative links are uniformly sampled from negative link space, we have:

$$
\mathbb{E}(score_k) = \mathbb{E}\left(\frac{|E^o| \cdot |E_k^-|}{2 \cdot \deg(k) \cdot |E^-|} \right) \approx \frac{|E^o| \cdot 2}{2 \cdot \deg(k) \cdot N} = \frac{\overline{\deg}(G^o)}{\deg(k)}, \quad (7)
$$

where $\overline{\deg}(G^o)$ is the average degree of all nodes in the observed graph. Thus, node unbalance is in inverse proportion to node degree, which means the lower the node degree is the harder is for models to predict potential neighbors.

4.2 Harmonic Weighting

To improve the performance on low-degree nodes, we need to design a negative sample weighting method to lower the unbalanced score of low-degree nodes. However, lowering the unbalanced score also makes negative samples have a lower weight than positive samples. Thus, we first design a weighting formula that balances the scores while keeping positive and negative samples balanced. Then we seek for an approximation that reduces the effect on high-degree nodes.

Given a weight strategy $W_{i,j}$ for negative samples, unbalanced score of node "k" becomes:

$$score_k = \frac{|E^o| \cdot \sum_{e^-(i,j)\in|E_k^-|} W_{i,j}}{2 \cdot \deg(k) \cdot |E^-|}. \tag{8}$$

In order to keep positive and negative sample balance in the loss function, $W_{i,j}$ needs to satisfy the following circumstances:

$$\sum_{e^-(i,j)\in E^-} W_{i,j} = |E^-|. \tag{9}$$

Note that when the unbalanced scores of all nodes are equal to one, we have the following equation set:

$$\begin{cases} |E^o| \cdot \sum_{e^-(i,j)\in E_1^-} W_{i,j} = 2 \cdot \deg(1) \cdot |E^-| \\ \quad\quad\quad\vdots \\ |E^o| \cdot \sum_{e^-(i,j)\in E_N^-} W_{i,j} = 2 \cdot \deg(N) \cdot |E^-| \end{cases}. \tag{10}$$

By adding all the equations, we get:

$$|E^o| \cdot 2 \cdot \sum_{e^-(i,j)\in E^-} W_{i,j} = 2 \cdot \sum_{k=1}^N \deg(k) \cdot |E^-| \tag{11}$$

$$\Rightarrow \sum_{e^-(i,j)\in E^-} W_{i,j} = |E^-|. \tag{12}$$

This means when unbalanced scores all equal one, positive and negative samples always balance. Thus we only need to design a weighting formula that satisfies N equations in (10). To simplify the design of the weighting strategy, we decompose $W_{i,j}$ by $W_{i,j} = W_i W_j$, which reduces the number of variables equals to the number of equations. As E^- is sampled from negative links, we calculate the expected value of the unbalanced score and can get the following result:

$$\begin{aligned} \mathbb{E}(score_k) &= \frac{|E^o|\cdot\mathbb{E}(\sum_{e^-(i,j)\in E_k^-} W_{i,j})}{2\cdot\deg(k)\cdot|E^-|} \\ &\approx \frac{|E^o|\cdot\frac{\mathbb{E}(|E_k^-|)}{N-1}(\sum_{i=1...N,i\neq k} W_i W_k)}{2\cdot\deg(k)\cdot|E^-|} \\ &\approx \frac{\overline{\deg(G^o)}\cdot W_k \sum_{i=1...N,i\neq k} W_i}{(N-1)\deg(k)}. \end{aligned} \tag{13}$$

Considering linked node pairs are a very small portion of all the node pairs, we have the following approximation based on (12):

$$\sum_{i=1}^{N} W_i = \sqrt{(\sum_{i=1}^{N} W_i)^2}$$

$$\approx \sqrt{\sum_{e^-(i,j) \in E_{all}^-} W_i W_j}$$

$$= \sqrt{\sum_{e^-(i,j) \in E_{all}^-} W_{i,j}}$$

$$= \sqrt{|E_{all}^-|}$$

$$\approx N. \tag{14}$$

E_{all}^- is the set of all negative samples. When W_k is far smaller than $\sum_{i=1}^{N} W_i$, we have:

$$\mathbb{E}(score_k) \approx \frac{\overline{\deg}(G^o) \cdot W_k \sum_{i=1}^{N} W_i}{(N-1)\deg(k)} \approx \frac{\overline{\deg}(G^o) \cdot W_k}{\deg(k)}. \tag{15}$$

We expect the score equal to one, thus we can set $W_k = \frac{\deg(k)}{\overline{\deg}(G^o)}$. However direct using this weighting strategy will influence nodes with higher degrees significantly, so we need to smooth the weight scores.

Laplace smoothing adds one when counting the number of each member to smooth the frequency of them. Inspired by that, we smooth the weighting score by averaging with one. We choose the harmonic average for smoothing so as to make the weighting score closer to the theoretical result for low-degree nodes and keep the score closer to one for others. Thus we call our weighting method harmonic weighting. The mathematical expression of HAW is:

$$HAW_k = \frac{2}{\frac{\overline{\deg}(G^o)}{\deg(k)} + 1} = \frac{2 \cdot \deg(k)}{\overline{\deg}(G^o) + \deg(k)}. \tag{16}$$

By using HAW to weight every negative sample term in the loss function we get the following HAW-based loss function:

$$L_{HAW} = \frac{1}{|E^o|} \sum_{e(i,j) \in E^o} \mathcal{L}(\bar{A}(i,j), 1)$$

$$+ \frac{1}{|E^-|} \sum_{e^-(i,j) \in E^-} \frac{4 \deg(i)\deg(j)}{(\overline{\deg}(G^o) + \deg(i))(\overline{\deg}(G^o) + \deg(j))} \mathcal{L}(\bar{A}(i,j), 0). \tag{17}$$

We use the HAW-based loss function for link prediction tasks to enhance existing link prediction methods.

Table 1. Details of datasets. Std is short for standard deviation.

Dataset	Cora	Citeseer	Pubmed	Texas	Wisconsin	Cornell
# nodes	2708	3327	19717	183	251	183
# edges	5278	4552	44324	279	450	277
average degree	3.90	2.74	4.50	3.05	3.59	3.03
std of degrees	5.23	3.38	7.43	7.83	7.94	7.04

5   Experiments

5.1   Experimental Setup

We conduct experiments on six datasets: Cora [13], Citeseer [6], Pubmed [14], Texas [5], Wisconsin [5], and Cornell [5]. Their details are shown in Table 1. Following the common settings [11,15,16], we randomly sample 85% links as a training set, 5% links as a validation set, and consider the rest 10% links as a testing set. We call this kind of sampling as link-wise uniform sampling.

However, this dataset partition will make the evaluation dominated by high-degree nodes. As the links are sampled uniformly for testing, there are fewer links to be predicted for low-degree nodes. In order to better reflect the performance of low-degree nodes, the number of links for low-degree nodes should be enlarged in the testing set. Therefore, we design a node-wise uniform sample method that samples a comparable number of links for nodes with different degrees. Specifically, the sampling probability of a link (m, n) is defined as follows:

$$p(m, n) = \frac{\frac{1}{degree_m \cdot degree_n}}{\sum_{e(i,j) \in E} \frac{1}{degree_i \cdot degree_j}}. \tag{18}$$

By using the node-wise uniform sampling, we can get a new testing set and a new validation set. By considering the rest of the links as the training set, we get new datasets for evaluation. For a comprehensive comparison, we evaluate performance by using AUC and AP on both kinds of sampling settings.

To show the effectiveness of HAW, we combine it with three auto-encoder based link prediction methods(GAE [11], VGAE [11], and ARVGA [16]) and the state-of-the-art subgraph classification based method called WALKPOOL [15][1]. To be more reliable, we run our experiments with 5 different random seeds and report the average performance and the corresponding standard deviation.

5.2   Experimental Results

Table 2, 3, 4 and 5 show the link prediction performance of AUC and AP under datasets constructed with link-wise uniform sampling and node-wise uniform sampling. The best and suboptimal performances are marked in bold and under-lined, respectively. We can draw the following conclusions:

---

[1] We use WP as a shorthand.

Table 2. Results measured by AUC under link-wise uniform sampling.

	Cora	Citeseer	Pubmed	Texas	Wisconsin	Cornell
GAE	91.78±0.97	91.09±1.45	97.04±0.15	67.24±17.13	66.37±10.13	77.15±4.50
VGAE	91.29±0.54	90.60±1.84	96.47±0.12	71.36±7.92	70.71±8.41	72.21±13.68
ARVGA	92.50±0.33	91.84±0.89	96.73±0.12	71.33±8.64	75.34±6.05	77.15±3.61
WP	94.62±0.55	92.51±0.84	**98.62±0.10**	74.84±4.63	82.95±6.54	85.19±5.44
HAW w/ GAE	92.83±0.52	92.24±0.68	96.77±0.09	73.72±6.88	72.12±7.99	78.03±2.81
HAW w/ VGAE	92.83±0.53	91.80±2.07	96.14±0.16	**77.56±4.18**	72.93±8.57	77.92±3.78
HAW w/ ARVGA	93.39±0.90	93.46±0.53	96.29±0.18	75.94±2.14	76.13±6.22	77.78±4.34
HAW w/ WP	**94.81±0.49**	**93.51±1.19**	98.52±0.12	76.33±3.97	**83.21±6.63**	**86.30±6.07**

Table 3. Results measured by AUC under node-wise uniform sampling.

	Cora	Citeseer	Pubmed	Texas	Wisconsin	Cornell
GAE	86.47±0.85	82.64±1.87	93.91±0.36	57.23±7.70	52.33±7.88	60.06±10.32
VGAE	86.32±0.86	82.10±1.42	92.41±0.59	54.49±10.37	57.93±5.32	57.48±11.57
ARVGA	88.03±0.66	84.00±1.32	92.67±0.38	62.50±5.20	53.89±2.77	66.23±7.94
WP	88.20±1.66	85.15±3.05	**94.81±0.17**	59.94±8.75	67.82±5.39	66.58±6.99
HAW w/ GAE	89.31±0.50	85.81±1.22	94.01±0.24	64.69±3.92	61.58±3.15	63.29±7.09
HAW w/ VGAE	89.46±0.86	85.59±0.63	92.66±0.60	58.52±7.13	67.96±2.65	60.12±10.64
HAW w/ ARVGA	**90.13±0.47**	**88.76±1.61**	92.93±0.42	**65.32±3.65**	59.91±7.82	66.75±7.30
HAW w/ WP	89.00±1.74	86.15±2.62	94.46±0.14	60.93±9.25	**68.05±5.41**	**67.74±5.57**

Table 4. Results measured by AP under link-wise uniform sampling.

	Cora	Citeseer	Pubmed	Texas	Wisconsin	Cornell
GAE	92.19±0.90	92.07±1.12	97.23±0.16	71.48±14.22	72.33±7.82	81.72±3.72
VGAE	92.59±0.40	91.92±1.31	96.58±0.07	78.15±6.29	75.18±8.80	78.23±11.59
ARVGA	93.11±0.39	92.77±0.55	96.82±0.15	78.31±7.27	79.11±5.21	82.61±2.86
WP	95.01±0.54	93.60±0.83	**98.58±0.10**	80.43±4.04	86.08±5.09	88.20±4.92
HAW w/ GAE	93.57±0.52	92.88±0.64	96.96±0.02	80.47±4.65	76.32±8.12	82.13±2.56
HAW w/ VGAE	93.91±0.44	93.19±1.54	96.28±0.13	**82.66±3.43**	77.04±8.11	82.81±3.31
HAW w/ ARVGA	93.50±1.26	94.02±0.19	96.29±0.20	81.56±2.74	79.64±5.01	82.35±3.24
HAW w/ WP	**95.12±0.62**	**94.23±1.03**	98.52±0.12	80.84±4.75	**86.39±5.27**	**88.71±5.79**

Table 5. Results measured by AP under node-wise uniform sampling.

	Cora	Citeseer	Pubmed	Texas	Wisconsin	Cornell
GAE	85.75±1.11	80.89±1.82	93.55±0.44	62.91±5.54	56.97±4.48	58.74±8.43
VGAE	86.05±0.95	81.49±1.84	91.74±0.66	61.48±7.41	60.36±4.55	56.39±7.93
ARVGA	87.07±1.10	82.82±1.65	91.87±0.43	66.92±3.18	56.65±5.28	67.82±5.39
WP	88.68±1.26	84.61±2.92	**94.75±0.11**	63.45±7.42	72.71±2.74	70.90±7.11
HAW w/ GAE	88.93±0.79	85.00±1.79	93.68±0.31	65.23±5.92	63.28±4.77	60.63±6.31
HAW w/ VGAE	89.07±1.25	85.87±0.89	92.09±0.65	64.60±6.04	70.53±3.63	58.46±8.67
HAW w/ ARVGA	**89.27±0.72**	**88.52±2.07**	92.73±0.56	**68.36±2.56**	61.10±7.78	68.05±5.41
HAW w/ WP	89.13±1.55	85.77±2.74	94.47±0.21	63.35±8.69	**72.90±2.93**	**72.20±5.50**

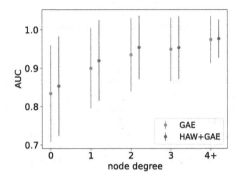

Fig. 3. The figure shows the performance of GAE and HAW+GAE on different degrees of nodes in Citeseer. Dots indicate the average score and lines indicate the scale of standard deviation. To show it clearly, red lines and red dots shift right a little bit. (Color figure online)

Firstly, methods with HAW achieve better performance on most datasets. For example, HAW improves the performance of all methods on Citeseer. Besides, the improvements of HAW are more obvious on difficult datasets(e.g., Texas and Wisconsin) where link prediction methods get worse results.

Secondly, HAW gets a larger performance gain in node-wise uniform sampling settings, which indicates HAW can improve the performance on low-degree nodes significantly. Specifically, on Cora and Citeseer, models with HAW get consistently better performance than the state-of-the-art WALKPOOL method. On the other hand, compared with link-wise uniform sampling, the performance of all basic models with node-wise uniform sampling drops severely, which means it is harder for models to predict potential neighbors for low-degree nodes.

Thirdly, our HAW has not achieved better results on Pubmed. We think this is because our HAW is designed to improve the performance of low-degree nodes. However, Pubmed has a higher average degree than other datasets, which indicates there exist more high-degree nodes.

In order to further show the difference of HAW on nodes of different degrees, we compared the performance of GAE and HAW+GAE on Citeseer in Fig. 3. It can be concluded that performance on low-degree nodes is significantly improved, without damage to the performance of high-degree nodes.

5.3   Visualization

To understand the difference between basic methods and methods with HAW, we visualize the embeddings learned by GAE and HAW+GAE using 2D t-SNE [12]. As shown in Fig. 4, the embeddings of low-degree nodes learned by GAE are almost randomly distributed in the embedding space. In contrast, when combined with HAW, the embeddings of low-degree nodes gather around high-degree nodes. That means HAW enables models to preserve similarity relations of low-degree nodes with others and thus helps models predict potential neighbors.

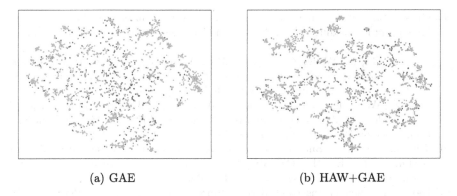

(a) GAE   (b) HAW+GAE

Fig. 4. Visualization of the embeddings of Citeseer. Zero-degree nodes are colored red, one-degree nodes are colored orange, and the other nodes are colored blue. (Color figure online)

6 Conclusion and Future Work

In this work, we first find and analyze the performance unbalance on link prediction methods for nodes with different degrees, and then we propose a reweighting strategy called HAW to preserve potential neighbors for low-degree nodes in link prediction. By assigning lower weights to negative samples that are related to low-degree nodes, HAW can help the model focus on the information of positive samples and keep the loss of positive and negative samples balanced. Experimental results show that link prediction methods can achieve better performance when combined with HAW. This work raises attention to the performance of low-degree nodes in link prediction. In the future, we will explore data augmentation methods to further improve the performance of low-degree nodes.

Acknowledgement. This work was supported in part by No.XDC02050200 and No.E110101114 program.

References

1. Brin, S., Page, L.: The anatomy of a large-scale hypertextual web search engine. Comput. Netw. ISDN Syst. **30**(1–7), 107–117 (1998)
2. Cai, L., Ji, S.: A multi-scale approach for graph link prediction. In: Proceedings of the AAAI Conference on Artificial Intelligence, vol. 34, pp. 3308–3315 (2020)
3. Cai, L., Li, J., Wang, J., Ji, S.: Line graph neural networks for link prediction. IEEE Trans. Pattern Anal. Mach. Intell. **44**(9), 5103–5113 (2021)
4. Chen, X., et al.: Imagine by reasoning: a reasoning-based implicit semantic data augmentation for long-tailed classification. In: Proceedings of the AAAI Conference on Artificial Intelligence, vol. 36, pp. 356–364 (2022)
5. Craven, M., et al.: Learning to extract symbolic knowledge from the world wide web. In: AAAI/IAAI, pp. 509–516 (1998)

6. Giles, C.L., Bollacker, K.D., Lawrence, S.: Citeseer: an automatic citation indexing system. In: Proceedings of the third ACM Conference on Digital Libraries, pp. 89–98 (1998)

7. Guo, Y., Gu, X., Wang, Z., Fan, H., Li, B., Wang, W.: Rcs: an attributed community search approach based on representation learning. In: 2021 International Joint Conference on Neural Networks (IJCNN), pp. 1–8. IEEE (2021)

8. Guo, Z., Wang, F., Yao, K., Liang, J., Wang, Z.: Multi-scale variational graph autoencoder for link prediction. In: Proceedings of the Fifteenth ACM International Conference on Web Search and Data Mining, pp. 334–342 (2022)

9. Jeh, G., Widom, J.: Simrank: a measure of structural-context similarity. In: Proceedings of the Eighth ACM SIGKDD International Conference on Knowledge Discovery and Data Mining, pp. 538–543 (2002)

10. Katz, L.: A new status index derived from sociometric analysis. Psychometrika **18**(1), 39–43 (1953)

11. Kipf, T.N., Welling, M.: Variational graph auto-encoders. arXiv preprint arXiv:1611.07308 (2016)

12. Van der Maaten, L., Hinton, G.: Visualizing data using t-sne. J. Mach. Learn. Res. **9**(11), 2579–2605 (2008)

13. McCallum, A.K., Nigam, K., Rennie, J., Seymore, K.: Automating the construction of internet portals with machine learning. Inf. Retr. **3**, 127–163 (2000)

14. Namata, G., London, B., Getoor, L., Huang, B., Edu, U.: Query-driven active surveying for collective classification. In: 10th International Workshop on Mining and Learning with Graphs, vol. 8, pp. 1–8 (2012)

15. Pan, L., Shi, C., Dokmanić, I.: Neural link prediction with walk pooling. In: International Conference on Learning Representations (2022)

16. Pan, S., Hu, R., Long, G., Jiang, J., Yao, L., Zhang, C.: Adversarially regularized graph autoencoder for graph embedding. In: Proceedings of the Twenty-Seventh International Joint Conference on Artificial Intelligence, pp. 2609–2615 (2018)

17. Qian, M., Gu, X., Chu, L., Dai, F., Fan, H., Li, B.: Flexible order aware sequential recommendation. In: Proceedings of the 2022 International Conference on Multimedia Retrieval, pp. 109–117 (2022)

18. Stanfield, Z., Coşkun, M., Koyutürk, M.: Drug response prediction as a link prediction problem. Sci. Rep. **7**(1), 1–13 (2017)

19. Wang, Z., Tan, Y., Zhang, M.: Graph-based recommendation on social networks. In: Proceedings of the 12th Asia-Pacific Web Conference, pp. 116–122. IEEE (2010)

20. Wu, X., Cheng, Q.: Stabilizing and enhancing link prediction through deepened graph auto-encoders. In: Proceedings of the Thirty-First International Joint Conference on Artificial Intelligence. vol. 2022, pp. 3587–3593. NIH Public Access (2022)

21. Yang, Y., Gu, X., Fan, H., Li, B., Wang, W.: Multi-granularity evolution network for dynamic link prediction. In: Pacific-Asia Conference on Knowledge Discovery and Data Mining, pp. 393–405. Springer, Heidelberg (2022). https://doi.org/10.1007/978-3-031-05933-9_31

22. Zhang, M., Chen, Y.: Link prediction based on graph neural networks. Adv. Neural Inf. Process. Syst. **31**, 5171–5181 (2018)

23. Zhou, T., Lü, L., Zhang, Y.C.: Predicting missing links via local information. Eur. Phys. J. B **71**, 623–630 (2009)

24. Zitnik, M., Leskovec, J.: Predicting multicellular function through multi-layer tissue networks. Bioinformatics **33**(14), i190–i198 (2017)

6D Object Pose Estimation with Attention Aware Bi-gated Fusion

Laichao Wang[1,2], Weiding Lu[1,2], Yuan Tian[3], Yong Guan[1,2],
Zhenzhou Shao[1,2(✉)], and Zhiping Shi[1]

[1] College of Information Engineering, Capital Normal University, Beijing 100048,
China
{2211002004,2211002065,guanyong,shizp}@cnu.edu.cn
[2] Beijing Key Laboratory of Light Industrial Robot and Safety Verification, Capital
Normal University, Beijing 100048, China
zshao@cnu.edu.cn
[3] Industrial and Commercial Bank of China Limited Beijing Branch, Beijing 100032,
China
tianyuan2727@126.com

Abstract. Accurate object pose estimation is a prerequisite for successful robotic grasping tasks. Currently keypoint-based pose estimation methods using RGB-D data have shown promising results in simple environments. However, how to fuse the complementary features from RGB-D data is still a challenging task. To this end, this paper proposes a two-branch network with attention aware bi-gated fusion (A2BF) module for the keypoint-based 6D object pose estimation, named A2BNet for abbreviation. A2BF module consists of two key components, bidirectional gated fusion and attention mechanism modules to effectively extract information from both RGB and point cloud data, prioritizing crucial details while disregarding irrelevant information. Several A2BF modules can be embedded in the network to generate complementary texture and geometric information. Extensive experiments are conducted on the public LineMOD and Occlusion LineMOD datasets. Experimental results demonstrate that the average accuracy using the proposed method on both datasets can reach 99.8% and 67.6% respectively, outperforms the state-of-the-art methods.

Keywords: Object pose estimation · Gated fusion · Attention
mechanism

1 Introduction

Robotic grasping in 3D scenes is a fundamental task that has a wide range of
applications, including object sorting, palletizing, material handling, and object

Supported by the Natural Science Foundation of China (62272322, 62002246, 62272323)
and the Project of Beijing Municipal Education Commission (KM202010028010) and
Applied Basic Research Project of Liaoning Province(2022JH2/101300279).

manipulation [1]. Accurate 6D pose estimation of objects is one of the prerequisites for successful robotic grasping. Although numeral methods have been developed to address the problem of robotic grasping, the complex environment in the real world often poses a great challenge to achieving high accuracy in 6D pose estimation. For example, changes in lighting conditions, background interference, lack of textures on objects, and occlusion usually reduce the performance of 6D pose estimation significantly.

The research on pose estimation started with methods based on RGB images. In recent years, with the significant development of deep learning techniques, researchers have used convolutional neural networks (CNNs) to address the challenges caused by the limitations of RGB images, such as sensor noise, changes in lighting conditions, background interference [2–4]. However, compared to RGB-D-based methods, these methods have inferior pose estimation accuracy due to the absence of geometric information. Application of RGB-D-based pose estimation methods has attract attention recently [5,6,16,17].

Generally, RGB-D based 6D pose estimation methods are mainly divided into two categories, including end-to-end methods and keypoint detection methods. The former ones have a higher computational cost since they need to take into account the decoupling problem of forecasting rotation and translation and involve direct regressing on both rotation and translation [2,5,6]. The latter based on regressing keypoints has a lower computational cost compared to the end-to-end approach. It also performs well in challenging conditions such as varying lighting and object occlusions [3,6,8,10]. Therefore, in this paper, we develop a keypoint-based pose estimation method.

How to integrate the characteristics and advantages of the features extracted from the two modalities is still a noteworthy question for RGB-D based methods. Current fusion strategies used in the RGB-D based pose estimation methods rely heavily on direct concatenation or nearest neighbor-based fusion, which still has room for improvement [3,5]. Since multi-modality features may contribute differently to performance gain, focusing on the crucial information of RGB and depth features during fusion may has the potential to further improve the accuracy of 6D pose estimation.

To address the above-mentioned issues, we propose an attention-aware bi-gated fusion Network (A2BNet) to extract representative features from RGB and point cloud images. The flowchart of the proposed method is shown in Fig. 1. The proposed method uses a bidirectional feature fusion module to selectively fuse texture and geometric information using gate fusion. In addition, a dual-path attention feature extraction module is designed to further enhance the effectiveness of feature extraction. In summary, the main contributions of the proposed method are three-fold:

- We propose a distinctive gate mechanism. This method selectively integrates texture and geometric features from both RGB and point cloud data, resulting in a unique feature fusion. This novel fusion approach not only enhances the accuracy of pose estimation but also underscores our innovative contribution in information integration.

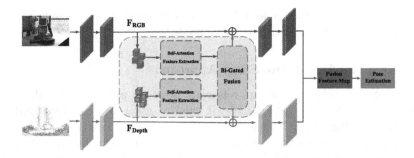

Fig. 1. The flowchart of the proposed 6D pose estimation based on attention-aware bi-gated fusion method (A2BNet).

- In the feature extraction stage, a simple and efficient feature selection method is proposed based on an attention mechanism. The network powered by the attention mechanism tends to extract useful and informative features while disregarding redundant ones, improving both feature extraction efficiency and subsequent pose estimation accuracy.
- To validate the feasibility of our approach, we conducted experiments on the LineMOD dataset and the Occlusion LineMOD dataset. The experimental results demonstrate the advantages of our method over existing state-of-the-art approaches.

2   Related Work

This section introduces two groups of methods for pose estimation, i.e., the end-to-end one-stage methods based on the holistic approach and the two-stage methods based on keypoint detection. The holistic approach usually use an end-to-end structure where the network directly outputs predictions for the rotation and translation of the target object. The keypoint detection approach usually estimates the rotation and translation first with algorithms such as least squares, and then the keypoints of the target object are predicted by the network.

2.1   End-to-End One-Stage Approach

The one-stage method exhibits faster processing speed, however, its accuracy is relatively lower, especially when detecting small objects and occluded objects, where the detection performance is unsatisfactory. CDPN [14] disentangles the pose by separately predicting rotation and translation, addressed the issue of inconsistent accuracy of Translation when facing different objects.PoseCNN [2] introduces a new loss function for symmetric objects and performs well in handling occlusion and symmetric objects. DeepIM [21] utilizes a separate 3D position and 3D orientation representation to predict the relative pose transformation. In recent years, methods such as G2L-Net [9] extract coarse point clouds from RGB-D images, incorporate the point clouds into the network for 3D instance segmentation and object translation prediction, and then convert the refined point clouds into local canonical coordinates for object rotation prediction.

2.2 Keypoint-Based Two-Stage Approach

The Two-Stage method, although relatively slower in processing speed, demonstrates higher accuracy. Additionally, it exhibits fewer false positives and false negatives, resulting in a reduced number of both false detections and missed detections. Hu *et al.* [8] proposed a method which performs better in scenarios where objects are occluded. PVNet [3] predicts the direction from each pixel to each keypoint. Even if a keypoint is not visible, it can be located using information from another visible part, which helps in predicting the pose of occluded objects. DenseFusion [5] fuses 2D images with 3D point cloud information to obtain more accurate and robust object pose estimation results. Hybrid-Pose [19] employs a network to predict different intermediate representations, and utilizes regression module to filter out outliers in the predicted intermediate representations. CRT-6D [7] employ a lightweight deformable Transformer that is connected iteratively to optimize the proposed pose estimation on sampled OSKFs(Object Spatial Keypoints Features). PVN3D [6] is built upon the DenseFusion network, predicting the directions of pixels and keypoints enhances local features and significantly improves the accuracy of 6D pose estimation. PR-GCN [16] combines point refinement networks and multimodal fusion graph convolutional networks to address ineffective representations of depth data and insufficient integration of different modalities. PAV-Net [17] can effectively mitigate external interferences and enhance the efficiency of utilizing influential point features through point-wise attention voting. FFB6D [10] proposes a bidirectional fusion strategy based on neighboring points and pixels. It performs bidirectional fusion in each convolutional layer during the encoding and decoding stages.

3   Method

3.1   Overview

The proposed method, outlined in Fig. 2a, involves two branches. The initial branch utilizes ResNet-34 [11] and PSPNet [12] to establish the encoder and decoder, respectively, extracting features from RGB images. The second branch, employing RandLA-Net [24], constructs both encoder and decoder for depth image representation. Incorporated within convolutional layers (as depicted in Fig. 2a), A2BF modules enrich the fusion process. Extracted texture and geometric features then combine, proceeding to the pose estimation module for generating 3D object keypoints. Ultimately, the object's rotation and translation transformations are computed through the least squares method for pose estimation.

The A2BF module integrates a self-attention module and a gate fusion module, as shown in Fig. 2b. The self-attention module computes attention weights

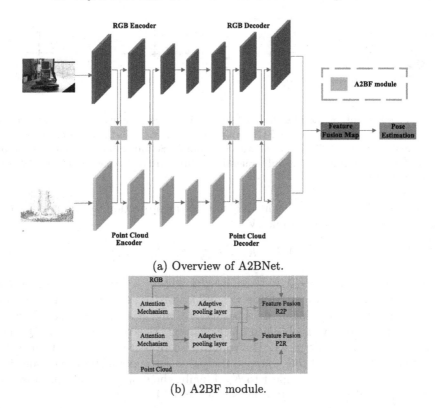

(a) Overview of A2BNet.

(b) A2BF module.

Fig. 2. Overview of A2BNet and A2BF module.

for both RGB and point cloud branches, highlighting critical information across modalities for performance enhancement. The feature fusion module retains complementary details while excluding irrelevant data. Subsequently, features from the gate fusion module proceed to the next feature layer. The ensuing sections detail the introduction of the two A2BF sub-modules.

3.2 Self-attention Mechanism Module

This module utilizes self-attention mechanism to explore the correlations between feature vectors in both RGB and point cloud data. By examining the relevance among these vectors, it identifies the texture and geometry features that are worth focusing on during the feature extraction stage. This enables the module to perform feature selection from both RGB and point cloud data, which then serve as the inputs to the subsequent feature fusion module.

As shown in Fig. 3, for RGB data, the module processes the input tensor into a sequence consisting of discrete tokens. Each token is represented by a feature vector. This module represents the input sequence of RGB data as $F_{rgb}^{in} \in \mathbb{R}^{N \times D_{frgb}}$, F_{rgb}^{in} represents the input sequence, N represents the number of tokens in the sequence, and each token is represented by a feature vector

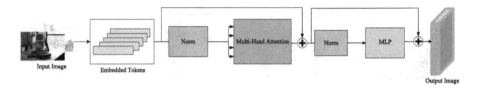

Fig. 3. Self-attention module in RGB and point cloud branches. Input features of dimensions $640 \times 480 \times C$ in the RGB branch ($1 \times 12288 \times C$ in the point cloud branch) are transformed into 307,200 tokens with C channels. Self-attention computes attention scores for each feature vector. Subsequently, features are resized back to $640 \times 480 \times C$ ($1 \times 12288 \times C$).

of dimensionality D_{frgb}. This module uses linear projection to compute a set of queries(Q), keys(K), and values(V).

$$Q_r = F_{rgb}^{in} M^q, \quad K_r = F_{rgb}^{in} M^k, \quad V_r = F_{rgb}^{in} M^v, \tag{1}$$

where $M^q \in \mathbb{R}^{D_{frgb} \times D_q}, M^k \in \mathbb{R}^{D_{frgb} \times D_q}, M^v \in \mathbb{R}^{D_{frgb} \times D_q}$ represents the weight matrix. Next, the attention weights for texture feature vectors are computed using the scaled dot product between Q_r and K_r. The similarity between the vectors Q_r and K_r is obtained by calculating the score vector using Qr multiplied by the transpose of Kr. The resulting score vector is divided by a scalar represented by D_k, and then mapped through a softmax function to obtain similarity scores of Q_r with respect to K_r. Finally, the scores are multiplied by the vector V_r, resulting in the attention score vector A_r.

$$A_r = softmax(\frac{Q_r K_r{}^T}{\sqrt{D_k}}) \tag{2}$$

Finally, the network utilizes a multi-layer perception (MLP) with non-linear transformations to compute the output features F_{rgb}^{out}. The storage of the output features is the same as the input features.

$$F_{rgb}^{out} = MLP(A_r) + F_{rgb}^{in} \tag{3}$$

Similarly, this module utilizes attention mechanism to calculate the correlation of point cloud data that contains geometric information, resulting in F_{point}^{out}.

3.3   Gate Fusion Module

After using the self-attention module to filter out the texture features and geometric features that are worthy of attention from both modalities, this section employs gate fusion module to fusion the extracted features from the two modalities.

The proposed gate fusion module is applied simultaneously to the RGB and point cloud branches, named the Point to RGB(P2R) fusion module and RGB to Point(R2P) fusion module. as shown in Fig. 4. They have identical structures with the same number of channels, denoted as C, but different feature sizes.

These two modules utilize reset and update gates to control the information filtering process.

To avoid neglecting the importance of texture and geometric information in the gating information, a gating state weight calculation step is proposed in this paper, which introduces the learning of weights and error calculation, enables more accurate filtering and fusion of information by the reset gate and update gate.

$$r = \sigma(W_r(U_{rgb} \odot f^{rgb} + U_{point} \odot f^{p2r} + b)), \qquad (4)$$

$$z = \sigma(W_z(U_{rgb} \odot f^{rgb} + U_{point} \odot f^{p2r} + b)), \qquad (5)$$

where U_{rgb} and U_{point} refer to the weighted generation part of f^{rgb} and f^{point}. b represents the error calculation. Afterwards, each data in the feature tensor is transformed into a value ranging from 0 to 1 using convolutional layers and an activation function, acting as gate signals.

To retain the memory information f^{rgb} from RGB and concatenate it with another input feature f^{point}, a convolution operation is performed, followed by an activation function to obtain the current candidate feature. The element values of the candidate feature range from -1 to 1.

$$f^f = tanh(W_h(f^{p2r}, f^{rgb'})). \qquad (6)$$

Subsequently, the proposed method use the update gate to filter and update the relevant information. $z \odot f^f$ represents how much information extracted from RGB data is retained, while $(1 - z) \odot f^f$ represents how much of the current candidate information, f^f, should be added. $(1 - z)$ is used as a selection of information from f^f or as the forgetting of unimportant information in f^f, resulting in the feature selection outcome.

$$F^{rgb} = (1 - z) \odot f^f + z \odot f^{rgb}. \qquad (7)$$

Similarly, for the depth modality, the feature fusion module for geometric features is used to obtain F^{point} for subsequent feature extraction.

4   Experimental Results and Discussion

In this section, two sets of experiments are conducted to verify the proposed method A2BNet. In the first set of experiment, A2BNet is compared with the

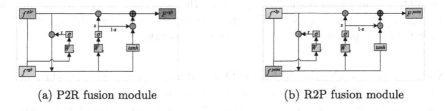

(a) P2R fusion module   (b) R2P fusion module

Fig. 4. Bi-gated fusion modules for RGB and Point cloud features.

state-of-the-art methods, and the second experiment is the ablation study to evaluate the effects of attention mechanism and bi-gated fusion for 6D object pose estimation.

Two public benchmark datasets, LineMOD and Occlusion LineMOD, are used in the experiments. In order to achieve good accuracy in the LineMOD dataset with 13 classes of objects, there are a number of difficulties to overcome, i.e., the pose estimation for textureless objects, cluttered backgrounds, and varying lighting conditions. The Occlusion LineMOD dataset is derived from the LineMOD dataset, including 8 classes of objects with severe occlusion in the scenes. In the experiments, the training data do not include instances with occlusion, testing on the Occlusion LineMOD dataset poses a great challenge. We split the training and testing sets following previous works [2,3] and generated synthesis images for network training [3,6].

ADD [11] is used as an evaluation metric, which is calculated by taking the average distance between corresponding points on the object's vertices in the predicted and ground truth poses, considering the pose transformation from the camera coordinate system to the object coordinate system. In particular, the metric ADD-0.1d is adopted, which defines the predicted vertices with an error accuracy within 10% of the object's diameter as accurate predictions.

During network training, both synthetic data and fusion data are used as input for auxiliary training. The batch size is set to 1 during training and testing, and the number of iterations is 25. The RGB images and depth images in the dataset have a size of 640×480. Prior to training, the depth images are processed into point cloud data with the size of $1 \times 12,288$ using camera intrinsic parameters, which are then combined with RGB data of size 640×480 as inputs to the model. The training strategy for optimization is stochastic gradient descent.

4.1 Comparison with the State-of-the-art Methods on Two Benchmark Datasets

Comparison on LineMOD Dataset. In the experiments, A2BNet is trained using LineMOD dataset only, and tested on both LineMOD and Occlusion LineMOD datasets. Table 1 shows the testing results on the LineMOD dataset, compared with Pose-CNN DeepIM [21], PVNet [3], CDPN [14], DPOD [4], Point-Fusion [15], Dense-Fusion [5], G2L-Net [9], PVN3D [6], FFB6D [10], PR-GCN [16] and PAV-Net [17]. The average accuracy of A2BNet reach 99.8%, outperforms the above methods. And ADD-0.1d for all 13 objects is greater than 99%. The proposed method performs the best on 7 classes of objects. For the remaining 6 classes, the results are acceptable, with a tiny difference of less than 0.3% compared to the best-performing network.

Comparison on Occlusion LineMOD Dataset. The experimental results on the Occlusion LineMOD dataset are shown in Table 2, compared with state-of-the-art methods. The accuracy of pose estimation is significantly lower compared to the results on LineMOD dataset, indicating that the Occlusion LineMOD dataset is much more challenging. Specifically, A2BNet performs the best on 4 out of 8 classes in the dataset. The pose estimation of the object ape demonstrating significant improvement by 25.2%. For the remaining 4 classes, the results were not significantly inferior. The overall average accuracy reach 67.6%, showing a promising improvement of 1.3% compared to the current state-of-the-art pose estimation method.

In summary, our network performs well in pose estimation accuracy on the LineMOD and Occlusion LineMOD datasets. Specifically, our network demonstrates competitive performance in estimating the poses of objects with low texture and irregular shapes. It also performs well for large-sized and irregular-shaped objects, even in occluded scenarios. However, it shows limitations in estimating the poses of textureless and regular-shaped objects in occluded scenarios, where occlusion can hinder the selection of reliable keypoints.

4.2 Ablation Study

To validate the effectiveness of the two sub-modules in the proposed A2BF module, the object ape is chosen from the LineMOD dataset, and the following three pose estimation networks are compared:

- Pose estimation utilizing only the backbone network.
- Pose estimation employing the dual self-attention mechanism.
- Pose estimation incorporating the bidirectional gated fusion module into the network.
- Pose estimation using the complete A2BF module(Attension + Bi-Gated).

From Table 3, it can be observed that the inclusion of the attention mechanism module improved the performance of the pose estimation network, resulting in a 2.1% increase in accuracy for object pose detection. Furthermore, compared to recent 6D pose estimation networks, it still achieved satisfactory results. This demonstrates that the method of modifying the weights of feature maps to focus

Table 1. Comparison results with state-of-the-art methods on the LineMOD dataset.

ObjectMethod	RGB				RGB-D								
	Pose- CNN DeepIM [21]	PVNet [3]	CDPN [14]	DPOD [4]	Point- Fusion [15]	Dense- Fusion [5]	G2L- Net [9]	PVN3D [6]	FFB6D [10]	PR- GCN [16]	PAV- Net [17]	Ours
ape	77.0	43.6	64.4	87.7	70.4	92.3	96.8	97.3	98.4	99.2	-	**100.0**
benchvise	97.5	99.9	97.8	98.5	80.7	93.2	96.1	99.7	100.0	99.8	-	**100.0**
camera	93.5	86.9	91.7	96.1	60.8	94.4	98.2	99.6	99.9	100.0	-	99.9
can	96.5	95.5	95.9	99.7	61.1	93.1	98.0	99.5	99.8	99.4	-	**100.0**
cat	82.1	79.3	83.8	94.7	79.1	96.5	99.2	99.8	99.9	99.8	-	**100.0**
driller	95.0	96.4	96.2	98.8	47.3	87.0	99.8	99.3	100.0	99.8	-	99.7
duck	77.7	52.6	66.8	86.3	63.0	92.3	97.7	98.2	98.4	**98.7**	-	98.4
eggbox	97.1	99.2	99.7	99.9	99.9	99.8	100.0	99.8	100.0	99.6	-	**100.0**
glue	99.4	95.7	99.6	96.8	99.3	100.0	100.0	100.0	100.0	**100.0**	-	99.8
holep-uncher	52.8	82.0	85.8	86.9	71.8	92.1	99.0	99.9	99.8	-	**99.9**
iron	98.3	98.9	97.9	100	83.2	97.0	99.3	99.7	99.9	99.5	-	**100.0**
lamp	97.5	99.3	97.9	96.8	62.3	95.3	99.5	99.8	99.9	100.0	-	99.9
phone	87.7	92.4	90.8	94.7	78.8	92.8	98.9	99.5	99.7	**99.7**	-	99.6
Avg	88.6	86.3	89.9	95.2	73.7	94.3	98.7	99.4	99.7	99.6	99.3	**99.8**

Table 2. Comparison results with state-of-the-art methods on the Occlusion LineMOD dataset.

Object Method	Pose-CNN [2]	Hu- et al. [8]	pix2-Pose [18]	PVNet [3]	DPOD [4]	Hu- et al. [20]	Hybird-Pose [19]	PVN3D [6]	FFB6D [10]	PR-GCN [16]	CRT-6D [7]	Ours
ape	9.6	12.1	22.0	15.8	-	19.2	20.9	33.9	47.2	40.2	53.4	**78.6**
can	45.2	39.9	44.7	63.3	-	65.1	75.3	88.6	85.2	76.2	92.0	**93.3**
cat	0.9	8.2	22.7	16.7	-	18.9	24.9	39.1	45.7	**57.0**	42.0	43.6
driller	41.1	45.5	44.7	65.7	-	69.0	70.2	78.4	81.4	82.3	81.4	**82.5**
duck	19.6	17.2	15.0	25.2	-	25.3	27.9	41.9	53.9	30.0	44.9	**55.0**
eggbox	22.0	22.1	25.2	50.2	-	52.0	52.4	**80.9**	70.2	68.2	62.7	49.6
glue	38.5	35.8	32.4	49.6	-	51.4	53.8	68.1	60.1	67.0	**80.2**	54.9
holepuncher	22.1	36.0	49.5	39.7	-	45.6	54.2	74.7	85.9	**97.2**	74.3	83.7
Avg	24.9	27.0	32.0	40.8	47.3	43.3	47.5	63.2	66.2	65.0	66.3	**67.6**

on important information while ignoring non-essential information can more efficiently extract data features and thereby improve pose estimation accuracy. The pose estimation network incorporating both the dual self-attention mechanism and the bidirectional gate fusion module achieved the optimal pose estimation results. It exhibited a 6.5% improvement in pose estimation accuracy for the ape object compared to the backbone network without the modules.

The effectiveness of pose estimation heavily depends on the ability of the extracted feature maps to facilitate the selection of reliable keypoints. By incorporating A2BF modules, our proposed network captures relevant information and enhances the quality of keypoints, thus improving the accuracy of pose estimation, particularly in the presence of occlusions. As shown in Fig. 5, it is obvious that the network with the A2BF modules can accurately estimate the poses of occluded objects, while the network without these modules exhibits significant rotational prediction errors when dealing with occlusions.

Table 3. The efficiency of each component in A2BF module.

Fusion Stage		Pose Result
Attention	Bi-Gated Fusion	Add-0.1d
		93.5
✓		95.6
	✓	97.2
✓	✓	100

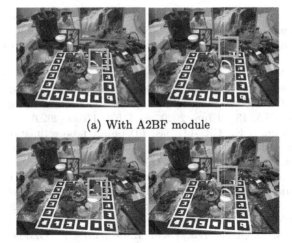

(a) With A2BF module

(b) Without A2BF module

Fig. 5. Visual comparison of A2BF module.

5   Conclusion

This paper introduces an attention-aware bi-gated fusion approach, achieving a balance between accuracy and computational complexity through a two-stage keypoint-based pose estimation backbone. It features two branches for texture features from RGB images and geometric features from point cloud data. A gate feature fusion module enhances object pose estimation accuracy by selectively filtering and fusing features due to texture-geometry complementarity. Furthermore, an attention module heightens network representation by emphasizing scene-important data. Experiments confirm method performance enhancement, albeit with limitations. Future directions include reducing label dependency via unsupervised 6D pose networks and integrating few-shot learning into 6D pose estimation while minimizing training data usage.

References

1. Zhang, H., Tang, J., Sun, S., et al.: Robotic grasping from classical to modern: a survey. arXiv preprint arXiv:2202.03631 (2022)
2. Xiang, Y., Schmidt, T., Narayanan, V., Fox, D.: Posecnn: a convolutional neural network for 6d object pose estimation in cluttered scenes. arXiv preprint arXiv:1711.00199 (2017)
3. Peng, S., Liu, Y., Huang, Q., Zhou, X., Bao, H.: Pvnet: pixel-wise voting network for 6dof pose estimation. In: Proceedings of the IEEE Conference on Computer Vision and Pattern Recognition, pp. 4561–4570 (2019)
4. Zakharov, S., Shugurov, I., Ilic, S.: Dpod: dense 6d pose object detector in RGB images. arXiv preprint arXiv:1902.11020 (2019)

5. Wang, C., et al.: DenseFusion: 6D object pose estimation by iterative dense fusion. In: Proceedings of the 2019 IEEE/CVF Conference on Computer Vision and Pattern Recognition (CVPR), Long Beach, CA, USA, 15–20 June 2019, pp. 3338–3347 (2019)

6. He, Y., Sun, W., Huang, H., Liu, J., Fan, H., Sun, J.: PVN3D: a deep point-wise 3D keypoints voting network for 6DoF pose estimation. In: Proceedings of the 2020 IEEE/CVF Conference on Computer Vision and Pattern Recognition (CVPR), Seattle, WA, USA, 13–19 June 2020, pp. 11629–11638 (2020)

7. Castro, P., Kim, T.K.: CRT-6D: fast 6D object pose estimation with cascaded refinement transformers. In: 2023 IEEE/CVF Winter Conference on Applications of Computer Vision (WACV), pp. 5735–5744 (2022)

8. Hu, Y., Hugonot, J., Fua, P., Salzmann, M.: Segmentation-driven 6D object pose estimation. In: Proceedings of the 2019 IEEE/CVF Conference on Computer Vision and Pattern Recognition (CVPR), Long Beach, CA, USA, 15–20 June 2019, pp. 3380–3389 (2019)

9. Chen, W., Jia, X., Chang, H.J., Duan, J., Leonardis, A.: G2L-net: global to local network for real time 6D pose estimation with embedding vector features. In: Proceedings of the 2020 IEEE/CVF Conference on Computer Vision and Pattern Recognition (CVPR), Seattle, WA, USA, 13–19 June 2020, pp. 4232–4241 (2020)

10. He, Y., Huang, H., Fan, H., Chen, Q., Sun, J.: FFB6D: a full flow bidirectional fusion network for 6D pose estimation. In: Proceedings of the IEEE Computer Society Conference on Computer Vision and Pattern Recognition, Virtual, 19–25 June 2021, pp. 3002–3012 (2021)

11. He, K., Zhang, X., Ren, S., Sun, J.: Deep residual learning for image recognition. In: Proceedings of the IEEE Conference on Computer Vision and Pattern Recognition, pp. 770–778 (2016)

12. Zhao, H., Shi, J., Qi, X., Wang, X., Jia, J.: Pyramid scene parsing network. In: Proceedings of the IEEE Conference on Computer Vision and Pattern Recognition, pp. 2881–2890 (2017)

13. Hu, Q., et al. Randla-net: efficient semantic segmentation of large-scale point clouds. In: Proceedings of the IEEE/CVF Conference on Computer Vision and Pattern Recognition, pp. 11108–11117 (2020)

14. Li, Z., Wang, G., Ji, X.: CDPN: coordinates-based disentangled pose network for real-time rgb-based 6-dof object pose estimation. In: Proceedings of the IEEE International Conference on Computer Vision, pp. 7678–7687 (2019)

15. Xu, D., Anguelov, D., Jain, A.: Pointfusion: deep sensor fusion for 3d bounding box estimation. In: Proceedings of the IEEE Conference on Computer Vision and Pattern Recognition, pp. 244–253 (2018)

16. Zhou, G., Wang, H., Chen, J., Huang, D.: PR-GCN: a deep graph convolutional network with point refinement for 6D pose estimation. In: 2021 IEEE/CVF International Conference on Computer Vision (ICCV), Montreal, QC, Canada, pp. 2773–2782 (2021). https://doi.org/10.1109/ICCV48922.2021.00279

17. Huang, J., Xia, C., Liu, H., Liang, B.: PAV-Net: point-wise attention keypoints voting network for real-time 6D object pose estimation. In: 2022 International Joint Conference on Neural Networks (IJCNN), Padua, Italy, pp. 1–8 (2022). https://doi.org/10.1109/IJCNN55064.2022.9892089

18. Park, K., Patten, T., Vincze, M.: Pix2Pose: pixel-wise coordinate regression of objects for 6D pose estimation. In: 2019 IEEE/CVF International Conference on Computer Vision (ICCV), Seoul, Korea (South), pp. 7667–7676 (2019). https://doi.org/10.1109/ICCV.2019.00776

19. Song, C., Song, J., Huang, Q.: Hybridpose: 6d object pose estimation under hybrid representations. In: Proceedings of the IEEE/CVF Conference on Computer Vision and Pattern Recognition, pp. 431–440 (2020)
20. Hu, Y., Fua, P., Wang, W., Salzmann, M.: Single-stage 6d object pose estimation. In: Proceedings of the IEEE/CVF Conference on Computer Vision and Pattern Recognition, pp. 2930–2939 (2020)
21. Li, Y., Wang, G., Ji, X., et al.: Deepim: deep iterative matching for 6d pose estimation. In: Proceedings of the European Conference on Computer Vision (ECCV), pp 683–698 (2018)

Author Index

Printed in the United States
by Baker & Taylor Publisher Services